ISBN 978-0-365-31358-8
PIBN 10288480

STATE OF NEW YORK.

No. 24.

IN SENATE.

JANUARY 19, 1886.

SECOND ANNUAL REPORT

OF THE NEW YORK STATE DAIRY COMMISSIONER.

To the Legislature of the State of New York;

The New York State Dairy Commissioner respectfully submits the following report:

The following is a statement in detail of the assistant commissioners, experts, chemists, agents, and counsel employed by the Commissioner, with their compensation, expenses and disbursements:

Date.		Name of payee	Compensation.	Expenses and disbursements.
1885.				
January	20.	W. G. Spence, expert	$60 00	$23 20
	30.	P. E. Eysaman, expert	42 00	67 40
	30.	Geo. L. Flanders, assistant com'r	100 00	....
	30.	Geo. B. Fellows, expert and agent	100 00	....
February	11.	B. F. Van Valkenburgh, asst. com'r.	166 66	72 98
	11.	Edward W. Martin, chemist	104 00	49 46
	11.	Elwyn Waller, chemist	40 00	....
	11.	Charles M. Stillwell, chemist	150 59	....
	11.	J. R. Wheeler, expert and agent	80 00	48 62
	11.	Thomas R. Gray, expert and agent.	80 00	46 57
	11.	Wm. W. Mecteer, expert and agent.	80 00	42 33
	11.	Arthur C. Salmon, attorney	165 00	3 00
	11.	Samuel J. White, expert and agent.	60 00	31 23
	11.	Walter Moeller, chemist and expert.	100 00	....
	11.	E. S. Wilson, expert and agent	105 00	92 04
	11.	W. G. Spence, expert and agent	60 00	24 91
	11.	E. G. Love, chemist	119 37	....
	11.	Francis V. S. Oliver, attorney	120 00	....
	11.	Charles Sears, expert and agent	60 00	14 60
	11.	T. C. DuBois, expert and agent	80 00	107 94
	11.	Chas. S. Kellogg, expert and agent	39 00	9 75
	11.	Marcus A. Perry, assistant com'r	100 00	41 68

Date.		Name of payee.	Compensation.	Expenses and disbursements.
1885.				
February	11.	R. D. Clark, chemist	$100 00	....
	18.	R. A. Witthaus, chemist	76 00	....
	18.	J. J. Sorogan, expert and agent	60 00	$33 15
	27.	Geo. L. Flanders, assistant com'r	100 00	....
	27.	Geo. B. Fellows, expert and agent	100 00	....
	28.	Edward W. Martin, chemist and expert	42 00	7 04
	28.	Charles M. Stillwell, chemist	103 61	3 20
	28.	Arthur C. Salmon, attorney	95 00	3 87
	28.	Wm. W. Meeteer, expert and agent	80 00	19 39
	28.	E. S. Wilson, expert and agent	105 00	99 32
	28.	J. J. Sorogan, expert and agent	60 00	14 96
	28.	Charles Sears, expert and agent	60 00	8 85
	28.	Walter Moeller, chemist and expert	100 00	....
	28.	J. R. Wheeler, expert and agent	80 00	54 08
	28.	B. F. Van Valkenburgh, asst. com'r	166 66	19 20
	28.	T. C. DuBois, expert and agent	80 00	116 48
	28.	F. V. S. Oliver, attorney	150 00	....
	28.	Horace W. White, expert and agent	31 50	12 85
	28.	Samuel J. White, expert and agent	60 00	21 65
	28.	R. D. Clark, chemist	100 00	12 41
	28.	Thomas R. Gray, expert and agent	80 00	35 51
	28.	M. A. Perry, assistant com'r	100 00	18 64
	28.	R. A. Witthaus, chemist	164 00	....
March	2.	W. G. Spence, expert and agent	60 00	13 92
	9.	A. C. Salmon, attorney	30 00	....
April	1.	Geo. L. Flanders, assistant com'r	100 00	....
	1.	Geo. B. Fellows, expert and agent	100 00	....
	7.	R. D. Clark, chemist	100 00	9 93
	9.	Walter Moeller, chemist and expert	100 00	....
	9.	F. V. S. Oliver, attorney	90 00	....
	9.	R. A. Witthaus, chemist	224 00	....
	9.	Charles Sears, agent and expert	60 00	13 28
	9.	Arthur C. Salmon, attorney	120 00	....
	9.	Charles M. Stillwell, chemist	147 50	....
	9.	Thomas R. Gray, expert and agent	80 00	43 52
	9.	Elwyn Waller, chemist	50 00	....
	9.	Francis E. Englehardt, chemist	15 00	....
	9.	E. G. Love, chemist	86 25	....
	9.	J. R. Wheeler, expert and agent	80 00	60 25
	9.	E. S. Wilson, expert and agent	105 00	150 24
	9.	T. C. DuBois, expert and agent	80 00	155 14
	9.	B. F. Van Valkenburgh, assist. com'r	166 66	125 48
	9.	Horace W. White, expert and agent	43 50	76 32
	9.	Edward W. Martin, chemist	192 00	26 92
	9.	J. J. Sorogan, expert and agent	60 00	41 75
	9.	W. G. Spence, expert and agent	60 00	36 52
	9.	Wm. P. Mason, chemist	60 00	....
	9.	M. A. Perry, assistant commissioner	100 00	90 73
	9.	Samuel J. White, expert and agent	60 00	36 10

Date.		Name of payee.	Compensation.	Expenses and disbursements.
1885.				
April	13.	Wm. W. Meeteer, expert and agent.	$80 00	$34 22
May	1.	Geo. L. Flanders, assistant com'r ..	100 00	....
	1.	Geo. B. Fellows, expert and agent...	100 00	....
	6.	B. F. Van Valkenburgh, assist. com'r	166 66	40 05
	6.	Edward W. Martin, chemist	136 00	10 50
	6.	R. D. Clark, chemist	100 00	4 20
	6.	Walter Moeller, chemist and expert.	100 00	1 13
	6.	E. S. Wilson, expert and agent	105 00	45 40
	7.	Marcus A. Perry, assist. com'r.....	100 00	14 70
	7.	A. C. Salmon, attorney	255 00	8 23
	7.	E. G. Love, chemist	70 00	....
	7.	Stillwell & Gladding, chemists	266 66	....
	7.	T. C. DuBois, expert and agent....	80 00	70 66
	7.	W. G. Spence, expert and agent...	60 00	36 64
	7.	J. J. Sorogan, expert and agent....	60 00	33 23
	7.	Samuel J. White, expert and agent.	60 00	20 60
	7.	J. R. Wheeler, expert and agent...	80. 00	60 08
	7.	Thomas R. Gray, expert and agent.	80 00	65 71
	7.	Wm. W. Meeteer, expert and agent.	80 00	40 32
	7.	Charles Sears, expert and agent ...	60 00	18 00
	7.	Horace W. White, expert and agent.	12 00	5 46
	7.	R. A. Witthaus, chemist	50 00	....
	7.	Wm. Manlius Smith, chemist	15 00	....
June	1.	Geo. L. Flanders, asst. com'r	150 00	....
	1.	Geo. B. Fellows, expert and agent .	100 00	....
	5.	R. D. Clark, chemist............	100 00	26 90
	5.	M. A. Perry, assistant commissioner.	100 00	2 10
	5.	Edward W. Martin, chemist and expert	88 00	2 35
	5.	Walter Moeller, chemist and expert.	100 00	....
	6.	T. C. DuBois, expert and agent...	80 00	121 36
	6.	E. G. Love, chemist	101 25	....
	6.	Stillwell & Gladding, chemists	164 63	....
	6.	Arthur C. Salmon, attorney	60 00	....
	6.	F. V. S. Oliver, attorney	210 00	....
	6.	Thomas R. Gray, expert and agent.	80 00	54 48
	6.	J. R. Wheeler, expert and agent ..	80 00	77 15
	6.	W. W. Meeteer, expert and agent..	80 00	54 18
	6.	W. G. Spence, expert and agent...	60 00	31 66
	6.	J. J. Sorogan, expert and agent ...	60 00	27 16
	6.	Samuel J. White, expert and agent.	60 00	18 45
	6.	B. F. Van Valkenburgh, assistant commissioner	166 66	54 52
	6.	E. S. Wilson, expert and agent....	105 00	84 25
	6.	H. W. White, expert and agent....	24 00	7 12
	11.	Charles Sears, expert and agent....	60 00	20 80
	12.	Charles B. Evers, expert and agent.	40 00	....
	15.	Geo. N. Loveridge, expert and agent.	75 00	....
	24.	Samuel Hand, attorney	1,000 00	31 00

Date.		Name of payee.	Compensation.	Expenses and disbursements.
1885.				
July	1.	Geo. L. Flanders, assist. com'r	$150 00	....
	1.	Geo. B. Fellows, expert and agent .	100 00	....
	1.	R. D. Clark, chemist............	150 00	....
	2.	M. A. Perry, assist. commissioner..	100 00	$33 86
	6.	Charles Sears, expert and agent ...	60 00	18 04
	6.	Samuel J. White, expert and agent	60 00	19 30
	6.	J. J. Sorogan, expert and agent....	60 00	29 10
	6.	Edward W. Martin, chemist	266 00	....
	6.	Wm. G. Spence, expert and agent..	60 00	31 94
	7.	B. F. Van Valkenburgh, asst. com'r	166 66	29 58
	7.	Geo. C. Hodges, chemist	5 00	....
	7.	Alvin W. Barry, attorney.........	180 00	....
	7.	Walter Moeller, expert and agent..	100 00	....
	7.	R. D. Clark, chemist.............	....	95 84
	7.	Stillwell & Gladding, chemists	26 37	....
	7.	H. W. White, expert and agent....	30 00	14 50
	7.	Thomas R. Gray, expert and agent.	80 00	33 30
	7.	T. C. DuBois, expert and agent ...	80 00	14 58
	7.	J. R. Wheeler, expert and agent...	80 00	63 39
	7.	Wm. W. Meeteer, expert and agent.	80 00	48 09
	9.	New York Stencil Works, for cheese brands	....	250 00
	13.	E. S. Wilson, expert and agent....	105 00	37 30
	13.	F. V. S. Oliver, attorney	120 00	....
	14.	Charles B. Evers, expert and agent.	53 33	....
	28.	Risley, Quin & Perry, attorneys and counselors	900 00	19 25
	28.	Wm. G. Spence, expert and agent..	42 50	19 28
	31.	Geo. L. Flanders, assist. com'r	150 00	....
August	1.	Geo. B. Fellows, expert and agent.	100 00	....
	1.	R. D. Clark, chemist............	150 00	....
	10.	H. W. White, expert and agent....	75 00	9 63
	10.	B. F. Van Valkenburgh, asst. com'r	166 66	140 27
	10.	J. R. Wheeler, expert and agent ..	80 00	56 80
	10.	Stillwell & Gladding, chemists	127 50	....
	10.	T. C. DuBois, expert and agent....	80 00	69 11
	10.	Wm. W. Meeteer, expert and agent.	80 00	58 96
	10.	Samuel J. White, expert and agent.	60 00	25 10
	10.	J. J. Sorogan, expert and agent...	60 00	45 36
	10.	E. S. Wilson, expert and agent....	105 00	30 35
	10.	Thomas R. Gray, expert and agent.	80 00	36 34
	10.	F. V. S. Oliver, attorney..........	60 00	....
	10.	Walter Moeller, expert and agent..	100 00	....
	10.	Charles Sears, expert and agent....	60 00	20 90
	10.	Howard J. Babcock, chemist and expert	40 00	....
	12.	Edward W. Martin, chemist and expert	276 00	....

Date. 1885.		Name of payee.	Compensation.	Expenses and disbursements.
August	12.	M. A. Perry, assistant com'r......	$150 00	$59 06
	12.	F. P. Vandenburgh, chemist and expert........................	66 00	2 00
	13.	R. D. Clark, chemist and expert...	....	15 14
	14.	E. G. Love, chemist	18 75	....
September	1.	Geo. L. Flanders, assistant com'r ...	150 00	....
	1.	R. D. Clark, chemist	150 00	....
	1.	Geo. B. Fellows, expert and agent..	100 00	....
	5.	Howard J. Babcock, chemist and expert	40 00	....
	5.	D. Magone, attorney and counselor.	200 00	....
	5.	Charles Sears, expert and agent....	60 00	29 25
	5.	E. S. Wilson, expert and agent....	105 00	29 39
	5.	F. V. S. Oliver, attorney and agent.	120 00	....
	5.	Thomas R. Gray, expert and agent.	80 00	38 84
	5.	Stillwell & Gladding, chemists.....	143 12	2 50
	5.	Samuel J. White, expert and agent.	60 00	26 15
	5.	Edward W. Martin, chemist	237 00	20 00
	5.	T. C. DuBois, expert and agent ...	80 00	50 68
	5.	Wm. W. Meeteer, expert and agent.	80 00	43 39
	5.	J. R. Wheeler, expert and agent...	80 00	36 67
	8.	Norman A. Lawlor, att'y and agent.	75 00	....
	8.	H. W. White, expert and agent....	75 00	6 74
	8.	J. J. Sorogan, expert and agent....	60 00	43 76
	8.	Walter Moeller, chemist and expert.	100 00	....
	14.	B. F. Van Valkenburgh, ass't com'r.	166 66	61 14
	17.	R. A. Witthaus, chemist	490 00	....
	17.	M. A. Perry, assistant com'r	150 00	48 98
	22.	Risley, Quin & Perry, attorneys and counselors....................	1,290 00	68 10

In addition to the above, this department has expended for necessary apparatus, for stationery, blanks, furniture for the office, expenses and salary of the Commissioner, the sum of $4,093.28, making the expenses up to and including the 30th day of September, $30,-133.97; during this period we have collected in fines and penalties, $2,617.45, of which sum we have paid over to the State Treasurer $2,288.81, the balance having been used to pay the expenses of trials, etc., as provided by chapter 202, Laws of 1884.

Since September 30, there has been collected in fines and penalties the sum of $1,000.

The year just closed has been a very eventful one to this department.

The case which arose in the city and county of New York, under section 6 of chapter 202 of the Laws of 1884, and which is known as

The People v. *Marx*, was appealed to the Court of Appeals by the defendant, who had been unsuccessful in all the courts below, and that court decided that the particular section under which the defendant was convicted was unconstitutional and void.

The sixth section of the law of 1884 provided in effect that no oleaginous substance, other than that made from pure unadulterated milk or cream, and designed to take the place of butter, should be made or sold. The Court of Appeals held that the prohibiting of the making or selling of any article "designed to take the place of butter" must be construed to mean that the manufacture or sale of any and every oleaginous substance, to be used as a *substitute* for butter, was prohibited, and that such a prohibition would not only cover every article of the kind now known, but also any such article which might be hereafter produced, whether the same was similar in appearance, or in any other respect, to natural butter, or not, and would prevent the manufacture or sale of any oleaginous substitute for natural butter, although such substitute might be perfectly wholesome, very desirable, and so totally dissimilar to natural butter as not to deceive any one, and that such a prohibition was so sweeping in its terms and effects as to be unconstitutional and void. This single point is all the court decided.

The learned judge who wrote the opinion of the court says: "All the witnesses who have testified as to the quality of oleomargarine may be in error; still that would not change a particle the nature of the question, or the principles by which the validity of the act is to be tested. Section 6 is broad enough in its terms to embrace not only oleomargarine, but any other compound, however wholesome, valuable or cheap, which has been or may be discovered or devised, for the purpose of being used as a substitute for butter. Every such product is rigidly excluded from manufacture or sale in this State.

"One of the learned judges who delivered opinions at the General Term endeavored to sustain the act, on the ground that it was intended to prohibit the sale of any artificial compound, as butter or cheese, made from unadulterated milk or cream. That it was that design to deceive which the law rendered criminal. If that was a correct interpretation of the act, we should concur with the learned judge in his conclusion as to its validity." We do not think that section 6 is capable of the construction claimed. The prohibition is not of the manufacture or sale of an article designed as an imitation of dairy butter or cheese, or intended to be passed off as such, but of an article designed to take the place of dairy butter or cheese. The artificial product might be green, red or white, instead or yellow, and totally dissimilar in appearance to ordinary dairy butter, yet it might be designed as a substitute for butter, and if so, would fall within the prohibition of the statute. Simulation of butter is not the act prohibited. There are other statutory provisions fully covering that subject.

Chapter 215 of the Laws of 1882, by its first section prohibits the introduction of any substance into imitation butter or cheese, for the purpose of imparting thereto a color resembling that of yellow butter or cheese. The second section prohibits the sale of oleomargarine, or imitation butter, thus covered, etc.

Chapter 238 of the Laws of 1882, provides (section 1) that every person who shall manufacture for sale, or offer for sale, or expose any article in semblance of butter or cheese, not the legitimate product of the dairy, must distinctly and durably stamp on the side of every cheese and on the top and side of every tub, firkin or package, the words 'Oleomargarine butter,' or if containing cheese, 'Imitation cheese,' and chapter 246 of the Laws of 1882 makes it a misdemeanor to sell at wholesale or retail any of the above articles, representing them to be butter.

These enactments seem to cover the entire subject of fraudulent imitation of butter and of sales of other compounds as dairy products and they are not repealed, etc.

The provisions of this last act are covered by one of the acts of 1882 above cited, and the provisions of the repealed acts in relation to the dairy products are covered by substituted provisions, in the act of 1884, but the statutes directed against fraudulent simulations of butter and the sale of such simulations as dairy butter are left to stand. Further statutes to the same effect were enacted in 1885. Consequently if the provisions of section 6 should be held invalid, there would still be ample protection in the statutes against fraudulent imitations of dairy butter, or sales of such imitations as butter.

Notwithstanding all this, the venders and dealers, in imitation butter, have deliberately and persistently represented in every possible way that there is now no law in our State to prevent the open manufacture and sale of these adulterated goods, in the face of the fact as stated in the opinion of the court, that there are several unrepealed statutes relating to this subject, besides the Laws of 1885.

The adverse decision of the Court of Appeals imposed very grave responsibility and a vast amount of work upon the department. It became necessary to examine with great care not only our new law of 1885, but also all of the statutes relating to this subject which had previously been passed by our Legislature. Many laws had, at different times, been enacted, designed to prevent frauds in the manufacture and sale of dairy products and their imitation. Frequent changes had been made by amendments, to meet the decisions of the court, and the discoveries of the many new methods employed to cheapen imitation goods and deceive the public. Some of these laws have been expressly repealed; others repealed by necessary implication, and the whole list, possibly, in some respects modified or affected by the recent decision of our highest court.

Legal counsel were employed to do this work, and after days and even weeks spent in a most critical examination of all those statutes, and careful analysis of all the authorities bearing upon the questions involved, a plan of aggressive operations and work was agreed upon which has since been closely followed. Full and carefully arranged briefs were prepared and furnished to the district attorneys of the counties in which it was expected most of our work would be done, and such attorneys as we employed in different localities to assist us were also put in possession of our views of the laws, and thus prepared for the prosecution of the work of the department.

Early in the summer a case arose in Albany, involving the constitutionality of that important section of our present law, creating, as we supposed and still believe, a proper milk standard. The Supreme Court, at General Term, however, by a majority of its members rendered an adverse decision, holding substantially that while it was doubtless within the power of the Legislature to create a milk standard by proper enactment, yet in our statute it had simply attempted to prescribe a rule of evidence, and having undertaken to make conclusive evidence of the guilt of one indicted for committing a crime, that which in its very nature could not be conclusive, our Legislature exceeded its authority, and the enactment, in that respect, was unconstitutional and void. So it will be seen that whenever we have attempted to draw these laws up tight in what was intended to be their most valuable and vital sections, a strand has parted somewhere and we have suffered a fall.

At the time of the decision by the Court of Appeals in *The People v. Marx*, there were pending a very large number, I believe more than a hundred, indictments and other proceedings against violators of the law of 1884. Although we believe that in the great majority of those cases the accused could and ought to be convicted and punished under the unrepealed old statutes and the provisions of the Penal Code, and have so advised the district attorneys of the proper counties, yet nothing has thus far been done in those cases, excepting in the county of Erie, where some of them have been presented to the grand jury, and as I am informed ten or twelve indictments have been found.

After the argument of this case in the Court of Appeals and before it was decided, chapters 183 and 458 of the Laws of 1885 were passed and became operative, to which reference is made by that court in the opinion spoken of.

Under these statutes and others and some of the provisions of the Penal Code we have been proceeding against those engaged in the manufacture and the sale of the forbidden imitation and adulterated goods.

We have commenced criminal proceedings in more than one hundred different cases, several persons have been convicted and fined, and indictments and other proceedings are now pending in a very large number of cases. Our experts, agents, chemists and attorneys are before the courts almost every day now, and we are making progress quite as satisfactorily as can reasonably be expected, considering the dilatory hindrances interposed by accused parties and the many vexatious delays incident to the attempt to enforce such statute before the police courts of such large cities as New York and Brooklyn.

We have been able to try a large number of our cases in Brooklyn, for the reason that in that city all misdemeanors are tried before the court issuing the warrant, unless a judge of a higher court gives a written certificate that it is a proper case to be tried by indictment, when the case is removed to another tribunal. Such certificates are seldom given, and our cases are there speedily moved on to trial. In the city and county of New York the practice is altogether different. A warrant is issued by a police magistrate who can only impose fines in cases where the defendant pleads guilty. In all but about ten of our cases in New York the defendants have elected to be tried at General Sessions. In such cases the evidence is presented to a grand jury, and if indictments are found bail is given, which practically pigeon-holes the proceeding, because in that great city there is always a very great number of indictments found for misdemeanors which cannot be tried promptly, if at all, for the reason that it is made by law the duty of the district attorney to first try those cases where the prisoner is in confinement, no bail having been given. The district attorney of New York has courteously and promptly done what he could for us, consistent with his other duties, and has caused some of our indictments to be tried, which has greatly facilitated our work. I fear we shall never be able to have our cases promptly disposed of in New York under the present system of procedure in criminal prosecutions in that city.

There are now pending about fifty civil cases brought in the Supreme Court to recover the penalty prescribed for violations of the several sections of our law, and which are in addition to the fines which the criminal court in criminal proceedings may impose upon those convicted of a misdemeanor. In the cases so brought for violation of the sections against selling impure, unwholesome and adulterated milk, we were suddenly confronted with the decision in the Albany milk case. That case has been appealed to and argued in the Court of Appeals and a favorable decision there is confidently looked for. We are trying hard to get the civil causes for penalties for selling oleomargarine to trial, but not being preferred causes, it will probably be several months before they will be reached and disposed of.

We have recently brought to trial one of our civil cases for the penalty prescribed for selling adulterated milk in the county of Westchester, which resulted in a verdict for the defendant.

The reports of the assistant commissioners and of our chemists are herewith presented and are of great value.

Mr. Edward W. Martin's report on milk, it is believed, will set at rest all question as to the milk standard created by chapters 202 of the Laws of 1884, and 183 of the Laws of 1885. That report shows the work done during the year 1885, the amount of milk produced for market in several milk-producing counties, gives the results of analyses of adulterated milk, the best methods of analysis, and some methods of testing milk. Great care has been taken to obtain samples of milk from a number of different counties of the State, from herds of different breeds of cattle upon farms in different localities in some of the same counties, upon table-land and low-land; also from cows fed upon different kinds of feed, including brewers' grains and other feed supposed by some people to produce an inferior quality of milk. The milk of a very large number of cows has thus been subjected to every known test, and it appears that not a single herd nor an individual cow has been found whose milk contained more than eighty-eight per cent of water, nor less than twelve per cent of milk solids, which contained not less than three per cent of fat. The wisdom of establishing the milk standard created by the Legislature cannot now be doubted.

Our experts have accompanied those who deliver milk to families in cities and have taken samples from time to time just as regular customers were served, with a view of ascertaining whether there is any appreciable difference in the quality of the milk so delivered by milkmen. We have also taken full cans of milk and dipped from them at intervals, as the groceryman would probably do in serving his customers. These samples were all tested and analyzed and the results are given. The report also treats upon the subject of feeding milch cows.

A large number of analyses of cream and condensed milk were made and the results are here given.

The report closes with a bibliography of milk, which will be of great value to investigators and others, as it gives a complete reference to everything that has been written upon this subject during the past twenty years, together with a list of the books published.

The report of Mr. B. F. Van Valkenburg, assistant commissioner at New York, is very complete and gives a very full account of our work under his immediate care. He states that our prosecutions during the last quarter of 1884 had greatly reduced the amount of sales of imitation butter; that the Court of General Sessions in New York then determined not to try any more of our cases, including those where the defendant was charged with fraudulently selling imitation for pure dairy butter, until after the decision of the case of *The People* v. *Marx* should be announced. Dealers thus encouraged commenced selling again, and more than double the amount of these forbidden goods were sold during the first quarter of 1885 than had been sold during the preceding quarter year. In Brooklyn, Kings county, however, the local courts proceeded to promptly try and dispose of our cases, and the result was that but very little oleo was sold outside of the city of New York. Thirty-nine arrests were made between December 31, 1884, and April 30, 1885, when the new law took effect. Ten cases were made in New York in May, 1885, and none in Kings county.

As usual the approach of hot weather and the low price of butter caused a suspension of sales of imitation goods, and about June 1, our butter experts began to give all their spare time to the work of protecting consumers against frauds in the sales of impure and adulterated milk, and continued until about September 1, when they and some of the milk experts again commenced operations to prevent sales of bogus butter. A large number of arrests have been made and indictments found. Our cases in Brooklyn have been promptly tried and disposed of as will appear from the statement of Mr. Salmon, our attorney there, which is filed herewith. In the city and county of New York we have not been able to proceed much further than to obtain indictments, and for the reason which has before been stated. The effect of our prosecutions and proceedings has been very salutary. For the last two months a large majority of respectable dealers in the cities of New York and Brooklyn who have heretofore sold imitation butters have quit selling it and now deal altogether in the genuine article. Public sentiment has been aroused and much good has been done by the articles in favor of pure dairy products which have of late appeared in the press. The large and powerful organization known as the "New York Retail Grocers' Union," at a meeting held at Grocers' Hall, No. 213 East Twenty-third street, New York city, December 14, 1885, unanimously adopted and promulgated the following preamble and resolutions:

"WHEREAS, The New York Retail Grocers' Union has put itself on record to encourage the sale of all pure goods, and to discourage and try to prevent all deceptions that are and may be practiced upon our customers by the sale of imitation or impure goods; and

"WHEREAS, We are informed that the laws of the State of New York prohibit the sale of all imitations of butter in the way they are at present manufactured and offered for sale; and

"WHEREAS, The sale of all imitation butter has been a detriment to the legitimate business of the retail dealer, inasmuch as that it has been a constant temptation to the dealers therein to sell it for butter, and that such fraudulent sale has created a prejudice and fear among the consumers in regard to the purchase of all butters. Therefore be it

"*Resolved*, That we discourage the sale of all imitation of butter and urge all our members and the trade in general not to handle it in any manner or form, until such a time when the manufacturers thereof will produce and offer for sale to us an article that will be distinct in appearance and different in color to that of genuine butter; that will be free from all temptation to fraud, and that will be manufactured and sold to us in strict accordance with the laws of this State. And be it further

"*Resolved*, That we most respectfully petition the State Dairy Commission to use all power that is invested in them to stop the fraudulent sale of all imitation of butter.

"C. F. BUSSING, *President*.

"H. TONJES, *Secretary*."

Mr. Van Valkenburg says, however, that there is another and a secret organization of retail grocers consisting of about two hundred members, who have raised a fund and employed attorneys to defend its members, when prosecuted by this department, but he says that three-fourths of the retail grocers now refuse to handle imitation butter upon the ground that it is a fraud.

The charge, so frequently made, that our farmers are using "oleo" oils to mix with dairy butter has been thoroughly investigated by Mr. Van Valkenburg and he reports that there is no foundation whatever for the statement and that it is not true; the sources of his information are given. And I will here add that during all the time I have been Commissioner, but one single case of alleged adulteration by a farmer has been brought to our notice. That case was carefully and completely investigated, and although there were some circumstances which, unexplained, were somewhat suspicious, yet we were unable to find that the charge was true.

His report further shows that while sales of imitation butter within this State have been very materially reduced, sales in the United States have increased very much since 1883. The inference can be fairly drawn that the vigorous enforcement of our law has reduced sales and consumption of these goods in this State to about one-fourth the amount they have heretofore been, but the manufacturers and venders of imitation butter have sought and found markets elsewhere. Valuable figures and statistics are given showing the amount of pure butter and of the imitations which have been sold, the extent to which consumers have been imposed upon, and the dairy interests of this and other States injured.

He also furnishes a detailed statement of work performed, number of days' attendance at court and elsewhere, samples obtained, etc., by the different chemists, employees, agents, attorneys, etc., of this department, under his immediate supervision.

Many other facts are given and suggestions made which will be found to be of importance.

The report of Mr. M. A. Perry, assistant commissioner at Buffalo, gives a very full statement of our work in that part of the State. It is doubtful if another city can be found within our State where the people are so grossly imposed upon by the dealers in milk as they are in Buffalo. His statement shows that the adulteration of milk has been extensively practiced there by wholesalers and retailers of that article. Out of 254 samples of milk obtained in the city of Buffalo, the analyses showed that 186 were adulterated, and he says that a large majority of the dealers from whom samples were obtained, intentionally adulterate that important article of food, and that nothing but the rigid enforcement of the law will cure the evil. A large number of arrests have been made and seven cases have been brought to trial and convictions obtained. Here, too, offenders have organized and raised funds for defense, but thus far Buffalo courts and juries have sustained us in our effort to enforce the law. Other trials will be had as fast as they can be brought on, and many other arrests will soon be made. Previous to the decision of the case of *The People* v. *Marx*, several samples of imitation butter were obtained in Buffalo, but no cases were brought to trial. Recently indictments have been found against persons who had sold these goods. Evidence has been obtained against several dealers for selling imitation butter since the Laws of 1885 became operative, and such cases will be brought to trial as soon as possible. We are well sustained by the press and public sentiment in the western portion of the State. The board of supervisors of Erie county, at a meeting held on the 30th day of December last, unanimously adopted the following preamble and resolutions:

"WHEREAS, The manufacture and sale of oleomargarine and butterine are being brought into direct competition with and undermining the greatest agricultural interest of the State; and

"WHEREAS, Said goods are sold in many instances for genuine dairy products in the county, and we believe their use, together with that of adulterated milk, to be injurious to the health and lives of its inhabitants; and

"WHEREAS, There have been a number of cases brought before the grand jury and indictments found against several parties on the testimony produced before said grand jury. Therefore

"*Resolved*, That it is the sense of this Board that all parties indicted for selling such adulterated food, representing the same to be genuine dairy product, contrary to law, be prosecuted, and that the district attorney of this county be and is hereby requested to make strenuous efforts to have all persons who are or may hereafter be indicted for such offenses punished to the full extent of the law. Adopted."

Upon the whole it gives me pleasure to be able to report that we are succeeding as well in our efforts to enforce the law in the State as could be expected considering the many vexatious hindrances and delays with which we have to contend.

The manufacturers of these deceptive imitations of dairy butter have made great progress in the art of producing an article designed and well calculated to deceive consumers, and we have been compelled to do a considerable amount of experimental work in order to keep close to them and be able to detect their counterfeits. A large number of analyses have, therefore, been made, all the known and some new methods have been thoroughly tested and verified, so that I feel confident there is now no imitation product manufactured and put upon the market which cannot be certainly detected by methods and means now in use by this department.

The report of Dr. Elwyn Waller and Mr. Edward W. Martin of the School of Mines, Columbia College, New York, which is herewith presented, is one of the most valuable contributions to the literature of this subject which has been produced in many years, so full and complete is that report, that I shall be able to give but a very imperfect synopsis of it. A brief description of the mode of making butter is given; the constituents of butter are furnished in the form of tables. A list of the various adulterants is given, and also methods of testing for them. Methods of manufacturing imitation butter from the original process down to a late date are described. A list of materials used, as stated in and called for by various patents which have been obtained is also mentioned. Rough-and-ready tests are given to determine the character of the fat used, some of which are believed to be entirely new. A description of the coloring matter used is also given, with the tests for them.

This report of Dr. Waller and Mr. Martin can only be appreciated after that careful study which it will surely receive by all those who are interested in this subject.

The growth and development of the sentiment of hostility to all these counterfeits of pure dairy products have been rapid and marvelous. I have written the proper officer of every State and organized Territory of the United States asking for copies of all their recent laws upon this subject. I have received prompt replies and find that twenty-seven States and two Territories have already enacted restrictive or prohibitory statutes. A summary of all these laws will be found among the appendices to this report.

The fact that sales of these deceptive imitations of butter are considerable, and in many instances apparently on the increase in some of those States, is evidence that the people there are congratulating themselves that since the passage of their laws forbidding such sales very little, if any, imitation butter is now sold. We are confident that these sales will not be very materially reduced until those States and Territories which have not already done so, provide means for the enforcement of their anti-oleo laws by officers specially designated and appointed for that purpose.

Last winter a law was enacted (chap. 193, Laws of 1885), providing for the adoption of a State brand for full-cream cheese, and making it the duty of the Dairy Commissioner to procure such brands and issue them to such manufacturers of full-cream cheese as should apply for them and conform to the regulations and requirements prescribed. Such brands were obtained and issued to four hundred and fifty-two different manufactories. This enactment and the work done in pursuance of its provisions has been of very great value to dairymen. The excellent reputation of our full-cream cheese has been re-established, and the advance obtained in the price of such cheese over those not thus branded has been from one-half cent to one cent per pound, which amounts to a very large sum of money in the aggregate. I have received a communication which is herewith presented from Mr. B. D. Gilbert, who is the secretary of the board of trade at Utica, N. Y. Mr. Gilbert is as competent to speak upon the subject of the value to dairymen of this brand as any man in the State. His communication shows how, to what extent, and for what reason this brand has proved advantageous, and will be read with much interest. Doubtless a much greater number of these brands will be issued during the coming year.

During the year a chemist was secured and suitable experts employed to assist in the enforcement of our law in the great dairy county of St. Lawrence. I am assured that this proceeding has been of great value to the dairymen in that region. Such plants ought to be established at different points throughout the State so as to be available to dairymen in the several dairy sections, but so long as this enormous and expensive litigation continues necessary in order to prevent manufacture and sales of counterfeit butter, we shall not be able to extend that branch of the service without a considerable increase in the amount appropriated for the use of this department.

The report of Dr. R. D. Clark, who is a practicing physician, and is also the chemist of this department at Albany, is of inestimable value. It is in fact an exhaustive treatise upon the subjects under consideration and is herewith submitted. It is proper to say that very soon after my first annual report was presented I began the work of investigating the subject of the character of these counterfeit butters as articles of food. A considerable evidence as to the unwholesomeness of these goods had been adduced by and before the Senate Committee on Public Health just prior to the enactment of chapter 202 of the Laws of 1884, but it was not generally accepted as conclusive. I had several interviews upon this subject with Dr. Clark very early last winter, and he then determined to undertake this important work. It was decided that a series of experiments in artificial digestion should be undertaken; analytical and microscopical investigations were to be utilized; a thorough and careful examination of standard works and authorities upon physiological subjects, bearing upon the matters under consideration, was to be made, and every thing done which gave any promise of assisting in arriving at a correct conclusion. The Doctor undertook this task in addition to all the other work which he was to perform as chemist for this department, although it was entirely clear that he would be obliged to surrender a considerable of his practice as a physician to others, that he might be able to accomplish all the work to be done. In order to aid him in this investigation I procured from the Patent Office at Washington certified copies of the several patents obtained by manufacturers, and some of the ingredients covered and provided for by these patents he mentions. This report of Dr. Clark's first takes up the general work of the department intrusted to him and treats of several matters of much importance.

A brief history of butter is given. He mentions various substances designated butter by the alchemists; he enumerates some of the circumstances which vary the amount of fat in milk, such as the breed, age, food of cattle, etc.; gives the best butter-producing age of the cow, the influence of temperature, of foods, etc., upon the quality and quantity of milk, and the governing principles and physiological laws appertaining. He mentions the established fact that fat is manufactured in the body of the animal from substances which contain no fat in themselves, and that the fat taken in food is itself probably broken up, reformed and not merely stored in the body; and relates many of the experiments upon which these discoveries are based. The subject of cream is taken up and the various influences which affect the rising of cream are fully discussed; and various processes of churning cream are considered. The chemical constituents of butter are given and the process and manner of producing rancidity in butter are described.

The report then takes up the subject of adulteration of butter, and much valuable information is given; and then proceeds to discuss the all-important question, whether these artificial compounds, made in imitation of and sold for pure dairy butter, are wholesome or not. I shall not attempt to give any synopsis of this portion of his report, for obvious reasons. It is sufficient for me to say that he has devoted himself with exhaustless patience and energy to the faithful performance of this work, and that he submits with great confidence, that the facts shown establish, beyond reasonable doubt, the unwholesomeness of these artificial butters, upon the following grounds:

First. On account of their indigestibility;

Second. On account of their insolubility when made from animal fats;

Third. On account of their liability to carry germs of disease into the human system; and

Fourth. On account of the probability of their containing, when made under certain patents, or in a careless manner, unhealthy ingredients.

Some valuable experiments are now in progress which will not be completed in time for this report. It is our purpose to continue this experimental work and other investigations of imitation butter as articles of food, the results of which will from time to time be made known.

The results of analyses of several samples of full-cream cheese will also be found in the report of Dr. Clark, together with the methods of the manufacture of such cheese in use by some prominent cheesemakers in the State.

The assistant commissioners and all the employees of this department have very satisfactorily performed their duties and have at all times exhibited that zeal, energy and spirit which has been absolutely essential in order that we might be able to accomplish what has been done during the year.

ALBANY, *January* 15, 1886

J. K. BROWN,
New York State Dairy Commissioner.

APPENDIX.

New York, *December* 31, 1885.

Hon. Josiah K. Brown, *New York State Dairy Commissioner, New Capitol, Albany, N. Y.:*

Dear Sir — I have the honor of submitting the following report for the year ending December 31, 1885:

The vigorous prosecutions of the violators of the law prohibiting the sale of oleomargarine during the last quarter of the year 1884, had, at the opening of the present year, greatly reduced the sale of spurious goods in the city of New York, but at about that time the Court of General Sessions of this city decided not to try any cases in relation to the sale of imitations of butter (although all such cases were where oleomargarine had been sold as butter), until the case of *The People* v. *Marx* was decided by the Court of Appeals. Consequently the dealers in fraudulent goods were encouraged to renewed efforts in pushing the sale of these counterfeits to their customers for butter. To such an extent was this done that the sales for the first three months of this year were, no doubt, more than double that of the last quarter of the previous year.

But in Kings county the judges having decided to continue to try all cases brought before them in which there was fraud, all prosecutions for selling oleomargarine for butter were promptly tried by the courts and a large number of convictions obtained. Consequently there was very little of these compounds sold during the winter and early spring months in Kings and other counties outside of New York.

The State experts were actively engaged in obtaining evidence of violations of the law, in attending court, in examinations and trials during the winter and early spring months.

But in April it was found that the usual low price of butter in that season had greatly reduced the sale of spurious goods in New York and other cities. All prosecutions up to and including April 30th, were brought under chapter 202, Laws of 1884. Thirty-nine arrests had been made between December 31, 1884, and the above-named date.

Twenty-three cases were disposed of by trial, resulting in the conviction of sixteen offenders, who were fined $100 each; and seven cases were discharged on trial; leaving sixteen cases not disposed of, to which add fifty-two of the prosecutions commenced previous to December 31, 1884, making sixty-eight cases brought under chapter 202, not disposed of by trial.

On May 1st I proceeded to obtain evidence of violations of chapter 183, Laws of 1885, and during the month ten cases were made in New

York against parties for selling oleomargarine for butter, while in Kings county we could find no violations of the law during the month.

I was aware from former experience that a very limited amount of imitation butter could be sold during the heated term, owing to the fact that the flavors of raw animal, vegetable and nut oils combined with chemicals now used in the manufacture of these spurious goods called oleomargarine, are, under high temperature, liable to decompose; and that the flavor of the oils can be so readily detected by the consumer as to render the article unsalable or nearly so. By reason of this and other facts in a conference with you, it was deemed best on the 1st of June to use the spare time of the butter experts looking after violations of the law by the sale of bogus butter, and to assign them to duty in looking after violations of the law relating to milk, as the summer months covered the period during which the largest amount of milk was handled, consequently the greater danger to the health of consumers from the adulteration of this necessary article of food during the heated term.

About September 1st, butter having advanced and the weather being cool, oleomargarine again appeared upon the market; consequently the experts who had been put upon milk duty during the summer months, and part of the force originally on milk inspection, were assigned to the duty of looking after violations of the law by the sale of counterfeit butter for the real article, since which time they have obtained evidence against a large number of persons who have been arrested and held to bail for trial, and several parties have been tried and convicted. During September and October the wholesale and retail dealers in spurious goods seemed to think that the prosecutions did not amount to much. But for the last two months they have evidently realized that it is an unsafe business, and a large majority of those who formerly handled the counterfeits are now doing a legitimate trade in pure butter. In fact very few of the dealers, except those that are retailing large quantities, are now handling the stuff. As they do not desist after having been repeatedly arrested, they have apparently concluded and, as a matter of fact, say that they are able to pay all the fines that can be imposed on them and still make money out of this nefarious business.

We have met with many obstacles during the year, and it has required a vast amount of labor to watch the manufacturers and wholesale dealers, as well as the retail grocers, and obtain sufficient evidence to warrant making arrests. For like all violators of the law they have moved with secrecy and have handled this oleaginous compound under cover of every style of package known to the butter trade, and branded the goods under the name of many well-known brands of creamery butter. Also shipped them as butter, and the manufacturers of and wholesale dealers in these vile compounds have stood ready at all times to go on the bail bond of the retailers when prosecuted. The New York Retail Grocers' Union, an organization representing eight hundred grocers, held a meeting December 3, 1885, called for the purpose of discussing the oleomargarine question, the manufacturers and dealers in these goods were invited, also the wholesale and commission dealers in butter, and requested to take part in the discussion of the subject. During the discussion, a grocer remarked that he had been informed by the assistant State dairy commissioner that the imitation goods, as then manufactured, could not be sold without violating the law, even if sold for oleomargarine, as they were an imitation and semblance of natural butter. Several manufacturers and dealers present offered to put up from one to five hundred dollars each to defend him and pay his fine if found guilty on trial, if he would handle the goods as then made in violation of law.

At the above meeting the following preamble and resolutions were offered and discussed, then the meeting adjourned to December 14, when they were unanimously adopted, and forwarded to me by C. F. Bussing, Esq., president of the organization, and I embody them in this report to show the spirit in which all reputable dealers look upon these counterfeits.

The following is the preamble and resolutions adopted:

"WHEREAS, The New York Retail Grocers' Union has put itself on record to encourage the sale of all pure goods, and to discourage and try to prevent all deceptions that are and may be practiced upon our customers by the sale of imitation or impure goods; and

"WHEREAS, We are informed that the laws of the State of New York prohibit the sale of all imitation of butter, in the way they are at present manufactured and offered for sale; and

"WHEREAS, The sale of all such imitation butter has been a detriment to the legitimate business of the retail dealer, inasmuch as that it has been a constant temptation to the dealers therein to sell it for butter, and that such fraudulent sale has created a prejudice and fear among the consumers in regard to the purchase of all butters. Therefore be it

"*Resolved,* That we discourage the sale of all imitations of butter and urge all our members and the trade in general not to handle it in any manner or form, until such a time when the manufacturers thereof will produce and offer for sale to us an article that will be distinct in appearance and different in color to that of genuine butter; that will be free from all temptation to fraud, and that will be manufactured and sold to us in strict accordance with the laws of this State. And be it further

"*Resolved,* That we most respectfully petition the State Dairy Commission to use all power that is invested upon them to stop the fraudulent sale of all imitations of butter."

There is a secret organization called the Grocers' Protective Association, representing about two hundred grocers in New York and Brooklyn. They are organized under the pretense of protecting its members from unlawful prosecutions. But the attorney for the association appears as counsel for a large majority of the dealers who are prosecuted for selling this deleterious compound for butter; showing that they are associated together in order to the more effectually defy the law, and share the costs and penalties when a member is convicted. But it gives me great pleasure to state that at least three-fourths of the retail grocers refuse to handle oleomargarine for the reason that it cannot be sold unless palmed off on their customers for butter, which they will not do, although they are compelled to see their trade drifting into the hands of their unscrupulous neighbors, who do not hesitate to traffic on the confidence of their patrons and deal out this deleterious compound to them, well knowing that they would not knowingly buy it as an article of food. The handling of these fraudulent goods seems to so degrade the retailers, that after deceiving their customers by selling it for butter, in nearly every instance, when prosecuted, they go on the witness stand and swear that they always sell it for what it is.

It having been persistently asserted, by manufacturers of and dealers in oleomargarine, that a very large percentage of the New York State dairymen were adulterating their butter by adding animal and vegetable oils to their product, I have during the year taken great pains to investigate the matter and have had a large number of samples of dairy butter analyzed and in every instance have found them pure. I have also tried to ascertain if any oils have been shipped into the dairy sections of the State, but failed to learn of a single shipment. I have also inquired of the dealers in oils used by the manufacturers to adulterate butter, and every one of them say they have never sold any oil that they have any reason to think has been for the use of dairymen, but that their sales were invariably to regular manufacturers of oleomargarine.

I have also made inquiries of a large number of commission houses each of which handle butter for from five hundred to one thousand farmers. Each and every one without exception say they have never found a single instance where they have reason to think that the dairymen have adulterated their butter.

I do not say there is not a dairyman in the State who adulterates his product, but I do say that I have failed to find any evidence of it, and am fully convinced that it is not practiced to an extent worth noticing, if at all.

It is also claimed by manufacturers and dealers in imitation butter that they desire to have their goods sold on their merits and for what they are.

If they are honest in their assertions, why is it that they make it to resemble butter so closely and pack it in every style of package known to the butter trade, and nearly always way-bill it butter when shipped?

Why have they not devised some new style of package to handle it in, and adopted some color distinct from butter?

Believing the consumers to be the greatest sufferers from these spurious goods and that the evil could be more effectually eradicated by commencing where deception begins, I have directed my attention largely to the retail dealers, who knowingly and fraudulently sell these imitations representing them to be butter.

It is an undisputed fact that the demand for these goods, unlike any other article of food, is not created by the desire of the consumer or his preference for it; but it is entirely built up by the avarice and rapacity of the dealers who deceive their customers in every sale they make.

After commencing prosecutions against a large number of grocers, I turned my attention to the manufacturers and wholesale dealers in these oleaginous compounds and have commenced prosecutions against every manufacturer, and a large number of the wholesale dealers, who have continued to manufacture and sell in violation of law. Although their business has been greatly reduced they still continue to defy the law.

The greatest difficulty in dealing with these goods arises from their being shipped into the State from several of the Eastern and Western States as butter. Consequently we cannot detect them until they are in the hands of the retailer. If any way can be devised to prevent these goods being shipped into this State unless the contents of the packages are plainly marked on the outside of each package, it would be of great service to this department in enforcing the law.

The entire receipts of both butter and oleomargarine from Chicago for the year have been only 33,135 packages, and at least 8,135 of this amount were butter, leaving the estimated amount of oleomargarine from Chicago, 2,500 packages, and as it is conceded that Chicago furnishes fully one-third of the bogus goods handled in New York city, therefore it is safe to say that not over 75,000 packages have been handled during the year, and not less than half of this amount has been shipped out of the State again, consequently the consumption of these goods has been reduced to a minimum of the quantity made in the United States, which is variously estimated at from 800,000 to 1,000,000 packages for the year.

It was estimated that there was manufactured in the United States during the year 1885, 600,000 packages of all kinds of butter imitations, and that about one-third of this amount was handled in New York; showing that the enforcement of the law had reduced the amount handled sixty per cent, while the quantity manufactured in the United States has increased over fifty per cent.

And the prosecution having greatly reduced the quantity of oleomargarine sold in the State during the year, as compared with former years, the manufacturers and dealers in spurious goods, realizing that they could not sell them in this State, without great risk, have looked for a market for their product in other States, and as some of the States having no law regulating the sale of counterfeits of butter, and most of those that have laws on the subject have no special provisions for enforcing them, there has been very little done to prevent the market of such States from being flooded with compounds that cost from ten to fourteen cents per pound. To such an extent has this been done that it has resulted in nearly excluding New York State dairy butter from all the markets that were formerly supplied with the product of our dairies. Consequently we find ourselves with an unprecedentedly large crop and no outlet for it. Therefore we have lower prices than for many years past. The following statistics will show the depreciation of that portion of the crop handled in New York city for the past four years.

The receipts and value of butter handled in New York city for the four years ending November 30, 1882–1885, were as follows:

For the year ending November 30, 1882:

Eastern....	44,215,990 lbs., valued at 29½ cts. per lb.	$13,043,617 05
Western....	35,648,850 lbs., valued at 28 cts. per lb..	9,981,678 00
Total....	79,864,840 lbs........................	$23,025,295 05

For the year ending November 30, 1883:

Eastern....	44,804,060 lbs., valued at 26 cts. per lb..	$11,649,055 60
Western....	45,743,850 lbs., valued at 24 cts. per lb..	10,978,524 00
Total....	90,547,910 lbs........................	$22,627,579 60

For the year ending November 30, 1884:

Eastern....	38,263,820 lbs., valued at 24½ cts. per lb.	$9,374,635 90
Western....	49,853,350 lbs., valued at 22 cts. per lb..	10,967,737 00
Total....	88,117,170 lbs........................	$20,342,372 90

For the year ending November 30, 1885:

Eastern....	39,480,350 lbs., valued at 22 cts. per lb..	$8,685,677 00
Western....	54,086,500 lbs., valued at 20 cts. per lb..	10,817,300 00
Total....	93,566,850 lbs........................	$19,502,977 00

These figures show by comparing 1885 with 1882 that the receipts for 1885 were 13,702,010 pounds in excess of 1882, but that the marketable value was $3,522,318.05 less.

In order to show the real difference, take the receipts of 1885 — 93,566,850 pounds, and value eastern at 29½ cts. and western at 28 cts., the same as the receipt of 1882, we have a total value of	$26,790,923 25
Deduct the actual value of 1885....................	19,502,977 00
Showing a loss on amount handled thus compared of	$7,287,946 25

This loss has been sustained by the dairy interest that depends on New York city for a market, and as not over ten per cent of the population of the United States depend upon New York city for butter, the total loss to the dairy interest of the United States cannot be less than $70,000,000 per annum on butter, while milk and cheese are greatly reduced in value by indirect competition with the most gigantic fraud of the age. If the consumer were benefited by this depreciation in the value of the dairy product, it would be some argument in favor of this compound misnamed oleomargarine. But the fact is that it cannot be sold to consumers except as butter, and is always retailed at nearly the price of good dairy butter. If offered at a low price, the purchaser would at once suspect that it was spurious and would not buy it. Consequently while these counterfeits are being purchased by the retailer at ten to fourteen cents per pound, and dealt out to his unsuspecting customers for butter at twenty to thirty cents per pound, making not less than one hundred per cent on his purchases, dairy butter awaits a market until it becomes old flavored and then sells to exporters for ten to fifteen cents per pound. The result being that no one except the manufacturer and dealer receives any benefit, while the consumer is defrauded and the dairy interest ruined.

We have investigated all the largest cities and villages along the Hudson, and several in the central part of the State, but have found very few parties selling oleomargarine in violation of law, and very few violations in any city or village, and in many of the towns we could find no spurious goods. No arrests have been made outside of New York, Kings and Westchester counties under my direction since April, but I hope to be able to commence the prosecution of those against whom we have evidence of having violated the law, very soon.

During the year the State experts have purchased 8,072 samples and have appeared in court to obtain warrants, attend examinations before police magistrates, before the grand juries, and in attending trials at Sessions, 1,773 times.

The following is a detailed report of work performed by certain employees of the department, during the year, in prosecuting violators of the laws relating to the sale of oleomargarine:

1. *Jedadiah R. Wheeler, "Expert," employed in above service, 277 days.*

Number of days in court	121
Number of days obtaining evidence	139
Number of days inspecting stores	17
Number of samples purchased	1,387
Number of samples delivered to chemist	35
Number of evenings obtaining samples	124
Number of complaints made	26
Number of appearances in cases	179
Number of stores visited	1,720
Number of stores inspected	518

2. *William W. Meeteer, "Expert," employed in above service, 255 days.*

Number of days in court	129
Number of days obtaining evidence	105
Number of days inspecting stores	21
Number of samples purchased	1,332
Number of samples delivered to chemist	55
Number of evenings obtaining samples	137
Number of complaints made	31
Number of appearances in cases	203
Number of stores visited	1,842
Number of stores inspected	693

3. *Thomas R. Gray, "Expert," employed in above service, 254 days.*

Number of days in court	128
Number of days obtaining evidence	106
Number of days inspecting stores	20
Number of samples purchased	1,414
Number of samples delivered to chemist	46
Number of evenings obtaining samples	111
Number of complaints made	35

Number of appearances in cases	225
Number of stores visited	1,795
Number of stores inspected	558

4. *Thomas C. DuBois, "Expert," employed in above service,* 289 *days.*

Number of days in court	71
Number of days obtaining evidence	120
Number of days inspecting stores	59
Number of days on special duty	39
Number of samples purchased	1,625
Number of samples delivered to chemist	23
Number of evenings obtaining samples	119
Number of complaints made	11
Number of appearances in cases	103
Number of stores visited	2,329
Number of stores inspected	652

5. *Edmond S. Wilson, "Expert," employed in above service,* 249 *days.*

Number of days in court	104
Number of days obtaining evidence	97
Number of days inspecting stores	48
Number of samples purchased	1,547
Number of samples delivered to chemist	32
Number of evenings obtaining samples	99
Number of complaints made	16
Number of appearances in cases	143
Number of stores visited	2,340
Number of stores inspected	598

6. *Charles Sears, "Expert," employed in above service,* 66 *days.*

Number of days in court	31
Number of days obtaining evidence	35
Number of days inspecting stores	....
Number of samples purchased	298
Number of samples delivered to chemist	17
Number of evenings obtaining samples	41
Number of complaints made	11
Number of appearances in cases	35
Number of stores visited	396
Number of stores inspected	....

7. *Joseph J. Sorogan, "Expert," employed in above service,* 74 *days.*

Number of days in court	21
Number of days obtaining evidence	53
Number of days inspecting stores	....
Number of samples purchased	291

Number of samples delivered to chemist	15
Number of evenings obtaining samples	51
Number of complaints made	6
Number of appearances in cases	37
Number of stores visited	346
Number of stores inspected	....

8. *Archibald D. Clark, "Expert," employed in above service,* 92 *days.*

Number of days in court	28
Number of days obtaining evidence	24
Number of days on special duty	40
Number of samples purchased	178
Number of samples delivered to chemist	5
Number of evenings obtaining samples	26
Number of complaints made	5
Number of appearances in cases	34
Number of stores visited	157
Number of stores inspected	....

9. *Charles M. Stillwell, A. M., and Thomas S. Gladding, A. M.*

Number of analyses of butter samples	141
Number of appearances in cases	196
Number of days in court	102

10. *Edward G. Love, Ph. D.*

Number of analyses of butter samples	89
Number of appearances in cases	113
Number of days in court	61

11. *E. H. Bartley, M. D.*

Number of analyses of butter samples	4
Number of appearances in cases	13
Number of days in court	9
Number of days investigating lard	5

12. *J. F. Geisler, Ph. C.*

(Employed from November 15, 1885.)

Number of analyses of butter samples	12
Number of appearances in cases	1
Number of days in court	1

13. *A. C. Salmon, Counsel.*

Number of cases prosecuted	53
Number of cases disposed of	46

Number of convictions	29
Number of dismissals	17
Number of appearances in cases	212
Number of days in court	87
Number of cases removed to General Sessions	2
Number of appeals taken	12
Amount of fines imposed	$2,975

14. *F. V. S. Oliver, Counsel.*

Number of cases prosecuted	66
Number of cases disposed of	8
Number of convictions	2
Number of dismissals	4
Number of appearances in cases	217
Number of days in court	80
Number of cases bailed to General Sessions	44
Number of cases bailed to Special Sessions	14
Amount of fines imposed	$200

15. *Le Roy S. Gove, Counsel, employed from October* 27, 1885.

Number of cases prosecuted	18
Number of cases disposed of	2
Number of convictions	1
Number of dismissals	2
Number of appearances in cases	62
Number of days in court	27
Number of cases bailed to General Sessions	8
Number of cases bailed to Special Sessions	7
Amount of fines imposed	$100

16. *Total.*

Number of days	1,928
Number of days in court	1,005
Number of days obtaining evidence	679
Number of days inspecting stores	165
Number of days on special duty	79
Number of samples purchased	8,072
Number of samples delivered to chemist	228
Number of evenings obtaining samples	708
Number of complaints made	141
Number of appearances in cases	1,773
Number of stores visited	10,925
Number of stores inspected	3,018
Number of analyses, "butter samples"	228
Number of analyses, butter samples delivered by G. F. Gadley,	2
Number of analyses, butter samples delivered by G. Palmer	2
Number of analyses, butter samples delivered by asst. com'r.	2
Number of analyses, butter samples delivered by New York health officer	1
Number of analyses, lard from diseased hog	1

Number of analyses, cheese samples	10
Number of cases bailed to General Sessions	58
Number of cases bailed to Special Sessions	83
Number of appeals taken	13
Number of cases disposed of	54
Number of cases dismissed on examination and trials	23
Number of convictions	31
Number of cases not disposed of	87
Amount of fines imposed	$3,175

When it is considered that at every appearance in court we have to be ready with counsel, chemist and witnesses, and that cases are pending in not less than twenty different courts, you can readily see that it requires a great deal of labor and care to see that our witnesses are on hand in the proper courts. And it often happens that we have cases in three or four courts the same day.

It is my opinion that the course of procedure in the courts of New York county is too complicated. In order to bring a case of simple misdemeanor to trial we have to obtain a warrant from a police magistrate, before whom the accused can have a preliminary examination. If sufficient cause is shown to hold him, he is then allowed or required to give bail. The case then has to be brought before the grand jury, and if an indictment is found it is placed on the calendar for trial for the Court of General Sessions. To bring an offender to trial in New York under this system causes three times the labor that it does in Brooklyn, where all misdemeanors are tried before the court issuing the warrant (with or without a jury as the accused may elect), except that where sufficient reason is shown, any one of the judges of a higher court may remove the case to another tribunal.

It seems to me that in the proper administration of justice there should be a tribunal so created or organized in the city of New York, that in all cases of misdemeanor the defendant can be speedily tried before the court or magistrate issuing the warrant, with or without a jury, and having also the right to have his case removed to another and higher court whenever a judge of such higher court shall certify that it is a proper case for the consideration of the grand jury. Such court should be distinct from the police courts, in that particular differing from the practice in Brooklyn. The tribunal thus constituted and having only charges of misdemeanors brought before it, could proceed to try, dispose of them promptly and in regular order on the calendar. Under the practice so established, proceedings for punishing misdemeanors would not be subjected to delay as they now are whenever prison cases are ready for trial.

Since April 30, 1885, there has been evidence obtained of one hundred and thirty-two violations of law and one hundred and two arrests have been made, leaving thirty cases in which arrests have not yet been made. Of the one hundred and two arrests made, thirty-one have been disposed of by trial, resulting in fifteen convictions, and sixteen have been discharged on trial for various reasons, principally by showing they were not owners of the store when the sale was made. Sixty-three cases are still pending that were brought under chapter 183, Laws of 1885.

And there have been ten arrests made under chapter 215, Laws of 1882, for violating the hotel clause in that act, and two have been convicted. During the year under all of the laws governing the sale of imitations of dairy products there have been one hundred and forty-one arrests made, and thirty-one convictions, and twenty-three discharges on examination or trial, leaving eighty-seven cases not disposed of.

The convictions have been mostly in Kings county. The only reason why this has been so is the difference in the manner of prosecuting as before stated.

I cheerfully embrace this opportunity to acknowledge on behalf of this department the valuable assistance received from the officers of the New York and Brooklyn Health Boards, the officers of the Retail Grocers' Union, the committee representing the butter trade, and many private citizens, all of which is heartily appreciated.

The district attorneys, in the counties in which actions have been brought, have given the business of the department prompt attention and forwarded the prosecutions in every case as fast as the great press of business in their respective offices would permit. As most of the prosecutions have been in New York and Kings counties, the department is especially indebted to the district attorneys of those counties, and their able assistants for their prompt attention to all matters connected with this department.

The prosecutions instituted and the actions brought under my immediate supervision in this section of the State have received prompt attention in nearly every instance from the judges of the court in which they were brought.

In conclusion, I am greatly pleased to be able to say that all of the counsel, chemists, experts and agents appointed by you for the department in this vicinity have done their full duty, taking pride in the success of the department, and have greatly assisted me in accomplishing what I have the honor to report.

Respectfully submitted,
B. F. VAN VALKENBURGH,
Assistant New York State Dairy Commissioner.

BUFFALO, N. Y., *December 31st*, 1885.

HON. JOSIAH K. BROWN, *New York State Dairy Commissioner, Albany, N. Y.:*

DEAR SIR — I beg leave to submit the following report:

On the 1st day of December, 1884, I was employed by you as an expert and agent at Buffalo, N. Y., to assist in the proper enforcement of the laws of the State relating to dairy products, with instructions from you to take such steps as were necessary for that purpose.

I immediately entered upon the execution of this trust by taking a general survey of the field in which I was to operate, and in reporting to you the condition of affairs as I found them at Buffalo.

During the months of December, January, February and March ensuing I purchased from wholesale and retail dealers at Buffalo, some seventy-five samples of suspected butter, which in every case was represented and sold to me for a genuine article. And in further pursuance of your instructions I delivered these samples properly sealed in glass jars, marked and numbered, to Prof. R. A. Witthaus, city chemist, at Buffalo, for chemical analysis; and taking from him at the time of delivery of each and every such sample a proper receipt therefor. These samples were all carefully analyzed by the chemist, who subsequently reported to me the result of his labors. Of the number thus analyzed twenty-five were found to be adulterated and in some of the cases as much as seventy-five per cent of foreign matter was discovered. Many of the other samples were found to have been adulterated, but not to so great an extent. The analyses of these samples were all properly certified by the chemist and the certificates thereof forwarded to you.

In the early part of April, 1885, after consultation with the district attorney of Erie county, and receiving from him the assurance of his support and co-operation, and in furtherance of your directions, I began prosecutions against some of the violators of these laws that come under the sixth section of the act of 1884. Informations were filed and warrants of arrest issued. These offenders, on being brought into court and arraigned, pleaded not guilty. Several continuances were had at the instance and request of the defendants. Finally it was, with your concurrence, agreed that the further prosecution of these cases should be postponed until after a decision had been arrived at in the case of *The People* v. *Marx*, which was then pending in the Court of Appeals, touching the constitutionality of the sixth section of this act. The adverse decision by the Court of Appeals in the Marx case was soon thereafter made known, and all further attempts to bring these cases to trial were abandoned.

Prior to the passage of the act of 1884, the manufacture of oleomargarine and butterine was extensively carried on in the city of Buffalo. Its passage was the means of accomplishing a complete suspension of their manufacture here, and materially aided in suppressing the sale of the goods. At the time when the decision in the Marx case was made known this compound had almost wholly disappeared from the counters of both wholesale and retail dealers, and the community was encouraged in the hope and belief that imposition in the purchase and consumption of butter was nearly at an end.

On the 1st day of July, 1885, I received my commission from you as assistant State dairy commissioner.

On or about the middle of April of the current year various complaints were made to me by our citizens in relation to the poor quality of the milk that was being sold by grocerymen and milk dealers. I then proceeded to investigate the condition of the milk trade in Erie and adjoining counties Samples of milk were obtained, which were likewise subjected to chemical analysis by Prof. Witthaus. It was soon discovered that the adulteration of milk was extensively practiced by both the wholesale and retail dealer. Adulteration and deceptive practices seem to have been the rule and honesty the exception. Out of two hundred and fifty-four samples of milk obtained in the city of

Buffalo, from April 15 to the 25th day of November, 1885, one hundred and eighty-six were found to be adulterated. The adulteration consisted principally in the admixture of water. The percentage of this convenient and seductive fluid, as shown by the cases that have been analyzed, ranged from one to twenty-five per cent. I have no hesitation in saying and find, as the result of my labors in this department, that a large majority of the dealers in milk in this vicinity from whom samples have been obtained, disrespect and intentionally violate the provisions of this law, and that nothing short of its rigid enforcement will cure the evil.

The prosecutions of these cases have been delayed for prudential reasons until the decision of the General Term of the Third Department of the Supreme Court, in which the constitutionality of that part of the act of 1884, relating to milk, then under consideration, should become known. Immediately thereafter, and on the 4th day of the present month, I began prosecutions against these offenders in the Court of Special Sessions at Buffalo. Upwards of seventy suits have been begun, and fifty-four warrants of arrest issued. Of this number forty-four have been arraigned. In four of the cases the defendants, after their plea of not guilty had been entered, and upon a demand for an immediate trial on the part of the people, without further proceedings withdrew their pleas of not guilty and were each fined the sum of $25. In two cases the evidence on behalf of the people being presented, the defendants, by advice of their attorneys, withdrew their pleas of not guilty and were likewise fined the minimum sum fixed in the act, making thus far six cases of conviction.

Soon after the above-mentioned six cases had been disposed of, I learned that a powerful combination had been formed among the grocerymen who deal in milk, and among the milkmen in this city, having for its sole object resistance to the enforcement of this law; that large sums of money had been assessed and collected from the members of these combinations in furtherance of this object. Able lawyers have been employed by them. Upon the trial of the seventh of this class of cases, which occurred on the 29th and 30th days of the present month, the defendant demanded a trial by jury. Two eminent and skillful lawyers appeared for the defense. The case was hotly and stubbornly contested, but ended in a verdict of guilty. Hence the people have been successful in this very important test case by the conviction of the defendant. No decision adverse to the people has thus far been rendered, and the cases now undetermined will be brought to trial as rapidly as possible. The energetic and persistent prosecution of these offenders is earnestly recommended, as being the only means afforded for the protection of the people from fraud and imposition in one of the principal essentials of human consumption, comfort and health.

During the latter part of the past summer, and during the fall, I made strenuous efforts to procure a hearing before a grand jury, for the purpose of securing indictments against the persons from whom were purchased the samples of spurious butter last winter and spring. My efforts in this direction have been attended with many perplexities and delays, owing to the fact that our criminal courts have been, and still are, overloaded with important criminal business; but I am pleased to report that my efforts to obtain a hearing before a grand jury were finally successful, and on or about the 13th day of November, 1885, I succeeded, through the kind and considerate assistance of the district attorney, in going before a grand jury of the Court of Oyer and Terminer and obtaining ten indictments against these offenders. I regret not being able to say at what time these cases will be heard, but stringent efforts will continue to be made for a speedy trial. It is my intention to bring other cases for presentment at the sitting of the next grand jury, or as soon thereafter as a hearing can be obtained, and seek to procure indictments in all the remaining cases against the persons from whom I obtained samples of spurious butter last winter and spring.

During the months of November and December, 1885, I have caused samples of butter suspected of being spurious to be obtained in the city of Buffalo, and submitted them for chemical analysis to Prof. Witthaus. The whole number of such samples so delivered since the 1st of November last is eighteen, of which twelve have been declared by the chemist to be adulterated. The certificates of analysis of these samples will be forwarded to you with this report.

Of the two hundred and ninety samples of milk secured by me, two hundred and nineteen have been pronounced by the chemist adulterated; the adulteration consisting principally in the admixture of water. Out of the ninety-four cases of suspected butter forty have been found to contain foreign substances, and upwards of thirty were largely composed of compounds other than butter made from pure milk or cream.

I am happy to report that the people throughout the western section of the State have of late manifested a much greater interest than heretofore in the suppression and extinguishment of this overshadowing evil; and that there is an earnest desire for the energetic and rigid enforcement of the dairy laws intended for their protection. Public sentiment in this respect has been stimulated and strengthened in the hearty support given it by the entire press of Western New York. While the manufacturers of bogus butter have disappeared from our midst, there remains the manufacturer of oleo oils in large quantities, and the importation of spurious butter is still carried on to that extent as greatly to depreciate in value the legitimate products of the dairy.

On the 30th day of the present month the supervisors of Erie county, at their annual session, passed a resolution condemnatory of these impositions upon the people and industries of the State, and on account of which the greatest agricultural interests are made to suffer; and instructed the district attorney to put forth renewed efforts for their suppression and the punishment of all persons engaged in the traffic.

The following is a statement of the action of the board of supervisors, as reported:

"WHEREAS, The manufacture and sale of oleomargarine and butterine are being brought into direct competition with and undermining the greatest agricultural interest of the State; and

"WHEREAS, Said goods are sold in many instances for genuine dairy products in the county, and we believe their use, together with that of adulterated milk, to be injurious to the health and lives of its inhabitants; and

"WHEREAS, There have been a number of cases brought before the grand jury, and indictments found against several parties on the testimony produced before said grand jury; therefore

"*Resolved,* That it is the sense of this board that all parties indicted for selling such adulterated food, representing the same to be genuine dairy products, contrary to law, be prosecuted, and that the district attorney of this county be and is hereby requested to make strenuous efforts to have all persons who are, or may hereafter be, indicted for such offenses punished to the full extent of the law. Adopted."

The following is a detailed report of work performed by certain employees of the department during the year, in prosecuting violators of the laws relating to the sale of milk:

I take pleasure in saying that all persons who have been employed with me for this department have faithfully performed their duties, and have greatly assisted me in accomplishing the work which I have the honor now to report.

Respectfully submitted,
MARCUS A. PERRY,
Assistant Dairy Commissioner, Buffalo, N. Y.

REPORT ON MILK.

By Edward W. Martin.

B. *Charles Sears, "Expert," employed in above service,* 241 *days.*

Number of days inspecting milk	191
Number of days inspecting dairies	36
Number of days attending court	14
Number of samples delivered to chemist	106
Number of specimens examined	7,297
Number of milk inspections	4,946
Number of creameries inspected	11
Number of condenseries inspected	3
Number of dairies inspected	28
Number of stables inspected	28
Number of cows inspected	670
Number of miles traveled, about	28,000

D. *Joseph J. Sorogan, "Expert," employed in above service,* 233 *days.*

Number of days inspecting milk	207
Number of days inspecting dairies	5
Number of days attending court	21
Number of samples delivered to chemist	42
Number of specimens examined	4,413
Number of milk inspections	3,953
Number of creameries inspected	8
Number of condenseries inspected	......
Number of dairies inspected	2
Number of stables inspected	10
Number of cows inspected	102
Number of miles traveled, about	21,000

E. *Samuel J. White, "Expert," employed in above service,* 306 *days.*

Number of days inspecting milk	183
Number of days inspecting dairies	38
Number of days attending court	25
Number of days on special duty	60
Number of samples delivered to chemist	126
Number of specimens examined	3,692
Number of milk inspections	3,151

Number of creameries inspected	11
Number of condenseries inspected	1
Number of dairies inspected	34
Number of stables inspected	34
Number of cows inspected	851
Number of miles traveled about	24,000

H. *William W. Meeteer, "Expert," employed in above service,* 52 *days.*

Number of days inspecting milk	42
Number of days inspecting dairies	......
Number of days attending court	......
Number of days on special duty	10
Number of samples delivered to chemist	25
Number of specimens examined	652
Number of milk inspections	536
Number of creameries inspected	......
Number of condenseries inspected	......
Number of dairies inspected	......
Number of stables inspected	2
Number of cows inspected	15
Number of miles traveled, about	4,500

I. *Jedediah R. Wheeler, "Expert," employed in above serviece,* 27 *days.*

Number of days inspecting milk	15
Number of days inspecting dairies	12
Number of days attending court	......
Number of samples delivered to chemist	9
Number of specimens examined	330
Number of milk inspections	170
Number of creameries inspected	......
Number of condenseries inspected	......
Number of dairies inspected	2
Number of stables inspected	2
Number of cows inspected	64
Number of miles traveled. about	3,500

G. *Thomas R. Gray, "Expert," employed in above service,* 50 *days.*

Number of days inspecting milk	29
Number of days inspecting dairies	......
Number of days attending court	......
Number of days on special duty	21
Number of samples delivered to chemist	19
Number of specimens examined	1,613
Number of milk inspections	723
Number of creameries inspected	1
Number of condenseries inspected	1
Number of dairies inspected	......
Number of stables inspected	14

Number of cows inspected	282
Number of miles traveled	2,700

J. *Thomas C. DuBois, "Expert," employed in above service,* 18 *days.*

Number of days inspecting milk	18
Number of days inspecting dairies	......
Number of days attending court	......
Number of samples delivered to chemist	5
Number of specimens examined	450
Number of milk inspections	333
Number of creameries inspected	......
Number of condenseries inspected	3
Number of dairies inspected	......
Number of stables inspected	......
Number of cows inspected	......
Number of miles traveled	1,313

F. *Edward S. Wilson, Agent, employed in above service,* 53 *days.*

Number of days inspecting milk	35
Number of days inspecting dairies	......
Number of days attending court	......
Number of days on special duty	18
Number of samples delivered to chemist	9
Number of specimens examined	620
Number of milk inspections	395
Number of creameries inspected	1
Number of condenseries inspected	......
Number of dairies inspected	......
Number of stables inspected	25
Number of cows inspected	509
Number of miles traveled	2,800

K. *Archibald D. Clark, "Expert," employed in above service,* 9 *days.*

Number of days inspecting milk	9
Number of days inspecting dairies	......
Number of days attending court	......
Number of samples delivered to chemist	......
Number of specimens examined	342
Number of milk inspections	235
Number of creameries inspected	......
Number of condenseries inspected	......
Number of dairies inspected	......
Number of stables inspected	......
Number of cows inspected	......
Number of miles traveled	352

Total.

Number of days	989
Number of days inspecting milk	729
Number of days inspecting dairies	91
Number of days attending court	60
Number of days on special duty	109
Number of samples delivered to chemist	341
Number of specimens examined	19,409
Number of milk inspections	14,442
Number of creameries inspected	32
Number of condenseries inspected	8
Number of dairies inspected	66
Number of stables inspected	115
Number of cows inspected	2,496
Number of miles traveled, about	88,165

Daily average per week of cans, of forty quarts each, of milk, condensed milk and cream.

RAILROADS.	Week ending January 7.			Week ending January 14.			Week ending January 21.			Week ending January 28.		
	Milk.	Condensed milk.	Cream.	Milk.	Condensed milk.	Cream.	Milk.	Condensed milk.	Cream.	Milk.	Condensed milk.	Cream.
Lake Erie and Western	3,283	27	74	2,972	23	76	3,077	26	72	3,000	23	70
Harlem and New York Central	2,404	144	12	2,529	183	13	2,412	194	13	2,389	199	16
Ontario and Western	1,435	....	33	1,458	....	38	1,471	....	37	1,501	....	33
New York City and Northern.......	1,074	....	....	1,067	....	....	1,062	....	....	1,074	....	.....
New York, New Haven and Hartford.	1,038	....	....	1,023	....	....	1,026	....	....	1,032	....	.....
New York, Susquehanna and West'n.	962	....	14	970	....	10	971	....	14	976	....	11
New Jersey Central...............	402	....	6	408	....	6	414	....	5	423	....	5
Miscellaneous	462	....	....	412	....	....	455	....	...	455	....	.....
Total	11,060	171	139	10,839	206	143	10,888	220	140	10,850	222	134
Surplus disposed of on platform per forty-quart can................	$1 50			$1 60			$1 70			$1 75		

Daily average per week of cans, of forty quarts each, of milk, condensed milk and cream.

RAILROADS.	Week ending February 4.			Week ending February 11.			Week ending February 18.			Week ending February 25.		
	Milk.	Condensed milk.	Cream.	Milk.	Condensed milk.	Cream.	Milk.	Condensed milk.	Cream.	Milk.	Condensed milk.	Cream.
Lake Erie and Western	3,129	25	69	3,116	22	77	3,161	20	80	3,252	17	74
Harlem and New York Central	2,378	197	17	2,380	202	15	2,371	202	20	2,413	193	24
Ontario and Western	1,541	....	40	1,558	....	36	1,426	....	36	1,761	....	43
New York City and Northern	1,073	....	....	1,074	....	....	1,069	....	...	1,062	....	....
New York, New Haven and Hartford	1,029	....	....	1,043	....	....	1,011	....	...	1,082	....	....
New York, Susquehanna and West'n	976	....	9	983	....	12	989	....	16	1,058	....	13
New Jersey Central	423	....	5	427	....	5	429	....	5	430	....	4
Miscellaneous	451	....	....	457	....	....	453	....	....	449	....	....
Total	11,000	222	140	11,038	224	145	10,909	222	571	11,507	210	158
Surplus disposed of on platform per forty-quart can	$1 [illegible]5			$1 65			$1 60			$2 00		

Daily average per week of cans, of forty quarts each, of milk, condensed milk and cream.

RAILROADS.	Week ending March 4.			Week ending March 11.			Week ending March 18.			Week ending March 25.		
	Milk.	Condensed milk.	Cream.	Milk.	Condensed milk.	Cream.	Milk.	Condensed milk.	Cream.	Milk.	Condensed milk.	Cream.
Lake Erie and Western	4,658	14	47	3,247	16	71	3,302	20	68	3,283	21	65
Harlem and New York Central......	2,426	203	24	2,408	190	26	2,415	188	20	2,424	189	18
Ontario and Western..............	1,696	....	49	1,672	....	57	1,751	....	52	1,767	....	52
New York City and Northern.......	1,090	....	....	1,103	....	....	1,112	....	....	1,120	....	.....
New York, New Haven and Hartford.	1,079	....	....	1,080	....	....	1,079	....	....	1,102	....	.....
New York, Susquehanna and West'n.	997	....	17	970	....	12	965	....	12	956	....	13
New Jersey Central................	381	....	5	443	....	6	440	...	5	438	....	5
Miscellaneous....................	469	....	....	458	...	....	457	....	....	459	..	.....
Total..........	12,791	217	142	11,381	206	172	11,521	208	157	11,549	210	158
Surplus disposed of on platform per forty-quart can...............	$1 50			$1 60			$1 55			$1 65		

Daily average per week of cans, of forty quarts each, of milk, condensed milk and cream.

ILROADS.	Week ending April 1.			Week ending April 8.			Week ending April 15.			Week ending April 22.		
	Milk.	Condensed milk.	Cream.	Milk.	Condensed milk.	Cream.	Milk.	Condensed milk.	Cream.	Milk.	Condensed milk.	Cream.
Lake Erie and Western	3,346	22	71	3,549	24	88	3,492	22	94	3,573	24	104
Harlem and New York Central	2,409	167	21	2,309	59	19	2,312	92	22	2,327	88	21
Ontario and Western	1,778	....	61	1,783	....	146	1,758	....	153	1,667	....	155
New York City and Northern	1,144	....	....	1,179	....	....	1,176	....	....	1,198	....	....
New York, New Haven and Hartford	1,124	....	....	1,100	....	....	1,108	....	....	1,106	....	....
New York, Susquehanna and West'n	990	....	8	1,036	....	18	1,173	....	6	1,111	....	29
New Jersey Central	437	....	5	418	....	5	432	....	5	429	....	6
Miscellaneous	477	....	....	476	....	....	469	....	....	469	....	....
Total	11,705	189	166	11,850	83	276	11,920	114	280	11,880	112	315
Surplus disposed of on platform per forty-quart can	$1 65			$2 00			$3 00			$3 50		

Daily average per week of cans, of forty quarts each, of milk, condensed milk and cream.

RAILROADS.	Week ending April 29.			Week ending May 6.			Week ending May 13.			Week ending May 20.		
	Milk.	Condensed milk.	Cream.	Milk.	Condensed milk.	Cream.	Milk.	Condensed milk.	Cream.	Milk.	Condensed milk.	Cream.
Lake Erie and Western	3,920	23	135	3,697	23	96	3,622	25	90	3,893	26	153
Harlem and New York Central	2,368	93	22	2,390	67	19	2,344	70	19	2,351	69	21
Ontario and Western	1,865	....	215	1,789	91	109	1,723	109	84	1,766	107	84
New York City and Northern	1,231	....	....	1,276	....	....	1,278	....	....	1,317	....	.....
New York, New Haven and Hartford	1,146	....	....	1,154	....	....	1,129	....	....	1,156	....	.....
New York, Susquehanna and West'n	1.218	....	33	1,183	....	22	1,170	....	14	1,218	....	52
New Jersey Central	449	....	6	430	....	7	444	....	5	471	....	7
Miscellaneous	479	....	....	485	....	....	476	....	....	479	...	.....
Total	12,676	116	411	12,404	181	253	12,186	204	212	12,651	202	317
Surplus disposed of on platform per forty-quart can	$3 50			$1 20			$1 25			$1 32		

Daily average per week of cans, of forty quarts each, of milk, condensed milk and cream.

RAILROAD.	Week ending May 27.			Week ending June 3.			Week ending June 10.			Week ending June 17.		
	Milk.	Condensed milk.	Cream.	Milk.	Condensed milk.	Cream.	Milk.	Condensed milk.	Cream.	Milk.	Condensed milk.	Cream.
Lake Erie and Western	3,746	24	153	4,087	27	157	4,068	30	165	4,228	30	218
Harlem and New York Central	2,412	70	22	2,302	61	17	2,354	55	13	2,376	59	24
Ontario and Western	2,014	109	147	1,993	108	141	2,005	113	153	2,006	105	199
New York City and Northern	1,378	....	....	1,265	....	....	1,407	....	....	1,388	....	....
New York, New Haven and Hartford	1,229	....	....	1,229	....	....	1,255	....	....	1,285	....	....
New York, Susquehanna and West'n	1,310	....	45	1,325	....	53	1,304	....	53	1,369	....	73
New Jersey Central	483	...	6	474	....	6	470	....	9	486	....	6
Miscellaneous	499	....	....	507	....	....	493	....	....	466	....	....
Total	13,066	203	373	13,182	196	374	13,356	198	393	13,604	194	521
Surplus disposed of on platform per forty-quart can	$1 32			$1 30			$1 20			$2 00		

Daily average per week of cans, of forty quarts each, of milk, condensed milk and cream.

RAILROADS.	Week ending June 24.			Week ending July 1.			Week ending July 8.			Week ending July 15.		
	Milk.	Condensed milk.	Cream.	Milk.	Condensed milk.	Cream.	Milk.	Condensed milk.	Cream.	Milk.	Condensed milk.	Cream.
Lake Erie and Western............	4,702	25	200	4,244	27	206	4,163	23	221	4,440	23	216
Harlem and New York Central.....	2,352	66	27	2,312	66	26	2,144	27	33	2,303	63	33
Ontario and Western..............	2,261	176	174	2,211	99	210	2,138	111	197	2,402	122	187
New York City and Northern.......	1,369	...	....	1,341	....	....	1,295	....	....	1,241	....	....
New York, New Haven and Hartford	1,314	...	...	1,288	....	....	1,245	....	....	1,269	....	....
New York, Susquehanna and West'n	1,403	...	65	1,339	....	66	1,307	....	75	1,405	....	49
New Jersey Central...........	488	....	7	480	....	9	466	....	7	499	....	6
Miscellaneous.....................	487	....	10	469	....	12	447	....	10	429	....	12
Total...........	14,376	267	303	13,686	192	529	13,205	161	543	13,988	208	503
Surplus disposed of on platform, per forty-quart can........	0 75			$1 20			$1 12			$3 00		

Daily average per week of cans, of forty quarts each, of milk, condensed milk and cream.

RAILROADS.	Week ending July 22.			Week ending July 29.			Week ending August 5.			Week ending August 12.		
	Milk.	Condensed milk.	Cream.	Milk.	Condensed milk.	Cream.	Milk.	Condensed milk.	Cream.	Milk.	Condensed milk.	Cream.
Lake Erie and Western	4,466	21	220	4,468	22	192	4,793	23	182	4,010	20	170
Ontario and Western	2,571	113	224	2,537	115	215	2,622	126	195	2,388	117	140
Harlem and New York Central	1,956	72	33	1,952	92	42	2,301	88	46	2,024	36	28
New York, Susquehanna and West'n.	1,456	....	96	1,473	....	73	1,580	....	66	1,316	....	58
New York City and Northern	1,213	....	....	1,222	....	....	1,426	....	....	1,267	....	.....
New York, New Haven and Hartford	1,266	....	....	1,249	....	....	1,222	....	....	1,179	....	.....
New Jersey Central	505	....	4	511	....	3	477	....	3	480	....	4
Miscellaneous	412	....	9	410	....	10	413	....	9	422	....	12
Total	13,845	206	586	13,822	239	535	14,834	237	501	13,086	353	412
Surplus disposed of on platform per forty-quart can	$3 00			$1 12			$1 50			$1 60		

Daily average per week of cans, of forty quarts each, of milk, condensed milk and cream.

RAILROADS.	Week ending August 19.			Week ending August 26.			Week ending Sept. 2.			Week ending Sept. 9.		
	Milk.	Condensed milk.	Cream.	Milk.	Condensed milk.	Cream.	Milk.	Condensed milk.	Cream.	Milk.	Condensed milk.	Cream.
Lake Erie and Western	3,876	23	181	3,775	28	154	3,768	25	151	3,752	23	148
Ontario and Western	2,164	113	143	2,062	115	131	2,109	112	128	2,099	110	125
Harlem and New York Central	1,967	30	27	1,824	31	26	1,823	29	24	1,821	30	23
New York, Susquehanna and West'n.	1,250	....	53	1,265	....	53	1,268	....	50	1,256	....	48
New York City and Northern	1,243	....	....	1,245	...	.....	1,240	....	....	1,237	....	.....
New York, New Haven and Hartford.	1,103	....	....	1,109	....	.....	1,098	....	....	1,083	....	...
New Jersey Central	445	....	3	444	....	3	437	....	2	438	....	3
Miscellaneous	415	....	5	418	....	7	412	....	5	415	....	6
Total	12,463	166	412	12,142	169	374	12,155	166	360	12,101	163	353
Surplus disposed of on platform, per forty-quart can	80 cents.			90 cents.			$1 50			$1 52		

Daily average per week of cans, of forty quarts each, of milk, condensed milk and cream.

RAILROADS.	Week ending Septemb'r 16.			Week ending Septemb'r 23.			Week ending Septemb'r 30.			Week ending October 7.		
	Milk.	Condensed milk.	Cream.	Milk.	Condensed milk.	Cream.	Milk.	Condensed milk.	Cream.	Milk.	Condensed milk.	Cream.
[illegible] Erie and [illegible]	3,850	22	178	3,781	24	160	3,691	21	133	3,657	24	93
Ontario and [illegible]	2,161	110	137	2,102	128	135	2,112	109	122	1,956	110	66
Harlem and New York Central	1,948	28	28	1,829	33	30	1,798	32	23	2,130	54	7
New York, Susquehanna and West'n.	1,230	....	55	1,271	....	57	1,262	....	41	1,000	....	18
New York City and N[illegible]rn	1,238	....	....	1,258	....	....	1,247	....	....	1,149	....	....
New York, New Haven and [illegible]ford.	1,100	....	...	1,102	....	....	1,110	....	....	1,086	....	....
New Jersey Central	438	....	4	436	....	5	440	....	3	424	....	3
[illegible]als	410	....	6	420	....	4	416	....	6	310	....	4
Total	12,375	160	378	12,199	185	391	12,076	162	328	11,712	188	191
Surplus disposed of on platform per forty-quart can	$1 60			$1 65			$1 75			$2 00		

Daily average per week of cans, of forty quarts each, of milk, condensed milk and cream.

RAILROADS.	Week ending October 14.			Week ending October 21.			Week ending October 28.			Week ending November 4.		
	Milk.	Condensed milk.	Cream.	Milk.	Condensed milk.	Cream.	Milk.	Condensed milk.	Cream.	Milk.	Condensed milk.	Cream.
New York, Lake Erie and Western	3,492	27	78	3,539	24	84	3,587	27	70	3,424	22	70
New York, Ontario and Western	1,821	115	44	1,810	112	49	1,940	115	52	1,808	122	39
New York Central and Harlem	2,161	61	10	2,198	46	18	2,204	71	14	2,200	50	14
New York City and Northern	1,157	....	....	1,162	....	...	1,156	...	....	1,132	....	.....
New York, Susquehanna and West'n.	1,259	...	14	1,254	..	17	1,153	....	21	1,124	....	18
New York, New Haven and Hartford.	1,079	....	....	1,069	....	....	1,052	....	...	1,048	....	....
New Jersey Central	404	....	4	407	....	4	414	.. .	5	393	....	.....
Miscellaneous	455	....	9	464	...	7	452	....	6	441	....	8
Total	12,028	203	159	11,903	182	179	11,958	213	168	11,570	194	149
Surplus disposed of on platform per forty-quart can	$1 00			$1 25			$2 00			$2 50		

Daily average per week of cans, of forty quarts each, of milk, condensed milk and cream.

RAILROAD.	Week ending November 11.			Week ending November 18.			Week ending November 25.			Week ending December 2.		
	Milk.	Condensed milk.	Cream.	Milk.	Condensed milk.	Cream.	Milk.	Condensed milk.	Cream.	Milk.	Condensed milk.	Cream.
New York, Lake Erie and Western..	3,450	20	86	3,490	23	98	3,391	26	97	3,184	30	87
New York, Ontario and Western	1,788	129	46	1,731	130	35	1,730	165	36	1,570	166	31
New York Central and Harlem......	2,212	96	20	2,260	115	32	2,292	105	30	2,259	98	22
New York City and Northern.......	1,105	...	....	1,113	....	....	1,080	....	....	1,065	..	.....
New York, Susquehanna and Western.	1,125	....	23	1,128	....	25	1,083	....	22	998	....	15
New York, New Haven and Hartford	1,044	....	....	1,041	...	....	1,025	....	....	985	....	.....
New Jersey Central...............	375	....	3	392	....	5	385	....	3	375	....	3
Miscellaneous....................	443	...	6	438	....	8	427	....	5	416	...	3
Total........................	11,552	245	184	11,593	268	183	11,418	296	193	10,852	294	161
Surplus disposed of on platform, per forty-quart can.	$1 50			$1 65			$2 75			$3 25		

Daily average per week, of cans of forty quarts each, of milk, condensed milk and cream.

RAILROADS.	Week ending December 9.			Week ending December 16.			Week ending December 22.			Week ending Dec. 30 and including Dec. 31.		
	Milk.	Condensed milk.	Cream.	Milk.	Condensed milk.	Cream.	Mil	Condensed milk.	Cream.	Milk.	Condensed milk.	Cream.
New York, Lake Erie and Western..	3,158	21	64	3,218	21	68	3,263	22	70	3,669	24	82
New York Central and Harlem......	2,319	76	22	2,325	74	16	2,418	92	18	2,568	84	19
New York, Ontario and Western ...	1,470	120	25	1,560	135	27	1,562	118	32	1,771	120	40
New York City and Northern.......	1,050	....	....	1,071	....	....	1,074	....	....	1,112	....	....
New York, Susquehanna and West'n.	1,004	....	17	1,032	....	14	1,040	.. .	16	1,060	....	17
New York, New Haven and Hartford	No daily	report	rece'd.	... ..	....	...	.. ...	....	....	......	....	....
New Jersey Central...............	374	....	5	386	....	9	390	....	7	414	....	9
Miscellaneous....................	451	....	10	448	....	7	460	....	10	495	....	12
Total	9,826	217	143	10,040	230	141	10,207	232	153	11,089	228	179
Surplus disposed of on platform per forty-quart can...............	$1 50			$3 00			$1 25			$2 50		

Total receipts over all roads for the year ending December 31, 1885.

MONTHS.	Milk (cans).	Cream (cans).	Condensed milk (cans).	Estimated cost (freight included).	Exchange prices.	Averaged market prices.	Platform prices.
					cents.	cents.	
January	370,846	4,304	6,308	$638,150	3 1-4	3 1-3	$1 55
February	344,163	4,331	6,482	626,900	3	3 1-4	1 62
March	390,155	4,963	6,948	663,500	3	3 1-16	1 52
April	405,998	9,006	3,864	751,000	2 3-4	2 7-8	1 47
May	446,670	9,702	7,072	642,250	2 1-4	2 3-8	1 10
June	465,344	13,803	7,025	642,800	2	2 2-16	1 20
July	478,904	16,637	7,126	740,500	2 1-4	2 5-16	1 65
August	444,578	11,203	5,935	750,800	2.27-31	2 7-8	1 30
September	400,558	6,424	6,113	595,500	2 1-2	2 5-8	1 25
October	413,125	5,296	6,631	629,750	2 1-2	2 5-8	1 45
November	385,589	4,858	6,643	710,450	3 1-4	3 5-16	2 05
December	384,529	4,331	7,061	752,600	3 26-31	3 5-8	1 97
Total	4,930,459	94,868	77,208	$8,169,300	33 57-124	34 1-4	Equal to 36 cts.

By averaging these prices we have 34 3-5 cents, the true market value.

Freight on 4,930,459 cans, less 240,000 brought by dealers, is at an average of 31 cents	$1,454,000
Freight on 172,076 cans of cream and condensed milk, at 45 cents is	77,434
	$1,531,434

Analyses of condensed milk.

Number of inspection.	Name of sample.	Water.	Total solids.	Fat.	Sugar and caseine.	Ash.
B. 7839	Condensed milk	55.74	44.26	13.03	28.83	2.40
B. 7840	Evaporated milk	51.59	48.41	14.25	31.22	2.55
E. 185	Condensed milk	26.73	73.27	9.42	52.11	2.12
G. 633	Condensed milk	54.36	45.64	13.57	28.91	2.26
J. 60	Preserved milk	24.61	75.39	9.96	63.53	1.90
J. 61	Preserved milk	25.14	74.86	9.86	63.11	1.89
J. 123	Condensed milk	58.49	41.51	11.94	27.40	2.17

METHODS OF ANALYSIS.

A careful study of the question during the past year leads me to conclude that Waller's method of milk analysis, as given in the report of 1884, is the best.

The American Society of Public Analysts inquired into the matter and found that Waller's method was generally used in this country.

An experience of some six years has led me to conclude that this method is in every way satisfactory; as the whole operation is conducted with the minimum amount of manipulation.

In regard to the analysis of condensed milk the method given in the report of 1884 seems to answer all purposes and to be a rapid and accurate one.

No new methods of detecting other adulterations have come under my notice. A test for the presence of water is described in the Analyst, vol. X, p. 146, J. Uffelman, and consists in placing in a porcelain capsule a few drops of a solution of diphenylamine in concentrated sulphuric acid; upon adding a few drops of the suspected milk a blue color will form if the milk contains water. This arises from the fact that all water contains more or less nitrates. These nitrates form with the diphenylamine a deep blue color.

This test is too delicate, however, for practical use; if we merely rinse out a can with water, allow it to drain say for twenty minutes, and then fill it with pure milk, a few drops of this milk will show a decided reaction with the diphenylamine.

Testing Milk.

Many experiments were made on the methods of testing milk during the past year in order to arrive at some practical method for the use of milk producers and buyers. The lactometer and lactoscope (Fesers) seemed to be the instruments best fitted for the use of practical men. The lactometer to determine the specific gravity and the lactoscope to determine the percentage of fat. The usefulness of this last test is shown in the tables of the dairy inspections, for it will be seen that in each dairy there are a number of cows giving milk rich in fat, and it can readily be understood that with a little attention the cows in the dairy could be separated into cheese, butter and milk cows.

Methods of Testing.

The fact that the specific gravity of milk is an indicater of the per cent of its constituents, more particularly the amount of water, has been a subject of discussion for some time past. That a large percentage of cream will reduce the specific gravity, provided that the solids not fat, viz.: caseine, sugar and salts, remain the same, is undoubtedly a fact, but it must be taken into consideration that milk rich in fatty matters is usually rich in solids not fat. It must be distinctly understood that we speak of the average milk and not of isolated cases.

The testimony given in the Schrumpf case shows conclusively that the average milk has a specific gravity greater than one hundred degress on the lactometer whose one hundred point indicates a specific gravity of 1.029. The endeavor to disprove this fact by the testimony for the defense showed the extreme difficulty in finding isolated cases of a single cow giving milk of a specific gravity less than 1.029. From reliable sources it has come to my knowledge that the experts employed for the defense spared neither time nor money in the endeavor to find milk, no matter of what quality, whose specific gravity was less than 1.029.

The results given below of many hundred tests made by the experts of this commission show conclusively that not only does the average milk of a dairy have a specific gravity greater than 100 degrees at 60 degrees Fahrenheit, but that milk from single cows nearly always has a specific gravity greater than 100 degrees at 60 degrees Fahrenheit.

A chemist who pretends to learn or know any thing about milk by study in his laboratory alone must be classed as one who still has to learn the rudiments of the subject. The only possible way to obtain a knowledge of this subject is for the investigator to spend a large portion of his time among the milk producers.

The practical knowledge thus obtained enables him then to theorize on the subject. To my mind the practical information to be obtained from the farmers of this State on the subject of milk cannot be overestimated, and the opinions of any chemist or expert on many matters relating to milk are absolutely worthless without such practical knowledge. It has become a matter of considerable importance for the milk producer to determine the quality of the milk produced, not only from his dairy, taken as a whole, but from each individual of it.

The little attention paid to this subject is shown on reference to the results of the inspection of dairies by this commission in various sections of this State, for out of each herd whose milk was tested it is seen that a few of the cows at least produced milk whose quality was almost equal to the average Alderney.

Now a farmer who would test the milk of each cow and by this means separate the herd into two classes, first, butter cows, second, milk cows, would certainly be the gainer.

The instruments designed for this purpose are simple in construction and easily used.

1. *The Lactometer.*

This should be used as follows: Cool the milk to 60 degrees Fahrenheit. Float the lactometer in the liquid, being careful not to wet that part of the stem of the instrument above the surface of the milk, and note the point at which the instrument floats.

2. *The Lactoscope.*

The lactoscope is an instrument, see First Annual Report, page 90, which indicates the per cent of fat in the milk; and the results obtained by it depend upon the fact that the color of the milk is due to the globules of fat which the watery part holds in suspension. See plate 1 of First Annual Report.

Now it can be readily understood that if we add water to milk the opacity will grow less and less as the amount of water is increased, because the fat globules are spread further and further apart; or in other words, if we add a quart of water to a quart of milk, the fat globules contained in the quart of milk are now mixed through the two quarts of milk and water, and therefore the milk is only half as opaque as before the water was added. In the same way enough water could be added so that the milk would become almost transparent.

The lactoscope depends then on this principle and is used as follows: Add the milk from the measuring pipette in the manner directed; then add small quantities of water, shaking each time. After the addition of the water hold out at arm's length and determine if the black lines on the white glass stem can be seen. Continue this addition of water until the lines can be seen through the mixture of milk and water. Then note the per cent of fat indicated on the side of the instrument at the point to which the mixture of milk and water has risen.

This will be the per cent of fat contained in the sample of milk tested. A few examples of the amount of fat found by the lactoscope compared with that found by analysis will be of interest.

Number of sample.	Per cent of fat by lactoscope.	Per cent of fat by analysis.
B. 7540	4.00	4.00
B. 7552	3.50	3.49
E. 2953	4.01	4.00
E. 2966	4.00	3.99
B. 7614	4.00	4.17
B. 7622	3.00	2.63
B. 7653	4.25	4.33
B. 7655	3.50	3.44
B. 7764	3.25	3.24
E. 3051	3.75	3.60
E. 3072	4.00	4.34
B. 7708	4.50	4.40
B. 7716	1.75	1.82
B. 7748	3.00	3.16
E. 3123	4.00	4.18
E. 3157	4.50	4.60
B. 7418	2.50	2.52
E. 2835	3.75	3.56
E. 2848	3.75	4.04
H. 489	3.50	3.42
E. 3124	3.50	3.33
E. 2822	3.55	3.43
E. 2805	3.50	3.42
E. 2777	2.25	2.37
E. 2773	3.50	3.54
H. 422	2.50	2.52
E. 2736	2.50	2.53
H. 389	2.75	2.78

Number of sample.	Per cent of fat. by lactoscope	Per cent of fat. by analysis.
G. 621	1.25	1.10
F. 335	2.00	1.90
G. 609	3.00	2.93

From these data, viz., specific gravity and per cent of fat, the per cent of the various constituents of the sample of milk, viz., water, sugar and caseine and salts, may be calculated, as follows:

1. The lactometric standing of the milk taken at 60 degrees Fahrenheit.

2. The per cent of fat by the lactoscope. (Fesers.)

3. Sixty-eight hundredths of one per cent of ash is taken as the average amount of ash or salts in milk. Experiments show that in a mixture of fat and water, every per cent of butter fat reduces the specific gravity by .001, while every per cent of solids not fat, viz.: sugar, caseine and salts increases the specific gravity by 0·00375.

So that knowing the specific gravity at 60 degrees Fahrenheit and the per cent of fat the following simple calculation will give approximately the water, fat, sugar and caseine and salts.

The specific gravity of water at 60 degrees Fahrenheit = 1·000. Now if we add to this the per cent of fat = F. found by the lactoscope we will lower the specific gravity of the mixture by the number F × .001. That is 1·000 (the specific gravity of water) less F × 0·001 = the specific gravity of the mixture of butter fat and water A, but the specific gravity of the milk in question was S.

Then S — A = increase in specific gravity of the milk due to solids not fat. But each per cent of solids not fat, increase the specific gravity by 0·00375.

Therefore, $\frac{S - A}{.00375}$ = per cent of solids not fat.

Value of Lactometer Degrees in Specific Gravity.

Lactometer.	Gravity.	Lactometer.	Gravity.	Lactometer.	Gravity.
0	1·00000	18	1·00522	36	1·01044
1	1·00029	19	1·00551	37	1·01073
2	1·00058	20	1·00580	38	1·01102
3	1·00087	21	1·00609	39	1·01131
4	1·00116	22	1·00638	40	1·01160
5	1·00145	23	1·00667	41	1·01189
6	1·00174	24	1·00696	42	1·01218
7	1·00203	25	1·00725	43	1·01247
8	1·00232	26	1·00754	44	1·01276
9	1·00261	27	1·00783	45	1·01305
10	1·00290	28	1·00812	46	1·01334
11	1·00319	29	1·00841	47	1·01363
12	1·00348	30	1·00870	48	1·01392
13	1·00377	31	1·00899	49	1·01421
14	1·00406	32	1·00928	50	1·01450
15	1·00435	33	1·00957	51	1·01479
16	1·00464	34	1·00986	52	1·01508
17	1·00493	35	1·01015	53	1·01537

Lactometer.	Gravity.	Lactometer.	Gravity.	Lactometer.	Gravity.
54......	1·01566	77......	1·02233	99......	1·02871
55......	1·01595	78......	1·02262	100......	1·02900
56......	1·01624	79......	1·02291	101......	1·02929
57......	1·01653	80......	1·02320	102......	1·02958
58......	1·01682	81......	1·02349	103......	1·02987
59......	1·01711	82......	1·02378	104......	1·03016
60......	1·01740	83......	1·02407	105......	1·03045
61......	1·01769	84......	1·02436	106......	1·03074
62......	1·01798	85......	1·02465	107......	1·03103
63......	1·01827	86......	1·02494	108......	1·03132
64......	1·01856	87......	1·02523	109......	1·03161
65......	1·01885	88......	1·02552	110......	1·03190
66......	1·01914	89......	1·02581	111......	1·03219
67......	1·01943	90......	1·02619	112......	1·03248
68......	1·01972	91......	1·02630	113......	1·03277
69......	1·02001	92......	1·02668	114......	1·03306
70......	1·02030	93......	1·02697	115......	1·03335
71......	1·02059	94......	1·02726	116......	1·03364
72......	1·02088	95......	1·02755	117......	1·03393
73......	1·02117	96......	1·02784	118......	1·03422
74......	1·02146	97......	1·02813	119......	1·03451
75......	1·02175	98......	1·02842	120......	1·03480
76......	1·02204				

Example.

A sample of milk was tested with the lactometer and lactoscope.

Lactometer = 108 at 60 degrees Fahrenheit.

Lactoscope = 3·25 per cent of fat.

Upon referring to the table showing the actual specific gravity of lactometer degrees we find 108 = a specific gravity of 1·03132.

Specific gravity of water equals 1·000. Each per cent of fat reduces this specific gravity by 0·001. Then 3·25 per cent of fat would reduce the specific gravity by 0·001 × 3·25 = 0·00325, or 1·0000 — 0·00325 = 0·99675.

This number would then represent mixture of water with 3·25 per cent of butter fat. Now the real specific gravity of the milk was 1·03132, and this increase above 0·99675 is due to the solids not fat, then 1·03132 — 0·99675 = 0·03457, but 1 per cent of the solids, not fat, increases the specific gravity by 0·00375.

Therefore, if we divide 0·03457 by 0·00375, we obtain the per cent of solids, not fat, or 9·21.

We now have,

1. The per cent of fat....................................	3·25
2. The per cent of solids not fat..........................	9·21
	12·46

The sum of the above must be equal to the total solids.

Then $100-12\cdot46=87\cdot54=$ per cent of water. Allowing 0·68 per cent for the salts we finally obtain the following results:

Water	87·54
Fat	3·25
Sugar — caseine	8·53
Salts	0·68
	100·00

Putting the formula in the form of a rule we have :

Multiply the per cent of fat found by the lactoscope by 0·001, and subtract the result from 1·0000. Call this number A. Find the real specific gravity from the table of lactometer degrees. Subtract A from this and divide the result by 0·00375, quotient will equal the solids not fat.

To illustrate how nearly the percentage of the constituents of milk can be determined from the specific gravity and per cent of fat by the lactoscope, I give below some analyses of milk compared with the calculated analyses.

	Constituents from actual analyses.		Constituents calculated from specific gravity and per cent of fat.	
	No. 1.	No. 2.	No. 1.	No. 2.
Water	87·27	86·62	87·13	86·37
Total solids	12·73	13·38	12·87	13·65
Fat	3·60	4·34	3·75	4·25
Sugar and caseine	8·41	8·37	8·44	8·72
Solids not fat	9·13	9·04	9·12	9·40
Salts	0·72	0·67	0·68	0·68

Milk Standard.

Before the chemist can decide whether the milk under examination is adulterated or not, that is as far as the addition of water or the removal of cream is concerned, he must know the greatest variation of the constituents of milk and endeavor to fix some standard of purity in order to determine whether water has been added or cream removed.

This important fact can only be determined by a thorough inspection and examination of the milk of cows of all ages and breeds, and kept under the most varying conditions.

Too high a standard would injure the producer ; because it would require him to keep a certain kind or particular breed of cows in order that his product should meet the requirements of the law, while too low a standard would injure both comsumer and producer. The consumer, because if the standard was much below that of pure milk, the law would be merely an incentive to the dishonest producer to adulterate his product, thereby not only injuring the consumer, but by increasing the quantity and decreasing the quality, injure the producer, as the price of such milk would undoubtedly be lower. Every quart of water added to milk represents just so much money taken from the pocket of the honest producer.

In the report of the Dairy Commissioner for 1884, a large number of authorities were quoted and a large number of analyses were given, besides the actual analyses and observations of the chemists and experts of the commission, and it was clearly shown that average milk never contained more than eighty-eight per cent of water, nor less than three per cent of fat, nor less than nine per cent of solids not fat.

The tables given here are the results of the observations of the chemists and experts of this commission, during the year 1885.

It will be noticed that not a single cow gave milk that fell below the standard in any particular.

The method adopted for the estimation of the fat, etc., in milk from single cows is described under the head of testing (page 54).

The examinations of the milk were conducted in the following manner: The cows were milked in the presence of the inspector. The contents of the pail were thoroughly mixed and a sample was taken, cooled to 60 degrees Fahrenheit, and tested by means of the lactometer and lactoscope (Fesers).

The morning's and evening's milk of at least ten cows, taken at random in each herd, was examined in this way.

The average of the A. M. and P. M. milk from at least ten cows was taken, brought to the chemist and analyzed.

It will be seen that the dairies visited were both upland and lowland; and that the cows were fed on various kinds of food.

Notice particularly that even those cows fed largely on brewers' grains gave milk above the standard. If then, the law requires that milk should not contain more than eighty-eight per cent of water, nor less than three per cent of fat, certainly no one can complain of too high a standard.

In Massachusetts the law requires at least thirteen per cent of milk solids, and courts of competent jurisdiction have sustained this standard. In looking over the tables it will be noticed that the specific gravity of milk even from single cows never falls below 100 degrees on the lactometer at a temperature of 60 degrees Fahrenheit = specific gravity 1·029, except in those instances when the milk tested was that of strippers.

The question of the use of the lactometer was fully discussed in the report for 1884. Attention is called to the results for 1885, merely to substantiate the proposition laid down in the 1884 report, viz., that normal milk from healthy cows will never have a specific gravity of less than 100 degrees of the lactometer at 60 degrees Fahrenheit = a specific gravity of 1·029, at 60 degrees Fahrenheit.

In conclusion, attention is called to that part of the report for 1884, discussing the probable spread of epidemics by the use of milk adulterated with contaminated water.

The epidemic of typhoid fever at Plymouth, Pa., during the past year illustrates the danger of this kind of adulteration.

A proper standard for milk may prevent epidemics of this kind; and past experience has shown that the control of the milk supply saves thousands of lives annually.

Analyses of the Average Milk of the Different Dairies Inspected.

[a Morning's milk. b Evening's milk. c Noon's milk.]

Number of inspection and analysis.	Number of cows.	Water.	Fat.	Caseine and sugar.	Salts.	Solids not fat.	Total solids.
B. 3298	a37	85·35	5·57	8·25	0·73	8·28	14·65
B. 3307	a1	87·31	3·60	8·42	0·67	9·09	12·69
B. 3308	a45	87·56	3·21	8·56	0·67	9·23	12·44
B. 3359	a24	86·38	4·46	8·41	0·75	9·16	13·62
B. 3369	a15	86·11	4·42	8·72	0·77	9·40	13·91
B. 3731	b36	86·44	4·09	8·75	0·72	9·47	13·56
B. 5871	a37	86·71	3·86	8·68	0·75	9·43	13·29
B. 5887	b37	87·42	3·18	8·67	0·73	9·40	12·58
B. 6464	a37	86·81	3·23	9·23	0·73	9·94	13·19
B. 6480	b37	87·51	3·05	8·73	0·71	9·44	12·49
B. 7043	a37	86·75	3·60	8.90	0·75	9·65	13·25
B. 7058	b37	87·38	3.49	8·39	0·74	9·13	12·62
B. 7540	a32	85·93	4·00	9·42	0·65	10·07	14·07
B. 7552	b32	86·59	3·49	9·22	0·70	9·92	13·41
B. 7855	a32	85·49	4·35	9·53	0·63	10·16	14·51
B. 7869	b32	86·01	3·78	9·56	0·65	10·21	13·99
B. 7881	b11	86·93	3·59	8·76	0·72	9·48	13·07
B. 7893	a11	86·78	3·87	8·65	0·70	9·35	13·22
B. 7909	b36	86·63	3·84	9·48	0·66	10·14	13·37
B. 7925	a36	86·81	3·55	9·00	0·74	9·74	13·19
B. 7935	b20	86·04	4·24	9·14	0·72	9·86	13·96
B. 7942	16	84·27	5·86	8·65	0·93	9·58	15·73
B. 7958	b53	85·42	4·39	9·46	0·73	10·19	14·58
B. 5926	a62	84·80	5·55	8·83	0·82	9·65	15·20
B. 5937	b62	85·63	4·52	9·08	0·77	9·85	14·37
E. 1661	a6	85·80	4·88	8·55	0·77	9·32	14·20
E. 1662	b6	85·50	5·08	8·63	0·79	9·42	14·50
E. 2279	a6	85·76	4·01	9·48	0·75	10·23	14·24
E. 2280	b6	84·97	5·09	9·18	0·76	9·94	15·03
E. 2805	b24	87·10	3·42	8·79	0·69	9.48	12·90
E. 2822	a24	87·30	3·43	8·58	0·69	9·27	12·70
E. 2835	b14	87·08	3·56	8·61	0·75	9·36	12·92
E. 2848	a14	86·22	4·04	9·01	0·73	9·74	13·78
E. 2868	b30	87·81	3·08	8·37	0·74	9·11	12·19
E. 2874	a1	86·02	3·16	10·09	0·73	10·82	13·98
E. 2888	a30	87·29	3·36	8·67	0·68	9·35	12·71
E. 2953	b17	86·32	4·01	8·89	0·78	9·67	13·68
E. 2966	a17	86·27	3·99	9·04	0·70	9·74	13·73
E. 2998	b32	86·85	4·08	8·34	0·63	9·07	13·15
E. 3030	a32	85·91	4·96	8.44	0·69	9·13	14·09
E. 3051	b23	87·27	3·60	8·41	0·72	9·13	12·73
E. 3072	a23	86·62	4·34	8·37	0·67	9·04	13·38

ANALYSES OF THE AVERAGE MILK OF THE DIFFERENT DAIRIES INSPECTED — *Continued.*

Number of inspection and analysis.	Number of cows.	Water.	Fat.	Caseine and sugar.	Salts.	Solids not fat.	Total solids.
E. 3102........	b300	86·52	4·19	8·66	0·63	9·29	13·48
E. 3111........	b1	85·86	4·85	8·47	0·62	9·29	14·14
E. 3123........	b1	86·03	4·18	9·04	0·75	9·79	13·97
E. 3124........	b300	87·44	3·33	8·50	0·73	9·23	12·56
E. 3145........	b300	86·41	4·02	8·80	0·78	9·58	13·59
E. 3151........	a6	83·60	6·60	9·80	0·76	9·04	16·40
E. 3157........	b6	84·81	4·60	9·80	0·79	10·59	15·19
E. 3173........	a13	86·88	3·24	9·22	0·66	9·88	13·12
E. 3184........	b13	86·84	3·28	9·23	0·65	9·88	13·16
E. 3196........	b14	86·49	3·12	9·78	0·61	10·39	13·51
E. 3207........	a14	87·07	3·68	8·58	0·67	9·25	12·93
E. 3231........	a13	87·03	3·35	8·86	0·76	9·62	12·97
E. 3312........	a6	83·82	5·74	9·56	0·78	10·34	16·18
E. 3318........	b6	83·35	6·32	9·56	0·77	10·33	16·65
E. 3404........	a6	84·16	6·33	8·69	0·82	9·51	15·84
E. 3410........	b6	85·37	5·56	8·25	0·82	9·07	14·63
E. 3415........	a5	84·91	6·18	8·22	0·69	8·91	15·09
E. 3420........	b5	86·43	5·72	6·76	1·09	6·85	13·57
I. 52..........	b20	86·19	4·38	8·70	0·73	9·43	13·81
I. 53..........	a20	87·38	3·24	8·65	0·73	9·48	12·61
I. 79..........	b20	87·54	3·18	8·57	0·71	9·28	12·46
I. 89..........	a20	85·69	4·84	8·77	0·70	9·47	14·31
I. 166.........	a20	87·19	3·81	8·33	0·69	9·02	12·83
I. 167.........	b20	87·33	3·60	8·39	0·68	9·07	12·67
C. 1886........	a	87·03	4·07	7·21	0·69	8·90	12·97
C. 2196........	....	87·47	3·14	8·62	0·77	9·39	12·53
D. 5926........	a10	84·80	5·55	8·83	0·82	9·65	15·20
D. 5937........	b10	85·63	4·52	9·08	0·77	9·85	14·47

ANALYSES OF MILK OF SINGLE COWS.

Number of inspection.	Water.	Total solids.	Fat.	Sugar and caseine.	Salts.
B. 3298..............	85·35	14·65	5·57	8·35	0·73
B. 3307..............	87·31	12·69	3·60	8·42	0·67
B. 3308..............	87·56	12·44	3·21	8·56	0·67
B. 3309..............	86·09	13·91	3·24	9·90	0·77
B. 3310..............	86·78	13·22	3·59	8·96	0·67
B. 3314..............	87·55	12·45	3·01	8·71	0·73
E. 2281..............	86·48	13·52	4·00	8·77	0·75
E. 3111..............	85·86	14·14	4·85	8·47	0·82
E. 3145..............	86·41	13·59	4·02	8·80	0.78

Analyses of adulterated milk.

Number of inspection.	Water.	Total solids.	Fat.	Sugar and caseine.	Salts.
A. 17	88·28	11·72	2·84	8·88	0·77
A. 75	89·34	10·66	2·48	8·16	0·64
A. 166	88·24	11·76	3·24	7·84	0·68
A. 171	89·32	10·68	2.69	7·51	0·48
A. 257	88·88	11·12	2·60	8·43	....
A. 267	88·03	11·97	2·58	9·39	....
A. 286	88·38	11·62	2·27	9·35	....
A. 424	91·68	8·32	0·95	7·27	....
A. 476	90·54	9·46	2·17	7·28	....
A. 765	88·52	11·48	3·12	7·69	0·67
A. 768	88·33	11·67	3·27	7·78	0·62
A. 1399	88·86	11·14	2·69	7·77	0·68
A. 1400	89·35	10·65	3·13	6·92	0·60
A. 1401	88·98	11·02	2·64	7·70	0·68
A. 1403	89·12	10·88	2·88	7·34	0·66
A. 1470	88·28	11·72	3·33	7·73	0·66
A. 1537	89·11	10·89	3·21	7·12	0·56
A. 1546	89·10	10·90	3·05	7·28	0·57
A. 1547	88·44	11·56	4·17	6·68	0·71
A. 1581	88·65	11·35	2·98	7·78	0·59
A. 1582	90·78	9·22	2·41	6·36	0·45
A. 1616	90·40	9·60	2·34	6·69	0·57
A. 1617	89·26	10·74	3·08	7·08	0·58
A. 2350	88·44	11·56	3·47	7·66	0·43
A. 2351	88·74	11·26	2·87	7·87	0·52
A. 2373	88·71	11·29	2·87	7·84	0·58
A. 2422	88·51	11·49	3·20	7·74	0·55
A. 2562	88·48	11·52	3·38	7·54	0·60
A. 2563	88·35	11·65	3·23	7·84	0·58
A. 2564	89·13	10·87	3·28	6·97	0·62
A. 2565	89·38	10·62	3·36	6·72	0·54
A. 2572	89·45	10·55	0·73	9·08	0·74
A. 2575	88·54	11·46	1·56	9·90	....
B. 1638	88·46	11·54	2·98	7·91	0·65
B. 1653	88·88	11·12	2·72	7·73	0·57
B. 1655	89·25	10·75	2·51	7·63	0·61
B. 2155	89·11	10·89	2·98	7·25	0·66
B. 2157	88·28	11·72	2·56	8·53	0·63
B. 2187	91·40	8·60	2·54	5·59	0·47
B. 2263	90·80	9·20	3·01	5·74	0·45
B. 2268	89·80	10·20	2·53	7·11	0·56
B. 2527	89·26	10·74	2·88	7·21	0·65
C. 272	88·83	11·17	4·01	6·45	0·71
C. 917	89·93	10·07	2·16	7·32	0·59

Analyses of adulterated milk.—(Continued).

Number of inspection.	Water.	Total solids.	Fat.	Sugar and caseine.	Salts.
C. 1492	90·41	9·59	2·94	6·20	0·45
D. 183	90·57	9·43	2·72	6·24	0·47
D. 266	90·00	10·00	2·59	6·86	0·55
D. 955	88·60	11·40	2·21	8·56	0·63
E. 798	89·75	10·25	2·46	7·18	0·61
E. 892	88·24	11·76	2·81	8·34	0·61
A. 3169	89·54	10·46	3·12	6·65	0·69
A. 3315	88·66	11·34	2·90	7·76	0·68
A. 3361	88·30	11·70	2·64	8·33	0·73
A. 3370	89·25	10·75	2·46	7·80	0·49
A. 3392	87·79	12·21	2·14	9·36	0·71
A. 3401	87·39	12·61	2·59	9·33	0·69
A. 3405	88·07	11·93	2·56	8·56	0·73
A. 3422	87·76	12·24	1·52	8·63	2·09
A. 3478	88·56	11·54	2·42	8·38	0·64
A. 3481	88·62	11·38	2·91	7·75	0·72
A. 3579	88·55	11·45	2·75	8·11	0·59
A. 3586	87·83	12·17	2·63	8·87	0·67
B. 3054	88·75	11·25	3·25	7·34	0·66
B. 3562	89·42	10·58	3·52	6·51	0·55
B. 3610	88·47	11·53	3·83	7·09	0·61
B. 4747	88·95	11·05	2.41	8·03	0·61
B. 4750	90·92	9·08	2·39	6·24	0·45
B. 4751	91·15	8·85	2·31	6·11	0·43
B. 4752	89·00	11·00	2·11	8·31	0·58
B. 4757	89·54	10·46	2·35	7·54	0·57
B. 6422	88·45	11·55	2·60	8·37	0·68
B. 6759	88·26	11·74	2·05	9·01	0·68
B. 6876	92·50	7·50	1·20	5·64	0·66
B. 7100	89·30	10·70	2·10	8·02	0·58
B. 7128	88·82	11·18	2·89	7·52	0·67
B. 7418	88·84	11·16	2·52	8·11	0·53
B. 7419	89·69	10·31	3·20	6·51	0·60
B. 7664	88·63	11·37	3·24	7·54	0·59
B. 7716	90·66	9·34	1·82	7·02	0·50
B. 7748	88·91	11·09	3·16	7·40	0·53
C. 1885	88·45	11·55	3·50	7·39	0·66
D. 2007	90·90	9·10	2·60	5·94	0·56
D. 2392	88·50	11·50	3·15	7·72	0·63
D. 5767	89·12	10·88	1·36	8·65	0·93
D. 5768	89·64	10·36	·26	9·35	0·75
D. 2760	89·76	10·24	2·48	7·15	0·61
D. 1644	91·79	8·29	2·03	5·77	0·49
D. 2769	88·96	11·04	3·26	7·11	0·67
D. 3125	90·34	9.46	2·84	6·07	0·55

Analyses of adulterated milk — (Continued).

Number of inspection.	Water.	Total solids.	Fat.	Sugar and caseine.	Salts.
D. 3152	89·34	10·66	2·85	7·22	0·59
D. 3676	89·10	10·88	2·59	7·60	0·69
D. 4362	90·04	9·96	2·06	7·33	0·57
D. 4812	89·00	11·00	3·36	7 04	0·60
E. 666	88·78	11·22	2·66	7·87	0·69
E. 768	88·55	11·45	3·09	7·73	0·63
E. 798	89·75	10·25	2·46	7·18	0·61
E. 2178	88·23	11·77	1·67	9·33	0·77
E. 2269	89·11	10·89	2·59	7·65	0·65
E. 2634	89·65	10·35	2·02	7·70	0·63
E. 2714	88·76	11·14	2·69	7·79	0·66
E. 2726	90·48	9·52	0·49	8·29	0·74
E. 2735	88·74	11·26	2·39	8·18	0·69
E. 2736	89·93	10·07	2·53	6·92	0·62
E. 2773	88·73	11·27	3·54	7·14	0·59
E. 2777	89·04	10·96	2·37	7·98	0·61
F. 122	89·40	10·60	2·84	7·19	0·57
F. 244	89·81	10·19	2·55	7·08	0·56
F. 246	89·33	10·67	3·05	7·06	0·56
F. 335	91·67	8·33	1·90	5·98	0·45
F. 352	91·11	8·89	2·20	6·16	0·53
F. 354	90·47	9·53	2·60	6·38	0·55
G. 451	88·65	11·35	2·67	8·09	0·59
G. 474	89.06	10·94	2.90	7·47	0·57
G. 561	89·20	10·80	2·57	7·62	0·61
G. 589	88·73	11·27	2·91	7·69	0·67
G. 609	88·48	11.52	2·93	7·96	0·63
G. 612	88·74	11·26	2·72	7·86	0·68
G. 621	90·56	9·44	1·10	7·73	0·61
G. 630	89·35	10·65	2·85	7·20	0·60
G. 666	89·17	10·83	3·03	7·16	0·64
H. 184	88·44	11·56	2·82	8·05	0·69
H. 229	90·19	9·81	1·85	7·35	0·61
H. 231	89·05	10·95	3·31	7·05	0·59
H. 240	88·97	11·03	3·27	7·18	0·58
H. 372	89·75	10·25	1·72	7·94	0·59
H. 475	90·52	9·48	2·37	6·64	0·49
H. 378	91·06	8·94	1·11	7·33	0·50
H. 389	89·86	10·14	2·78	6·79	0·57
H. 422	89·28	10·72	2·62	7·43	0·67
H. 435	89·18	10·82	3·29	7·06	0·53
I. 7	89·99	10·01	2·67	6·74	0·60
J. 91	89·77	10·23	2·48	7·24	0·51

Morning of January 2, 1885; farm of H. N. Smith, Montgomery, Orange county; cows in herd, 16; treatment, kind; housing, good; food, wheat-bran, grains and corn-meal.

Number of inspection.	Name of cow.	Breed.	Age of cow.	Number of times calving.	Time of last calving.	Number of quarts given.	Lactometer at 60 deg. Fahr.	Per cent of cream.
			years.		mos.			
B. 3013 ..	Fanny	Dutch ...	10	8	4	6	110	18
B. 3014 ..	Jenny	Native ...	10	8	4	6	112	18
B. 3015 ..	Dutch	Dutch ...	13	10	3	7	108	12
B. 3016 ..	Daisy	Native ...	10	8	5	4	104	24
B. 3017 ..	Fanny	Native ...	13	7	3	8	114	16
B. 3018 ..	Dolly	Native ...	8	6	6	4	110	18
B. 3019 ..	Molly	Native ...	9	7	5	4	112	20
B. 3020 ..	Bertha	Native ...	9	6	5	3	108	22
B. 3021 ..	Average ..	....	..	..	..	..	110	20

Morning of January 17, 1885; farm of John D. Mould; Montgomery, Orange county; cows in herd, 37; treatment, kind; housing, good; food, wheat middlings, bran and grains.

Number of inspection.	Name of cow.	Breed.	Age of cow.	Number of times calving.	Time of last calving.	Number of quarts given.	Lactometer at 60 deg. Fahr.	Per cent of cream.
			years.		mos.			
B. 3281	Blossom ...	Native	9	7	1	10	106	20
B. 3282	Mooley	Native	4	2	1	9	112	16
B. 3283	Spot	Native	10	8	4	5	110	22
B. 3284	Kennedy	Native	8	6	3	4	100	24
B. 3285	Alderney ...	Alderney.	5	3	2	6	112	18
B. 3286	Brindle	Native ...	9	7	2	7	108	16
B. 3287	Morrissey ...	Native ...	3	1	4	3	118	14
B. 3288	Thief	Native ...	8	6	2	5	110	20
B. 3289	Crow	Native ...	7	5	1	5	112	18
B. 3290	Crazy	Native ...	10	8	4	3	114	16
B. 3291	Bob. Tit	Native ...	12	10	4	4	114	30
B. 3292	Blue	Native ...	10	8	3	5	116	22
B. 3293	Degraw	Native ...	6	4	5	2	100	24
B. 3294	Big Horn ...	Native ...	8	6	6	3	114	10
B. 3295	Maggie	Native ...	3	1	4	2	108	18
B. 3296	Brindle	Native ...	4	2	6	2	112	20
B. 3297	Star	Native ...	10	8	7	3	102	14
B. 3298	Average ..	....	..	..	..	..	108	20

Morning of January 20, 1885; farm of B. C. Sears, Blooming Grove, Orange county; cows in herd, 45; treatment, good; housing, good; food, hay, corn-meal and sprouts.

Number of inspection.	Name of cow.	Breed.	Age of cow.	Number of times calving.	Time of last calving.	Number of quart given.	Lactometer at 60 deg. Fahr.	Per cent of cream.
			years.		mos.			
B. 3299....	B. Sears....	Holstein..	5	3	2	8	114	16
B. 3300....	Shot.......	Holstein..	9	7	2	9	102	16
B. 3301....	Morris. ...	Native...	7	5	1	10	112	14
B. 3302....	Hannah....	½ Aldern'y	9	6	3	9	102	16
B. 3303....	Daisy.......	Alderney.	9	7	2	9	114	12
B. 3304....	Cherry.....	½ Holstein	10	8	8	10	112	20
B. 3305....	Button.....	Jersey....	9	7	1	6	116	26
B. 3306....	B. Alderney.	Alderney.	12	10	3	6	110	24
B. 3307....	Jennie.....	Ayrshire..	2	1	3	5	110	12
B. 3308....	Average..	....	..	..	..	..	110	18

Morning of January 22, 1885; farm of C. Newkirk, Montgomery, Orange county; cows in herd, 26; treatment, kind; housing, good; food, wheat-bran and corn-meal.

Number of inspection.	Name of cow.	Breed.	Age of cow.	Number of times calving.	Time of last calving.	Number of quarts given.	Lactometer at 60 deg. Fahr.	Per cent of cream.
			years.		mos.			
B. 3316....	Ketz	Native...	9	7	2	10	106	24
B. 3317....	Weighman .	Native...	4	2	2	4	116	18
B. 3318....	Black.......	Native...	7	5	2	5	118	24
B. 3319....	White.......	Native...	9	6	2	6	116	14
B. 3320....	James O...	Native...	12	10	2	3	106	18
B. 3321....	Red	Native...	2	1	2	4	118	32
B. 3322....	Little Spot.	Native...	2	1	1	4	118	24
B. 3323....	CrumpHorn	Native...	5	3	1	5	118	26
B. 3324....	White Romp	Native...	10	7	2	4	116	10
B. 3325....	O. Spot....	Native...	12	9	6	4	108	14
B. 3326....	Gentle.....	Native...	10	7	2	5	114	18
B. 3327....	Bowman ...	Native...	10	8	4	3	114	22
B. 3328....	Spot.......	Native...	7	5	3	4	118	36
B. 3329....	Bowen.....	Native...	12	9	1	6	114	20
B. 3330....	Morrisen...	Native...	10	8	1	6	116	22
B. 3331....	Hill... ...	Native...	7	5	2	5	116	24
B. 3332....	Snyder......	Native...	9	6	3	5	114	22
B. 3333....	Daisy......	Native...	6	4	2	6	114	22
B. 3334....	Dill	Native...	10	7	3	6	118	18
B. 3335....	Graham ...	Native...	9	6	2	4	114	24
B. 3336....	Average..	....	..	..	..	..	112	22

Morning of January 23, 1885; farm of Henry Bergen, Montgomery, Orange county; cows in herd, 24; treatment, kind; housing, good; food, corn-meal and wheat-bran.

Number of inspection.	Name of cow.	Breed.	Age of cow.	Number of times calving.	Time of last calving.	Number of quarts given.	Lactometer at 60 deg. Fahr.	Per cent of cream.
			years.		mos.			
B. 3337....	Mollie........	Native...	3	1	3	3	100	24
B. 3338....	Big Mollie...	Native...	6	4	3	8	110	16
B. 3339....	B. Beef....	Native...	7	4	2	9	112	14
B. 3340....	B. Face....	Native...	8	6	3	7	114	18
B. 3341....	Roney........	Native...	8	6	4	6	112	22
B. 3342....	Black	Native...	5	3	6	9	108	24
B. 3343....	Betsey.....	Native...	6	4	2	9	112	14
B. 3344....	Roxey	Native...	10	8	1	8	114	14
B. 3345....	W. Face....	Native...	7	4	6	3	114	10
B. 3346 ...	Strawberry..	Native...	12	9	6	3	108	14
B. 3347....	Hasbrouck...	Native...	7	5	6	4	110	14
B. 3348....	Ayrshire....	Native...	11	9	11	3	110	20
B. 3349....	Rapid	Native...	6	4	9	3	110	24
B. 3350....	Betsey......	Native...	6	3	9	3	108	16
B. 3351....	White	Native...	5	3	6	4	118	24
B. 3352....	Black........	Native...	9	6	9	3	110	22
B. 3353....	Scrub........	Native...	8	6	4	3	110	10
B. 3354....	B. Red......	Native...	8	5	6	4	106	18
B. 3355....	Blue.........	Native...	6	4	3	6	110	22
B. 3356....	Sallie........	Native...	8	6	3	6	112	22
B. 3357....	Mollie........	Native...	10	7	2	7	114	18
B. 3358....	Daisy......	Native...	6	4	2	8	116	18
B. 3359....	Average...	....	..	..	..	..	110	20

Morning of January 24, 1885; farm of J. H. Van Keuren, Hamptonburgh, Orange county; cows in herd, 15; treatment, kind; housing, good; food, hay, cotton-seed meal, wheat middlings.

Number of inspections.	Name of cow.	Breed.	Age of cow.	Number of times calving.	Time of last calving.	Number of quarts given.	Lactometer at 60 deg. Fahr.	Per cent of cream.
			years.		mos.			
B. 3360....	B. Wallace..	Native...	7	5	3	10	114	18
B. 3361....	One Horn..	Native...	8	5	2	4	118	28
B. 3362....	Ackerly....	Native...	13	10	2	8	110	24
B. 3363....	Line Back..	Native...	12	10	3	6	120	14
B. 3364....	L. Wallace.	Native...	5	3	2	8	116	20
B. 3365....	Blue........	Native...	7	5	1	9	110	20
B. 3366....	White face..	Native...	8	6	4	8	115	20
B. 3367....	White.. ..	Native...	8	5	3	6	118	18
B. 3368....	Brindle.....	Native...	7	4	3	6	112	24
B. 3369....	Average..	....	..	..	..	..	112	22

Evening of January 29, 1885; farm of J. B. Mould, Montgomery, Orange county; cows in herd, 20; treatment, kind; housing, good; food, corn stalks, hay, sprouts, wheat-bran and middlings.

Number of inspections.	Name of cow.	Breed.	Age of cow.	Number of times calving.	Time of last calving.	Number of quarts given.	Lactometer at 60 deg. Fahr.	Per cent of cream.
			years.		mos.			
B. 3461....	Lill........	Native...	9	7	3	7	110	22
B. 3462....	Nell........	Native...	6	5	2	10	102	12
B. 3463....	Crumple ...	Dutch ...	7	5	4	5	116	16
B. 3464 ...	Brennan....	Native...	3	1	3	5	114	22
B. 3465....	Massey.....	Native...	8	6	2	6	102	8
B. 3466....	Jenney.....	Native...	14	..	5	3	112	12
B. 3467....	Hart........	Native...	8	..	6	3	116	8
B. 3468....	Susie	Native...	4	2	3	5	110	20
B. 3469....	B. Lill......	Native...	4	2	1	4	114	20
B. 3470 ...	Alderney....	Alderney..	10	8	6	3	100	30
B. 3471....	O. Nell.....	Native...	12	10	6	2	110	12
B. 3472....	Average..	....	..	..	..	..	110	16

Evening of January 30, 1885; farm of M. Shafer, Montgomery, Orange county; cows in herd, 40; treatment, kind; housing, good; food, corn-meal and middlings.

Number of inspection.	Number of cow.	Breed.	Age.	Number of times calving.	Time of last calving.	Number of quarts given.	Lactometer at 60 deg. Fahr.	Per cent of cream.
			years.		mos.			
B. 3473....	1	Native...	6	4	2	8	110	12
B. 3474....	2	Native...	8	6	3	6	110	22
B. 3475....	3	Native...	5	3	3	4	112	22
B. 3476....	4	Native...	9	6	2	6	108	14
B. 3477....	Bradley	Native...	7	4	3	7	106	12
B. 3478....	6	Native...	10	9	1	7	110	18
B. 3479....	7	Native...	4	2	3	6	110	30
B. 3480....	8	Native...	12	9	3	6	106	24
B. 3481....	9	Native...	6	4	3	5	112	10
B. 3482....	10	Native...	7	5	3	5	112	16
B. 3483....	11	Native...	7	4	3	6	110	18
B. 3484....	12	Native...	6	4	3	4	118	20
B. 3485....	13	Native...	9	6	4	4	110	18
B. 3486....	14	Native...	10	7	4	4	112	16
B. 3487....	15	Native...	12	9	4	5	114	18
B. 3488....	16	Native...	4	2	3	4	116	14
B. 3489....	17	Native...	6	4	3	5	112	16
B. 3490....	18	Native...	7	5	3	6	112	18
B. 3491....	Average..	....	..	..	..	..	110	18

Morning of February 9, 1885; *farm of N. J. Quackenbos, Montgomery, Orange county; cows in herd,* 18; *treatment, kind; housing, good; food, hay, corn-meal, wheat-bran.*

Number of inspection.	Number of cow.	Breed.	Age.	Number of times calving.	Time of last calving.	Number of quarts given.	Lactometer at 60 deg. Fahr.	Per cent of cream.
			years.		mos.			
B. 3623....	1	Native...	10	7	3	5	114	20
B. 3624....	2	Native...	5	3	1	8	116	20
B. 3625....	3	Native...	6	4	2	6	120	16
B. 3626....	4	Native...	8	6	3	5	118	20
B. 3627...	5	Native...	6	4	3	5	120	20
B. 3628....	6	Alderney..	16	..	3	4	118	16
B. 3629....	7	Native...	7	5	1	6	110	22
B. 3630....	8	Native...	6	4	3	7	114	18
B. 3631....	9	Native...	6	4	1	8	110	18
B. 3632....	10	Native...	11	8	2	5	118	24
B. 3633....	11	Native...	5	3	2	6	116	22
B. 3634....	12	Native...	6	3	3	5	112	18
B. 3635....	13	Native...	8	5	2	6	114	20
B. 3636....	14	Native...	10	8	1	6	112	18
B. 3637....	15	Native...	9	7	1	6	118	18
B. 3638....	16	Native...	7	4	1	6	118	20
B. 3639....	Average..	....	..	..	..	..	114	20

Evening of February 21, 1885; farm of S. J. Morris, Montgomery, Orange county; cows in herd, 17; treatment, kind; housing, good; food, hay, wheat-bran and middlings, cotton-seed meal.

Number of inspection.	Number of cows.	Breed.	Age of cow.	Number of times calving.	Time of last calving.	Number of quarts given.	Lactometer at 60 deg. Fahr.	Per cent of cream.
			years		mos.			
B. 3819....	1	Native...	6	4	3	8	116	18
B. 3820....	2	Native...	14	11	1	7	116	30
B. 3821....	3	Native...	8	5	1	9	114	20
B. 3822....	4	Native...	14	12	½	6	116	20
B. 3823....	5	Native...	8	6	12	5	120	26
B. 3824....	6	Native...	16	13	1	6	118	18
B. 3825....	7	Native...	14	11	½	6	108	18
B. 3826....	8	Native...	6	4	6	4	118	24
B. 3827....	9	Native...	9	6	½	6	110	28
B. 3828....	10	Native...	6	4	6	4	118	18
B. 3829....	11	Native...	6	4	5	4	112	22
B. 3830....	12	Native...	7	4	½	6	118	20
B. 3831....	13	Native...	6	3	4	5	114	22
B. 3832....	14	Native...	9	6	3	6	116	24
B. 3833....	15	Native...	8	6	4	6	114	22
B. 3834....	16	Native...	10	7	4	5	114	22
B. 3835....	17	Native...	7	5	4	6	116	24
B. 3836....	Average..	....	..	..	..	..	114	22

Evening of February 27, 1885; *farm of G. O. Smith, Montgomery, Orange county; cows in herd,* 21; *treatment, kind; housing, good; food, hay, wheat-bran, middlings, and cotton-seed meal.*

Number of inspection.	Name of cow.	Breed.	Age of cow.	Number of times calving.	Time of last calving.	Number of quarts given.	Lactometer at 60 deg. Fahr.	Per cent of cream.
			years.		mos.			
B. 3961....	Spot........	Native...	9	7	1	7	110	20
B. 3962....	Alderney...	Native...	5	3	1	7	108	16
B. 3963....	Roan	Native...	15	12	3	3	110	14
B. 3964 ...	Bright.....	Native...	6	4	1	6	110	20
B. 3965....	Cherry.....	Native...	8	5	½	6	112	14
B. 3966....	Black......	Native...	6	4	2	6	110	16
B. 3967....	Red........	Native...	8	6	7	4	106	20
B. 3968....	R. Heifer...	Native...	2	1	6	4	110	18
B. 3969....	White......	Native...	8	6	1	8	108	18
B. 3970....	Yellow.....	Native...	7	5	3	5	114	30
B. 3971....	Kicker.....	Native...	7	5	2	6	106	20
B. 3972....	Blue.......	Native...	8	6	2	6	110	20
B. 3973....	Bell.......	Native...	10	8	3	5	110	18
B. 3974....	Lill........	Native...	8	6	2	6	108	20
B. 3975....	Average..	....	..	..	..	..	108	19

Evening of February 28, 1885; *farm of William Y. Lodge, Montgomery, Orange county; cows in herd,* 2; *treatment, good and kind; housing, very good; food, hay and corn-meal.*

Number of inspection.	Name of cow.	Breed.	Age of cow.	Number of times calving.	Time of last calving.	Number of quarts given.	Lactometer at 60 deg. Fahr.	Per cent of cream.
			years.		mos.			
B. 3976....	Leland.....	Alderney.	8	6	2	5	112	26
B. 3977....	Lill........	Alderney.	10	8	3	5	110	22
B. 3978....	Average..	....	..	..	..	..	110	24

Evening of March 2, 1885; farm of H. Tower, Hamptonburgh, Orange county; cows in herd, 24; treatment, kind; housing, good; food, hay, wheat, middlings and grains.

Number of inspection.	Number of cow.	Breed.	Age of cow.	Number of times calving.	Time of last calving.	Number of quarts given.	Lactometer at 60 deg. Fahr.	Per cent of cream.
			years.		mos.			
B. 3979....	1	Native...	7	5	2	5	112	20
B. 3980....	2	Native...	7	6	3	5	110	22
B. 3981....	3	Native...	4	2	4	4	116	18
B. 3982....	4	Native...	4	2	2	5	104	16
B. 3983....	5	Native...	4	2	2	6	106	20
B. 3984....	6	Native...	7	5	1	7	110	16
B. 3985....	7	Native...	6	8	2	7	108	22
B. 3986....	8	Native...	9	6	1	7	106	16
B. 3987....	9	Native...	4	2	4	3	112	16
B. 3988....	10	Native...	7	5	1	7	104	24
B. 3989 ...	11	Native...	4	2	6	3	114	12
B. 3990....	12	Native...	8	6	5	5	106	22
B. 3991....	13	Native...	5	3	10d.	7	118	12
B. 3992....	14	Native...	5	2	1m.	7	116	12
B. 3993....	15	Native...	9	7	1m.	6	106	18
B. 3994....	16	Native...	9	6	10d.	9	104	18
B. 3995....	17	Native...	2	1	4	4	112	16
B. 3996....	18	Native...	2	1	10d.	5	114	16
B. 3997....	19	Native...	7	5	2m.	5	112	18
B. 3998...	20	Native...	7	5	3m.	5	110	20
B. 3999....	21	Native...	4	3	3m.	4	114	18
B. 4000....	22	Native...	7	5	1m.	7	112	18
B. 4001....	Average..	....	..	..	..	..	112	18

Morning of March 4, 1885; farm of H. J. Comfort, Montgomery, Orange county; cows in herd, 22; treatment, kind; housing, good; food, hay, corn-meal and wheat-bran.

Number of inspection.	Number of cow.	Breed.	Age of cow.	Number of times calving.	Time of last calving	Number of quarts given.	Lactometer at 60 deg. Fahr.	Per cent of cream.
			years.		mos.			
B. 4016....	1	Native...	7	5	3	7	110	16
B. 4017....	2	Native...	10	7	3	7	110	16
B. 4018....	3	Native...	6	3	3	6	108	12
B. 4019....	4	Native...	7	4	4	2	110	16
B. 4020....	5	Native...	6	4	3	8	108	16
B. 4021....	6	Native...	11	8	2	9	106	12
B. 4022....	7	Native...	11	8	2	6	110	16
B. 4023....	8	Native...	5	3	5	3	114	18
B. 4024 ...	9	Native...	9	6	3	3	112	20
B. 4025....	10	Native...	6	4	1	8	108	12
B. 4026....	11	Native...	7	5	6	4	110	18
B. 4027....	12	Native...	8	5	9	4	114	16
B. 4028....	Average..	....	..	..	..	..	110	16

Evening of March 19, 1885; farm of C. H. Bonne, Montgomery, Orange county; cows in herd, 23; treatment, kind; housing, good; food, hay, wheat-bran, cotton-seed meal and corn-meal.

Number of inspection.	Number of cow.	Breed.	Age of cow.	Number of times calving.	Time of last calving.	Number of quarts given.	Lactometer at 60 deg. Fahr.	Per cent of cream.
			years.		mos.			
B. 4357 ...	1	Native...	8	6	2	7	108	12
B. 4358....	2	Native...	8	6	5	6	104	22
B. 4359....	3	Native...	8	5	6	4	106	20
B. 4360....	4	Native...	8	6	5	3	110	22
B. 4361....	5	Native...	10	8	4	4	110	16
B. 4362....	6	Native...	10	7	1	9	112	14
B. 4363....	7	Native...	8	6	2	9	108	14
B. 4364....	8	Native...	10	8	2	9	108	14
B. 4365....	9	Native...	6	4	2	9	108	14
B. 4366....	10	Native...	7	5	4	5	110	16
B. 4367....	11	Native...	12	10	4	3	106	24
B. 4368 ...	12	Native...	7	5	3	3	110	18
B. 4369....	13	Native...	5	3	4	4	114	12
B. 4370....	14	Native...	5	3	2	6	108	18
B. 4371....	15	Native...	6	4	2	8	110	16
B. 4372 ...	16	Native...	7	5	3	5	112	16
B. 4373....	Average..	....	..	..	..	..	108	17

Evening of March 20, 1885; farm of David Sparks, Montgomery, Orange county; cows in herd, 24; treatment, kind; housing, good; food, hay, wheat-middlings, corn-meal and cotton-seed meal.

Number of inspection.	Name of cow.	Breed.	Age of cow.	Number of times calving.	Time of last calving.	Number of quarts given.	Lactometer at 60 deg. Fahr.	Per cent of cream.
			years.					
B. 4374....	Stump.....	Native...	8	6	8d.	5	120	14
B. 4375....	R. Lee.....	Native...	13	10	2d.	6	118	12
B. 4376....	Van Wagner	Native...	3	1	1m.	3	118	16
B. 4377....	Star.......	Native...	4	2	4m.	5	120	16
B. 4378....	Cherry.....	Native...	8	6	12m.	5	118	16
B. 4379....	White......	Native...	8	6	1m.	6	116	16
B. 4380....	Black......	Native...	7	5	2m.	6	110	18
B. 4381....	Dippie.....	Native...	6	4	4m.	4	114	18
B. 4382....	Flora......	Native...	9	5	2m.	7	116	14
B. 4383....	Benedict...	Native...	3	1	6m.	2	114	18
B. 4384....	White, 2d..	Native...	3	1	8d.	4	118	14
B. 4385....	Stuffy......	Native...	8	6	2d.	6	130	18
B. 4386....	Rosie......	Native...	14	10	3d.	6	118	12
B. 4387....	Average..	....	..	..	..	..	116	16

Morning of May 27, 1885; James Vandereef & Son, gentlemen in village; cows in herd, 1; treatment, kind; housing, good; food, pasture, middlings, wheat-bran, cotton-seed meal twice a day; residence, Montgomery, Orange county.

Number of inspection.	Name of cow.	Breed.	Age of cow.	Number of times calving.	Time of last calving.	Number of pounds given.	Lactometer at 60 deg. Fahr.	Per cent of cream.
			years.		1885.			
B. 5855..	Lola...	Alderney.	2	1	March.	16 6oz.	112	22
			July 1.					
	Lola...	Alderney.	2	1	March.	15	112	24

Evening of February 14, 1885; farm of J. De Witt Krebs, Montgomery, Orange county; cows in herd, 38; treatment, kind; housing, good; food, wheat-bran, corn-meal and ground oats.

Number of inspection.	Number of cow.	Breed.	Age of cow.	Number of times calving.	Time of last calving.	Number of quarts given.	Lactometer at 60 deg. Fahr.	Per cent of cream.
			years.		mos.			
B. 3712..	1	R. Ayrshire.	4	3	4	10	112	28
B. 3713..	2	R. Ayrshire.	2	1	4	8	118	14
B. 3714..	3	R. Ayrshire.	2	1	2	12	114	20
B. 3715..	4	R. Ayrshire.	2	1	4	12	116	14
B. 3716..	5	R. Ayrshire.	2	1	1	13	112	16
B. 3717..	6	R. Ayrshire.	2	1	3	10	118	16
B. 3718..	7	R. Ayrshire.	3	2	2	11	110	10
B. 3719..	8	R. Ayrshire.	5	3	2	14	110	26
B. 3720..	9	R. Ayrshire.	9	6	3	12	112	18
B. 3721..	10	R. Ayrshire.	7	5	2	15	112	24
B. 3722..	11	R. Ayrshire.	6	4	4	11	114	18
B. 3723..	12	½ Ayrshire .	6	4	2	15	110	22
B. 3724..	13	½ Ayrshire .	7	5	2	15	110	16
B. 3725..	14	½ Ayrshire .	3	2	2	12	114	18
B. 3726..	15	Native.....	8	6	1	14	116	16
B. 3727..	16	Native.....	8	6	15d	15	114	18
B. 3728..	17	¾ Ayrshire .	9	7	10d	4	122	20
B. 3729..	18	R. Ayrshire.	7	5	10m	4	110	20
B. 3730..	19	Ayrshire ...	14	11	5	2	102	28
B. 3731..	Average..	....	..	..	....	..	112	20

Evening of May 16, 1885; farm of J. De Witt Krebs, 150 acres, Montgomery, Orange county; cows in herd, 37; treatment, kind; housing, good; food, wheat-bran, cotton-seed meal three times a day.

Number of inspection.	Name of cow.	Breed	Age of cow.	Number of times calving.	Time of last calving.	Number of quarts given.	Lactometer at 60 deg. Fahr.	Per cent of cream.
			yrs.		1884.			
B. 5603.....	Lulie Douglas.........	Ayrshire.	4	3	Aug. 28	4	104	18
B. 5604.....	Mrs. McGregor..... ...	Ayrshire.	2	1	Sept. 22	3	112	18
B. 5605.....	Beauty of Wallkill......	Ayrshire.	2	1	Oct. 5	5	108	18
B. 5606.....	Thanksgiving..........	Ayrshire.	2	1	Dec. 10	5	110	18
					1885.			
B. 5607.....	Favorite of Brookside..	Ayrshire.	3	1	Jan. 4	5	108	20
					1884.			
B. 5608.....	Favorite, 6th...........	Ayrshire.	3	1	Oct. 26	5	114	16
B. 5609.....	Favorite, 5th.	Ayrshire.	4	2	Nov. 21	4	110	18
B. 5610.....	Maid of Wallkill.	Ayrshire.	9	6	Nov. 3	6	110	16
B. 5611....	Bessie.................	Ayrshire.	5	3	Dec. 3	6	108	20
					1885.			
B. 5612.....	Favorite, 3d...........	Ayrshire.	8	6	Mch. 31	10	112	16
B. 5613.....	Favorite, 2d...........	Ayrshire.	14	12	April 13	8	112	20
					1884.			
B. 5614.....	M. Dolly...............	Ayrshire.	6	4	Sept. 16	5	114	18
B. 5615.....	Mary..................	½ Ayrshire.	3	2	Dec. 12	5	112	18
B. 5616.....	Gowdy	½ Ayrshire.	7	5	Dec. 10	7	112	18
B. 5617.....	Dinah..................	½ Ayrshire.	6	4	Dec. 6	7	114	16
					1885.			
B. 5618.....	Peggie.................	Native ..	8	6	Jan. 25	7	112	16
B. 5619.....	Sallie..................	Native ..	8	6	Jan. 29	7	114	16
B. 5620.....	Average...........	....	..	....	....	..	110	18

Morning of May 30 and evening of May 30, 1885; farm of J. DeWitt Krebs, [illegible], Orange county; cows in herd, 37; treatment, kind; housing, [illegible]; food, pasture, wheat middlings and bran.

Number of inspection.	Name of cow.	Breed.	Age of cow.	Number of times calving.	Time of last calving.		Number of pounds given.	Lactometer at 60 deg. Fahr.	Per cent of cream.	Number of inspection.	Number of pounds given.	Lactometer at 60 deg. Fahr.	Per cent of cream.
			yrs.		1884.								
B. 5856	[illegible]ie [illegible]glas	Ayrshire	4	8	August	28	6 6 oz.	112	20	B. 5872	7 12 oz.	108	18
B. 5857	Mrs. [illegible]Gregor	Ayrshire	2	1	September	22	6	114	18	B. 5873	7 4 oz.	114	18
B. 5858	[illegible]ely of Wallkill	Ayrshire	2	1	[illegible]er	5	8	110	20	B. 5874	9 6 oz.	110	18
B. 5859	Thanksgiving	Ayrshire	2	1	[illegible]er	10	10	112	18	B. 5875	12	112	16
					1885.								
B. 5860	Favorite of [illegible]ookville	Ayrshire	8	1	January	4	10	110	20	B. 5876	12 6 oz.	110	18
					1884.								
B. 5861	Favorite, 6th	Ayrshire	8	1	[illegible]er	26	8 6 oz.	14	20	B. 5877	10 6 oz.	114	21
B. 5862	Favorite 5th	Ayrshire	4	2	November	[illegible]	8	110	20	B. 5878	8 12 oz.	110	20
B. 5863	[illegible] of [illegible]	Ayrshire	9	6	[illegible]er	3	18	112	22	B. 5879	15 6 oz.	112	22
B. 5864	Bessie	Ayrshire	5	8	December	8	11 6 oz.	108	20	B. 5880	[illegible] 12 [illegible]	108	20
B. 5865	Favorite, [illegible]	Ayrshire	8	6	[illegible]ch	31	19 6 oz.	110	20	B. 5881	22	1[illegible]	18
B. 5866	Favorite, [illegible]	Ayrshire	14	12	April	13	19	112	22	B. 5882	21 6 oz.	114	20
B. 5867	[illegible] Dolley	½ Ayrshire	6	4	September	16	9 6 oz.	14	20	B. 5883	11 6 oz.	112	20
B. 5868	Mary	½ Ayrshire	3	2	[illegible]er	12	10	12	22	B. 5884	[illegible] 4 oz.	114	20
B. 5869	Gowdy	½ Ayrshire	7	5	[illegible]er	[illegible]	10	14	24	B. 5885	[illegible] 8 oz.	112	22
B. 5870	[illegible]nah	½ Ayrshire	6	4	December	6	12	110	18	B. 5886	16 4 oz.	1[illegible]	18
B. 5871	Average	....	....	....	....		....	112	20	B. 5887	....	112	19

Morning and evening of June 29, 1885; *farm of J. De Witt Krebs, Montgomery, Orange county; cows in herd,* 37; *treatment, kind; housing, good; food, pasture and four quarts wheat-bran per day.*

Number of inspection.	Name of cow.	Breed.	Age of cow.	Number of times calving.	Time of last calving.	Number of pounds given.	Lactometer at 60 deg. Fahr.	Per cent of cream.	Number of inspection.	Number of pounds given.	Lactometer at 60 deg. Fahr.	Per cent of cream.
			years.		1884.							
B. 6449	[illegible]	Ayrshire	4	8	August 28	8 3-4	118	[illegible]	B. 6465	4 1-4	114	16
B. 6450	[illegible] McGregor	Ayrshire	2	1	[illegible]er 22	5	116	14	B. 6466	6	116	[illegible]
B. 6451	Beauty of Wallkill	Ayrshire	2	1	[illegible]er 5	7 1-4	116	14	B. 6467	8 1-3	116	12
B. 6452	Thanksgiving	Ayrshire	2	1	[illegible]er 10	9	120	14	B. 6468	11 1-4	116	14
					1885.							
B. 6453	Favorite of [illegible]	Ayrshire	3	1	January 4	8 1-2	118	12	B. 6469	11 1-4	114	16
					84.							
B. 6454	Favorite, 6th	Ayrshire	3	1	October 21	7 1-2	118	18	B. 6470	9 1-4	[illegible]	18
B. 6455	Favorite, 5th	Ayrshire	4	2	M[illegible] 21	5 3-4	116	12	B. 6471	6 1-2	112	12
B. 6456	[illegible] of [illegible]allkill	Ayrshire	9	6	[illegible]er 8	9	112	14	B. 6472	18	14	14
B. 6457	Bessie	Ayrshire	5	8	[illegible]er 8	11 3-4	116	16	B. 6473	10 1-2	14	16
B. 6458	[illegible]ite, 3d	Ayrshire	8	6	March 31	15 1-2	114	[illegible]	B. 6474	18 1-4	112	14
B. 6459	Favorite, 2d	Ayrshire	14	12	April 18	15 3-4	114	16	B. 6475	18 3-4	112	16
B. 6460	[illegible]ss Dolley	Ayrshire	6	4	[illegible] [illegible]	5 3-4	118	20	B. 6476	7 1-2	[illegible]	20
					1885.							
B. 6461	[illegible]avorite of Walden	Ayrshire	2	1	June 6	12 1-2	118	18	B. 6477	14	118	16
					1884.							
B. 6462	[illegible]sa	Ayrshire	2	1	[illegible]er 25	6 1-2	120	14	B. 6478	7 3-4	118	14
					1885.							
B. 6463	[illegible]y Beauty	Ayrshire	3	1	April 10	10 1-4	118	16	B. 6479	12 3-4	116	16
B. 6464	Average	....	....	....	....	....	114	16	B. 6480	....	114	15

Morning and evening of August 3, 1885; farm of J. DeWitt Krebs, Montgomery, Orange county; cows in herd, 37; treatment, kind; housing, good; food, wheat middlings, bran and cotton-seed meal.

Number of inspection.	Name of cow.	Breed.	Age of cow.	Number of times calving.	Time of last calving.	Number of pounds given.	Lactometer at 60 deg. Fahr.	Per cent of cream.	Number of inspection.	Number of pounds given.	Lactometer at 60 deg. Fahr.	Per cent of cream.
			years.		1884.							
B. 7029.	Mistress McGregor	Ayrshire.....	2	1	September 32	2 3-4	114	20	B. 7044....	8 3-4	114	20
B. 7030.....	Thanksgiving..............	Ayrshire.....	2	1	December 10	6 3-4	112	20	B. 7045....	5 3-4	114	18
B. 7031.....	Beauty of Wallkill.........	Ayrshire.....	2	1	October 5	3 1-2	110	14	B. 7046....	4	108	14
					1885.							
B. 7032.....	Favorite of Brookville........	Ayrshire,....	3	1	January 4	6 3-4	108	14	B. 7047....	7	110	14
					1884.							
B. 7033.... .	Favorite, 6th..............	Ayrshire.....	3	1	October 26	6	112	22	B. 7048....	6 1-2	114	22
B. 7034.....	Favorite, 5th..............	Ayrshire.....	4	2	November 21	2 3-4	106	16	B. 7049....	3 1-4	108	16
B. 7035.....	Bessie.....................	Ayrshire.....	5	3	December 3	5 1-4	106	16	B. 7050....	5 1-2	108	16
B. 7036.....	Maid of Wallkill............	Ayrshire.....	9	6	November 3	6 3-4	108	16	B. 7051....	7 3-4	108	14
					1885.							
B. 7037.....	Favorite, 3d	Ayrshire.....	8	6	March 31	13 1-4	106	18	B. 7052 ...	15 1-4	106	18
B. 7038.....	Favorite, 2d	Ayrshire.....	14	12	April 13	11 1-4	104	14	B. 7053....	13 3-4	106	13
					1885.							
B. 7039.....	Favorite of Walden	Ayrshire.....	2	1	June 6	9 1-2	114	16	B. 7054....	11	112	16
					1884.							
B. 7040.....	Rosa Bruce.................	Ayrshire.....	2	1	December 25	5 1-4	112	12	B. 7055....	5 1-2	112	11
B. 7041.....	May Beauty.................	Ayrshire.....	3	2	April ..	8 1-2	114	14	B. 7056....	9	114	13
B. 7042.....	Mistress Dolley.............	Ayrshire.....	6	4	September 16	4 1-4	110	14	B. 7057....	4 1-2	106	14
B. 7043.....	Average....	....	....	....	....	....	110	18	B. 7058....	....	110	16

Morning and evening of September 7, 1885; farm of J. DeWitt Krebs, Montgomery, Orange county; cows in herd, 32; treatment, kind; housing, good; food, wheat-bran and corn-fodder.

Number of inspection.	Name of cow.	Breed.	Age of cow.	Number of times calving.	Time of last calving.		Number of pounds given.	Lactometer at 60 deg. Fahr.	Per cent of cream.	Number of inspection.	Number of pounds given.	Lactometer at 60 deg. Fahr.	Per cent of cream.
			years.										
B. 7529	[illegible]ty of Wallkill	Ayrshire	2	1	1884. October	5	4	102	16	B. 7541	4	102	16
B. 7530	Thanksgiving	Ayrshire	2	1	[illegible]er	10	5	120	20	B. 7542	6 1-4	120	20
B. 7531	Favorite of Brookville	Ayrshire	3	1	1885. January	4	7	112	20	B. 7543	9	110	18
B. 7532	Maid of Wallkill	Ayrshire	3	1	1884. [illegible]r	22	6	116	24	B. 7544	9	114	24
B. 7533	Bessie	Ayrshire	9	6	[illegible]r	3	7	112	13	B. 7545	7	112	20
B. 7534	Favorite, 6th	Ayrshire	5	3	[illegible]er	3	4 1-2	112	20	B. 7546	4 3-4	110	18
B. 7535	Favorite, 3d	Ayrshire	3	6	1885. [illegible]	31	13	110	[illegible]	B. 7547	17	106	16
B. 7536	Favorite, 2d	Ayrshire	14	12	April	13	12 1-4	114	20	B. 7548	14 1-2	10	18
B. 7537	Favorite of Walden	Ayrshire	2	1	June	6	8 1-2	112	20	B. 7549	9 3-4	[illegible]	18
B. 7538	[illegible]sa Bruce	Ayrshire	2	1	1884. [illegible]er	25	6	112	18	B. 7550	6	112	18
B. 7539	[illegible]y Blossom	Ayrshire	3	2	1885. April	7	7 3-4	110	20	B. 7551	8	110	20
B. 7540	[illegible]rage	....	....	....	....	....	....	112	20	B. 7552	....	110	19

Morning and evening of October 5, 1885; farm of J. DeWitt Krebs, Montgomery, Orange county; cows in herd, 32; treatment, kind; housing, good; food, corn-fodder and wheat-bran.

Number of inspection.	Name of cow.	Breed.	Age of cow.	Number of times calving.	Time of last calving.	Number of pounds given.	Lactometer at 60 deg. Fahr.	Per cent of cream.	Number of inspection.	Number of pounds given.	Lactometer at 60 deg. Fahr.	Per cent of cream.
			years.		1885.							
B. 7842.....	Lelia Douglas..................	Ayrshire.....	5	4	October 1	18	114	28	B. 7856....	12 1-2	114	26
B. 7843.....	Favorite of Walden..........	Ayrshire.....	2	1	June 6	8 1-4	116	18	B. 7857....	8 1-2	118	18
					1884.							
B. 7844.....	Thanksgiving................	Ayrshire.....	2	1	December 10	8	114	20	B. 7858....	2 1-2	122	18
					1885.							
B. 7845.....	Favorite of Brookville........	Ayrshire.....	3	1	January 4	6 1-2	116	20	B. 7859....	5 3-4	114	18
B. 7846.....	Favorite, 6th................	Ayrshire.....	3	1	October 22	4 3-4	110	20	B. 7860....	4 1-4	110	20
					1884.							
B. 7847.....	Bessie....	Ayrshire.....	5	3	December 8	8	104	22	B. 7861....	8	106	20
					1885.							
B. 7848.....	Maid of Wallkill....	Ayrshire.....	9	6	March 8	4 1-2	116	20	B. 7862....	4	116	20
B. 7849. ..	Favorite, 2d..................	Ayrshire.....	14	12	April 18	12	114	18	B. 7863....	11 3-4	112	16
B. 7850.....	Favorite, 3d..................	Ayrshire.....	8	6	March 31	11 3-4	112	16	B. 7864....	10	112	14
B. 7851.....	May Blossom	Ayrshire.....	3	1	April 2	7	110	14	B. 7865....	6 1-4	112	14
B. 7852.....	Mistress Dolley.............	Ayrshire.....	6	5	September 25	17	116	24	B. 7866....	16	114	20
					1884.							
B. 7853.....	Rosa Bruce.................	Ayrshire.....	2	1	December 25	4	116	24	B. 7867....	4	112	22
					1885.							
B. 7854.....	Hugo Lassie.................	Ayrshire.....	2	1	September 21	18 1-2	112	16	B. 7868....	13 1-4	112	14
B. 7855.....	Average.................	....	....	....	....	....	114	20	B. 7869....	....	115	19

Evening of November 9, 1885, *and morning of November* 10, 1885; *farm of J. De Witt Krebs, Montgomery, Orange county; treatment, kind; housing, good; food, wheat-bran and middlings.*

Number of inspection.	Name of cow.	Breed.	Age of cow.	Number of times calved.	Time of last calving.	Number of pounds given.	Lactometer at 60 deg. Fahr.	Per cent of cream.	Number of inspection.	Number of pounds given.	Lactometer at 60 deg. Fahr.	Per cent of cream.
			years.		1885.							
B. 7870.....	Lulie Douglas..............	Ayrshire.....	5	4	October 1	11 3-4	104	20	B. 7882....	10	106	22
B. 7871.....	Favorite of Walden..........	Ayrshire.....	2	1	June 6	7	112	20	B. 7883....	7 1-4	110	22
B. 7872.....	Baty of Wallkill...........	Ayrshire.....	3	2	October 20	11 1-2	110	16	B. 7884....	11 3-4	110	16
B. 7873.....	Favorite of Brookville........	Ayrshire.....	2	1	January 4	4 1-2	112	20	B. 7885....	3 3-4	112	20
					1884.							
B. 7874.....	Favorite, 8d................	Ayrshire.....	8	6	March	8	110	18	B. 7886....	9 1-4	112	18
					1885.							
B. 7875.....	Favorite, 2d................	Ayrshire.....	14	13	April 13		112	20	B. 7887....	10	112	22
B. 7876.....	May Bim................	Ayrshire.....	3	2	April 7	4 3-4	112	16	B. 7888 ..	5 1-4	110	18
B. 7877.....	Lise Cherry...............	Ayrshire.....	4	3	October 4	11 1-4	112	14	B. 7889....	12 1-2	112	16
B. 7878.....	Miss Dolley...............	Ayrshire.....	6	5	September 25	14 1-2	108	16	B. 7890....	13 3-4	110	18
B. 7879.....	Hugo Lassie................	Ayrshire.....	2	1	September 21	12 1-2	108	14	B. 7891....	12 3-4	110	16
B. 7880.....	[illegible], 7th................	Ayrshire.....	2	1	October 15	12 1-2	112	16	B. 7892....	12 3-4	112	18
B. 7881.....	Average................	....	....	....	....	....	110	18	B. 7893....	....	111	20

Evening of December 3, 1885, and morning of December 4, 1885; farm of J. De Witt Krebs, Montgomery, Orange county; cows in herd, 32; treatment, kind; housing good; food, wheat-bran, middlings and corn-fodder.

Number of inspection.	Name of cow.	Breed.	Age of cow.	Number of times calving.	Time of last calving.	Number of pounds given.	Lactometer at 60 deg. Fahr.	Per cent of cream.	Number of inspection.	Number of pounds given.	Lactometer at 60 deg. Fahr.	Per cent of cream.
			years.		[illegible]85.							
B. 7894	[illegible] [illegible]uglas 4th	Ayrshire	5	4	October 1	10 1-4	04	18	B. 7910	10 3-4	08	18
B. 7895	Favorite of [illegible] W[illegible]	Ayrshire	2	1	June 6	6 1-4	10	22	B. 7911	6 3-4	12	22
B. 7896	[illegible]uty of Wallkill	Ayrshire	3	2	[illegible]er 20	10 1-2	08	18	B. 7912	11 1-2	10	20
B. 7897	[illegible]vorite of [illegible] [illegible]e	Ayrshire	3	1	January 4	3 1-4	12	22	B. 7913	3	14	22
B. 7898	Favorite, 6th	Ayrshire	3	2	No[illegible]ember 19	11	13	16	B. 7914	12 3-4	18	18
B. 7899	[illegible]avorite, 5th	Ayrshire	4	3	[illegible]er 29	16	20	16	B. 7915	16 1-2	24	16
B. 7900	[illegible]e	Ayrshire	6	4	[illegible]er 11	17	08	18	B. 7916	15 3-4	12	18
B. 7901	Favorite, 3d	Ayrshire	8	6	[illegible]h 31	5 3-4	10	16	B. 7917	6	14	18
B. 7902	Favorite, 2d	Ayrshire	14	12	April 13	8	12	24	B. 7918	8	14	24
B. 7903	[illegible]y [illegible]dy	Ayrshire	3	2	April 7	3 1-2	10	18	B. 7919	3 1-2	12	20
B. 7904	[illegible]e [illegible]y	Ayrshire	3	3	October 4	11 1-4	12	16	B. 7920	11	13	18
B. 7905	[illegible]ing	Ayrshire	3	2	[illegible]er 25	11 1-2	14	20	B. 7921	13	18	20
B. 7906	Mistress [illegible]lley	Ayrshire	7	5	[illegible]er 25	12 1-4	10	16	B. 7922	15 1-4	10	18
B. 7907	Hugo [illegible]e	Ayrshire	2	1	[illegible]r 21	10 3-4	06	16	B. 7923	10	06	18
B. 7908	Favorite, 7th	Ayrshire	2	1	[illegible]er 15	11	08	18	B. 7924	12	10	16
B. 7909	Eve[illegible]ing [illegible]erage	...	...	...	...	...	110	18	...	...	...	...
	[illegible]ing average	...	...	...	...	...	...	...	B. 7925	...	112	19

Evening of December 4, 1885; farm of Abner Bookstaver, Montgomery, Orange county; cows in herd, 20; treatment, kind; housing, good; food, corn-fodder, corn-stalks, and brewers' grains.

Number of inspection.	Name of cow.	Breed.	Age of cow.	Number of times calving.	Time of last calving.	Number of quarts given.	Lactometer at 60 deg. Fahr.	Per cent of cream.
			years.		1885.			
B. 7926	New	Native	8	6	July.	6	116	10
B. 7927	Cherry	Native	7	5	Aug.	6	118	12
B. 7928	Black	Native	6	3	Nov.	8	118	14
B. 7929	Bodine	Native	11	8	Aug.	7	110	28
B. 7930	Spot	Native	10	7	Nov.	9	108	16
B. 7931	Brownie	Native	7	5	April.	4	116	20
B. 7932	Wooley	Native	13	10	June.	8	114	18
B. 7933	Sleepy	Native	12	9	Nov.	9	108	12
B. 7934	Brindle	Native	8	6	Oct.	5	120	26
B. 7935	Average	....	..	..	....	..	114	18

December 4, 1885; farm of R. A. Fisher, Montgomery, Orange county; cows in herd, 16; treatment, kind; housing, good; food, four quarts cob corn and four quarts ground oats per day.

Number of inspection.	Name of cow.	Breed.	Age of cow.	Number of times calving.	Time of last calving.	Number of quarts given.	Lactometer at 60 deg. Fahr.	Per cent of cream.
			years.		mos.			
B. 7936	Sylvina	Jersey	7	5	6	3	112	28
B. 7937	Hattie	Jersey	5	3	5	4	112	38
B. 7938	French Queen of Upholme.	Jersey	3	2	8	2	114	20
B. 7939	Lucy of Lee	Jersey	8	6	7	3	110	24
B. 7940	Young Rosette.	Jersey	6	3	7	3	106	26
B. 7941	Fairy Stone	Jersey	2	1	8	2	110	46
B. 7942	Average	....	..	..	....	..	110	28

The following is the pedigree of the above herd:

Sylvina, No. 8299; sire, Rex, No. 1330; dam, Robema, No. 3840; Sylvina was dropped September 22, 1879. She is finely bred, tracing to Comlis Lily, Albert, No. 44, and Jack Dasher. She has made fourteen pounds of butter per week for Dr. E. D. Newell of New Brunswick, N. J. She dropped the 5th day of April, 1885, a heifer calf by Stokes Pogis Perfeeline; she is a very promising heifer.

Queen of Upholme, No. 1883; sire, Delaware Darling, No. 3461; dam, Pet of Clifton, No. 1386. She is a three-year-old heifer; dropped a fine heifer calf March 2, 1885; by Minnie's Duke of Darlington; fine udder and well-placed teats.

Hattie, No. 14640; sire, Young Piene, No. 6518; dam, One Eyed May, No. 14164; dropped 1880. Hattie is a superior milker and fine breeder; milk rich and a good butter-maker testing fourteen pounds per week.

Lucy of Lee, No. 6593; sire, Pine Cliff, No. 1106; dam, Jewel Beauty 2d, No. 1701; dropped March 15, 1877. Lucy of Lee's grand-dam was Jewel Beauty, imported by Mr. Sharpless of Philadelphia, Pennsylvania; he considered her one of the best in his herd; she is a persistent milker and a regular breeder; she has never been tested but has made as high as fourteen pounds of butter per week.

Young Rosette, No. 19656; sire, Willie F. S., No. 245; dam, Rosette F. S., No. 1818; Young Rosette is six years old, bred by Rev. Edward G. Renouf, St. Martin's, Island of Jersey. She is a splendid cow and has made for her owner fifteen pounds of butter in one week on grass, and feed, three quarts of corn-meal night and morning. She is a regular breeder.

Fairy Stone, No. 24459; sire, Lemon Peel P. — 320 — C.; dam, Rather Pretty F., No. 5933 C. She is an imported heifer and has in her veins the blood of Bobby Verhunnus, Young Rose and Coomassie, and other noted strains, and gives great promise for a fine butter maker.

Bertha M., No. 31339; sire, Fishu Polonius, No. 5931; dam, Hattie French, No. 14640. This heifer dropped a heifer calf April 12, 1885, when only nineteen months old, and made eight and one-half pounds of butter in one week three months after calving.

Evening of December 12, 1885; *farm of A. Koffman, Sunny Side, Montgomery, Orange county; cows in herd,* 53; *treatment, kind; housing, good; food, corn-meal, wheat-bran and beets.*

Number of inspection.	Name of cow.	Breed.	Age of cow.	Number of times calving.	Time of last calving.	Number of quarts given.	Lactometer at 60 deg. Fahr.	Per cent of cream.
			years.	mos.	1885.			
B. 7943 ..	Essie............	Alderney	4	8	Sept. 15	5	118	24
B. 7944 ..	Big Nellie........	Alderney	4	8	Sept. 16	4	114	30
B. 7945 ..	Pussie...........	Jersey	4	8	May 2	4	112	30
B. 7946 ..	Susan, 1st.......	Holstein.....	8	1	June 10	6	114	20
B. 7947 ..	Fannie K........	Holstein.....	5	3	July 16	6	120	18
B. 7948 ..	Jennie K........	Holstein.....	4	2	Oct. 15	7	118	18
B. 7949 ..	Susan, 2d........	Durham	7	5	Aug. 10	6	116	24
B. 7950 ..	Stocken Leg.....	Dutch	5	3	Sept. 25	6	114	22
B. 7951 ..	Brindle..........	Native	6	4	Oct. 2	6	110	28
B. 7952 ..	Alderney	Alderney	17	15	May 8	4	110	32
B. 7953 ..	Goat	Native	6	4	Oct. 2	6	110	28
B. 7954 ..	Black Bet	Holstein.....	8	1	May 1	5	116	18
B. 7955 ..	Sarah K.........	Jersey	4	2	Oct. 16	5	110	28
B. 7956 ..	Clara B..........	Alderney	4½	3	Aug. 10	5	118	28
B. 7957 ..	Rosa Lee.... ...	Alderney	8	1	Aug. 4	7	118	26
B. 7958 ..	Average......	....	....	....	....	....	114	24

Morning of May 30, 1885; *Wallkill, Orange county; farm of Lewis S. Wisner; cows in herd,* 6; *treatment, kind: housing, good; food, lowland pasture.*

Number of inspection.	Name of cow.	Breed.	Age of cow.	Number of this calving.	Time of last calving.	Number of quarts given.	Lactometer at 60 deg. Fahr.	Per cent of cream.
			years.		1884.			
E. 1656 ..	Bertie Hart......	A. J. C. C. 10,208	5	4	Nov. ..	6	108	30
					1885.			
E. 1657 ..	Georganna, 2d....	A. J. C. C. 22,126	8	2	April..	9	114	22
E. 1658 ..	Belle of Orange..	A. J. C. C. 26,744	8	7	Feb'ry.	8	110	33
E. 1659 ..	Carrie Mitchell...	A. J. C. C. 26,882	5	4	March.	11	112	22
					1884.			
E. 1660 ..	Bobby's Rose....	A. J. C. C. 24,424	4	2	Dec....	5	112	25
E. 1661 ..	Morn'g's average	....	....	....	......	....	112	26
E. 1662 ..	Even'g's average	....	....	....	......	....	108	26

Morning of June 30, 1885; *Wallkill, Orange county; farm of Lewis S. Wisner; cows in herd,* 6; *treatment, kind; housing, good; food, lowland pasture.*

Number of inspection.	Name of cow.	Breed.	Age of cow.	Number of this calving.	Time of last calving.	Number of quarts given.	Lactometer at 60 deg. Fahr.	Per cent of cream.
			years.		1884.			
E. 2274 ..	Bertie Hart......	A. J. C. C. 10,208	5	4	Nov...	4	114	30
					1885.			
E. 2275 ..	Georganna, 2d...	A. J. C. C. 22,126	8	2	April..	7	114	21
E. 2276 ..	Belle of Orange..	A. J. C. C. 26,744	8	7	Feb'ry.	7	116	35
E. 2277 ..	Carrie Mitchell...	A. J. C. C. 26,882	5	4	March.	8	112	24
					1884.			
E. 2278 ..	Bobby's Rose....	A. J. C. C. 24,424	4	2	Dec....	8	116	25
E. 2279 ..	Morn'g's average	....	....	....	......	....	118	27
E. 2280 ..	Even'g's average	....	....	....	......	....	114	30

Morning and evening of September 26, 1885; Wallkill, Orange county, farm of Lewis S. Wisner; cows in herd, 6; treatment, kind; housing, good; food, lowland pasture.

Number of inspection.	Name of cow.	Breed.	Age of cow.	Number of times calving.	Time of last calving.	Number of quarts given.	Morning. Lactometer at 60 deg. Fahr.	Morning. Per cent of cream.	Morning. Per cent of fat.	Number of inspection.	Number of quarts given.	Evening. Lactometer at 60 deg. Fahr.	Evening. Per cent of cream.	Evening. Per cent of fat.
			years.		85.									
E. 8146.....	Carrie Mill..........	A. J. C. C..	5	4	March.........	2	110	24	4.50	E. 8152....	8	110	24	4.50
E. 8147.....	...a, 2d..........	A. J. C. C..	3	2	April.........	3	116	30	4.25	E. 8153....	4	116	28	4.00
E. 8148.....	...by Rose...	A. J. C. C..	4	2	84. ...er ..... 1885.	2	102	33	5.50	E. 8154....	8	106	30	5.25
E. 8149.....	Mitchaline..............	A. J. C. C..	8	6	July	8	114	18	3.50	E. 8155....	7	114	16	3.25
E. 8150.....	Belle of Orange..........	A. J. C. C..	8	7	February......	4	110	25	4.75	E. 8156....	8	110	23	4.50
E. 8151.....	Average..............	....	....	....	....	....	109	25	4.25	....	....	....	....	.....
	Average.	....	....	....	....	....	....	....	....	E. 3157....	....	116	23	4.00

Calculated analysis, September 26, 1885. Five cows owned by Lewis S. Wisner, Wallkill, Orange county. Morning's milking.

Number of inspection.	Water.	Total solids.	Fat.	Solids not fat.	Sugar caseine.	Salts.
E. 3146	85·78	14·22	4·50	9·72	9·04	0·68
E. 3147	85·75	14·35	4·25	10·10	9·42	0·68
E. 3148	85·17	14·83	5·50	9·33	8·65	0·68
E. 3149	86·76	13·24	3·50	9·74	9·06	0·68
E. 3150	85·75	14·25	4·75	9·50	8·82	0·68
E. 3151	(Average, see page 62.)					

Calculated analysis, September 26, 1885. Five cows owned by Lewis S. Wisner, Wallkill, Orange county. Evening's milking.

Number of inspection.	Water.	Total solids.	Fat.	Solids not fat.	Sugar caseine.	Salts.
E. 3152	85·78	14·22	4·50	9·72	9·04	0·68
E. 3153	85·97	14·03	4·00	10·03	9·35	0·68
E. 3154	85·16	14·84	5·25	9·59	8·81	0·68
E. 3155	87·07	12.93	3·25	9·68	9·00	0·68
E. 3156	85·78	14·22	4·50	9·72	9·04	0·68
E. 3157	(Average, see page 62.)					

Morning and evening of October 31, 1885; farm of Lewis S. Wisner (Willow Farm), Wallkill, Orange county; cows in herd, 6; treatment, the very best; housing, good; food, lowland pasture.

MORNING.

Number of inspection.	Name of cow.	Breed.	Age of cow.	Number of times calving.	Time of last calving.	Number of quarts given.	Lactometer at 60 deg. Fahr.	Per cent of fat by lactoscope.
E. 3307.....	Bertie Heart..	A. J. C. C..	5	3	1884. Nov. 26	2	108	3.25
E. 3308.....	Mitchaline....	A. J. C. C..	8	6	1885. April 7	3	116	4.00
E. 3309.....	Georganna, 2d.	A. J. C. C..	3	2	Feb. 12	3	122	4.00
E. 3310.....	Bell of Orange.	A. J. C. C..	8	6	Feb. 12	2	121	5.25
E. 3311.....	Carrie Mitchell	A. J. C. C..	5	3	March 9	4	120	4.50
E. 3312.....	Average...	....	....	....	....	....	118	4.50

EVENING.

Number of inspection.	Name of cow.	Breed.	Age of cow.	Number of times calving.	Time of last calving.	Number of quarts given.	Lactometer at 60 deg. Fahr.	Per cent of fat by lactoscope.
E. 3313.....	Bertie Heart..	A. J. C. C..	5	3	1884. Nov. 26	1	116	3.25
E. 3314.....	Mitchaline....	A. J. C. C..	8	6	1885. April 7	2 1-2	114	3.75
E. 3315.....	Georganna, 2d.	A. J. C. C..	3	2	Feb. 12	2 1-2	120	3.75
E. 3316.....	Bell of Orange.	A. J. C. C..	8	6	Feb. 12	1 1-2	119	5.00
E. 3317.....	Carrie Mitchell	A. J. C. C..	5	3	March 9	3 1-2	116	4.25
E. 3318.....	Average...	....	....	....	....	....	118	4.25

Morning and evening of November 29, 1885 *; farm of Lewis S. Wisner* (*Wisner Farm*), *Wallkill, Orange county ; cows in herd,* 6 *; treatment, the very best ; housing, good ; food, hay.*

MORNING'S.

Number of inspection.	Name of cow.	Breed.	Age of cow.	Number of times calving.	Time of last calving.	Number of quarts given.	Lactometer at 60 deg. Fahr.	Per cent of fat.
			years.		1885.			
E. 3399.....	Georganna, 2d	A. J. C. C.	3	2	Feb'ry 12	2 1-2	107	5.50
E. 3400.....	Belle of Ora'ge	A. J. C. C.	8	6	Feb'ry 12	1 3-4	119	5.75
E. 3401.....	Carrie Mitchell	A. J. C. C.	5	3	March 9	3	116	5.25
					1884.			
E. 3402.....	Bertie Heart..	A. J. C. C.	5	3	Nov'ber 26	1 3-4	113	4.75
					1885.			
E. 3403.....	Mitchaline....	A. J. C. C.	8	6	April 7	3	117	5.
E. 3404.....	Average....	....	....	....	....	....	113	5.25

EVENING'S.

Number of inspection.	Name of cow.	Breed.	Age of cow.	Number of times calving.	Time of last calving.	Number of quarts given.	Lactometer at 60 deg. Fahr.	Per cent of fat.
E. 3405.....	Georganna, 2d	A. J. C. C.	3	2	Feb'ry 12	2 1-4	109	5.25
E. 3406.....	Belle of Ora'ge	A. J. C. C.	8	6	Feb'ry 12	1 1-2	110	5.50
E. 3407.....	Carrie Mitchell	A. J. C. C.	5	3	March 9	2 3-4	115	5.25
					1884.			
E. 3408.....	Bertie Heart..	A. J. C. C.	5	3	Nov'ber 26	2	112	4.50
					1885.			
E. 3409.....	Mitchaline....	A. J. C. C.	8	6	April 7	3	116	5.
E. 3410.....	Average....	....	....	....	....	....	112	5.25

Morning and evening of December 16, 1885*; farm of Lewis S. Wisner, Middletown, Orange county ; cows in herd,* 5 *; treatment, kind ; housing, good; food, hay.*

MORNING'S MILKING.

Number of inspection.	Name of cow.	Breed.	Age of cow.	Number of times calving.	Time of last calving.	Number of quarts given.	Lactometer at 60 deg. Fahr.	Per cent of fat by lactoscope.	Per cent of cream.
			yrs.		1885.				
E. 3411.....	Bertie Heart..	A. J. C. C..	5	4	November....	5	116	4.00	25
E. 3412.....	Mitchaline	A. J. C. C..	8	6	July	4	116	4.00	25
E. 3413.....	Georganna, 2d.	A. J. C. C..	3	2	April........	1	85	5.00	35
E. 3414.....	Belle of Ora'ge	A. J. C. C..	8	7	February ...	1	98	5.00	35
E. 3415.....	Average ...	....	..	..	....	..	100	4.50	30

EVENING'S MILKING.

Number of inspection.	Name of cow.	Breed.	Age of cow.	Number of times calving.	Time of last calving.	Number of quarts given.	Lactometer at 60 deg. Fahr.	Per cent of fat by lactoscope.	Per cent of cream.
					1885.				
E. 3416... .	Belle of Orange	A. J. C. C..	8	7	February	1	98	5.00	34
E. 3417.....	Bertie Heart ..	A. J. C. C..	5	4	November....	3	115	4.00	26
E. 3418.....	Georganna, 2d.	A. J. C. C..	3	2	April........	1	86	5.00	34
E. 3419.....	Mitchaline	A. J. C. C..	8	6	July	3½	116	4.00	24
E. 3420.....	Average ...	....	..	..	....	..	101	4.50	31

"Willow Farm;" herd of registered Jersey cattle; Lewis S. Wisner, proprietor, Middletown, Orange county, N. Y.

Cows in herd, six.

Treatment, kind. Each cow being curried once every day. In the months of May, June, July, August and September, if the weather is stormy, the cows are stabled. In the months of October, November, December, January, February, March and April, they are also stabled except when being driven to water, which is done three times each day. They are milked in the stable the year round.

Housing, the very best. Stables partly under ground, well ventilated, with large roomy stalls. The water for the use of the cattle is pumped from a spring about five hundred yards from the stable by means of a hydraulic ram to a large trough situated in the stable yard. The water from this spring stands at a temperature of fifty-five degrees Fahrenheit in the warmest weather in summer. The trough used in the yard for receiving the water is thoroughly cleaned once each day in warm weather and twice each week in cold weather. The stable yard is sheltered by large barns on the west side and with sheds on the north and south sides; the east side is not sheltered except by a fence.

Food. In the months of May, June, July, August and September they are pastured on low-land, except when stabled in cold and stormy weather, when they are fed the best timothy hay with a small quantity of corn-meal. In the months of October, November and December they are fed the best of timothy hay and corn-fodder. In the months of January, February, March and April they are fed corn-meal, wheat middlings and bran with hay.

The following is the pedigree of the above herd:

1. Belle of Orange, No. 26,744, A. J. C. C. Born March 12, 1877. Sire, Hazlehurst, No. 761. Dam, Carrie, 5th, No. 1016. Dropped last calf February 12, 1885. This cow is considered by Mr. Wisner to be the best in his herd. She has never been tested but has made as high as twenty pounds of butter per week.

2. Bertie Heart, No. 10,208, A. J. C. C. Born September 14, 1879. Sire, Pompus, No. 2881. Dam, Bertie Morton 2d, No. 9939. Dropped last calf September 25, 1885. This cow has never been tested, but she has made as high as sixteen pounds of butter per week.

3. Carrie Mitchell, No. 26,882, A. J. C. C. Born June 9, 1880. Sire, Tomnoddy, No. 3260. Dam, Carrie, 5th, No. 1016. Dropped last calf March 9, 1885. This cow has never been tested, but she has made as high as fourteen pounds of butter per week.

4. Micheline, No. 22,064, A. J. C. C. (imported). Born February 15, 1881. Sire, Cetewayo, P., No. 224, C. Dam, Micheline, F., No. 3128, C. Dropped last calf July 8, 1885. This cow has made as high as nineteen pounds of butter per week.

5. Bobby's Rose, No. 24,424 (imported). Born December 9, 1881. Sire, Bobby P., No. 208, H. C. Dam, Cauliflower F., No. 2543, C. Dropped last calf December 16, 1884. Now due. This cow has made as high as nineteen pounds of butter per week.

6. Georgiana 2d, No. 22,126, (imported). Born October 28, 1882. Sire, Leadu by Cetewayo, P., No. 224, C. Dam, Georgiana, F., No.

43 C. Dropped last calf April 7, 1885. This cow has made as high as eighteen pounds of butter per week.

Mr. Wisner has one of the finest two year old Jersey bulls in this section of the State. At the Orange county fair in September, 1884, he received first premium, the bull then being but one year old.

At the State fair held in Albany in September, 1885, Mr. Wisner received a second premium on the bull, there being several on exhibition from different sections of the State, showing that Mr. Wisner's exhibition was of no mean order.

The above herd was inspected by me once each month from April, 1885, to January, 1886. The inspections were made of the milk given by each cow at both the morning and evening milkings. The tests used were the lactometer at a temperature of sixty degrees Fahrenheit for the purpose of obtaining the amount of the specific gravity; the lacteoscope for the purpose of obtaining the amount of butter fat; the cream gauge for the purpose of obtaining the per cent of cream.

Respectfully submitted,
SAMUEL J. WHITE,
Inspector.

Inspection letter "E."

Evening of August 24, 1885; *Fallsburgh, Sullivan county; farm of Andrew B. Bax'er; cows in herd,* 24; *treatment, kind; housing, good; food, upland pasture.*

Number of inspection.	Name of cow.	Breed.	Age of cow.	Number of times calving.	Time of last calving.	Number of quarts given.	Lactometer at 60 deg. Fahr.	Per cent of fat.
			yrs.		1885.			
E. 2789......	Stag............	Native...	10	8	April........	5	108	3.25
E. 2790......	Spot............	Native..	9	7	March.......	5	102	3.00
E. 2791......	Roney..........	Native...	5	3	June........	6	110	3.50
E. 2792......	Peggy..........	Native...	6	4	March.......	4	110	3.50
					1884.			
E. 2793......	Bill Wood	Native...	7	5	October......	2	106	4.25
					1885.			
E. 2794......	White..........	Native...	9	7	April........	6	106	3.50
E. 2795......	Topsie..........	Native...	8	6	May	6	102	3.00
					1884.			
E. 2796......	Armstrong......	Native...	4	2	May	4	106	3.00
					1885.			
E. 2797......	Sammie	Native...	6	4	April........	3	100	3.25
					1884.			
E. 2798......	Duckey.........	Native...	8	6	March.......	2	112	3.25
					1885.			
E. 2799......	Nate	Native...	9	6	February....	4	108	3.00
					1884.			
E. 2800......	Miller..........	Native...	8	6	April........	3	108	3.75
E. 2801......	Hammond	Native...	8	6	December ...	2	100	3.00
					1885.			
E. 2802......	Beckey	Native...	6	4	March.......	4	104	3.50
E. 2803......	Collard.........	Native...	2	1	April........	3	104	4.50
					1884.			
E. 2804......	Jennie..........	Native...	7	5	October	2	105	4.50
E. 2805......	Average......	....	..	....	....	....	108	3.50

Calculated analyses, August 24, 1885 ; sixteen cows owned by Andrew B. Baxter, Fallsburgh, Sullivan county ; evening's milking.

Number of inspection.	Water.	Total solids.	Fat.	Solids not fat.	Caseine and sugar.	Salts.
E. 2789	87·54	12·46	3·25	9·21	8·53	0·68
E. 2790	88·66	12·34	3·75	8·59	7·91	0·68
E. 2791	87·06	12·94	3·50	9·44	8·76	0·68
E. 2792	87·06	12·94	3·50	9·44	8·76	0·68
E. 2793	86·42	13·58	4·25	9·33	8·65	0·68
E. 2794	87·37	12·63	3·50	9·13	8·45	0·68
E. 2795	88·95	12·05	3·50	8·55	7·87	0·68
E. 2796	88·01	11·99	3·00	8·99	8·31	0·68
E. 2797	88·15	11·85	3·25	8·60	7·92	0·68
E. 2798	87·33	12·77	3·25	9·25	8·84	0·68
E. 2799	88·49	12·51	3·50	9·01	8·33	0·68
E. 2800	87·90	13·10	3·75	9·35	8·57	0·68
E. 2801	89·26	11·74	3·50	8·24	7·56	0·68
E. 2802	87·43	12·47	3·50	8·97	8·29	0·68
E. 2803	86·00	14·00	4·50	9·50	8·82	0·68
E. 2804	86·18	13·82	4·50	9·32	8·64	0·68
E. 2805	(Aver	age, see	page	61.)		

Morning of August 25, 1885, *Fallsburgh, Sullivan county ; farm of Andrew B. Baxter ; cows in herd,* 24 ; *treatment, kind ; housing, good ; food, upland pasture.*

Number of inspection.	Name of cow.	Breed.	Age of cow.	Number of times calving.	Time of last calving.	Number of quarts given.	Lactometer at 60 deg. Fahr.	Per cent of fat.
			yrs.					
E. 2806 ..	Stag	Native	10	8	1885. April	6	105	3.25
E. 2807 ..	Spot	Native	9	7	March	9	108	3.00
E. 2808 ..	Roney	Native	5	3	June	6	112	3.50
E. 2809 ..	Peggy	Native	6	4	March	4	110	3.50
E. 2810 ..	Bill Wood	Native	7	5	1884. October	8	116	4.25
E. 2811 ..	White	Native	9	7	1885. April	5	110	3.00
E. 2812 ..	Topsy	Native	8	6	May	6	101	3.00
E. 2813 ..	Armstrong	Native	4	2	1884. May	5	108	3.50
E. 2814 ..	Sammie	Native	6	4	1885. April	5	105	3.25
E. 2815 ..	Duckey	Native	8	6	1884. March	3	114	3.75
E. 2816 ..	Nate	Native	9	6	1885. Feb	7	105	3.00
E. 2817 ..	Miller	Native	8	6	1884. April	5	110	3.75
E. 2818 ..	Hammond	Native	8	6	Dec.	1	100	3.00
E. 2819 ..	Beckey	Native	8	6	1885. March	4	106	3.25
E. 2820 ..	Collard	Native	6	4	April	4	107	3.50
E. 2821 ..	Jennie	Native	7	5	1884. October	1	105	4.50
E. 2822 ..	Average	....	......	......	....	......	106	3.75

Calculated analyses, August 25, 1885; sixteen cows owned by Andrew B. Baxter, Fallsburgh, Sullivan county. Morning's milking.

Number of inspection.	Water.	Total solids.	Fat.	Solids not fat.	Sugar-caseine.	Salts.
E. 2806	87·77	12·23	3·25	8·98	8·30	0·68
E. 2807	86·26	11·74	3·00	8·74	8·06	0·68
E. 2808	86·91	13·09	3·50	9·59	8·91	0·68
E. 2809	87·07	12·93	3·50	9·43	8·75	0·68
E. 2810	85·41	14·59	4·25	10·34	9·66	0·68
E. 2811	87·70	12·30	3·00	9·30	8·62	0·68
E. 2812	88·69	12·11	3·75	8·56	7·86	0·68
E. 2813	87·22	12·78	3·50	9.28	8.60	0·68
E. 2814	87·77	12·23	3·25	8·98	8·30	0·68
E. 2815	86·44	13·56	3·75	9·81	9·13	0·68
E. 2816	88·40	12·60	3·75	8·85	8·17	0·68
E. 2817	86·76	13·25	3.75	9·50	8·82	0·68
E. 2818	89·41	12·43	3·50	8·93	7.41	0·68
E. 2819	87·68	12·32	3·25	9·06	8·20	0·68
E. 2820	87·48	12·52	3·50	9·02	8·34	0·68
E. 2821	86·18	13·82	4·50	9·32	8·64	0·68
E. 2822	(Aver	age, see	page	61.)		

with porcelain, to remove bacterial organisms. Chemists filter the very reagents which they employ in their analyses, knowing well that no weakening of the solutions is thus caused, and physicians and pharmacists constantly use the same process in their practice.

In view of the uncertainty of both chemical and biological examinations in the present state of our knowledge, the greatest importance attaches to the purely physical inspection of the surroundings of a source of water-supply to ascertain if there be any possible danger of its contamination, and if this be discovered, and cannot be surely prevented, the water is to be shunned, irrespective of all analytical testimony in its favor. As Corfield observes, "the average quality of a drinking water supplied to a place is not the matter of most importance, and, indeed, is rather a fallacious guide. What we want to know is the quality of the worst sample that the public are likely to be supplied with at any time." Simon's statement, that "it ought to be an absolute condition for a public water supply that it should be uncontaminable by drainage," is indorsed by the opinion of Parkes (Quain's Dict.) that "the great point in choosing water is, in practice, its freedom from any change of contamination with ex-

Evening of August 27, and morning of August 28, 1885; Fallsburgh, Sullivan county; farm of Elias Newkirk; cows in herd, 14; treatment, kind; housing, bad; food, low-land pasture.

Number of inspection.	Name of cow.	Breed.	Age of cow.	Number of times calving.	Time of last calving.	Number of quarts given.	Lactometer at 60 deg. Fahr., evening's.	Per cent of fat, evenings.	Number of inspection.	Lactometer at 60 deg. Fahr., morning's.	Per cent of fat, morning's.
			years.								
E. 2823	Knapp	Native	8	5	1884. Mar.	8	106	3.25	E. 2836	108	4.00
E. 2824	Lilly	Native	10	7	1885. April	8	100	3.00	E. 2837	106	3.75
E. 2825	Cherry	Native	11	8	February	4	103	3.00	E. 2838	100	3.00
E. 2826	Alderney	Native	6	4	August	10	110	3.00	E. 2839	116	3.00
E. 2827	Bright	Native	6	3	June	5	109	4.25	E. 2840	108	5.00
E. 2828	[illegible]	Native	5	3	March	4	103	4.00	E. 2841	114	4.50
E. 2829	Hauxhurst	Native	13	10	March	4	100	3.00	E. 2842	101	3.00
E. 2830	Spot, 1st	Native	5	1	1883. April	3	112	3.00	E. 2843	113	3.25
E 2831	Black	Native	9	6	1885. [illegible]ry	3	105	3.25	E. 2844	108	3.25
E. 2832	Spot, 2d	Native	5	2	1883. April	3	108	3.00	E. 2845	105	3.50
E. 2833	Grey	Native	6	4	1884. August	8	113	3.00	E. 2846	114	3.00
E. 2834	Long Legs	Native	9	6	1885. [illegible]ly	2	103	3.00	E. 2847	104	3.25
E. 2835	Average	....	....	....	....	....	108	3.50	E. 2848	109	4.00

Calculated analysis, August 27, 1885; twelve cows owned by Elias Newkirk, Fallsburgh, Sullivan county; evening's milking.

Number of inspection.	Water.	Total solids.	Fat.	Solids not fat.	Sugar and caseine.	Salts.
E. 2823......	87·16	12·89	3·25	9·64	8·96	0·68
E. 2824......	88·46	11·54	3·00	8·54	7·86	0·68
E. 2825......	88·58	11·67	3·00	8·67	7·99	0·68
E. 2826......	87·60	12·40	3·00	9·40	8·72	0·68
E. 2827......	85·88	14·12	4·50	9·62	8·94	0·68
E. 2828......	86·59	13·41	4·00	9·41	8·73	0·68
E. 2829......	87·47	12·53	3·00	9·53	8·85	0·68
E. 2830......	88·46	11·54	3·00	8·54	7·86	0·68
E. 2831......	87·54	12·46	3·00	9·46	8·78	0·68
E. 2832......	87·85	12·15	3·00	9·15	8·47	0·68
E. 2833......	87·77	12·23	3·25	8·98	8·30	0·68
E. 2834......	87·85	12·15	3·00	9·15	8·47	0·68
E. 2835......	(Aver	age, see	page 61.)	......	......	.....

Calculated analysis, August 27, 1885; twelve cows owned by Elias Newkirk, Fallsburgh, Sullivan county; morning's milking.

Number of inspection.	Water.	Total solids.	Fat.	Solids not fat.	Sugar and caseine.	Salts.
E. 2836......	86·59	13·41	4·00	9·41	8·73	0·68
E. 2837......	87·06	12·94	3·75	9·19	8·51	0·68
E. 2838......	88·08	11·92	3·00	8·92	8·24	0·68
E. 2839......	87·23	12·77	3·00	9·77	9·09	0·68
E. 2840......	85·32	14·68	5·00	9·68	9·00	0·68
E. 2841......	85·50	14·50	4·50	10·00	9·32	0·68
E. 2842......	87·39	12·61	3·00	9·61	8·93	0·68
E. 2843......	88·08	11·92	3·00	8·92	8·24	0·68
E. 2844......	86·85	13·15	3·25	9·90	9·22	0·68
E. 2845......	87·54	12·46	3·25	9·21	8·53	0·68
E. 2846......	86·97	13·03	3·50	9·53	8·85	0·68
E. 2847......	87·85	12·15	3·25	8·90	8·22	0·68
E. 2848......	(Aver	age, see	page 61.)	......	......	.....

Evening of August 31, and morning of September 1, 1885; Wallkill, Orange county, farm of Increase C. Jordan, cows in herd, 30; treatment, kind; housing, good; food, brewers' grains and lowland pasture.

Number of inspection.	Name of cow.	Breed.	Age of cow.	Number of times calved.	Time of last calving.	Number of quarts given.	Lactometer at 60 deg. Fahr. evening.	Per cent of fat evenings.	Number of inspection.	Lactometer at 60 deg. Fahr. mornings.	Per cent of fat evenings.
					1885.						
E. 2849	Spotter	Native	4	2	July	5	108	4.00	E. 2869	109	4.25
E. 2850	Mooley No. 1	Native	6	3	July	10	107	3.00	E. 2870	110	3.50
E. 2851	Old cow	Native	11	3	April	5	104	3.00	E. 2871	106	3.25
E. 2852	Spot	Native	8	6	July	8	101	3.00	E. 2872	104	3.00
E. 2853	Mooley No. 2	Native	6	4	March	7	105	3.25	E. 2873	106	4.00
E. 2854	Red	Native	4	2	August	6	123	3.25	E. 2874	123	3.25
E. 2855	Stub	Native	6	3	April	5	104	3.00	E. 2875	106	4.00
E. 2856	Bob	Native	6	3	February	4	106	3.50	E. 2876	102	3.25
E. 2857	Brown	Native	7	4	March	5	107	3.00	E. 2877	106	3.00
E. 2858	Grey	Native	7	4	March	6	106	4.25	E. 2878	108	3.75
E. 2859	Pink	Native	6	3	March	6	101	3.25	E. 2379	104	3.75
E. 2860	Little Red	Native	8	5	April	6	106	3.00	E. 2880	104	3.00
E. 2861	Curley	Native	5	2	March	4	109	3.25	E. 2881	105	4.25
E. 2862	Line Back	Native	5	2	March	5	114	4.25	E. 2882	117	4.25
E. 2863	Feefy	Native	5	2	March	4	109	3.50	E. 2883	107	3.75
E. 2864	Durham	Native	6	3	June	6	105	3.50	E. 2884	106	3.25
E. 2865	Little Pink	Native	5	2	March	5	103	3.50	E. 2885	110	3.75
E. 2866	Kicker	Native	5	2	April	6	115	4.25	E. 2886	114	4.25
E. 2867	Brindle	Native	8	6	April	5	103	3.50	E. 2887	108	4.00
E. 2868	Average	....	....	....	....	....	109	3.50	E. 2888	109	3.75

Calculated analysis, August 31, 1885; nineteen cows owned by Increase C. Jordan, Wallkill, Orange county. Evening's milking.

Number of inspection.	Water.	Total solids.	Fat.	Solids not fat.	Sugar-caseine.	Salts.
E. 2849	86·59	13·41	4·00	9·41	8·73	0·68
E. 2850	88·20	11·80	3·00	8·80	8·12	0·68
E. 2851	88·16	11·84	3·00	8·84	8·16	0·68
E. 2852	88·71	11·54	3·00	8·54	7·86	0·68
E. 2853	87·54	12·46	3·25	9·21	8·53	0·68
E. 2854	86·61	13·39	3·25	10·14	9·46	0·68
E. 2855	88·16	11·84	3·00	8·84	8·16	0·68
E. 2856	87·34	12·66	3·50	9·16	8·48	0·68
E. 2857	87·93	12·07	3·00	9·07	8·39	0·68
E. 2858	86·42	13·58	4·25	9·33	8·65	0·68
E. 2859	87·81	12·19	3·25	8·94	8·26	0·68
E. 2860	88·08	11.92	3·00	8·92	8·24	0·68
E. 2861	87·46	12·54	3·25	9·29	8·61	0·68
E. 2862	85·81	14·19	4·25	9·94	9·26	0·68
E. 2863	87·20	12·80	3·50	9·33	8·65	0·68
E. 2864	86.97	13·03	3·50	9·53	8·85	0·68
E. 2865	87·62	12·38	3·50	8·88	8·20	0·68
E. 2866	84·57	15·43	4·50	10·93	10·25	0·68
E. 2867	87·62	12·38	3·50	8·88	8.20	0·68
E. 2868	(Aver	age, see	page	61.)	....	....

Calculated analysis, September 1, 1885, nineteen cows owned by Increase C. Jordan, Wallkill, Orange county. Morning's milking.

Number of inspection.	Water.	Total solids.	Fat.	Solid not fat.	Sugar-caseine.	Salts.
E. 2869	85·88	14·12	4·50	9·62	8·94	0 68
E. 2870	87·06	12·94	3·50	9·44	8·76	0·68
E. 2871	87·69	12·31	3·25	9·06	8·20	0·68
E. 2872	88·48	12·52	3·75	8·77	8·09	0·68
E. 2873	86·74	13·26	4·00	9·26	8·58	0·68
E. 2874	86·29	13·71	3·50	10·21	9·53	0·68
E. 2875	86·74	13·26	4·00	9·26	8·58	0·68
E. 2876	87·85	12·15	3·50	8·66	7·98	0·68
E. 2877	88·01	11·99	3·00	8·99	8·31	0·68
E. 2878	86·90	13·10	3·75	9·35	8·67	0·68
E. 2879	87·21	12·79	3·75	9·04	8·36	0·68
E. 2880	88·16	11·84	3·00	8·84	7·98	0·68
E. 2881	86·50	13·50	4·25	9·25	8·57	0·68
E. 2882	85·57	14·43	4·25	10·18	9·50	0·68
E. 2883	86·98	13·02	3·75	9·27	8·59	0·68
E. 2884	87·69	12 31	3 25	9·06	8·38	0·68
E. 2885	86·75	13·25	3·75	9·50	8·82	0·68
E. 2876	85·81	14 19	4 25	9·94	9·26	0 68
E. 2887	86·97	15·03	4·00	9·03	8·35	0·68
E. 2888	(Aver	age, see	page	61.)	....	.....

Evening of September 3, and morning of September 4, 1885; Mount Hope, Orange county; farm of Matthew Writer; cows in herd, 25; treatment, kind; housing, good; food, upland pasture.

Number of inspection.	Name of cow.	Breed.	Age of cow.	Number of times calving.	Time of last calving.	Number of quarts given.	Lactometer at 60 deg. Fahr., evening's.	Per cent of fat, evening's.	Number of inspection.	Lactometer at 60 deg. Fahr., morning's.	Per cent of fat, morning's.
			years.		1885.						
E. 2889	Red Craus	Native	10	7	August	5	109	3.25	E. 2915	110	3.25
E. 2890	Twin Heifer	Native	6	3	August	5	106	3.50	E. 2916	109	3.50
E. 2891	Roan Heifer	Native	10	7	April	3	106	4.25	E. 2917	103	3.75
E. 2892	Hurdle	Native	8	5	May	3	105	3.25	E. 2918	106	4.25
E. 2893	Young White	Native	4	2	June	4	108	6.00	E. 2919	110	6.00
E. 2894	Speckle	Native	7	4	May	4	107	4.25	E. 2920	100	3.25
E. 2895	Old White	Native	12	9	May	4	100	3.25	E. 2921	110	4.50
E. 2896	Old Brooks	Native	11	8	April	4	104	3.25	E. 2922	103	3.25
E. 2897	Young Brindle	Native	2	1	April	3	105	3.50	E. 2923	107	3.75
E. 2898	Spot	Native	2	1	April	3	06	4.50	E. 2924	106	4.25
E. 2899	Young Brooks	Native	8	2	April	4	107	3.50	E. 2925	109	3.50
E. 2900	Teiney	Native	5	3	May	4	105	3.50	E. 2926	103	4.00
E. 2901	Young Lilly	Native	8	2	April	3	106	4.25	E. 2927	104	4.25
E. 2902	Young Darby	Native	5	3	July	4	110	3.50	E. 2928	113	3.25
E. 2903	Sleepy	Native	12	9	January	4	103	3.00	E. 2929	105	3.00
E. 2904	Berk	Native	8	2	March	4	04	3.50	E. 2930	109	3.25
E. 2905	Van Inwegin	Native	8	5	March	5	102	3.75	E. 2931	100	4.25
E. 2906	Old Craus	Mise	12	9	May	4	108	3.25	E. 2932	102	3.75
E. 2907	Old Lilly	Native	8	5	April	4	109	3.50	E. 2933	106	3.50
					1884.						
E. 2908	Stamp	Native	6	2	May	3	102	3.25	E. 2934	105	3.25
E. 2909	Rosey	Mise	3	2	March	3	102	3.75	E. 2935	104	3.75
E. 2910	Black Knight	Native	8	5	April	3	100	3.50	E. 2936	103	3.75
E. 2911	Daisy	Native	4	2	April	3	108	3.75	E. 2937	108	3.50
E. 2912	Belt	Mise	2	1	April	3	106	3.50	E. 2938	104	3.75
E. 2913	Old Knight	Native	11	8	May	4	109	3.75	E. 2939	102	4.00
E. 2914	Average	......	......	......	......	......	106	3.75			
	Average	......	......	......	......	......	......	......	E. 2940	107	3.75

Calculated analysis, September 3, 1885 ; *twenty-five cows ; owned by Matthew Writer, Mount Hope, Orange county. Evening's milking.*

Number of inspection.	Water.	Total solids.	Fat.	Solids not fat.	Sugar-caseine.	Salts.
E. 2889	87·46	12·54	3·25	9·29	8·61	0·68
E. 2890	87·37	12·63	3·50	9·13	8·45	0·68
E. 2891	87·77	12·23	3·25	8·98	8·30	0·68
E. 2892	86·42	13·58	4·25	9·33	8·65	0·68
E. 2893	83·05	16·95	6·00	9·95	9·27	0·68
E. 2894	88·70	11·30	3·25	8·05	7·37	0·68
E. 2295	86·08	13·92	4·25	9·67	8·99	0·68
E. 2896	87·85	12·15	3·25	8·90	8·22	0·68
E. 2897	87·45	12·55	3·50	9·05	8·37	0·68
E. 2898	86·11	13·89	4·50	9·39	8·71	0·68
E. 2899	87·30	12·70	3·50	9·20	8·72	0·68
E. 2900	87·45	12·55	3·50	9·05	8·37	0·68
E. 2901	86·42	13·58	4·25	9·33	8·65	0·68
E. 2902	87·06	12·94	3·50	9·44	8·76	0·68
E. 2903	88·24	11·76	3·00	8·76	8·08	0·68
E. 2904	87·53	12·47	3·50	8·97	8·29	0·68
E. 2905	87·37	12·63	3·75	8·88	8·20	0·68
E. 2906	87·92	12·08	3·25	8·83	8·15	0·68
E. 2907	87·14	12·86	3·50	9·36	8·68	0·68
E. 2908	87·68	12·32	3·50	8·82	8·14	0·68
E. 2909	87·27	12·63	3·75	8·88	8·20	0·68
E. 2910	87·83	12·17	3·50	8·67	7·99	0·68
E. 2911	87·29	12·71	3·75	8·96	8·28	0·68
E. 2912	87·37	12·63	3·50	9·13	8·45	0·68
E. 2913	86·82	13·18	3·75	9·43	8·75	0·68

Calculated analysis, September 4, 1885; *twenty-five cows owned by Matthew Writer; Mount Hope, Orange county. Morning's milking.*

Number of inspection.	Water.	Total solids.	Fat.	Solids not fat.	Sugar-caseine.	Salts.
E. 2915	87·41	12·59	3·25	9·34	8·66	0·68
E. 2916	87·14	12·86	3·50	9·36	8·68	0·68
E. 2917	86·90	13·10	3·75	9·35	8·67	0·68
E. 2918	86·16	13·84	4·25	9·59	8·91	0·68
E. 2919	83·90	16·10	6·00	10·10	9·42	0·68
E. 2920	88·46	11·54	3·25	8·29	7·61	0·68
E. 2921	86·06	13·94	4·50	9·44	8·76	0·63
E. 2922	87·55	12·45	3·25	9·20	8·52	0·68
E. 2923	86·98	13·02	3·75	9·27	8·59	0·68
E. 2924	86·69	13·31	4·25	9·06	8·38	0·68
E. 2925	87·14	12·86	3·50	9·36	8·68	0·68
E. 2926	86·97	13·03	4·00	9·03	8·35	0·68
E. 2927	86·58	13·42	4·25	9·17	8·49	0·68
E. 2928	87·38	12·62	3·25	9·37	8·69	2·68
E. 2929	88·08	11·92	3·00	8·92	8·24	0·68
E. 2930	87·46	12·54	3·25	9·29	8·61	0·68
E. 2931	86·89	13·11	4·25	8·86	8·18	0·68
E. 2932	87·37	12·63	3·75	8·88	8·20	0·68
E. 2933	87·37	12·63	3·50	9·13	8·45	0·68
E. 2934	87·77	12·23	3·25	8·98	8·30	0·68
E. 2935	87·21	12·79	3·75	9·04	8·36	0·68
E. 2936	87·29	12·71	3·75	8·96	8·28	0·68
E. 2937	87·22	12·78	3·50	9·28	8·60	0·68
E. 2938	87·21	12·79	3·75	9·04	8·36	0·68
E. 2939	87·05	12·95	4·00	8·95	8·27	0·68

Evening of September 7, and morning of September 8, 1885, Walton, Delaware county; farm of William C. Doig; cows in herd, 17; treatment, kind; housing, good; food, lowland pasture and fodder corn.

Number of inspection.	Name of cow.	Breed.	Age of cow.	No. of times calving.	Time of last calving.	No. of quarts given.	Lactometer at 60 deg. Fahr., evening's.	Per cent of fat evening's.	Number of inspection.	Lactometer at 60 deg. Fahr., morning's.	Per cent of fat, morning's.
			yrs.		1885.						
E. 2941	White	Native	2	1	April	3	105	3.75	E. 2954	108	3.75
E. 2942	Line Back	Native	8	7	January	3	115	5.25	E. 2955	118	5.25
E. 2943	Young Grizzly	Native	2	1	April	3	102	3.25	E. 2956	103	3.50
E. 2944	Florence	Native	2	1	April	3	104	3.25	E. 2957	102	4.00
E. 2945	Brown	Native	9	8	January	4	108	4.00	E. 2958	106	4.00
E. 2946	Spot	Native	9	8	March	5	100	3.75	E. 2959	104	3.75
E. 2947	Little Red	Native	2	1	April	3	115	4.00	E. 2960	102	3.75
					1884.						
E. 2948	Big Yellow	Native	9	8	December	2	105	5.25	E. 2961	118	5.25
					1885.						
E. 2949	Star	Native	4	2	February	4	108	3.75	E. 2962	105	3.75
E. 2950	Young Black	Native	4	2	April	4	102	3.25	E. 2963	105	3.25
E. 2951	Old White	Native	18	10	January	3	102	4.50	E. 2964	103	4.50
E. 2952	Black	Native	8	6	February	2	107	4.50	E. 2965	107	4.75
E. 2953	Average	Native	......	......	....	......	108	4.00	E. 2966	109	4.25

Calculated analysis September 7, 1885; twelve cows owned by William C. Doig, Walton, Delaware county. Evening's milking.

Number of inspection.	Water.	Total solids	Fat.	Solids not fat.	Sugar-caseine.	Salts.
E. 2941	87·13	12·87	3·75	9·12	8·44	0·68
E. 2942	84·46	15·54	5·25	10·29	9·61	0·68
E. 2943	87·68	12·32	3·50	8·82	7·96	0·68
E. 2944	87·68	12·32	3·50	8·82	7·96	0·68
E. 2945	86·90	13·10	4·00	9·10	8·42	0·68
E. 2946	87·29	12·71	3·75	8·96	8·28	0·68
E. 2947	87·20	12·80	4·00	8·80	8·12	0·68
E. 2948	84·46	15·54	5·25	10·29	9·61	0·68
E. 2949	87·13	12·87	3·75	9·12	8·44	0·68
E. 2950	87·54	12·46	3·25	9·21	8·53	0·68
E. 2951	86·57	13·43	4·50	8·93	8·25	0·68
E. 2952	86·03	13·97	4·50	9·47	8·79	0·68
E. 2953	(Average	see page	61.)	....	....	....

Calculated analysis, September 8, 1885; Twelve cows owned by William C. Doig, Walton, Delaware county; morning's milking.

Number of inspection.	Water.	Total solids.	Fat.	Solids not fat.	Sugar-caseine.	Salts.
E. 2954	86·90	13·10	3·75	9·35	8·67	0·68
E. 2955	84·23	15·77	5·25	10·52	9·84	0·68
E. 2956	87·61	12·39	3·50	8·89	8·21	0·68
E. 2957	87·05	12·95	4·00	8·95	8·27	0·68
E. 2958	86·74	13·26	4·00	9·26	8·58	0·68
E. 2959	87·21	12·79	3·75	9·04	8·36	0·68
E. 2960	87·37	12·63	3·75	8·88	8·20	0·68
E. 2961	84·64	15·36	5·25	10·11	9·43	0·68
E. 2962	87·13	12·87	3·75	9·12	8·44	0·68
E. 2963	87·77	12·23	3·25	8·98	8·30	0·68
E. 2964	86·07	13·93	4·50	9·43	8·75	0·68
E. 2965	84·71	14·29	4·75	9·54	8·86	0·68
E. 2966	(Average	see page	61.)			

Evening of September 10, 1885, and morning of September 11, 1885; Walton, Delaware county; farm of E. Wood; cows in herd, 32; treatment, kind; housing, good; food, upland pasture.

Number of inspection.	ne of cow.	Breed.	Age of cow.	Number of times calving.	Time of last calving.	Number of quarts given.	Lactometer at 60 deg. Fahr., evening's.	Per cent of fat, evening's.	Number of inspection.	Lactometer at 60 deg. Fahr., morning's.	Per cent of fat, morning's.
			years.		1885.						
E. 2967	Cherry	Native	4	2	February	3	111	4.50	E. 2999	110	4.25
E. 2968	Pet	Native	6	4	March	6	102	3.50	E. 3000	103	3.75
E. 2969	Leopard	Native	10	8	April	6	103	3.00	E. 3001	103	3.00
E. 2970	Lop Horn	Native	3	1	August	8	108	4.25	E. 3002	108	4.25
D. 2971	Peggy	Native	10	8	May	9	101	3.75	E. 3003	106	4.00
D. 2972	Lilly	Native	12	10	January	2	106	4.00	E. 3004	101	3.75
D. 2973	Pike Horn	Native	7	5	January	3	116	5.00	E. 3005	115	5.00
D. 2974	Bob	Native	2	1	June	3	108	4.75	E. 3006	109	4.75
D. 2975	Annie	Native	6	4	March	6	106	4.75	E. 3007	105	4.75
D. 2976	Tussy	Native	8	6	March	5	102	5.00	E. 3008	103	5.00
D. 2977	Osborne	Native	12	10	January	3	106	6.00	E. 3009	108	6.90
I. 2978	Shaw	Native	4	2	April	4	105	4.00	E. 3010	105	4.00
I. 2979	Grey	Native	3	1	April	5	110	4.50	E. 3011	110	4.50
I. 2980	Dun	Native	6	4	March	4	107	4.00	E. 3012	107	4.00
I. 2981	Old Brockle	Native	12	10	March	6	101	3.50	E. 3013	102	3.50
I. 2982	Young Leopard	Native	3	1	January	6	106	4.25	E. 3014	106	4.25
I. 2983	Black	Native	6	4	March	2	100	5.75	E. 3015	100	5.75
I. 2984	Liney	Native	8	6	February	3	100	3.25	E. 3016	100	3.50
I. 2985	Daisy	Native	3	1	March	4	108	4.25	E. 3017	107	4.50
I. 2986	Banker	Native	3	1	April	4	113	4.25	E. 3018	113	4.75
I. 2987	Brin	Native	5	3	February	2	101	3.25	E. 3019	102	3.50
I. 2988	Young Brockle	Native	8	6	March	7	103	4.25	E. 3020	102	4.50
I. 2989	Star	Native	2	1	April	4	109	5.50	E. 3021	109	5.50
I. 2990	Bet	Native	2	1	April	4	109	5.50	E. 3022	109	5.50
I. 2991	Webb	Native	14	12	March	4	108	5.00	E. 3023	108	5.00
I. 2992	Line Back	Native	2	1	April	3	107	5.25	E. 3024	107	5.25

					1884.						
I. 2993......	Old White	Native	10	8	April..........	8	104	4.50	E. 3025....	104	4.50
I. 2994......	Topsie.......................	Native	14	12	December	8	108	4.75	E. 3026....	107	4.75
					1885.						
I. 2995......	Watrus	Native	8	1	March..........	5	109	4.25	E. 3027....	108	4.25
I. 2996......	Spot..........................	Native	8	1	March..........	5	108	4.50	E. 3028....	108	4.50
I. 2997......	White	Native	8	1	March..........	5	108	4.50	E. 3029....	108	4.50
I. 2998......	Average	....	......	......	... .	......	108	4.75			
	Average	....	......	......	....	......	......	......	E. 3030....	109	5.00

Calculated analysis, September 10, 1885 ; *thirty-one cows owned by Ebenezer Wood, Walton, Delaware county. Evening's milking.*

Number of inspection.	Water.	Total solids.	Fat.	Solids not fat.	Sugar-caseine.	Salts.
E. 2967	85·72	14·28	4·50	9.78	9·10	0·68
E. 2968	87·68	12·32	3·50	8·82	8·14	0·68
E. 2969	88·24	11·76	3·00	8·76	8·08	0·68
E. 2970	86·24	13·76	4·25	9·51	8·83	0·68
E. 2971	86·74	13·26	4·00	9·26	8·58	0·68
E. 2972	87·44	12·56	3·75	8·81	8·13	0·68
E. 2973	84·79	15·21	5·00	10·21	9·53	0·68
E. 2974	85·66	14·34	4·75	9·59	8·91	0·68
E. 2975	85·79	14·21	4·75	9·46	8·78	0·68
E. 2976	85·78	14·22	5·00	9·22	8·54	0·68
E. 2977	84·21	15·79	6·00	9·79	9·11	0·68
E. 2978	86·82	13·18	4·00	9·18	8·50	0·68
E. 2979	85·77	14·23	4·50	9·73	9·05	0·68
E. 2980	86·66	13·34	4·00	9·34	8·66	0·68
E. 2981	87·76	12·24	3·50	8·74	8.06	0·68
E. 2982	86·42	13·58	4·25	9·33	8·65	0·68
E. 2983	85·22	14·78	5·75	9·03	8·35	0·68
E. 2984	88·02	11·98	3·50	8·48	7.80	0·68
E. 2985	85·95	14·05	4·50	9·55	8.87	0·68
E. 2986	85·56	14·14	4·50	9·94	9·26	0·68
E. 2987	87·76	12·24	3·50	8·74	8·06	0·68
E. 2988	86·34	13·66	4·50	9·16	8·48	0·68
E. 2989	84·61	15·39	5·50	9·89	9·21	0·68
E. 2990	84·61	15·39	5·50	9·89	9·21	0·68
E. 2991	85·32	14·68	5·00	9·68	9·00	0·68
E. 2992	85·08	14·92	5·25	9·67	8·99	0·68
E. 2993	86·21	13·79	4·50	9·29	8·61	0·68
E. 2994	85·64	14·36	4·75	9·61	8·93	0·68
E. 2995	86·19	13·81	4·25	9.56	8·88	0·68
E. 2996	86·34	14·66	4·50	9·16	8·48	0·68
E. 2997	85·95	14·05	4·50	9·55	8·87	0·68
E. 2998	(Aver	age, see	page	61.)		

Calculated analysis, August 24, 1885; thirty-one cows owned by Ebenezer Wood, Walton, Delaware county; Morning's milking.

Number of inspection.	Water.	Total solids.	Fat.	Solids not fat.	Sugar and caseine.	Salts.
E. 2999	85·75	14·25	4·50	9·70	9·02	0·68
E. 3000	87·29	12·71	3·75	8·96	8·28	0·68
E. 3001	88·24	11·76	3·00	8·76	8·08	0·68
E. 3002	86·24	13·76	4·25	9·51	8·83	0·68
E. 3003	86·74	13·26	4·00	9·26	8·58	0·68
E. 3004	87·44	12·56	3·75	8·81	8·13	0·68
E. 3005	84·78	15·22	5·00	10·22	9·54	0·68
E. 3006	85·82	14·18	4·75	9·43	8·75	0·68
E. 3007	85·89	14·11	4·75	9·36	8·68	0·68
E. 3008	85·71	14·29	5·00	9·29	8·61	0·68
E. 3009	84·05	15·95	6·00	9·95	9·27	0·68
E. 3010	86·82	13·18	4·00	9·18	8·50	0·68
E. 3011	85·77	14·23	4·50	9·73	9·05	0·68
E. 3012	86·66	13·34	4·00	9·34	8·66	0·68
E. 3013	87·76	12·24	3·50	8·74	8·06	0·68
E. 3014	86·42	13·58	4·25	9·33	8·65	0·68
E. 3015	85·22	14·78	5·75	9·03	8·35	0·68
E. 3016	88·30	12·70	3·50	9·20	8·52	0·68
E. 3017	86·03	13·97	4·50	9·47	8·79	0·68
E. 3018	85·34	14·66	4·75	9·91	9·23	0·68
E. 3019	87·68	12·32	3·50	8·82	8·14	0·68
E. 3020	86·34	13·66	4·50	9·16	8·48	0·68
E. 3021	84·61	15·39	5·50	9·89	9·21	0·68
E. 3022	84·61	15·39	5·50	9·89	9·21	0·68
E. 3023	85·32	14·68	5·00	9·68	9·00	0·68
E. 3024	85·08	14·92	5·25	9·67	8·99	0·68
E. 3025	86·21	13·79	4·50	9·29	8·61	0·68
E. 3026	85·64	14·36	4·75	9·61	8·93	0·68
E. 3027	86·19	13·81	4·25	9·56	8·88	0·68
E. 3028	86·34	13·66	4·50	9·15	8·48	0·68
E. 3029	85·95	14·05	4·50	9·55	8·87	0·68
E. 3030	(Aver	age, see	page	61.)		

Evening of September 14, and morning of September 15, 1885; Wawarsing, Ulster county; farm of Samuel D. Smith, cows in herd, 23; treatment, kind; housing, good; food, brewers' grains and upland pasture.

Number of inspection.	Name of cow.	Breed.	Age of cow.	Number of times calving.	Time of last calving.	Number of quarts given.	Lactometer at 60 deg. Fahr., evening's.	Per cent of fat, evening's.	Number of inspection.	Lactometer at 60 deg. Fahr., morning's.	Per cent of fat, morning's.
			years.		1885.						
E. 3031	Judy	Native	7	5	January	3	102	3.75	E. 3052	103	3.75
E. 3032	[illegible]Bad Horn	Native	5	3	January	4	103	3.25	E. 3053	110	3.50
					1884.						
E. 3033	Line Back	Native	13	11	[illegible]ber	6	110	3.75	E. 3054	110	3.75
E. 3034	P[illegible]ir	Native	10	8	December	6	106	3.25	E. 3055	106	3.75
					1885.						
E. 3035	Small [illegible]k	Native	5	3	January	5	104	3.50	E. 3056	107	3.50
					1884.						
E. 3036	Rampy	Native	12	10	[illegible]ber	3	103	3.25	E. 3057	110	3.75
					1885.						
E. 3037	Vermilion	Native	13	11	February	2	102	3.75	E. 3058	106	3.50
					1884.						
E. 3038	Kicker	Native	4	2	[illegible]ber	2	100	3.50	E. 3059	101	3.75
					1885.						
E. 3039	Ay[illegible]hire	Native	7	5	[illegible]ary	2	108	4.00	E. 3060	113	4.00
E. 3040	Black	Native	7	5	February	5	103	3.75	E. 3061	102	4.00
E. 3041	Spot	Native	5	3	[illegible]uary	4	107	3.75	E. 3062	104	3.50
					1884.						
E. 3042	I[illegible]an	Native	10	8	[illegible]ber	5	112	4.25	E. 3063	107	4.50
					1885.						
E. 3043	Tuff	Native	7	5	April	3	109	4.00	E. 3064	109	4.00
E. 3044	[illegible]on	Native	9	7	May	4	106	4.25	E. 3065	103	4.25
E. 3045	Upright	Native	10	8	September	5	117	3.00	E. 3066	117	3.25
					1884.						
E. 3046	Durham	Native	10	8	December	5	100	3.25	E. 3067	112	4.00

E. 3047.....	Hill	Native	5	3	1885. April..........	3	111	4.00	E. 3068....	101	4.00
E. 3048.....	Dominie	Native	14	12	1884. September.....	5	103	4.00	E. 3069....	103	3.75
E. 3049.....	Depugh	Native	6	4	1885. March.........	2	112	3.75	E. 3070....	102	3.25
E. 3050.....	Brindle	Native	5	3	February......	4	106	3.50	E. 3071....	106	3.50
E. 3051.....											
	Average	....	......	......	....	......	105	3.75	E. 3072....	107	4.00

Calculated analysis, September 14, 1885; twenty cows owned by Samuel D. Smith, Wawarsing, Ulster county. Evening's milking.

Number of inspection.	Water.	Total solids.	Fat.	Solids not fat.	Sugar-caseine.	Salts.
E. 3031	87·39	12·61	3·75	8·86	8·18	0·68
E. 3032	87·92	12·08	3·25	8·83	8·15	0·68
E. 3033	86·69	13·31	3·75	9·56	8·88	0·68
E. 3034	87·37	12·63	3·50	9·13	8·45	0·68
E. 3035	87·55	12·45	3·50	8·95	8·27	0·68
E. 3036	87·92	12·08	3·25	8·83	8·15	0·68
E. 3037	87·83	12·17	3·50	8·67	7·99	0·68
E. 3038	87·39	12·64	3·75	8·86	8·18	0·68
E. 3039	86·58	13·42	4·00	9·42	8·74	0·68
E. 3040	87·60	12·40	3·50	8·90	8·22	0·68
E. 3041	86·36	13·04	3·75	9·29	8·61	0·68
E. 3042	85·95	14·05	4·25	9·80	9·12	0·68
E. 3043	86·50	13·50	4·00	9·50	8·82	0·68
E. 3044	86·42	13·58	4·25	9·33	8·65	0·68
E. 3045	87·15	12·85	3·00	9·85	9·17	0·68
E. 3046	88·00	12·00	3·25	8·75	8·07	0·68
E. 3047	86·35	13·65	4·00	9·65	8·97	0·68
E. 3048	86·96	13·04	4·00	9·04	8·36	0·68
E. 3049	86·59	13·41	3·75	9·66	8·98	0·68
E. 3050	87·35	12·65	3·50	9·15	8·47	0·68
E. 3051	(Average, see page 61.)					

Calculated analysis, September 15, 1885; *twenty cows owned by Samuel D. Smith, Wawarsing, Ulster county. Morning's milking.*

Number of inspection.	Water.	Total solids.	Fat.	Solids not fat.	Sugar-caseine.	Salts.
E. 3052	87·29	12·71	3·75	8·96	8·28	0·68
E. 3053	87·06	12·94	3·50	9·44	8·76	0·68
E. 3054	86·69	13·31	3·75	9·56	8·88	0·68
E. 3055	87·06	12·94	3·75	9·19	8·51	0·68
E. 3056	87·29	12·71	3·50	9·21	8·53	0·68
E. 3057	86.69	13·31	3·75	9·56	8·88	0·68
E. 3058	87·37	12·63	3·50	9·13	8·45	0·68
E. 3059	87·39	12·61	3·75	8·86	8·18	0·68
E. 3060	86·20	13·80	4·00	9·80	9·12	0·68
E. 3061	87·04	12·96	4·00	8·96	8·28	0·68
E. 3062	87·55	12·45	3·50	8·95	8·27	0·68
E. 3063	86·01	13·99	4·50	9·49	8·81	0·68
E. 3064	86·50	13·50	4·00	9·50	8·82	0·68
E. 3065	86·20	13·80	4·00	9·80	9·12	0·68
E. 3066	86·83	13·17	3·25	9·92	9·24	0·68
E. 3067	88·00	12·00	3·25	8·75	8·07	0·68
E. 3068	86·25	13.75	4·00	9·75	9·07	0·68
E. 3069	87·12	12·88	4·00	8·88	8.20	0·68
E. 3070	86·89	13·11	3·75	9·36	8·68	0·68
E. 3071	87·37	12·63	3·50	9·13	8·45	0·68
E. 3072	(Aver	age, see	page	61.)	....	....

Evenings of September 22, 23, 24, 1885; Bedford, Westchester county, farm of William Nelson; cows in herd, 300; treatment, fair; housing, good; food, brewers' grains, corn-fodder and upland pasture.

Number of inspection.	Number of cow.	Breed.	Age of cow.	Number of times calving.	Time of last calving.	Number of quarts given.	Evening, Sept. 22. Lactometer at 60 deg. Fahr.	Evening, Sept. 22. Per cent of fat.
E. 3084	1	Native.	*....	*....	*....	2	107	4.25
E. 3085	2	Native.	....	....	....	4	109	3.50
E. 3086	3	Native.	....	....	....	3	108	4.75
E. 3087	4	Native.	....	....	....	2	112	3.75
E. 3088	5	Native.	....	....	....	6	113	4.25
E. 3089	6	Native.	....	....	....	7	110	3.25
E. 3090	7	Native.	....	....	....	5	110	3.75
E. 3091	8	Native.	....	....	....	2	113	3.25
E. 3092	9	Native.	....	....	....	8	103	5.00
E. 3093	10	Native.	....	....	....	8	101	5.50
E. 3094	11	Native.	....	....	....	6	110	4.00
E. 3095	12	Native.	....	....	....	7	107	3.25
E. 3096	13	Native.	....	....	....	3	112	4.75
E. 3097	14	Native.	....	....	....	4	108	4.00
E. 3098	15	Native.	....	....	....	5	106	3.25
E. 3099	16	Native.	....	....	....	4	112	3.75
E. 3100	17	Native.	....	....	....	2	105	3.50
E. 3101	18	Native.	....	....	....	1	111	3.75
E. 3102	Average		....	....	....	....	110	4.00

Number of inspection.	Number of cow.	Number of quarts given.	Evening, Sept. 23. Lactometer at 60 deg. Fahr.	Evening, Sept. 23. Per cent of fat.
E. 3103....	19	3	110	4.75
E. 3104....	20	2	108	4.50
E. 3105....	21	4	108	4.50
E. 2106....	22	6	109	3.75
E. 3107....	23	5	108	4.50
E. 3108....	24	3	110	4.25
E. 3109....	25	2	110	4.50
E. 3110....	26	1	112	3.75
E. 3111....	27	6	112	4.75
E. 3112....	28	4	114	3.50
E. 3113....	29	5	120	3.50
E. 3114....	30	2	117	3.50
E. 3115....	31	4	112	3.25
E. 3116....	32	6	108	5.25
E. 3117....	33	5	107	5.00
E. 3118....	34	4	117	5.00
E. 3119....	35	3	118	4.50
E. 3120....	36	2	110	4.00
E. 3121....	37	4	113	4.75
E. 3122....	38	3	112	4.25
E. 3123....	39	2	110	3.75
E. 3124....	Average...		115	4.00

Number of inspection.	Number of cow.	Number of quarts given.	Evening, Sept. 24. Lactometer at 60 deg. Fahr.	Evening, Sept. 24. Per cent of fat.
E. 3125....	40	2	112	3.00
E. 3126....	41	4	105	5.00
E. 3127....	42	2	104	4.50
E. 3128....	43	3	108	5.00
E. 3129....	44	5	112	4.25
E. 3130....	45	3	108	4.75
E. 3131....	46	2	116	4.00
E. 3132....	47	1	112	4.00
E. 3133....	48	4	116	3.75
E. 3134....	49	5	114	4.50
E. 3135....	50	6	112	4.50
E. 3136....	51	3	108	5.50
E. 3137....	52	4	106	4.75
E. 3138....	53	2	110	5.00
E. 3139....	54	3	109	4.00
E. 3140....	55	2	105	3.75
E. 3141....	56	4	112	4.50
E. 3142....	57	1	110	4.50
E. 3143....	43	4	114	3.50
E. 3144....	59	3	114	4.25
E. 3145....	Average.		115	4.00

*Could not obtain any information in reference to the above.

Calculated analysis, eighteen cows owned by William Nelson, Bedford, Westchester county. Evening's milking.

Number of inspection.	Water.	Total solids.	Fat.	Solids not fat.	Sugar-caseine.	Salts.
E. 3084	86·08	13·92	4·25	9·67	8·99	0·68
E. 3085	87·14	12·86	3·50	9·36	8·68	0·68
E. 3086	85·64	14·36	4·75	9·61	8·93	0·68
E. 3087	86·59	13·41	3·75	9·66	8·98	0·68
E. 3088	85·60	14·40	4·25	10·15	9·47	0·68
E. 3089	87·37	12·63	3·25	9·38	8·70	0·68
E. 3090	86·69	13·31	3·75	9·56	8·88	0·68
E. 3091	87·25	12·75	3·25	9·50	8·82	0·68
E. 3092	85·60	14·30	5·00	9·30	8·62	0·68
E. 3093	87·89	12·11	5·50	6·61	7·93	0·68
E. 3094	86·40	13·60	4 00	9·60	8·92	0·68
E. 3095	87·61	12·39	3·25	9·14	8·46	0·68
E. 3096	85·25	14·75	4·75	10·00	9·32	0·68
E. 3097	86·36	13·44	4 00	9·44	8·76	0·68
E. 3098	87·69	12·31	3·25	9·06	8·38	0·68
E. 3099	86·59	13·41	3·75	9·66	8·98	0·68
E. 3100	87·09	12·91	3·50	9·41	8·73	0·68
E. 3101	86·70	13·30	3·75	9·55	8·87	0·68
E. 3102	(Aver	age, see	page 62.)			

Calculated analysis, September 23, 1885; twenty-one cows owned by William Nelson, Bedford, Westchester county. Evening's milking.

Number of inspection.	Water.	Total solids.	Fat.	Solids not fat.	Sugar-caseine.	Salts.
E. 3103.........	85·75	14·25	4·75	9·50	8·82	0·68
E. 3104.........	85·95	14·05	4·50	9·55	8·87	0·68
E. 3105.........	85·95	14·05	4·50	9·55	8·87	0·68
E. 3106.........	86·82	13·18	3·75	9·43	8·75	0·68
E. 3107.........	85·95	14·05	4·50	9·55	8·87	0·68
E. 3108.........	86·91	13·89	4·25	9·64	8·96	0·68
E. 3109.........	85·78	14·22	4·50	9·72	9·04	0·68
E. 3110.........	86·59	13·41	3·75	9·66	8·98	0·68
E. 3111.........	86·59	13·41	3·75	9·66	8·98	0·68
E. 3112.........	88·32	12·68	3·25	9·43	8·75	0·68
E. 3113.........	86·28	13·72	3·50	10·22	9·54	0·68
E. 3114.........	86·52	13·48	3·50	9·98	9·30	0·68
E. 3115.........	87·25	12·75	3·25	9·50	8·82	0·68
E. 3116.........	85·00	15·00	5·25	9·75	9·07	0·68
E. 3117........	85·40	14·60	5·00	9·60	8·92	0·68
E. 3118.........	84·64	15·38	5·00	10·38	9·70	0·68
E. 3119.........	85·20	14·80	4·50	10·30	9·62	0·68
E. 3120.........	86·40	13·60	4·00	9·60	8·92	0·68
E. 3121.........	85·25	14·75	4·75	10·00	9·32	0·68
E. 3122.........	85·99	14·01	4·25	9·76	9·08	0·68
E. 3123.........	86·69	13·31	3·75	9·56	8·88	0·68
E. 3124.........	(Average	see page	62.)			

Calculated analysis twenty cows owned by William Nelson, Bedford, Westchester county. Evening's milking.

Number of inspection.	Water.	Total solids.	Fat.	Solids not fat.	Sugar-caseine.	Salts.
E. 3125	87·88	13·02	3·75	9·37	8·69	0·68
E. 3126	85·55	14·45	5·00	9·45	8·77	0·68
E. 3127	86·27	13·73	4·50	9·26	8·58	0·68
E. 3128	85·29	14·71	5·00	9·71	9·03	0·68
E. 3129	85·99	14·01	4·25	9·76	9·08	0·68
E. 3130	85·64	14·36	4·75	9·61	8·93	0·68
E. 3131	85·97	14·03	4·00	10·03	9·35	0·68
E. 3132	86·25	13·75	4·00	9·75	9·07	0·68
E. 3133	86·28	13·72	3·75	9·97	9·29	0·68
E. 3134	85·49	14·51	4·50	10·01	9·33	0·68
E. 3135	85·64	14·36	4·50	9·86	9·18	0·68
E. 3136	84·69	15·31	5·50	9·81	9·13	0·68
E. 3137	85·79	14·21	4·75	9·46	8·78	0·68
E. 3138	85·14	14·86	5·00	9·86	9·18	0·68
E. 3139	86·50	13·50	4·00	9·50	8·82	0·68
E. 3140	87·13	12·87	3·75	9·12	8·44	0·68
E. 3141	85·64	14·36	4·50	9·86	9·18	0·68
E. 3142	85·78	14·22	4·50	9·72	9·04	0·68
E. 3143	86·76	13·24	3·50	9·74	9·06	0·68
E. 3144	85·81	14·19	4·25	9·94	9·26	0·68
E. 3145	(Aver	age, see	page 62.)			

Morning and evening of October 6, 1885; Bedford, Westchester county; farm of Daniel J. Smith; cows in herd, 32; treatment, kind; housing, bad; food, upland pasture and fodder-corn.

Number of inspection.	Name of cow.	Breed.	Age of cow.	Number of times calving.	Time of last calving.	Number of quarts given.	Lactometer at 60 deg. Fahr., morning's.	Per cent of fat, morning's.	Number of inspection.	Number of quarts given.	Lactometer at 60 deg. Fahr., evening's.	Per cent of fat, evening's.
			years.		1885.							
E. 3163.....	Black Heifer	Native	6	3	September	6	118	3.25	E. 3174....	5	116	3.25
E. 3164.....	Red	Native	6	4	June	6	116	3.00	E. 3175....	7	113	3.75
E. 3165.....	Spotted Heifer.........	Native	4	2	September ...	4	120	4.00	E. 3176....	3	116	4.25
E. 3166.....	Line Back	Native	6	4	June	5	115	3.75	E. 3177....	6	114	4.00
E. 3167.. .	Speckle	Native	6	4	March	6	107	3.50	E. 3178....	7	106	3.50
E. 3168.....	Light Red	Native	7	5	March	3	105	4.00	E. 3179....	2	108	4.25
E. 3169.....	Spotted Cow	Native	7	5	February	2	112	3.25	E. 3180....	3	110	4.25
E. 3170.....	Red Heifer	Native	5	3	March	2	114	4.50	E. 3181....	3	112	5.00
E. 3171.....	Roan....................	Native	6	4	March	2	115	5.00	E. 3182....	3	114	6.00
E. 3172.....	Old Cow................	Native	8	6	May	2	110	3.25	E. 3183....	3	108	3.50
E. 3173.....	Average	....	....	....	....	....	114	3.75	E. 3184....	....	115	4.25

Calculated analysis, October 6, 1885; ten cows owned by Daniel J. Smith, Bedford, Westchester county. Morning's milking.

Number of inspection.	Water.	Total solids.	Fat.	Solids not fat.	Sugar and caseine.	Salts.
E. 3163......	86·76	13·24	3·25	9·99	9·31	0·68
E. 3164......	87·23	12·77	3·00	9·77	9·09	0·68
E. 3165......	85·66	14·34	4·00	10·34	9·66	0·68
E. 3166......	86·39	13·61	3·75	9·86	9·18	0·68
E. 3167......	87·30	12·70	3·50	9·20	8·52	0·68
E. 3168......	86·82	13·18	4·00	9·18	8·50	0·68
E. 3169......	87·25	12·75	3·25	9·50	8·82	0·68
E. 3170......	85·49	14·51	4·50	10·01	9·33	0·68
E. 3171......	84·78	15·22	5·00	10·22	9·54	0·68
E. 3172......	87·38	12·62	3·25	9·37	8·69	0·68
E. 3173......	(Aver	age, see	page 62.)			

Calculated analysis, October 6, 1885; ten cows owned by Daniel J Smith, Bedford, Westchester county. Evening's milking.

Number of inspection.	Water.	Total solids.	Fat.	Solids not fat.	Sugar and caseine.	Salts.
E. 3174......	86·92	13·08	3·25	9·83	9·15	0·68
E. 3175......	86·52	13·48	3·75	9·73	9·05	0·68
E. 3176......	85·65	14·35	4·25	10·10	9·42	0·68
E. 3177......	86·12	13·88	4·00	9·88	9·20	0·68
E. 3178......	87·37	12·63	3·50	9·13	8·45	0·68
E. 3179......	86·27	13·73	4·25	9·48	8·80	0·68
E. 3180......	86·11	13·89	4·25	9·64	8·96	0·68
E. 3181......	85·01	14·99	5·00	9·99	9·31	0·68
E. 3182......	83·59	16·41	6·00	10·41	9·73	0·68
E. 3183......	87·22	12·78	3·50	9·28	8·60	0·68
E. 3184......	(Aver	age, see	page 62.)			

Evening of October 7, and morning of October 8, 1885; Bedford, Westchester county; farm of Enoch T. Avery; cows in herd, 14; treatment, kind; housing, good; food, middlings and bran and lowland pasture.

Number of inspection.	Name of cow.	Breed.	Age of cow.	Number of times calving.	Time of last calving.	Number of quarts given.	Evening.		Number of inspection.	Number of quarts given.	Morning.	
							Lactometer at 60 deg. Fahr.	Per cent of fat.			Lactometer at 60 deg. Fahr.	Per cent of fat.
			years.		1885.							
E. 3186	Brindle	Native	7	5	April	5	118	5.00	E. 3197	4	119	4.75
E. 3187	Alderney	Native	9	7	April	5	116	4.50	E. 3198	6	116	4.50
E. 3188	Betsey	Native	10	8	June	4	114	4.75	E. 3199	3	116	4.50
E. 3189	Susan	Native	10	8	June	5	113	5.25	E. 3200	5	113	5.00
E. 3190	Grey	Native	7	5	June	5	114	5.00	E. 3201	4	116	5.00
E. 3191	Judy	Native	4	2	April	7	116	4.75	E. 3202	6	117	4.75
E. 3192	Red	Native	11	9	June	8	114	5.50	E. 3203	7	115	5.25
E. 3193	Dick	Native	9	7	June	5	113	4.75	E. 3204	6	115	4.75
E. 3194	Short Horn	Native	4	2	September	4	116	3.75	E. 3205	5	117	4.00
E. 3195	Star	Native	5	3	October	7	118	4.00	E. 3206	6	118	3.75
E. 3196	Average	....	....	....	....	....	116	4.75	E. 3207	....	117	4.25

Calculated analysis, October 7, 1885; ten cows owned by Enoch T. Avery, Bedford, Westchester county. Evening's milking.

Number of inspection.	Water.	Total solids.	Fat.	Solids not fat.	Sugar-caseine.	Salts.
E. 3186......	84·55	15·45	5·00	10·45	9·78	0·68
E. 3187......	85·33	14·67	4·50	10·17	9·49	0·68
E. 3188......	84·64	15·36	4·75	10·61	9·93	0·68
E. 3189......	84·64	15·36	5·25	10·11	9·43	0·68
E. 3190......	84·85	15·15	5·00	10·15	9·47	0·68
E. 3191......	85·02	14·98	4·75	10·23	9·55	0·68
E. 3192......	84·22	15·78	5·50	10·28	9·60	0·68
E. 3193.....	85·25	14·75	4·75	10·00	9·32	0·68
E. 3194......	86·28	13·72	3·75	9·97	9·29	0·68
E. 3195......	85·81	14·19	4·00	10·19	9·51	0·68
E. 3196......	(Aver	age, see	page 62.)			

Calculated analysis, October 8, 1885; ten cows owned by Enoch T. Avery, Bedford, Westchester county. Morning's milking.

Number of inspection.	Water.	Total solids.	Fat.	Solids not fat.	Sugar-caseine.	Salts.
E. 3197	84·78	15·22	4·75	10·47	9·79	0·68
E. 3198	85·33	14·67	4·50	10·17	9·49	0·68
E. 3199	85·33	14·67	4·50	10·17	9·49	0·68
E. 3200	84·93	15·07	5·00	10·07	9·39	0·68
E. 3201	84·70	15·30	5·00	10·30	9·62	0·68
E. 3202	84·94	15·06	4·75	10·31	9·63	0·68
E. 3203	84·46	15·54	5·25	10·29	9·81	0·68
E. 3204	85.09	14·91	4·75	10·16	9·48	0·68
E. 3205	85·89	14·11	4·00	10·11	9·43	0·68
E. 3206	86.13	13·87	3·75	10·12	9·44	0·68
E. 3207	(Aver	age, see	page 62.)			

Evening of October 9, and morning of October 10, 1885; Bedford, Westchester county, farm of Benjamin Mead; cows in herd, 13; treatment, kind; housing, good; food, brewers' grains, apples and lowland pasture.

Number of inspection.	Name of cow.	Breed.	Age of cow.	Number of times calving.	Time of last calving.	Number of quarts given.	Evening. Lactometer at 60 deg. Fahr.	Evening. Per cent of fat.	Number of inspection.	Number of quarts given.	Morning. Lactometer at 60 deg. Fahr.	Morning. Per cent of fat.
			years.		1885.							
E. 3208	Brady	Native	10	8	July	5	110	3.50	E. 3220	6	112	3.50
E. 3209	Black	Native	10	8	July	5	108	3.00	E. 3221	6	109	3.00
E. 3210	Red	Native	7	5	July	8	106	3.00	E. 3222	7	106	3.25
E. 3211	Dolly	Native	8	6	May	8	110	3.50	E. 3223	8	109	3.50
E. 3212	Brown	Native	7	5	July	8	118	3.00	E. 3224	7	116	3.25
E. 3213	Durham	Native	8	6	June	8	107	3.25	E. 3225	9	108	3.25
E. 3214	Merrett	Native	10	8	February	4	114	3.00	E. 3226	5	118	3.00
E. 3215	Williams	Native	8	6	May	2	112	4.00	E. 3227	8	110	4.00
E. 3216	Brady Heifer	Native	4	2	February	5	112	4.00	E. 3228	4	111	4.25
E. 3217	Red Williams	Native	8	6	February	2	110	4.25	E. 3229	3	109	4.50
E. 3218	Wild Cow	Native	10	8	February	4	108	3.50	E. 3230	5	105	3.50
E. 3219	Average	....	....	....	....	....	112	3.25	E. 3231	....	113	3.50

Calculated analysis, October 9, 1885; eleven cows owned by Benjamin Mead, Bedford, Westchester county. Evening's milking.

Number of inspection.	Water.	Total solids.	Fat.	Solids not fat.	Sugar-caseine.	Salts.
E. 3208......	87·06	12·94	3·50	9·44	8·76	0·68
E. 3209......	88·79	12·21	3·25	8·96	8·28	0·68
E. 3210......	88·00	12·00	3·00	9·00	8·32	0·68
E. 3211......	87·06	12·94	3·50	9·44	8·76	0·68
E. 3212......	87·08	12·92	3·00	9·92	9·24	0·68
E. 3213......	87·61	12·39	3·25	9·14	8·46	0·68
E. 3214......	87·36	12·64	3·00	9·64	8·96	0·68
E. 3215......	86·27	13·73	4·00	9·73	9·05	0·68
E. 3216......	86·27	13·73	4·00	9·73	9·05	0·68
E. 3217......	86·11	13·89	4·25	9·64	8·96	0·68
E. 3218......	87·61	12·39	3·50	8·89	8·21	0·68

Calculated analysis, October 10, 1885; eleven cows owned by Benjamin Mead, Bedford, Westchester county. Morning's milking.

Number of inspection.	Water.	Total solids.	Fat.	Solids not fat.	Sugar-caseine.	Salts.
E. 3220	86·91	13·09	3·50	9·59	8·91	0·68
E. 3221	88·41	11·59	3·50	9·09	8·41	0·68
E. 3222	87·69	12·31	3·25	9·06	8·38	0·68
E. 3223	87·14	12·86	3·50	9·36	8·66	0·68
E. 3224	86·92	13·08	3·25	9·83	9·15	0·68
E. 3225	87·53	12·47	3·25	9·22	8·54	0·68
E. 3226	87·46	12·54	3·00	9·54	8·86	0·68
E. 3227	86·43	13·57	4·00	9·57	8·89	0·68
E. 3228	86·06	13·94	4·25	9·69	9·01	0·68
E. 3229	85·87	14·13	4·50	9·63	8·95	0·68
E. 3230	87·45	12·55	3·50	9·05	8·57	0·68
E. 3231	(Average	see page	62.)			

Morning, noon and evening of June 2, 1885; farm of John Mitchell (Meadow-brook Farm), Cornwall, Orange county; cows in herd, 36; treatment, groomed once daily; housing, good; food, wheat-bran, corn-meal, hay, roots and oil-meal; during summer, grass.

Number of inspection.	Name of cow.	Breed.	Age of cow.	Number of times calving.	Time of last ng.	Number of quarts given.	Lactometer at 60 deg. Fahr.	Per cent of cream.	Number of inspection.	Number of quarts given.	Lactometer at 60 deg. Fahr.	Per cent of cream.	Number of inspection.	Lactometer at 60 deg. Fahr.	Per cent of cream.
			years.		1884.										
D. 3879	Charity	Holstein.....	5	3	Sept. 20	8	108	10	D. 3890....	5 3-4	108	10	D. 3901....	110	12
D. 3880	Dainty Dot....	Holstein.....	5	3	Oct. 21	10 1-4	100	17	D. 3891....	7	110	16	D. 3902....	100	18
D. 3881	Celeste	Holstein.....	5	3	Nov. 13	6 1-4	110	12	D. 3892....	4 1-4	110	12	D. 3903....	100	16
					1885.										
D. 3882	Aggie Louise ..	Holstein.....	5	4	April 6	9 1-8	106	12	D. 3893....	7 3-4	100	10	D. 3904....	100	10
					1884.										
D. 3883	Sylvina	Holstein.....	5	3	April 26	4 1-4	120	14	D. 3894....	3 1-2	106	16	D. 3905....	100	16
D. 3884	Matron	Holstein.....	7	5	July 28	8	110	12	D. 3895....	5 1-4	108	13	D. 3906....	103	14
D. 3885	Fanchon	Holstein.....	5	3	Oct. 20	5	112	9	D. 3896....	3	110	10	D. 3907....	102	9
D. 3886	Oriana.........	Holstein.....	6	4	June 28	6	112	10	D. 3897....	4	110	12	D. 3908....	116	11
D. 3887	Hyerine........	Holstein.....	5	3	Aug. 23	8 1-2	108	12	D. 3898....	5 1-8	110	14	D. 3909....	104	13
					1885.										
D. 3888	Ilga	Holstein.....	5	3	April 15	10 1-2	106	12	D. 3899....	6	108	12	D. 3910....	100	12
D. 3889	Average ...	....	....	....	....	....	112	12	D. 3900	....	108	14	D. 3911....	102	14

Morning and evening of November 28, 1885; farm of G. B. Robbins (Clover Valley Dairy Farm), Martin and Smith, managers, Marcy, Oneida county; cows in herd, 62; treatment, kind; housing, good; food, corn-fodder and hay.

Number of inspection.	Name of cow.	Breed.	Age of cow.	Number of times calving.	Time of last calving.	Number of quarts given.	Lactometer at 60 deg. Fahr.	Per cent of cream.	Number of inspection.	Number of quarts given.	Lactometer at 60 deg. Fah.	Per cent of cream.
			yrs.		1885.							
D. 5916. ...	Hehay........	Grade.....	8	6	March.......	8	110	15	D. 5927....	8	112	14
D. 5917.....	Curley........	Durham...	8	6	March.......	4	108	14	D. 5928....	8	106	14
D. 5918.....	Polly........	Durham...	7	5	February....	2	108	18	D. 5929....	2	108	20
D. 5919.....	Jennie........	Durham...	5	2	February....	8	110	14	D. 5930....	2	110	17
D. 5920.....	Silkey........	Durham...	8	6	March.......	4	106	16	D. 5931....	4	108	18
D. 5921.....	Black........	Durham...	6	4	March.......	4	106	18	D. 5932....	4	107	18
D. 5922.....	Dutch........	Durham...	8	6	March.......	8	108	14	D. 5933....	8	102	16
D. 5923.....	Beauty........	Durham...	6	4	March.......	4	107	16	D. 5934....	4	108	18
D. 5924.....	Wild Eye........	Durham...	8	6	March.......	4	104	15	D. 5935....	4	109	15
D. 5925. ...	Cherry........	Durham...	8	6	March.......	4	106	14	D. 5936....	4	108	16
D. 5926.....	Average........	....	....	....	....	....	108	16	D. 5937....	....	107	14

Dairy inspection by Inspector Joseph J. Sorogan, November 30, 1885. Clover Valley Dairy Farm, George B. Robbins, proprietor; managed by Martin & Smith. Number of acres, 335; cows in herd, 62; treatment, ordinary. Housing, stable which is of stone; dimensions, 50 by 85 feet; the other 40 by 80 feet, built above ground, both of which are well ventilated and great care is taken to have them kept clean. Manure cellar in rear of each which are cleaned out once a week. Two large Stover wind-mills pump the water from wells in adjoining yard which is so arranged that no drainage from the stables can affect the water. The proprietor has tried several herds of stock and has found that the Durham crossed with Ayrshire are the best for dairy purposes. Upon the farm may be seen the best Ayrshire bull in the county, he having taken first premium at the county fair the past two seasons. The feed at the present time is hay and corn-fodder which is fed until about December 15, then hay alone until about March 1, then oats ground with corn in the ear is fed until turned to grass, which is usually May 20. Mr. Robbins' dairy this season has averaged 5400 pounds of milk per cow, which net him not to exceed 69 cents per hundred pounds.

Evening of June 5, and morning of June 6, 1885; farm of Truman Baker (Cedar Grove Farm), Lebanon, Madison county; cows in herd, 20; treatment, kind; housing, good; food, grass in summer, hay in winter, until before calving; corn-meal and linseed meal, after calving.

Number of inspection.	Name of cow.	Breed.	Age of cow.	Number of times calving.	Time of last calving.	Number of quarts given.	Lactometer at 60 deg. Fahr.	Per cent of cream.	Number of inspection.	Number of quarts given.	Lactometer at 60 deg. Fahr.	Per cent of cream.
			years.		1885.							
I. 34	Lulu, No. 3569	Devon	7	4	April 20	8	112	20	I. 43	6 1-4	112	20
I. 35	Helena, No. 3273	Devon	10	7	April 16	6 1-2	106	16	I. 44	5	112	16
I. 36	Edith, No. 3318	Devon	14	11	April 25	6	110	28	I. 45	4	114	20
I. 37	Dora 4th, No. 1047	Devon	8	5	May 2	7 1-2	108	12	I. 46	6 1-4	108	20
I. 38	Lovey, No. 1887	Devon	11	8	March 3	6	108	20	I. 47	4	106	26
I. 39	Dolly, No. 1005	Devon	13	10	April 6	6 1-2	106	14	I. 48	4 1-4	110	16
I. 40	Lizzie, No. 3257	Devon	8	6	May 9	6 3-4	112	14	I. 49	5	112	22
I. 41	Meadow Queen, No. 4030	Devon	4	3	March 16	5	110	16	I. 50	4	112	20
I. 42	Myra, No. 3368	Devon	6	4	March 18	5	108	20	I. 51	In heat	this	A. M.
I. 52	Average (see page 62 for	analyses).....	....	....	....	....	110	....	I. 53	....	114	20

Morning and evening of July 9, 1885; *farm of Truman Baker* (*Cedar Grove Farm*), *Lebanon, Madison county; cows in herd,* 20; *treatment, kind; housing, good; food, grass in summer, good hay and small quantity of meal in winter.*

Number of inspection.	Name of cow.	Breed.	Age of cow.	Number of times calving.	Time of last calving.	Number of quarts given.	Lactometer at 60 deg. Fahr.	Per cent of cream.	Number of inspection.	Number of quarts given.	Lactometer at 60 deg. Fahr.	Per cent of cream.
			years.		1885.							
I. 70	Lizzie	Devon	8	6	May 9	7½	112	14	I. 80	6	110	10
I. 71	Myra	Do.	6	4	May 18	6½	110	16	I. 81	7	114	12
I. 72	Dora, 4th	Devon	8	5	May 2	5½	110	12	I. 82	6	108	8
I. 73	Lovely	Devon	11	8	March 23	6	110	12	I. 83	5	106	12
I. 74	Lulu	Do.	7	4	April 20	5	112	16	I. 84	7	110	12
I. 75	Meadow Queen	Do.	4	8	March 16	5	112	16	I. 85	5	110	14
I. 76	Helena	Do.	10	7	March 16	5	112	10	I. 86	6	110	12
I. 77	Edith	Do.	14	11	April 25	4	102	10	I. 87	6	108	14
I. 78	Dolly	Devon	13	10	April 6	4	104	12	I. 88	6	104	10
I. 79	Average (see page 62 for analyses)		....	....	....	....	110	16	I. 89	....	110	12

Morning of August 24, 1885; *farm of Truman Baker, Lebanon, Madison county; cows in herd,* 20; *treatment, kind; housing, good; food, pasture.*

Number of inspection.	Name of cow.	Breed.	Age of cow.	Number of times calving.	Time of last calving.	Number of quarts given.	Lactometer at 60 deg. Fahr.	Per cent of cream.
			years.		1885.			
I. 157....	Lizzie	Devon.........	8	6	May 9	5½	108	10
I. 158 ..	Myra............	Devon.........	6	4	May 18	6	108	14
I. 159....	Dora, 4th........	Devon.........	8	5	May 2	4	106	9
I. 160....	Lovely	Devon.........	11	8	Mar. 28	5½	108	10
I. 161....	Lulu	Devon.........	7	4	April 20	5½	110	12
I. 162....	Meadow Queen...	Devon.........	4	3	Mar. 16	6	106	18
I. 163....	Helena..........	Devon.........	10	7	Mar. 16	5	108	12
I. 164 ..	Edith...........	Devon.........	14	11	April 25	5½	110	18
I. 165....	Dolley	Devon.........	13	10	April 6	7	100	11
I. 166....	Average morn	ing's (see page 62	for an	alysis)	....	....	108	14
I. 167....	Average eveni	ng's	....	....	....	....	106	14

Experiments on Dipping.

It has been asserted that in dipping from a can of milk, the last portion dipped out, even when care is taken to stir up the contents before each dipping, will be much poorer in fat than the first; and that in this way great injustice might be done to a perfectly honest dealer if the milk was examined after a certain amount had been dipped from the can.

A number of experiments were made to determine exactly what effect the dipping from a forty-quart can had on the quality of the milk, The tables given here show the result. Except in the case where the inspector was on the wagon, the samples were taken so as to avoid stirring the milk as much as possible, and even under these circumstances it will be seen that the quality of the milk, as far as the per cent of fat obtained was concerned, is nearly the same from the beginning to the end of the experiment.

These experiments were made with the can at rest. If the milk was being carried around on an ordinary milk wagon the contents, from the continual jar, would tend to become mixed. That is, the cream would certainly not separate out as rapidly as if the can was at rest.

These experiments seem to point out the fact that where the last portion of a can of milk is of so poor a quality as to be called adulterated, the seller must have taken considerable pains to dip off the first portion so as to skim the milk, and knowingly and deliberately defraud those who might buy the last portions.

MIDDLETOWN, ORANGE COUNTY,
October 19, 1885.

The following twenty-five inspections were made by Inspector Samuel J. White from a can containing forty quarts of milk. The first two samples, containing one pint each, were taken from the can at six o'clock A. M. The following twenty-two samples were taken two samples at a time every hour up to five o'clock P. M. The twenty-fifth or last sample was taken at six o'clock P. M., after the contents of the can had been thoroughly stirred.

Number of inspection.	Time of taking sample.	Lactometer at 60 deg. F.	Per cent of fat.
E. 3232	6 A. M.	115	4·19
E. 3233		117	3·07
E. 3234	7 A. M.	116	3·59
E. 3235		116	3·36
E. 3236	8 A. M.	116	3·78
E. 3237		115	3·44
E. 3238	9 A. M.	115	3·61
E. 3239		115	3·67
E. 3240	10 A. M.	116	3·11
E. 3241		115	3·76
E. 3242	11 A. M.	115	3·29
E. 3243		116	3·60
E. 3244	12 M.	115	3·42
E. 3245		115	3·67
E. 3246	1 P. M.	115	3·46
E. 3247		115	3·47
E. 3248	2 P. M.	115	3·45
E. 3249		115	3·76
E. 3250	3 P. M.	115	3·59
E. 3251		115	3·71
E. 3252	4 P. M.	115	3·72
E. 3253		115	3·77
E. 3254	5 P. M.	115	3·62
E. 3255		115	3·57
E. 3256	6 P. M.	115	3·84

The following is the complete analysis of twenty-five samples of milk taken on the 19th day of October, 1885, from a can containing forty quarts. The first two samples were taken at six o'clock A. M. The next twenty-two samples were taken two at a time each hour up to six o'clock P. M. The twenty-fifth sample was taken after the remaining contents of the can had been thoroughly stirred. The first twenty-four samples were taken without stirring the milk except by drawing dipper from bottom of can.

Number of inspection.	Water.	Total solids.	Fat.	Sugar-caseine.	Salts.	Lactometer at 60 deg. F.
E. 3232	85·94	13·06	4·19	7·07	0·80	115
E. 3233	87·04	12·96	3·07	9·14	0·75	117
E. 3234	87·63	12·37	3·59	8·20	0·78	116
E. 3235	86·97	13·03	3·36	8·06	0·72	116
E. 3236	87·23	12·77	3·78	8·27	0·67	116
E. 3237	86·92	13·08	3·44	8·98	0·73	115
E. 3238	86·80	13·20	3·61	8·86	0·74	115
E. 3239	86·32	13·68	3·67	9·27	0·61	115
E. 3240	86·56	13·44	3·11	9·72	0·76	116
E. 3241	86·37	13·63	3·76	9·11	0·73	115
E. 3242	86·89	13·11	3·29	9·09	0·72	115
E. 3243	86·76	13·24	3·60	8·96	0·68	116
E. 3244	86·91	13·09	3·42	9·01	0·66	115
E. 3245	86·57	13·43	3·67	9·02	0·73	115
E. 3246	86·87	13·13	3·46	9·02	0·65	115
E. 3247	86·70	13·30	3·47	9·17	0·66	115
E. 3249	86·54	13·46	3·45	9·35	0·66	115
E. 3249	86·42	13·58	3·76	9·12	0·70	115
E. 3250	86·53	13·47	3·59	9·14	0·74	115
E. 3251	86·28	13·72	3·71	9·23	0·78	115
E. 3252	86·46	13·54	3·72	9·05	0·77	115
E. 3253	86·40	13·60	3·77	9·07	0·76	115
E. 3254	86·66	13·34	3·62	9·06	0·66	115
E. 3255	86·57	13·43	3·57	9·24	0·62	115
E. 3256	86·59	13·41	3·84	8·83	0·74	115

NEW YORK CITY,
October 23, 1885.

The following twenty-three samples of milk were taken by Inspector Samuel J. White from two cans containing forty quarts each, owned by Bullock & Shafer, corner of Broadway and Fifty-second street. The first ten samples were taken from can number one, each sample being taken from the quantity of milk being dipped by the driver of wagon numbered two, for Bullock & Shafer's customers. The inspector accompanying the driver on his trip. The next thirteen samples were taken from can number two in the same manner as above. The twenty-fourth or last sample numbered E. 3280, was taken from a third can of milk by request of Bullock & Shafer.

Can No. 1.

Number of inspection.	Lactometer at 60 deg. Fahr.	Per cent of fat.
E. 3257	113	4·55
E. 3258	112	4·15
E. 3259	114	4·59
E. 3260	113	4·63
E. 3261	113	4·09
E. 3262	113	4·18
E. 3263	113	4·17
E. 3264	114	4·11
E. 3265	112	4·58
E. 3266	112	3·98
Can No. 2.		
E. 3267	104	3·46
E. 3268	105	3·83
E. 3269	103	3·25
E. 3270	104	3·54
E. 3271	108	3·49
E. 3272	104	3·82
E. 3273	105	3·57
E. 3274	109	3·24
E. 3275	107	3·58
E. 3276	107	3·65
E. 3277	106	3·49
E. 3278	107	3·34
E. 3279	109	3·51
Can No. 3.		
E. 3280	104	3·68

The following is the complete analysis of twenty-three samples of milk taken on the 23d day of October, 1885, from two cans containing forty quarts each.

The samples were taken from the quantity of milk delivered by Bullock & Shafer from wagon No. 2, to their customers in New York city; each sample of milk being taken as the driver of wagon dipped the milk for delivery to customer, as before stated.

Can No. 1.

Number of inspection.	Water.	Total solids.	Fat.	Sugar-caseine.	Salts.	Lactometer at 60 deg. Fahr.
1. E. 3257	86·36	13·64	4·55	8·35	0·74	113
2. E. 3258	86·38	13·62	4·15	8·76	0·71	112
3. E. 3259	86·43	13·57	4·59	8·15	0·74	114
4. E. 3260	86·48	13·52	4·63	8·15	0·74	113
5. E. 3261	86·45	13·55	4·09	8·94	0·72	113
6. E. 3262	86·31	13·69	4·18	8·88	0·73	113
7. E. 3263	86·27	13·73	4·17	8·88	0·70	113
8. E. 3264	86·36	13·64	4·11	8·80	0·73	114
9. E. 3265	86·49	13·51	4·58	8·20	0·73	112
10. E. 3266	86·37	13·63	3·94	9·15	0·74	112

Can No. 2.

Number of inspection.	Water.	Total solids.	Fat.	Sugar-caseine.	Salts.	Lactometer at 60 deg. Fahr.
11. E. 3267	87·54	*12·46	3·46	8·34	0·62	104
12. E. 3268	87·20	*12·80	3·83	8·27	0·70	105
13. E. 3269	87·92	*12·08	3·25	8·83	0·67	103
14. E. 3370	87·83	12·17	3·54	8·00	0·63	104
15. E. 3271	87·70	12·30	3·49	8·12	0·69	108
16. E. 3272	87·11	*12·89	3·82	8·38	0·69	104
17. E. 3273	87·81	12·19	3·57	7·92	0·70	105
18. E. 3274	87·72	12·28	3·24	8·33	0·71	109
19. E. 3275	87·85	12·15	3·58	7·88	0·69	107
20. E. 3276	87·89	12·11	3·65	7·79	0·67	107
21. E. 3277	87·83	12·17	3·49	8·00	0·68	106
22. E. 3278	87·79	12·21	3·34	8·22	0·65	107
23. E. 3279	87·77	12·23	3·51	8·11	0·61	109
24. E. 3280	87·95	12·05	3·68	7·69	0·68	104

Sample No. E. 3280 was taken from a third can.

* Calculated.

MIDDLETOWN, ORANGE COUNTY,
October 26, 1885.

The following twenty-six samples were taken from a can containing forty quarts of milk. The first sample was taken after the contents of the can had been thoroughly stirred. The next twenty-four samples were taken two samples at a time at one hour intervals. The twenty-sixth or last sample was taken after the remaining contents of the can had been thoroughly stirred.

Number of inspection.	Lactometer at 60 deg. F.	Per cent of fat by lactoscope.
E. 3281	109	3·75
E. 3282	113	4·00
E. 3283	112	4·00
E. 3284	112	3·75
E. 3285	113	3·75
E. 3286	111	3·75
E. 3287	112	3·50
E. 3288	113	3·50
E. 3289	113	3·50
E. 3290	114	3·50
E. 3291	113	3·50
E. 3292	113	3·50
E. 3293	112	3·50
E. 3294	112	3·50
E. 3295	112	3·50
E. 3296	112	3·50
E. 3297	112	3·50
E. 3298	112	3·50
E. 3299	112	3·50
E. 3300	112	3·75
E. 3301	112	3·75
E. 3302	113	3·75
E. 3303	110	3·75

The following forty samples were taken from a can containing forty quarts of milk. The forty samples were taken from the can two samples at a time at intervals of thirty minutes, the first sample being taken at six o'clock A. M., the last at half past three P. M. The contents of the can were not stirred except by drawing the dipper from bottom of can. Each sample contained one quart.

Number of inspection.	Time of taking sample.	Lactometer at 60 deg. F.	Per cent of fat by lactoscope.
E. 3319	6 A. M.	112	3·75
E. 3320		113	3·75
E. 3321	6.30 A. M.	114	3·75
E. 3322		113	3·50
E. 3323	7 A. M.	113	3·50
E. 2324		113	3·50
E. 3325	7.30 A. M	113	3·50
E. 3326		113	3·50
E. 3327	8. A. M.	113	3·50
E. 3328		114	3·25
E. 3329	8.30 A. M.	114	3·25
E. 3330		114	3·25
E. 3331	9. A. M.	114	3·25
E. 3332		114	3·25
E. 3333	9.30 A. M.	114	3·25
E. 3334		114	3·25
E. 3335	10 A. M.	114	3·25
E. 3336		113	3·50
E. 3337	10.30 A. M.	114	3·25
E. 3338		114	3·25
E. 3339	11 A. M.	115	3·00
E. 3340		114	3·25
E. 3341	11.30 A. M.	114	2·25
E. 3342		114	3·25
E. 3343	12 M.	114	3·25
E. 3344		114	3·25
E. 3345	12.30 P. M.	114	3·25
E. 3346		114	3·25
E. 3347	1 P. M.	114	3·25
E. 3348		114	3·25
E. 3349	1.30 P. M.	114	3·25
E. 3350		114	3·25
E. 3351	2 P. M.	115	3·00
E. 3352		114	3·25
E. 3353	2.30 P. M.	114	2·25
E. 3354		114	3·25
E. 3355	3 P. M.	115	3·00
E. 3356		115	3·00
E. 3357	3.30 P. M.	115	3·00
E. 3358		115	3·00

Analysis by calculation.

Number of inspection.	Water.	Total solids.	Fat.	Solids not fat.	Sugar-caseine.	Salts.
E. 3319	86·60	13·40	3·75	9·65	8.97	0·68
E. 3320	86·52	13·48	3·75	9·73	9.05	0·68
E. 3321	86·34	13·66	3·75	9·81	9·13	0·68
E. 3322	86·83	13·17	3·50	9·67	8·99	0·68
E. 3323	86·83	13·17	3·50	9·67	8·99	0·68
E. 3324	86·83	13·17	3·50	9·67	8·99	0·68
E. 3325	86·83	13·17	3·50	9·67	8·99	0·68
E. 3326	86·83	13·17	3·50	9·67	8·99	0·68
E. 3327	86·83	13·17	3·50	9·67	8·99	0·68
E. 3328	87·07	12·93	3·25	9·68	9·00	0·68
E. 3329	87·07	12·93	3·25	9·68	9·00	0·68
E. 3330	87·07	12·93	3·25	9·68	9·00	0·68
E. 3331	87·07	12·93	3·25	9·68	9·00	0·68
E. 3332	87·07	12·93	3·25	9·68	9·00	0·68
E. 3333	87·07	12·93	3·25	9·68	9·00	0·68
E. 3334	87·07	12·93	3·25	9·68	9·00	0·68
E. 3335	87·07	12·93	3·25	9·68	9·00	0·68
E. 3336	86·83	13·17	3·50	9·67	8·99	0·68
E. 3337	87·07	12·93	3·25	9·68	9·00	0·68
E. 3338	87·07	12·93	3·25	9·68	9·00	0·68
E. 3339	87·31	12·69	3·00	9·69	9·01	0·68
E. 3340	87·07	12·93	3·25	9·68	9·00	0·68
E. 3341	87·07	12·93	3·25	9·68	9·00	0·68
E. 3342	87·07	12·93	3·25	9·68	9·00	0·68
E. 3343	87·07	12·93	3·25	9·68	9·00	0·68
E. 3344	87·07	12·93	3·25	9·68	9·00	0·68
E. 3345	87·07	12·93	3·25	0·68	9·00	0·68
E. 3336	87·07	12·93	3·25	9·68	9·00	0·68
E. 3347	87·07	13·93	3·25	9·68	9·00	0·68
E. 3348	87·07	12·93	3·25	9·68	9·00	0·68
E. 3349	87·07	12·93	3·25	9·68	9·00	0·68
E. 3350	87·07	12·93	3·25	9·68	9·00	0·68
E. 3351	87·31	12·69	3·00	9·69	9·01	0·68
E. 3352	87·07	12·93	3·25	9·68	9·00	0·68
E. 3353	87·07	12·93	3·25	9·68	9·00	0·68
E. 3354	87·07	12·93	3·25	9·68	9·00	0·68
E. 3355	87·31	12·69	3·00	9·69	9·01	0·68
E. 3356	87·31	12·69	3·00	9·69	9·01	0·68
E. 3357	87·31	12·69	3·00	9·69	9·01	0·68
E. 3358	87·31	12·69	3·00	9·69	9·01	0·68

MIDDLETOWN, ORANGE COUNTY,
November 25, 1885.

The following forty samples were taken from a can containing forty quarts of milk. The first sample was taken from the can at six o'clock A. M., after the contents of the can had been thoroughly stirred. The next thirty-eight samples were taken two at one time at intervals of thirty minutes; the contents of the can not being stirred except by drawing the dipper from bottom of can. The last sample taken was the last quart in the can.

Number of inspection.	Time of taking sample.	Lactometer at 60 deg. Fahr.	Per cent of fat by lactoscope.
E. 3359	6 A. M.	112	3·75
E. 3360	6.30 A. M.	112	3·75
E. 3361		112	3·75
E. 3362	7 A. M.	112	3·75
E. 3363		112	3·75
E. 3364	7.30 A. M.	111	4·00
E. 3365		111	4·00
E. 3366	8 A. M.	111	4·00
E. 3367		111	4·00
E. 3368	8.30 A. M.	112	4·00
E. 3369		112	4·00
E. 3370	9 A. M.	112	4·00
E. 3371		111	3·75
E. 3372	9.30 A. M.	110	3·75
E. 3373		111	3·75
E. 3374	10 A. M.	112	3·75
E. 3375		112	4·00
E. 3376	10.30 A. M.	111	4·00
E. 3377		111	4·25
E. 3378	11 A. M.	112	4·00
E. 3379		112	4·00
E. 3380	11.30 A. M.	112	3·75
E. 3381		112	3·75
E. 3382	12 M.	110	3·75
E. 3383		110	3·75
E. 3384	12.30 P. M.	111	3·75
E. 3385		111	3·75
E. 3386	1 P. M.	110	4·00
E. 3387		111	4·00
E. 3388	1.30 P. M.	112	4·00
E. 3389		111	3·75
E. 3390	2 P. M.	111	3·75
E. 3391		112	4·00
E. 3392	2.30 P. M.	113	4·00
E. 3393		112	4·00
E. 3394	3 P. M.	113	3·75
E. 3395		112	4·00
E. 3396	3.30 P. M.	113	3·75
E. 3397		111	3·75
E. 3398	4 P. M.	112	3·75

The following is the analysis by calculation of forty samples of milk taken on the 25th day of November 1885, from a can containing forty quarts. The first sample was taken at six o'clock A. M., after the contents of the can had been thoroughly stirred. The next thirty-eight samples were taken two samples at each time every thirty minutes from half past six A. M., up to half past three P. M., without stirring the milk except by drawing the dipper from bottom of can. The last sample, No. E. 3398, was taken at four o'clock P. M. It was the last quart in bottom of can.

Number of inspection.		Water.	Total solids.	Fat.	Solids not fat.	Sugar-caseine.	Salts.	Lactometer at 60 deg. F.
A. M. 6	E. 3359.	86·59	13·41	3·75	9·66	8·98	0·68	112
6.30.	E. 3360.	86·59	13·41	3·75	9·66	8·98	0·68	112
	E. 3361.	86·59	13·41	3·75	9·66	8·98	0·68	112
7.00.	E. 3362.	86·59	13·41	3·75	9·66	8·98	0·68	112
	E. 3363.	86·59	13·41	3·75	9·66	8·98	0·68	112
7.30.	E. 3364.	86·35	13·65	4·00	9·65	8·97	0·68	111
	E. 3365.	86·35	13·65	4·00	9·65	8·97	0·68	111
8.00.	E. 3366.	86·35	13·65	4·00	9·65	8·97	0·68	111
	E. 3367.	86·35	13·65	4·00	9·65	8·97	0·68	111
8.30.	E. 3368.	86·27	13·73	4·00	9·73	9·05	0·68	112
	E. 3369.	86·27	13·73	4·00	9·73	9·05	0·68	112
9.00.	E. 3370.	86·27	13·73	4·00	9·73	9·05	0·68	112
	E. 3371.	86·67	13·33	3·75	8·58	8·90	0·68	111
9.30.	E. 3372.	86·75	13·25	3·75	9·50	8·82	0·68	110
	E. 3373.	86·67	13·33	3·75	9·58	8·90	0·68	111
10.00.	E. 3374.	86·59	13·41	3·75	9·66	8·98	0·68	112
	E. 3375.	86·27	13·73	4·00	9·73	9·05	0·68	112
10.30.	E. 3376.	86·35	13·65	4·00	9·65	8·97	0·68	111
	E. 3377.	86·04	13·96	4·25	9·71	9·03	0·68	111
11.00.	E. 3378.	86·27	13·73	4·00	9·73	9·05	0·68	112
	E. 3379.	86·27	13·73	4·00	9·73	9·05	0·68	112
11.30.	E. 3380.	86·59	13·41	3·75	9·66	8·98	0·68	112
	E. 3381.	86·59	13·41	3·75	9·66	8·98	0·68	112
M. 12.00.	E. 3382	86·75	13·25	3·75	9·50	8·82	0·68	110
	E. 3383.	86·75	13·25	3·75	9·50	8·82	0·68	110
P. M. 12.30.	E. 3384	86·67	13·33	3·75	9·58	8·90	0·68	111
	E. 3385.	86·67	13·33	3·75	9·58	8·90	0·68	111
1.00.	E. 3386.	86·43	13·57	4·00	9·57	8·89	0·68	110
	E. 3387.	86·35	13·65	4·00	9·65	8·97	0·68	111
1.30.	E. 3388.	86·27	13·73	4·00	9·73	9·05	0·68	112
	E. 3389.	86·67	13·33	3·75	9·58	8·90	0·68	111
2.00.	E. 3390.	86·67	13·33	3·75	9·58	8·90	0·68	111
	E. 3391.	86·27	13·73	4·00	9·73	9·05	0·68	112
2.30.	E. 3392.	86·20	13·80	4·00	9·80	9·12	0·68	113
	E. 3393.	86·27	13·73	4·00	9·73	9·05	0·68	112
3.00.	E. 3394.	86·51	13·49	3·75	9·74	9·06	0·68	113
	E. 3395.	86·27	13·73	4·00	9·73	9·05	0·68	112
3.30.	E. 3396.	86·51	13·49	3·75	9·74	9·06	0·68	113
	E 3397.	86·67	13·33	3·75	9·58	8·90	0·68	111
4.00.	E. 3398.	86·59	13·41	3·75	9·66	8·98	0·68	112

FEEDING.

The question of the food, in kind and quality, for milch cows is of the greatest importance to the milk producer, and yet very few practical experiments have been made.

In looking over the literature relating to this subject, we find considerable data, but not of a practical kind. Graded cows seem to have been taken for the experiment, and again, the duration of the experiments have been too limited.

In these experiments ordinary cows should be selected, and at least twenty should be taken. Also the duration should be at least two months, for it will take that length of time at least to determine the effect of the food given upon the health of the cow, and quantity and quality of the milk produced. Some cows will respond at once to a change in diet, while others do not seem to be affected even after a long continued trial.

It is claimed that brewers' grains produce milk unfit for use, and that the health of the cow is permanently impaired. That some cows are so constituted as to be made extremely unwell from the use of brewers' grains, there can be no doubt; but we cannot conclude from this fact that brewers' grains are to be classed as an unhealthy food. Most cows placed on a diet of grains and other food usually appear to be perfectly healthy.

During the past year I sent the following questions to those who I supposed would be interested in this matter:

Are brewers' grains fed to milch cows in your vicinity?
What particular breed or kind of cows are kept?
How much brewers' grains are fed to each cow per day?
What other foods are given per day, and how much of each?
Are the grains fresh or from silos?
Is the yield of milk per day greater or less when brewers' grains are fed?
If any analyses have been made of the grains as fed to the cows, please fill the following table:

Per cent of water.
Per cent of fat.
Per cent of nitrogenous.
Per cent of non-nitrogenous.
Per cent of cellulose.
Per cent of acid (and what kind).

If any analyses have been made of milk from cows fed as above, please fill in the following table:

(Sp. Gr. at 60 deg. Fahr.)
Per cent by volume of cream.
Per cent of water.
Per cent of fat.
Per cent of sugar.
Per cent of caseine.
Per cent of salts.

Are the keeping qualities of the milk as good as that from cows not fed on brewers' grains ?

Do you think the use of brewers' grains imparts any odor to the milk ?

Do you think the continued use of brewers' grains has any effect on the health of the cows ?

Do you consider brewers' grains a healthy food for milch cows ?

If not, why ?

I received answers of some kind to nearly all the circulars sent, some four hundred and fifty. Out of these four hundred and fifty answers, only forty-two knew any thing about the subject, and none of the forty-two agreed on all of the questions, except this one: "Do you think the use of brewers' grains imparts any odor to the milk ?" and this in thirty-two replies was in the affirmative.

The majority seemed to think:

1. That the use of brewers' grains imparts a slight odor to the milk.

2. That fed in moderate quantities with other food it *was not unhealthy*.

3. That milk from cows fed on brewers' grains could not be used to make butter, cheese or condensed milk.

I was unable to obtain any analyses of brewers' grains showing the per cent of acid they contained, or the kind of acid. A large number of experiments made during the years 1880, 1881, 1882 and 1883, upon brewers' grains milk, to determine whether the curd formed when the milk became sour was in any way abnormal, resulted as follows:

The milk was placed in a four-ounce bottle, allowed to sour and the character of the curd and time of souring noted. The curd formed from the milk of cows fed in part on brewers' grains was tougher and was broken up with difficulty. The curd formed from milk from cows fed on hard feed was not so tough and was readily broken up by shaking. The time of souring was not so characteristic, the samples of milk from cows fed on hard feed turning sour, in some instances, first. These tests were made in lots of six to ten samples of brewers' grains milk, together with one to four samples of milk from cows fed on hard feed.

Some five hundred tests in all were made. With the exception of a tougher curd being formed, there was nothing abnormal in "Brewers' Grains Milk."

Many books have been written on the proper method of feeding cows, but they seem to be too theoretical for the practical farmer.

The needs of the milch cow from the latest German experiments are one part of nitrogenous food to five parts of carbonaceous food.

The following table taken from "Hints on Dairying," by T. D. Curtis, of Syracuse, gives the proportion of nitrogenous to carbonaceous in many foods.

Foods.	Nitrogenous.		Carbonaceous.
Meadow hay, medium	1	to	8·0
Red clover, medium	1	to	5·9
Lucerne, good	1	to	2·8
Swedish clover (alsike)	1	to	4·9

Foods.	Nitrogenous.		Carbonaceous.
Orchard grass, in blossom	1	to	6·5
White clover, medium	1	to	5·0
Timothy	1	to	8·1
Blue grass, in blossom	1	to	7·5
Red top	1	to	5·4
Fodder rye	1	to	7·2
Italian rye grass	1	to	6·3
Hungarian grass	1	to	7·1
Rich pasture grass	1	to	3·6
Green maize, German	1	to	8·9
Fodder oats	1	to	7·2
Sorghum	1	to	7·4
Pasture clover, young	1	to	2·5
Red clover, before blossom	1	to	3·8
Red clover, in blossom	1	to	5·7
White clover, in blossom	1	to	4·2
Buckwheat, in blossom	1	to	5·1
Fodder cabbage	1	to	5·2
Rutabaga leaves	1	to	3·9
Fermented hay from maize	1	to	12·0
Fermented hay from beet leaves	1	to	4·0
Fermented hay from red clover	1	to	4·1
Winter wheat straw	1	to	45·8
Winter rye straw	1	to	52·0
Winter barley straw	1	to	40·5
Oat straw	1	to	29·9
Corn stalks	1	to	34·4
Seed clover	1	to	7·4
Wheat chaff	1	to	24·1
Rye chaff	1	to	32·6
Oat chaff	1	to	23·8
Barley chaff	1	to	30·4
Potatoes	1	to	10·6
Artichokes	1	to	8·7
Rutabagas	1	to	8·3
Sugar beets	1	to	17·0
Carrots	1	to	9·3
Turnips	1	to	5·8
Wheat, grain	1	to	5·8
Rye, grain	1	to	7·0
Barley, grain	1	to	7·9
Oats, grain	1	to	6·1
Maize, grain	1	to	8·6
Millet, grain	1	to	5·4
Peas, grain	1	to	2·9
Buckwheat, grain	1	to	7·4
Cotton seed	1	to	4·6
Pumpkins	1	to	18·4
Coarse wheat bran	1	to	5·6
Wheat middlings	1	to	6·9

Foods.	Nitrogenous.		Carbonaceous.
Rye bran	1	to	5·3
Barley bran	1	to	4·5
Buckwheat bran	1	to	4·1
Hempseed cake	1	to	1·5
Sunflower	1	to	1·3
Corn bran	1	to	10·3
Brewers' grains	1	to	3·0
Malt sprouts	1	to	2·2
Wheat meal	1	to	5·7
Rape cake	1	to	1·7
Rape meal, extracted	1	to	1·3
Barley, middlings	1	to	6·0
Oat bran	1	to	9·7
Linseed cake	1	to	2·0
Linseed meal, extracted	1	to	1·4
Cotton seed meal, decort	1	to	1·8
Cotton seed cake, undecort	1	to	1·7
Cow's milk	1	to	4·4
Buttermilk	1	to	2·6
Skimmed milk	1	to	1·9
Cream	1	to	30·5

It seems hardly possible to state exactly what amount of food a cow should have each day. Every cow would, in all probability, need a different amount of food. But remembering that a mixed diet is the best, and that the cow needs one part of nitrogenous to five parts of carbonaceous food, and referring to the table given, the proper diet of a cow can easily be fixed.

A practical experiment was tried during the summer of 1885, by Mr. James Robinson, of Walton, Delaware county, in relation to the effect of feeding cows a small quantity of hard feed while at the same time they had abundant pasture. Two cows were experimented on, each receiving one-half pound of corn meal and one-half pound of wheat middling morning and evening. The result is shown in the tables, and a simple calculation shows the gain in milk over and above the cost of the feed. The milk was tested from time to time by means of the lactoscope and was found to contain from one-half to three-fourths of a per cent more fat, while the cows were being fed the ration of meal and middlings.

Maxie.

	Number of quarts given, A. M.	Number of quarts given, P. M.	Total number of quarts given.	Weather.
August 11..	4 1-2	4 1-2	9	
August 12..	4 1-2	4 1-2	9	
August 13..	4 1-2	4 1-2	9	
August 14..	Commenced feeding.			
August 15..	4 1-2	5	9 1-2	
August 16..	4 2-3	5	9 2-3	Warm.
August 17..	5	6	11	Showery.
August 18..	5	6	11	
August 19..	5 1-2	6	11 1-2	
August 20..	6	6 3-4	12 3-4	
August 21..	6	7 1-2	13 1-2	
August 22.	7	8 1-4	15 1-4	
August 23..	6 1-2	7	13 1-2	
August 24..	7 1-2	8	15 1-2	Heavy.
August 25..	7 1-2	6 1-2	14	Rainy.
August 26..	7 1-2	6 1-2	14	
August 27..	7 1-2	6 3-4	14 1-4	Cold.
August 28..	7	6 1-4	13 1-4	
August 29..	6	7 1-4	13 1-4	Freezing.
August 30..	6 3-4	6	12 3-4	Warmer.
August 31..	6	6 1-2	12 1-2	
September 1..	6 1-4	5	11 1-4	
September 2..	6 1-4	6	12 1-4	
September 3..	6 3-4	6 1-2	13 1-4	
September 4..	6 1-2	7 1-2	14	Cold.
September 5..	6 1-2	6 3-4	13 1-4	
September 6..	6 1-4	5 3-4	12	
September 7..	6 1-2	7	13 1-2	
September 8..	5 1-4	6	11 1-4	
September 9..	6 1-2	6 1-4	12 3-4	Warmer.
September 10..	5 1-2	5 1-2	11	
September 11..	5 1-2	6 1-4	11 3-4	
September 12..	5 1-2	6	11 1-2	
September 13..	5 1-2	5 3-4	11 1-4	
September 14..	5 1-2	6	11 1-2	
September 15..	5	6 1-2	11 1-2	
September 16..	5 1-2	6	11 1-2	
September 17..	5 1-4	6 1-2	11 3-4	Quite warm.
September 18..	5 1-4	6 1-2	11 3-4	
September 19..	5	6 1-4	11 1-4	
September 20..	6	6	12	
September 21..	5 1-2	6 1-2	12	
September 22..	5	6 1-4	11 1-4	
September 23..	5 3-4	6 1-4	12	Cold.
September 24..	5 1-4	6	11 1-4	
September 25..	5 1-2	6 1-2	12	
September 26..	6	6	12	
September 27..	6	5 1-2	11 1-2	Warmer.

Topsie.

	Number of quarts given, A. M.	Number of quarts given, P. M.	Total number of quarts given.	Weather.
August 11..	....	7	7	
August 12..	6 1-2	7	13 1-2	
August 13..	6 1-2	7	13 1-2	
August 14..	6 1-2	7	13 1-2	
August 15..	....	6 1-2	6 1-2	
August 16..	7 1-4	7 1-2	14 3-4	Warm and showery.
August 17..	7	8 1-2	15 1-2	
August 18..	7 1-2	8	15 1-2	
August 19..	7 1-2	8 1-2	16	
August 20..	7 1-2	9	16 1-2	
August 21 .	7 3-4	8 1-2	16 1-4	
August 22..	8	9 1-2	17 1-2	
August 23..	8	9	17	Heavy rain.
August 24..	9 1-2	9	18 1-2	
August 25..	9	8 1-2	17 1-2	
August 26..	9	9	18	
August 27..	9	8 1-2	17 1-2	Cold, nearly freezing.
August 28..	8 1-2	9	17 1-2	
August 29..	8 3-4	9	17 3-4	
August 30..	9	8 1-2	17 1-2	
August 31..	8 1-2	9 3-4	18 1-4	Warmer.
September 1..	9	7	16	Cold.
September 2..	8 3-4	8 1-4	17	
September 3..	8 1-4	8 1-2	16 3-4	
September 4..	8	7	15	
September 5..	7	8 1-2	15 1-2	
September 6..	7 1-2	7 1-2	15	
September 7..	7 1-2	8 1-2	16	
September 8..	7 1-4	6 3-4	14	Warmer.
September 9..	8	8	16	
September 10..	6 1-4	8	14 1-4	
September 11..	7	7	14	
September 12..	7 1-2	8	15 1-2	
September 13..	7	8	15	
September 14..	6 1-2	8 1-2	15	
September 15..	7	8	15	
September 16..	6 1-2	8	14 1-2	
September 17..	7	8	15	
September 18..	6 1-4	7 1-2	13 3-4	
September 19..	7	8	15	
September 20..	6	7 1-2	13 1-2	
September 21..	5 3-4	6 1-2	13 1-4	Very warm.
September 22..	6 1-2	7	13 1-2	
September 23..	6	7 1-2	13 1-2	
September 24..	5 3-4	8	13 3-4	
September 25..	6	7 1-2	13 1-2	
September 26..	6	8	14	
September 27..	6 1-2	7	13 1-2	Warm.
September 28..	6	....	6	

Condensed milk, analysis of; the analysis of milk from different animals and cream analysis.

GOAT'S MILK.

No.	Remarks.	Date of investigation.	Per cent of water.	Per cent of caseine.	Per cent of albumen.	Per cent of fat.	Per cent of milk-sugar.	Per cent of ash.	In dry substance. Per cent of nitrogen	In dry substance. Per cent of fat.	Analyst.
1	Morning	?	87.24	4.62		3.76-	4.88	0.89	5.81	29.47	Gorup Besanez.
2	Evening	"	82.25	4.31		9.88	4.05	0.82	3.89	52.80	
3	....	"	87.28	3.89		3.45	4.62	0.76	4.87	27.80	E. Felhal and N. Joly.
4	....	"	87.30	3.50	1.35	4.40	3.10	0.85	6.14	34.64	Doyene.
5	Morning	1856	84.89	......	......	4.61			......	30.51	Wicke.
6	Noon	1856	84.96	......	......	4.95	4.42	0.74	......	32.91	
7	Night	1856	84.44	......	......	5.22			......	33.58	
8	....	1868	......	2.85	1.00	5.87	4.25	......	......	......	Nast.
9	....	1868	......	3.15	1.50	5.85	4.28	......	......	......	
10	35 days after [illegible]	?	86.75	2.98	0.94	3.94	4.30	0.74	4.75	29.73	Commaille
11	1 year after calving	1868	83.59	3.65	0.93	6.15	4.98	0.70	4.45	37.48	
12	1 [illegible]nth after [illegible]	1868	85.61	3.00	0.79	4.11	5.72	0.77	4.24	28.56	
13	[illegible]al food	1868	87.84	2.95		3.37	......	......	3.86	31.82	F. Stohmann, O. Barber, R. Lehde.
14	Normal food	1868	88.89	2.75		3.57	......	......	3.79	30.75	
15	Normal food	1868	88.45	2.76		3.36	4.56	0.87	3.84	29.84	
16	([illegible]on of oil	1868	88.01	2.87		3.71	4.52	0.89	3.83	30.94	
17	Fodder poor in fat	1868	89.10	2.93		2.87	4.00	1.10	4.81	26.34	
18	Addition of albumen	1868	89.11	3.34		2.52	3.82	1.21	4.86	23.14	
19	Normal food	1868	87.75	3.51		3.48	4.19	1.07	4.57	28.41	
20	Normal food	1868	87.65	3.78		3.44	3.77	1.36	4.86	27.85	
21	([illegible]on of oil	1868	87.42	4.12		3.43	3.97	1.06	5.24	27.26	
22	Normal food	1868	87.65	3.07		3.76	......	......	3.96	30.44	
23	Normal food	1868	87.81	2.86		3.67	......	......	3.77	30.10	
24	Addition of oil	1868	87.62	3.03		3.74	4.77	0.84	3.88	30.21	
25	Normal food	1868	88.13	3.06		3.89	4.55	0.87	4.12	28.56	
26	Normal food	1868	87.85	3.16		3.47	4.62	0.90	4.19	28.56	

GOAT'S MILK—(*Continued.*)

No.	Remarks.		Date of investigation.	Per cent of water.	Per cent of caseine.	Per cent of albumen.	Per cent of fat.	Per cent of milk-sugar.	Per cent of ash.	In dry substance. Per cent of nitrogen.	In dry substance. Per cent of fat.	Analyst.
27	Fodder poor in fat		1868	88.98	3.28		2.48	4.29	0.97	4.72	22.50	F. Stohman, O. Barber, R. Lehde.
28	Addition of albumen		1868	87.55	3.85		3.03	4.33	1.24	4.98	24.83	
29	Normal food		1868	87.22	4.09		3.28	4.25	1.16	5.08	25.66	F. Stohmann, R. Frühling, A. Post.
30	Large addition of starch		1868	87.00	4.88		3.29	4.41	0.96	5.81	25.31	
31	[illegible]ure		1870	88.53	2.38		3.77	3.00	0.76	3.32	32.87	
32	Pasture		1870	88.97	2.79		3.00	4.38	0.85	4.05	27.20	
33	Hay and starch	1 goat	1870	87.71	2.47		3.86	4.70	0.75	3.50	29.76	Ditto
34	Hay and starch	2 goats	1870	89.36	2.96		2.47	4.40	0.81	4.45	23.21	Ditto
35	Hay and oil	1 goat	1870	87.97	2.75		3.96	4.51	0.81	3.66	32.92	Ditto
36	Hay and oil	2 goats	1870	88.57	3.10		3.18	4.19	0.90	4.34	27.82	Ditto
37	Hay	1 goat	1870	86.24	3.08		5.23	4.59	0.87	3.58	38.01	Ditto
38	Hay	2 goats	1870	87.76	3.27		3.61	4.51	0.87	4.27	29.49	Ditto
39	Hay and sugar	1 goat	1870	86.66	3.27		4.60	5.54	0.92	3.92	34.48	Ditto
40	Hay and sugar	2 goats	1870	83.61	3.46		2.47	4.60	0.86	4.87	21.69	Ditto
41	Hay	1 goat	1870	85.81	3.65		5.61	4.48	0.91	3.98	38.19	Ditto
42	Hay	2 goats	1870	87.04	3.71		3.84	4.52	0.89	4.58	29.68	Ditto
43	One hundred parts hay and one hundred parts oil cake.	April 14	1873	87.68	3.25		3.67	4.61	0.84	4.20	29.67	F. Stohmann, R. Frühling, O. Claus, P. Peterson, V. Seebach.
44		April 15	1873	86.93	3.88		4.05	4.80	0.84	4.14	31.01	
45		April 16	1873	86.88	3.88		3.70	5.25	0.95	4.11	28.12	
46		April 18	1873	86.86	3.50		4.40	4.79	0.95	4.11	32.26	
47		April 19	1873	86.39	3.81		4.04	4.81	0.95	4.48	29.68	
48		April 20	1873	85.61	4.00		4.69	4.75	0.95	4.45	32.59	
49		April 21	1873	86.49	3.94		3.73	4.89	0.95	4.67	27.61	
50		April 28	1873	84.02	4.81		5.73	4.49	1.00	4.82	35.86	
51		April 30	1873	85.83	4.31		4.43	4.93	1.00	4.70	30.20	
52		May 2	1873	84.59	4.51		5.13	4.72	1.00	4.68	33.29	
53		May 3	1873	84.95	4.63		4.96	4.46	1.00	4.92	32.96	
54		May 4	1873	95.63	4.63		4.23	4.51	0.56	5.16	29.46	
55	Mean of several analyses		1869	84.53	6.59		3.35	4.99	0.60	6.84	21.73	Meymott Tidy.
56	Upper Egypt		1869	87.99	2.44	0.99	4.24	3.73	0.62	4.57	35.30	

57	Paris and suburbs	1869	84.49	5.52		5.63	3.69	0.62	5.69	36.62	Becquerel Vernois.
58	Samen	1869	85.95	2.66	1.18	5.38	4 21	0.65	4.37	38.28	
59	Samen	1869	89.23	2.41	1.53	3.00	3.19	0.60	5.85	27.85	
60	Schwyz	1869	87.81	2.45	1.60	3.84	3.69	0.84	5.32	31.50	
61	Thibet breed	1869	85.65	2.45	1.32	5.54	4.34	0.70	3.09	38.60	
62	Short haired	1879	82.02	4.87		7.02	5.28	1.01	4.33	33.48	A. Voelcker.
63	Long haired Pyrenees	1879	84.48	3.94		6.11	4.68	0.79	4.06	39.87	
64	Hornless goats	1879	83.51	3.19		7.34	5.19	0.77	3.10	44.51	
65	1.027 Sp. Gr.	1879	86.40	3.30		4.40	5.10	0.80	3.88	32.35	Siedam-grotzky, Hofmeister.
66	1.030 Sp. Gr.	1879	88.20	3.40		2.80	4.75	0.85	4.61	23.73	
67	1.031 Sp. Gr.	1879	89.16	2.79		2.57	4.55	0.93	4.12	23.71	
68	1.030 Sp. Gr.	1879	89.90	2.67		2.20	4.46	0.77	4.28	21.78	
69	1.032 Sp. Gr.	1879	90.04	2.76		2.10	4.40	0.80	4.43	20.08	
70	1.026 Sp. Gr.	1879	90.52	2.71		2.11	3.86	0.78	4.57	22.26	
71	1.030 Sp. Gr.	1879	85.80	3.23		5.77	4.40	0.80	3.64	40.63	
72	1.028 Sp. Gr.	1879	87.40	3.18		3.96	4.60	0.86	4.04	31.48	
73	1.031 Sp. Gr.	1879	88.60	2.79		3.07	4.75	0.79	3.92	26.98	
74	1.031 Sp. Gr.	1879	89.08	2.87		2.57	4.63	0.90	4.19	23.48	
75	1.032 Sp. Gr.	1879	89.60	2.66		2.18	4.65	0.91	4.09	20.96	
76	1.031 Sp. Gr.	1879	90.15	2.56		1.63	4.81	0.85	4.16	16.55	
77	1.0289 Sp. Gr.	1882	88.49	4.22	0.30	3.74	4.12	0.77	6.28	32.49	J. Moser.
78	1.0278 Sp. Gr.	1882	88.45	3 29	0.18	3.97	4.81	0.76	4.77	34.87	
79	1.0289 Sp. Gr.	1882	89.27	2.95	0.28	3.32	4.21	0.76	4.82	30 94	
80	1.0281 Sp. Gr.	1882	88.15	4.99	0.51	4.17	4.24	0.74	7 43	35.19	
81	1.0289 Sp. Gr.	1882	88.47	4.68	0.20	3.75	4.38	0.76	6.77	32.52	
82	1.0299 Sp. Gr.	1882	89 60	2.26	0.19	2.98	4.14	0.81	3.77	28.65	
83	1.0284 Sp. Gr.	1882	88.36	2.45	0.47	4.10	4.10	0.79	4.01	35.22	
84	1.0286 Sp. Gr.	1882	88.01	2.64	0.43	4.22	4.22	0.75	3.83	35.86	
85	1.0276 Sp. Gr.	1882	88.88	3.10	0.13	4.34	3.52	0.71	4 65	31-65	
86	1.0291 Sp. Gr.	1882	89.85	2.98	0.13	4.45	3.05	0.77	4.67	28.64	
87	1.0277 Sp. Gr.	1882	87.89	3.67	0.16	4.30	4.40	0.75	5.06	36.33	
88	1.0276 Sp. Gr	1882	87.92	2.79	0.32	4.55	4.48	0.72	4.12	36.67	
89	1.0276 Sp. Gr.	1882	88.19	3.14	0.15	4.59	4.09	0.75	4.46	34.63	
90	1.0284 Sp. Gr.	1882	88.43	3.00	0.16	4.52	3.98	0.73	4.37	34.40	
91	1.0279 Sp. Gr.	1878	90.16	2.95		2.29	3.97	0.73	4.60	23.27	N. Gerber and Radenhausen.
	Minimum	....	82.02	2.16 / 2.47	0.14	2.09	3.10	0.35	3.10	16.55	
	Maximum	....	90.52	5.33 / 5.88	1.80	6.68	6.19	1.40	7.43	52.80	
	Average	....	87.33	3.01 / 3.52	0.51	3.94	4.39	0.82	4.51	30.59	

SHEEP'S MILK.

No.	Remarks.	Date of investigation.	Per cent of water.	Per cent of caseine.	Per cent of albumen.	Per cent of fat	Per cent of milk-sugar.	Per cent of ash.	In dry substance. Per cent of nitrogen.	In dry substance. Per cent of fat.	Analyst.
1	....	Within 50 years.	81.60	4.00	1.70	7.50	4.30	0.90	4.09	40.76	Doyen, 1.
2	Dishley sheep............	Within 50 years.	81.00	7.70		5.00	5.80	0.70	6.32	26.82	Felhal and Joly
3	Dishley sheep............	Within 50 years.	82.50	7.90		8 70	5.35	0.55	7.22	21.14	Felhal and Joly
4	Southdown..................	Within 50 years.	84.20	9.02		7.60	4.37	0.61	6.68	35.19	Felhal and Joly
5	Merino.................. ...	Within 50 years.	78.40	6.50		4.00	4.61	0.69	6.58	25.82	Felhal and Joly
6	Laurajais	Within 50 years.	76.90	8.80		10.40	4.16	0.16	5.77	45.13	Felhal and Joly
7	Tarascon....................	Within 50 years.	77.28	8.05		10.40	4.16	0.16	5.69	45.67	Felhal and Joly
8	Sheep from neighb'd of Paris.	Within 50 years.	83.28	6.98		5.18	3.94	0.72	6.66	30.59	Vernois & Becquerel, 3.
9	Merino......................	Within 50 years.	82.40	4.50		8.29	3.31	0.64	4.09	47.10	Vernois & Becquerel, 3.
10	....	Within 50 years.	83.12	4.18	1.13	5.87	4.49	0.92	5.03	31.81	Commaille, 4.
11	Bergamaker	1875	82.41	5.97		6.89	4.21	0.52	5.42	39.17	Rosell, 5.
12	....	1857	84.01	5.67		4.74	4.83	0.75	5.67	29.64	Bourchardat & Quevenne, 6.
13	Sp. Gr. 1.0416	1861	87.02	4.83		2.86	5.41	0.89	5.95	18.18	H. Grouven, 7.
14	Sp. Gr. 1.0390	1861	82.24	5.88		6.34	5.05	0.91	5.80	35.70	H. Grouven, 7.
15	Three days after ahbing..	1865	76.70	13.87		1.20	7.10	1.68	9.18	5.15	A. Voelcker, 8.
16	Three days after ahbing..	1865	83.10	5.76		4.45	5.73	0.96	5.44	26.83	A. Voelcker, 8.
17	Mean of herd of 300. Sp. Gr. 1.0375.	1877	75.43	7.19	1.46	11.91	3.26	1.07	5.64	48.48	W. Fleischmann and P. Vieth, 9.
18	Mean of herd of 300. Sp. Gr. 1.0361.	1877	77.15	6.08	1.59	10.64	3.64	1.03	5.46	46.56	W. Fleischmann and P. Vieth, 9.
19	Mean of herd of 300. Sp. Gr. 1.0372.	1879	75.43	6.17	1.62	11.78	4.03	1.02	5.07	47.74	W. Fleischmann and P. Vieth, 9.
20	Mean of herd of 300. Sp. Gr. 1.0371.	1880	74.59	6.59	1.85	11.95	3.94	1.08	5.31	47.03	W. Fleischmann and P. Vieth, 9.
21	Ordinary milk from English sheep.	1879	83.70	5.16		4.45	5.73	0.96	5.07	27.30	A. Voelcker, 10.
22	Ordinary milk from English sheep.	1779	75.00	6.58		12.78	4.66	0.93	4.21	51.12	A. Voelcker, 10.
23	Ordinary milk from English sheep.	1879	86.70	4.44		3.67	4.00	1.19	5.34	27.60	A. Voelcker, 10.
24	Ordinary milk from English sheep.	1879	86.12	5.59		2.16	4.93	1.20	6.44	15.56	A. Voelcker, 10.
25	Ordinary milk from English sheep. Sp. Gr. 1.042..	1879	84.15	5.91		2.32	6.57	1.05	5.96	14.64	A. Voelcker, 10.
26	Ordinary milk from English sheep. Sp. Gr. 1.031..	1879	79.02	4.56		10.24	5.19	0.99	3.48	48.81	A. Voelcker, 10.
27	Ordinary milk from English sheep. Sp. Gr. 1.036..	1879	84.24	4.81		4.78	5.80	0.87	4.88	30.88	A. Voelcker, 10.
28	Ordinary milk from English sheep. Sp. Gr. 1.041..	1879	84.78	5.37		3.65	5.46	0.79	5.63	23.90	A. Voelcker, 10.
29	Colostrum, Sp. Gr. 1.063	1879	69.74	17.37		2.75	8.85	1.29	9.18	9.09	A. Voelcker, 10.

30	Merino, four days after lambing, Sp. Gr. 1.0338..	188	80.72	3.61	0.83	8.90	3.24	0.77	3.68	46.17	F. Sterhmer.
.	Minimum..............	....	74.59	3.54 4.06	0.79	2.74	2.48	0.23	3.48	14.64	
	Maximum	....	87.02	5.42 8.43	1.71	9.55	7.74	1.61	7.22	51.12	
	Average...............	....	81.31	5.28 6.31	1.03	6.83	4.73	0.82	5.41	36.55	
..	Sheep's milk colostrum......	......	73.22	15.87		1.97	7.88	1.46	9.18	7.12	
	LAMA'S MILK.										
1	Mean of three analyses	......	86.55	3.00	0.90	3.15	5.60	0.80	4.64	23.42	Doyen.
	CAMEL'S MILK.										
1	....	......	86.94	3.67		2.90	5.87	0.66	4.50	22.21	Dragendorff.
2	....	......	......	4.00		......	5.80	......	......	......	Chatin.
3	Sp. gr. 1.0353...............	1877	86.21	3.96	0.38	3.23	5.81	0.91	5.04	23.43	Marchetti.
..	....	......	86.57	4.00		3.07	5.59	0.77	4.77	22.82	
	ELEPHANT'S MILK.										
1	One month after calving......	1880	66.99	3.21		22.07	7.40	0.63	1.54	66.26	Doremus.
	MARE'S MILK.										
1	Steppe mare... } Kirghiz ...	1871	......	....		2.12	7.26	......	......	23.55	Stahlberg.
2	Working mare. } Kirghiz ...	1871	......	....		2.45	5.95	......	......	27.22	Stahlberg.
3	Mean of fourteen analyses....	1875	90.81	1.95		1.06	6.26	0.39	3.22	10.94	Cameron.
4	....	1872	91.47	0.70	1.40	0.55	5.50	0.40	4.09	6.45	Doyen.
5	Sp. gr. 1.040	?	92.20	1.99		0.50	4.20	1.20	3.90	6.41	Hering.
6	Tartar mares	1871	92.48	1.33	0.36	0.65	4.72	0.29	3.60	8.66	J. Moser.
7	Steppe mares.................	?	90.26	1.82	1.03	1.26	5.34	0 29	3.04	12.94	Biel.
8	Steppe mares.................	1871	90.62	1.82	0.96	1.11	5.21	0.28	4.74	11.33	Biel.
9	Steppe mares.................	1871	90.88	1.31	0.71	1.56	5.73	0.31	3.36	16.22	Biel.
10	....	1871	89.29	1.59	0.28	1.16	7.16	0.36	2.79	10.83	Landosisky.
11	English blooded stock, 1.0345,	1879	91.49	0.87	0.46	0.12	6.41	0.33	2.50	1.41	P. Vieth.
12	1.0276......................	1879	92.53	0.75	0.83	0.36	4.69	0.49	3.38	4.82	P. Vieth.
13	Mare, five years old..........	1879	91.15	1.50		1.27	5.75	0.37	2.71	14.35	M. Schrodt.
14	....	1856-8	89.05	3.00		2.15	5.20	0.60	4.38	19.63	Filhol and Joly.
15	Twelve-year-old brown mare, 1.036......................	1878	91.76	2.45		0.39	5.99	0.31	4.75	4.78	N. Gerber and Radenhausen.
	Average	......	91.00	1.32 2.08	0.76	1.18	5.31	0.43	3.57	13.11	

ASS' MILK.

Number.	REMARKS.	Date of investigation.	Per cent of water.	Per cent of caseine.	Per cent of albumen.	Per cent of fat.	Per cent of milk sugar.	Per cent of ash.	In dry substance. Per cent of nitrogen.	In dry substance. Per cent of fat.	Analyst.
1	....	1846	89.63	0.60	1.55	1.50	6.40	0.32	3.32	14.46	Doyen.
2	Mean of fourteen analyses....	1846	90.47	1.95		1.29	6.29		3.27	13.54	Peligot.
3	....	1846	90.70	1.67		1.21	6.23		2.87	13.01	Simon.
4	Mean of several analyses.....	1857	89.36	2.26		1.37	7.04		3.40	12.88	Bourchadat & Quevenne.
5	Milk of five asses.	1878	88.03	3.08		2.82	5.29	0.78	4.12	23.56	Frühling & Schultz.
	Average	......	89.64	0.67 2.22	1.55	1.64	5.99	0.51	3.39	15.49	

SOW'S MILK.

Number.	REMARKS.	Date of investigation.	Per cent of water.	Per cent of caseine.	Per cent of albumen.	Per cent of fat.	Per cent of milk sugar.	Per cent of ash.	In dry substance. Per cent of nitrogen.	In dry substance. Per cent of fat.	Analyst.
1	Native Essex after five weeks sucking	1856	85.49	8.45		1.93	3.03	1.09	9.32	13.30	Scheven.
2	Native Essex after five weeks sucking.........	1856	88.17	7.36		1.08	2.26	1.18	9.97	8.71	Scheven.
3	Six days after birth..........	1865	80.43	12.89		3.14	2.79	0.71	10.54	16.00	Th. V. Gothen.
4	Nineteen days after birth.. ..	1865	89.26	5.68		2.82	1.59	0.87	8.46	26.26	Th. V. Gothen.
5	....	1866	82.98	6.89		6.88	2.01	1.29	3.46	40.30	Lintner.
6	....	1866	81.80	5.80		6.00	6.07	0.83	4.66	32.97	Cameron.
7	Mean of two analyses	1866	81.76	6.18		5.38	5.84	0.89	5.42	29.50	Ditto.
8	....	1866	82.46	5.09		9.23	1.69	1.53	4.64	52.62	Jern.
	Average	......	84.04	7.23		4.55	3.13	1.05	7.33	27.6	

BITCH'S MILK.

Number.	REMARKS.	Date of investigation.	Per cent of water.	Per cent of caseine.	Per cent of albumen.	Per cent of fat.	Per cent of milk sugar.	Per cent of ash.	In dry substance. Per cent of nitrogen.	In dry substance. Per cent of fat.	Analyst.
1	....	1868	......	5.52	2.99	10.77	3.05	......	......	......	Tolmatscheff.
2	....	1868	......	3.94	3.97	12.84	3.37	......	......	......	Tolmatscheff.

3	Ten days after birth..........	1838	65.74	17.40		16.20	2.90	1.50	8.13	47.29	Simon.
4	Later	1838	68.20	14.60		13.80	3.00	1.43	7.40	41.82	
5	Food, bread, mat and bones..	1845	69.80	13.60		12.40	2.50	0.77	7.21	41.06	Dumas.
6	Food, horse eat	1845	77.14	11.15		7.32	3.39	0.57	7.80	32.02	
7	(M, horse meat	1845	74.74	....		5.15	......	......	......	20.39	Dumas.
8	Food, bread and meat........	1845	85.90	....		3.09	......	......	......	16.35	
9	Food, bed and meat........	1845	75.90	....		6.84	......	......	......	28.28	
10	Exclusively bread............	1845	78.40	14.50		7.90	4.20	......	8.72	24.70	
11	Large dog, eight days with meat	1847	75.54	....		10.75	......	......	......	43.95	Bensch.
12	M, five days fed with meat	1847	77.52	....		10.95	......	......	......	48.71	
13	Man on meat.................	1866	77.26	5.19	3.97	10.64	2.49	0.44	6.4	6.79	Scubatin.
14	Man on potato.................	1866	82.95	4.25	3.92	4.98	3.41	0.47	7.67	29.21	
15	Man on fat	1866	77.37	5.92	4.25	10.11	2.14	0.39	7.19	44.68	
16	One day without food........	1866	79.45	4.28	3.97	9.82	2.06	0.42	6.42	47.79	
	Average	......	75.44	6.10 11.17	5.05	9.57	3.09	0.73	7.36	38.49	

CAT'S MILK.

1	....	58.59	81.63	3.12	5.96	3.88	4.91	0.58	7.91	13.13	A. Commalle.

CONDENSED MILK, WITHOUT ADDITION OF CANE SUGAR.

No.	REMARKS.	Date of investigation.	Per cent of water.	Per cent of caseine and albumen.	Per cent of fat.	Per cent of milk-sugar.	Per cent of cane-sugar.	Per cent of ash.	In dry substance. Per cent of nitrogen.	In dry substance. Per cent of fat.	Analyst.
1	From [illegible]y, U. S. A.	1871	53.54	14.44	13.12	16.80	......	2.60	4.95	28.24	C. F. Chandler and Schweitzer.
2	From [illegible]y, U. S. A.	1871	51.50	13.61	14.51	17.47	......	2.91	4.49	29.92	
3	From Purdy, U. S. A.	1871	49.28	15.43	14.58	17.75	......	2.96	4.88	28.72	
4	American	1871	50.40	17.80	14.20	15.60	......	2.00	5.74	28.63	S. Percy.
5	American	1871	61.00	10.60	11.20	15.70	......	1.50	4.35	28.72	
6	[illegible]an	1871	46.40	19.10	19.80	12.50	......	2.20	5.70	36.94	
7	American	1871	36.20	30.30	20.50	10.80	......	2.20	7.60	32.13	
8	American	1871	41.20	28.20	13.60	14.00	......	3.00	7.67	23.13	
9	American	1871	40.50	26.50	17.70	12.80	......	2.50	7.13	29.75	
10	Alpine milk	1881	62.84	11.89	11.11	12.08	......	2.24	4.90	29.89	F. Strohmer.
	Average		53.59	14.62	14.07	15.38	......	2.34	5.03	30.25	

CONDENSED MILK, WITH ADDITION OF CANE SUGAR.

No.	REMARKS.	Date of investigation.	Per cent of water.	Per cent of caseine and albumen.	Per cent of fat.	Per cent of milk-sugar.	Per cent of cane-sugar.	Per cent of ash.	In dry substance. Per cent of nitrogen.	In dry substance. Per cent of fat.	Analyst.
1	From [illegible], Switzerland	1868	24.13	13.67	8.67	10.82	40.48	3.28	2.88	11.43	C. Karmrodt.
2	From Cham, Switzerland	1873	26.23	9.48	8.34	53.89		2.01	2.05	11.31	J. Mosera, F. Soxhlet.
3	Fom [illegible], Switzerland	1873	28.90	10.16	9.35	54.59		2.00	2.14	12.29	
4	From [illegible], Switzerland	1873	24.70	9.77	6.02	57.40		2.11	2.08	8.00	
5	Fom [illegible], Swi[illegible]and	1873	26.95	10.56	9.18	51.13		2.23	2.31	12.50	J. Forster.
6	From [illegible], Switzerland	1872	27.30	8.00	9.26	52.69		2.25	1.77	12.83	P. Wagner.
7	From Cham, Switzerland	1875	25.95	13.11	10.46	48.32		2.15	2.83	14.13	N. Gerber.
8	Fom [illegible], Switzerland	1876	28.24	9.41	8.64	51.56		2.13	2.10	12.04	
9	From [illegible], Switzerland	1876	26.86	10.85	9.99	49.58		1.71	2.37	13.66	
10	Fm Cl[illegible]m, Switzerland	1877	23.51	11.00	10.62	52.96		2.11	2.30	13.88	
11	From [illegible], Switzerland	1876	25.50	13.10	10.00	50.80		1.70	2.60	13.42	A. Hutschesen-Smee.
12	From [illegible], Switzerland	1876	22.50	12.30	10.50	52.90		1.80	2.54	13.55	
13	From [illegible], Swi[illegible]zland	1876	20.50	[illegible]	10.80	54.10		1.90	2.56	13.58	
14	From [illegible], Switzerland	1879	23.48	11.85	9.70	11.95	41.41	2.11	2.50	12.68	N. Gerbert.
	Average	......	25.02	11.03	9.39	11.21	41.32	2.08	2.36	12.59	

15	From Vivis, Kempen	1872	23.40	10.00	13.83	50.74		2.03	2.09	18.05	P. Wagner
16	From Vivis, Kempen	1877	22.07	18.00	11.93	45.26		2.74	3.70	15.31	Wendler.
	Average	....	22.74	14.00	12.88	48.00		2.38	2.89	16.68	
17	Luxburg Alpina..............	1873	28.88	8.81	12.61	46.55		2.23	1.97	17.61	E. Shultze.
18	Luxburg Alpina..............	1873	29.09	7.79	10.48	51.83		2.07	1.76	14.71	E. Kopp.
19	Luxburg Alpina..............	1872	24.70	8.81	12.45	51.87		2.17	1.87	16.53	P. Wagner.
20	Luxburg Alpina..............	1875	20.93	9.62	18.78	49.69		1.96	1.95	23.75	} N. Gerber.
21	Luxburg Alpina..............	1876	24 12	10.88	10.27	52.84		1.94	2.29	13.58	
	Average	....	25.44	9.98	12.91	50.40		2.07	1.97	17.23	
22	Vevey, Switzerland	1872	23.40	10.00	13.83	50.74		2.08	2.09	18.05	P. Wagner.
23	Erland Nestle...............	1879	25.28	10.25	8.62	53.82		2.08	2.19	154	F. Soxhlet.
24	Erland Nestle...............	1879	24.75	12.67	11.58	11.19	37.69	2.17	2.69	15.82	N. Gerber.
	Average.................	....	24.48	10.97	11.83	11.19	39.95	2.08	2.82	14.97	
25	Cham Nestle...............	1880	25.35	11.26	9.37	9.47	42.47	2.08	2.41	12.42	L. Janke.
26	Thun..................	1877	35.66	16.35	14.63	30.18		3.12	4.07	22.81	} N. Gerber.
27	Thun.........................	1878	26.05	12.46	10.42	11.04	38.19	1.89	2.70	14.09	
	Average	....	30.86	14.41	12.55	10.81	29.76	2.51	3.39	18.45	
28	Gruyere, Switzerland.........	1878	29.80	8.81	12.61	46.55		2.23	2.01	17.96	E. Stie.
29	Sassie	1868	12.48	17.59	18.81	48.14		3.58	3.21	20.91	V. Gohren.
30	Werchritz.......	1868	28.63	10.85	12.18	.....		......	2.48	7107	K. Eeckhorn.
31	Werchrilz.......	1867	21.50	10.20	12.90	52.90		2.50	2.08	15.52	E. Peters.
32	Ipk....	1873	24.53	10.97	11.17	50.06		3.27	2.35	14.80	} J. Mr, F. Soxhlet.
33	Innspruck....................	1873	27.38	9.24	9.61	17.25	34.40	2.12	2.04	13.23	
34	Vienna..........	1879	24.26	10.82	9.68	53.18		2.16	2.29	12.71	
35	Milan....	1879	26.88	11.07	8.67	51.12		2.26	2.42	11.86	F. Soxhlet.
36	Milan.......................	1878	25.21	14.67	9.21	18.42	35.48	2.08	3.18	12.81	N. Gerber.
37	Milan.......	1880	26.75	9.95	8.81	15.13	37.16	2.05	2.17	12.03	J. Martenson.
38	Moo	1878	21.67	15.86	9.15	13.48	36.23	2.61	3.24	11.68	N. Gerber.
39	Hamburg.....................	1871	15.45	12.76	11.52	16.17	34.65	2.45	3.74	13.63	Schaedler.
40	Norway......................	1875	32.80	8.13	9.80	41.25		3.01	3.13	14.58	} N. Gerber.
41	Norway......................	1876	35.66	16.35	14 68	30.18		3.12	4.07	22.82	
42	Norway......................	1879	30.08	9.02	7.54	51.85		2.01	2.06	10.79	F. Soxhlet.
43	London......................	1879	25.56	12.39	9.90	10.18	40.10	1.87	2.66	18.16	} N. Gerber.
44	Wiltshire....................	1879	24.89	13.08	10.64	13.31	35.47	2.61	2.79	14.17	

CONDENSED MILK — (*Continued*).

Number.	REMARKS.	Date of investigation.	Per cent of water.	Per cent of caseine and albumen.	Per cent of fat.	Per cent of milk-sugar.	Per cent of cane sugar.	Per cent of ash.	In dry substance. Per cent of nitrogen.	In dry substance. Per cent of fat.	Analyst.
45	English origin not further known.	1876	24.30	18.52	10.80	16.50	27.11	2.77	3.91	14.27	Arthur Hill, Hassall.
46		1876	27.00	17.20	11.30	12.00	29.59	2.91	3.77	15.48	
47		1876	26.50	16.39	9.50	17.54	27.06	3.09	3.55	12.93	
48		1876	29.94	15.36	9.50	15.36	32.14	2.43	3.27	12.66	
49	New York Gail Borden.......	1875	27.72	9.92	8.61	51.84		1.81	2.20	11.91	N. Gerber.
50	New York Alderney..........	1876	28.38	10.22	9.23	51.57		1.56	2.28	12.89	The Same.
51	New [illegible] Cond. Milk Co	1878	25.43	11.85	7.01	10.11	44.22	1.89	2.44	9.40	V. Goddefroy.
52	[illegible]ed goat milk.........	1880	20.98	17.00	16.95	15.72	26.75	2.60	3.44	21.45	
	Minimum..........		15.45	8.20	5.96	10.11	......	1.62	1.76	8.00	
	Maximum		30.08	18.96	17.01	17.77	......	3.62	4.07	22.82	
	[illegible] average.		25.43	12.15	10.78	13.48	35.89	2.27	2.61	14.47	

CREAM.

Number.	REMARKS.	Date of investigation.	Per cent of water.	Per cent of caseine and albumen.	Per cent of fat.	Per cent of milk-sugar.	Per cent of cane sugar.	Per cent of ash.	In dry substance. Per cent of nitrogen.	In dry substance. Per cent of fat.	Analyst.
1	....	1860	83.23	4.24	8.17	3.02	......	......	4.05	48.72	F. Hoppe Seyler.
2	....	1860	80.42	4.29	10.84	3.74	......	......	3.51	55.36	
3	....	1860	82.36	4.16	9.76	......	......	......	3.77	55.33	
4	....	1863	59.25	2.20	35.00	3.05	......	0.50	0.86	85.89	Alex. Müller.
5	....	1863	52.51	2.69	41.16	3.19	......	0.45	0.91	86.67	
6	....	1863	70.41	......	23.05	5.95	......	0.59	......	77.89	
7	....	1863	59.92	2.60	33.55	3.80	......	0.62	1.04	83.71	
8	....	1863	60.10	2.78	33.56	2.99	......	0.63	1.09	84.11	The Same.
9	....	1867	66.73	3.18	25.04	4.84	......	0.71	1.53	75.26	
10	....	1867	71.80	2.75	20.94	3.96	......	0.55	1.56	74.26	
11	....	1867	74.20	2.66	17.93	4.57	......	0.65	1.65	69.50	
12	Obtained by Dev'nshire system	1849	56.04	6.02	34.38	2.67	......	0.88	2.19	78.21	Orthmann.

CREAM — (*Continued*).

13		1849	22.82	4.10	70.20	2.31	......	0.56	0.85	90.97	Gerard.
14		1859	68.28	4.22	2946	2.08	0.56	0.40	1.86	81.88	Hamberg.
15		1863	74.46	2.69	18.18	4.08	......	0.59	1.69	71.18	A. Voelcker.
16		1863	4680	......	240	......	......	2.19	......	72.16	A. Voelcker.
17		1863	56.50	......	81.57	......	......	3.49	......	72.57	A. Voelcker.
18		1863	61.67	2.62	88.48	1.56	......	0.72	1.09	87.22	A. Voelcker.
19		1863	79.52	......	15.56	......	......	0.63	......	5798	Gerard.
20	Skimmed at 60 deg. after 16 h.	1875	77.45	2.86	14.81	......	5.88	......	2.03	68.46	U. Kreussler.
21	Skimmed at 60 deg. after 28 h.	1875	777	3.17	15.07	......	5.39	......	2.24	66.55	U. Kreussler.
22	Skied at 60 deg. after 40 h.	1875	75.59	3.01	17.41	......	3.99	......	1.97	71.82	U. Kreussler.
23	Sied at 8 deg. fter 16 h.	1875	77.46	3.58	13.24	......	5.72	......	2.54	58.74	U. Kreussler.
24	Skimmed at 8 deg. after 28 h	1875	75.73	4.20	16.27	......	3.80	......	2.77	67.04	U. Kreussler.
25	Skimmed at 8 deg. after 40 h.	1875	74.93	2.75	17.07	......	5.25	......	1.76	669	U. Kreussler.
26	Skimmed at 1 deg. after 16 h.	1875	75.89	3.47	15.25	......	5.89	......	2.30	63 25	U. Kreussler.
27	Skimmed at 1 deg after 28 h.	1875	74.82	2.54	17.61	......	5.03	......	1.61	69.94	U. Kreussler.
28	Ten degrees forty hours	1875	72.75	2.48	18.65	......	6.12	......	1.46	68.44	U. Kreussler.
29	Fifteen degrees sixteen hours.	1875	73.46	3.48	17.31	......	5.75	......	2.10	65.60	U. Kreussler.
30	Twenty-eight hours	1875	71.77	3.10	20.45	......	4.68	......	1.76	72.44	U. Kreussler.
31	....	1876	62.12	5.83	30.64	1.27	......	0.14	2.46	80.89	Arth Hill Hassall.
32	....	1876	61 50	5.14	32.22	0.74	......	0.40	2.14	83.69	Arth Hill Hassall.
33	....	1876	63.24	2.70	31.42	2.36	......	0.28	1.18	85.47	Arth Hill Hassall.
34	....	1876	49.10	5.20	42.82	2.46	......	0.42	1.63	84.13	Arth Hill Hassall.
35	....	1876	43.04	7.40	44.76	4.45	......	0.35	2.08	78.23	Arth Hill Hassall.
36	....	1876	45.82	6.38	44.33	2.92	......	0.50	1.88	81.82	Arth Hill Hassall.
37	Lefeld's centrifugal machine..	1878	29.55	1.42	67.68	2.25	......	0.12	0.82	96.00	W. Fleischmann.
38	Laval's.....................	1880	66.12	2.69	27.69	3.03	......	0.47	1.27	81.78	A. Völcker.
39	By Swartz method....... ...	1878	63.40	......	22.08	......	......	......	......	69.57	J. König.
	Maximum	......	22.83	1.33	16.80	0.66	......	0.13	0.85	48.72	
	Minimum................	......	83.23	8.12	83.41	6.15	......	2.79	3.77	96.00	
	Average................	......	65.51	3.61	26.75	3.52	......	0.61	1.66	77.59	

LITERATURE.

Allen, A. H., Commercial Organic Analysis.
Blyth, A. W., Foods, their Composition and Analysis.
Birnbaum, K., Einfache Methoden zur Pruefung wichtiger Lebensmittel auf Verfaelschungen.
Chevallier & Baudrimont, Dictionnaire des alterations et falsifications des substances alimentaries, médicamenteuses et commerciales, avec l'indication des moyens de les reconnaitre.
Dairy Commissioners of New York, Report of, for 1884.
Dietzsch, O., Die wichtigsten Nahrungsmittel und Getraenke, deren Verunreinigungen und Verfaelschungen.
Fleischmann, W., Das Molkereiwesen.
Feser, J., Werth der bestehenden Milchproben.
Frytag, Werner, Eisbaum, Fleischer, Havenstein, Die Kuhmilch, ihre Erzeugung und Verwerthung.
Goeppelraeder, F., Beitrag zur Pruefung der Kuhmilch.
Hassall, Foods, its Adulterations and the Methods for their Detection.
Husson, C., Le lait, la crême et le beurre.
Kirchner, W., Beitraege zur Kenntniss der Kulmilch und ihrer Bestandtheile nach den gegenwaertigen Standpunkten wissenschaftlicher Forschung.
Kronbaum, K., Lehrbuch der rationellen Praxis der landwirthschaftlichen Gewerbe.
Loebner, A., Massregeln gegen Verfaelschung der Nahrungsmittel.
Hartiny, B., Die Milch, ihr Wesen und ihre Erzeugung.
Otto, J., Lehrbuch der landwirthschaftlichen Gewerbe.
Smee, A. H., Milk in Health and Disease.
Vieth, D. P., Milchpruefungsmethoden und die Controle der Milch in Staedten und Sammel-Molkereien.
Wanklyn, J. A., Milk Analysis. A Practical Treatise on the Examination of Milk and its Derivatives, Cream, Butter and Cheese.

ADULTERATIONS.

Analyst, vol. 3, page 41, E. H. Cook; Adulterations, On.
Analyst, vol. 3, page 384, Dr. Bernays; Adulterations, On.
Analyst, vol. 3, page 330, J. W. Gatehouse; Adulterations, On.
Analyst, vol. 3, page 235, D. Muter; Adulterations, Ingenious.
Chemical News, vol. 33, page 95; Adulteration of.
Chemical News, vol. 33, page 135; Adulteration of.
W., vol. 19, page 608; A., (3) vol. 2, page 172; C. Ct. bl., vol. 1873, page 185, Kissmuller; Milchverfaelschungen.
W., vol. 18, page 616; A. Ph., vol. 200, page 45; D. J., vol. 205, page 278; C. Ct. bl., vol. 1872, page 496; Dt. Ind. zt., vol. 1872, page 345, Hirschberg; Aseptien.
Chemical News, vol. 47, page 224; Benzoic and Boracic Acids.
W., vol. 28, page 927; Z. a. Ch., vol. 1882, page 531, E. Meissl; Benzoesaeure und Borsaeure in der Milch.
W., vol. 29, page 968; B., vol. 1883, page 1754, Forster; Borsaeure als Conservationsmittel.
W., vol. 5, page 444, Kletzinsky; Borax, Pruefung auf.
B., vol. 2, page, 333, E. H. Moore; Brighton, Adulterations at.

Analyst, vol. 4, page 12; Cane Sugar, Adulteration with.
Analyst, vol. 4, page 32, T. Stevenson; Cane Sugar, Adulteration with.
Analyst, vol. 4, page 53, J. H. Collins; Cane Sugar, Adulteration with.
Analyst, vol. 4, page 72, O. Hehner; Cane Sugar, Adulteration with.
Analyst, vol. 4, page 194; Cane Sugar, Adulteration with.
Analyst, vol. 5, page 37; Cane Sugar, Estimation of.
Chemical News, vol. 51, page 37; Chloralhydrate, Detection of.
Analyst, vol. 2, page 185; Dangerous Modes of.
Chemical News, vol. 40, page 102; Detection of.
J. Am. C. S., vol. 1, page 57, W. H. Watson; Detection of.
Analyst, vol. 4, page 49, E. A. Cameron; Falsifications.
W., vol. 30, page 1057, Lawrence; Fettstoffen, Milch mit Allzureichen.
Analyst, vol. 1, page 149; How to Adulterate.
Analyst, vol. 4, page 88; Preservatives.
W., vol. 30, page 1057, Loeflund; Milchconserven.
Analyst, vol. 5, page 227; Rain-water, Adulteration by.
Analyst, vol. 7, page 87; Report on, in New York.
Analyst, vol. 2, page 33, J. C. Bell; Salford, Adulteration at.
Analyst, vol. 2, page 47, J. C. Bell; Salford, Adulteration at.
Analyst, vol. 1, page 193, J. Muter; Salicylic Acid, Detection and Estimation of.
Chemical News, vol. 34, page 142; Salicylic Acid.
W., vol. 22, page 897, and vol. 24, page 988, Wagner; Salicylsaeure in der Milch-Wirthschaft.
W., vol. 27, page 836, L. V., vol. 27, page 143; Salicylsaeure, Nachweis.
W., vol. 28, page 927, Z. A. Ch., 1882, page 548, Bachmeyer; Soda, Nachweis von.
Chemical News, vol. 47, page 84; Soda.
Chemical News, vol. 43, page 210; Starch, Detection of.
Chemical News, vol. 39, page 172; Starch in.
W., vol. 69, page 690; D. J., vol. 210, page 477, Hager; Staerkemehl, Pruefung der.
Chemical News, vol. 49, page 126; Watering of Milk, Researches on.
W., vol. 30, page 1054; F. G. d. Vwft, 1884, page 322, Vieth, Wasser-Zusatz zur Milch.
W., vol. 30, page 1054; J. Ph. Ch., vol. 5, page 95, Lambue, Wasser zusatz zur Milch.

Analyses of Milk.

Analyst, vol. 1, page 200; Analysis of Milk at Bottom of Can.
Analyst, vol. 6, page 75, C. A. Cameron; Analyses of Milk.
Analyst, vol. 6, page 59, B. Dyer; Analyses of Milk.
Analyst, vol. 6, page 62, J. C. Bell; Analyses of Milk.
Analyst, vol. 6, page 209; Analysis of Milk at Manchester.
Analyst, vol. 7, page 60; Analyst, vol. 7, page 129; Analyst, vol. 7, page 164, O. Hehner; Analyses of Milk.
Analyst, vol. 7, page 221; Analysis and the Local Governing Board.
Analyst, vol. 7, page 53; Analyst, vol. 7, page 213, Dr. Vieth; Analyses of Milk.

Analyst, vol. 8, page 33, Dr. Vieth; Analyses of Milk during 1882.
Analyst, vol. 8, page 248, Dr. Duprè; Analyses of Milk.
Analyst, vol. 8, page 253, O. Hehner; Analyses of Milk.
Analyst, vol. 8, page 138; Analyses of Milk in Boston.
Analyst, vol. 8, page 256, Dr. Voelcker; Analyses of Milk.
Analyst, vol. 9, page 48, A. Angell; Analyses of Milk.
Analyst, vol. 9, page 186, M. Dechan; Analyses of Milk.
Analyst, vol. 9, page 186, T. Mebane; Analyses of Milk.
Analyst, vol. 9, page 69; Analysis of Milk in New York.
Chemical News, vol. 50, page 176; Analyses of Milk.
Chemical News, vol. 51, page 94; Analyses of Milk.
Chemical News, vol. 43, page 37; Analyses of Milk.
Chemical News, vol. 44, page 52; Analyses of Milk.
Chemical News, vol. 31, page 78; Analyses of Milk.
Chemical News, vol. 35, page 30; Analyses of Milk.
Chemical News, vol. 35, page 44; Analyses of Milk.
Chemical News, vol. 38, page 112; Analyses of Milk.
Chemical News, vol. 38, page 173; Analyses of Milk.
Chemical News, vol. 41, page 230; Analyses of Milk.
W., vol. 2, page 277; J. D. Ch. Méd., vol. 1855, page 577; P. Ct. bl., vol. 1856, page 304, Laconte; Milch Analyses.
W., vol. 21, page 907; A. Ch., vol. 1875, page 150, Gerber; Analyse der Kuhmilch, Analysis of Cows' Milk.
W., vol. 23, page 820; M. Z., vol. 1877, page 370, Truchot; Milchanalysen Verschiedener Kinderrassen.
W., vol. 23, page 820; M. Z., vol. 1877, page 311; D. J., vol. 255, page 517, A. Leclerc; Runkelruebenblaettern Gefuetterten Kuehen.
W., vol. 23, page 822; Ch. Ct. bl., vol. 1877, page 602, G. Christern; Analysen der Milch.
Chemical News, vol. 1877, page 184 (Biedermann's Ct. bl. f. Agric. Ch., vol. 1877), C. Jenssen; Analysen der Milch.
B. Ct. bl. Ag. Ch., vol. 1877, F. Schmidt; Analysen der Milch.
W., vol. 30, page 1050; L. V. st., vol. 31, page 55, Schrodt; Aschen der Kuhmilch.
Ana yst, vol. 4, page 44; Analyst, vol. 4, page 72; W., vol. 25, page 924, O. Hehner; Condensed Milk, Analysis of.
Analyst, vol. 6, page 221, Dr. Voelcker; Condensed Milk, Analysis of.
Analyst, vol. 8, page 171; Condensed Milk, Analysis of.
Chemical News, vol. 50, page 162; Condensed Milk, Analysis — Chemical and Physical; also of Infants' Milk Foods.
Chemical News, vol. 42, page 96; Condensed Milk, Analysis of.
W., vol. 20, page 812; Bull. Soc. En., vol. 1874, page 217; Chemical News, vol. 1874, page 66, A. Muntz; Condensirte Milch, Analyse.
W., vol. 28, page 918; M. Sc., vol. 1882, page 516; Condensirte Milch, Analyse.
Analyst, vol. 9, page 56, Dr. Vieth; Cream, Notes on.
Chemical News, vol. 41, page 102; Woman's Milk, Analysis of.

METHODS OF ANALYSIS.

Analyst, vol. 2, page 206, E. L. Cleever; Analysis at Somerset House.
Analyst, vol. 2, page 225, E. L. Cleever; Analysis, on.
Analyst, vol. 6, page 149, A. W. Blythe; Analysis, Apparatus as Applied to.

Analyst, vol. 9, page 42, A. Angell; Analysis, Discussion of.
Analyst, vol. 9, page 116, Dr. Muter; Analysis and Standards.
Analyst, vol. 9, pages 26, 29 and 210; Analysis at Somerset House.
Analyst, vol. 10, page 30, C. W. Stevens; Analyses of Milk.
Analyst, vol. 10, page 46, M. A. Adams; Analysis, New Method of.
Analyst, vol. 10, page 85, Adams; Analysis, Discussion on A Process.
Chemical News, vol. 45, page 176; Analysis.
Chemical News, vol. 45, page 141; Analysis, Instruments for.
W., vol. 8, page 518; J. f. pr. Ch., vol. 86, page 380, Muller; Analyse der Milch and Butter.
W., vol. 33, page 823; M. Z., 1877, page 221; Z. f. a. Ch., 1877, page 397; D. J., vol. 225, page 405; C. Ct. bl., 1877, page 727; L. Hanetti, S. Musso; Milch Analyse.
W., vol. 20, page 825; J. f. pr. Ch., 1877, page 329; C. Ct. bl., 1877, page 520; D. J., vol. 226, page 418; Chemical News, 1877, page 184; H. Ritthausen; Milch Analyse.
W., vol. 24; page 991; C. R., vol. 87, page 290; B. S. Ch., 1878, page 146; Chemical News, 1878, page 112; C. Ct. bl., 1878, 663; A. Adams; Verfahren zur.
Analyst, vol. 10, page 95, Frenzel and Wey; Caseine, Determination of.
W., vol. 16, page 518; W., J. f. pr. Ph., vol. 19, page 207; D. J., vol., 196; page 161; P. Ct. bl., 1870, page 761; Hess. Gew. bl., 1870, page 228; Dt. Ind. Zt., 1870, page 186; Chemical News, 1870, page 252; Kopler; Condensirte Milch, Untersuchung in.
W., vol. 30, page 1053; M. Z., 1884, page 2; Gerber; Creamometer.
W., vol. 5, page 444; W. Ct. bl., vol. 2, page 408; Fuchs; Rahm, Pruefung auf.
Chemical News, vol. 49, page 214; Examination of Milk.
Chemical News, vol. 40, page 229; Examination of Milk.
W., vol. 1, page 253; J. d. Ph., 1854, page 214; D. J., vol. 134, page 424; P. Ct. bl., 1854, 1392; Rosenthal; Milchproben.
W., vol. 1, page 254; C. R., vol. 28, page 505; J. f. pr. Ch., vol. 47, page 134; D. J., vol. 112, page, 367; Ph. Ct. bl., 1849, page 453; L. & K. J. b., 1849, page 605; Poggiale; Milchpruefung.
W., vol. 1, page 255; P. Ct. bl., 1855, page 639; Ph. N. bl., 1855, page 36; Marchand; Milchprobe.
W., vol. 3, page 331; A. Ch. Ph., vol. 102; page 57; D. J., vol. 144, page 303; Ch. Ct. bl., 1857, page 558; Otto; Milchpruefung.
W., vol. 3, page 333; P. N. bl., 1857, page 365; P. Ct. bl., 1858, page 144; Schw. Zschr. Ph., 1857, page 367; L. Ladé; Milchpruefung, Neue Methode.
W., vol. 4, page 415; C. R., vol. 46, page 236; D J., vol. 147, page 452; P. Ct. bl., 1858, page 624; E. Monier; Milchpruefung.
W., vol. 4, page 416; D. J., vol. 149, page 60; I. Michaelson; Milchpruefung.
W., vol. 4, page 416; D. J., vol. 147, page 132; P. Ct. bl., vol. 1858, page 545; P. N. bl., vol. 1858, page 85; C. Ct. bl., vol. 1858, page 329, C. Brunner; Milchpruefung.
W., vol. 4, page 421; Ch. Tech. M. Berlin, vol. 1859, page 95, Elsner; Milchuntersuchung, Ladé's Methods.
W., vol. 10, page 487; A. Ph., vol. 114, page 127; F. Z., vol. 1864, page 159; J. f. Ph., vol. 21, page 173; P. N. bl., vol. 1864, page 353; Ch. Ct. bl., vol. 1864, page 336, Hoyerman; Milchpruefung.

W., vol. 10, page 518; W. J. sch., vol. 14, page 145; P. Ct. bl., vol. 1867, page 927; D. J., vol. 184, page 529; Z. a. Ch., vol. 1867, page 250; Dt. Ind. zt., vol. 1867, page 328, R. Pibram; Milchuntersuchung, Zur.
W., vol. 13, page 518; P. Ct. bl., vol. 1867, page 1506, Begemann; Milchuntersuchung, Zur.
W., vol. 21, page 908; Am. Ch., vol. 1875, page 412, Voelcker; Milchpruefung.
W., vol. 21, page 908; D. J., vol. 217, page 342; P. Ct. bl., vol. 1875, page 1240, Klinger; Milchuntersuchungen.
W., vol. 21, page 908; Am. Ch., vol. 1875, page 417, MacAdam; Milchuntersuchung.
W., vol. 21, page 908; Ch. News, vol. 1875, page 266, Brown; Examination of milk.
W., vol. 25, page 937; Arch. Ph., vol. 1879, page 211; Ind. bl., vol. 1879, page 401; Ch. Ct. bl., vol. 1879, page 697; B., vol. 1879, page 2100, B. Ohm; Milchpruefung.
W., vol. 25, page 943, Fischer; Milchuntersuchungen.
W., vol. 27, page 836, Soxhlet; Milchuntersuchungen.
W., vol. 28, page 927; F. G. Vht., vol. 1882, page 143, Craine & Baker; Milchpruefungen.
W., vol. 16, page 520; W. Jsch. Ph., vol. 19, page 365; D. J., vol. 197, page 409; Ch. Ct. bl., vol. 1870, page 488, S. Pibram; Milchuntersuchungen.
W., vol. 17, page 670; M. Sc., vol. 1871, page 82, Dubrunfaut; Milchuntersuchung.
W., vol. 29, page 974, Schmid-Mullheim; Milchuntersuchungen.
W., vol. 7, page 479; J. f. pr. Ch., 84, page 145, V. Baumhauer; Milchuntersuchungen.
W., vol. 7, page 479; A. d. Ph., (2) vol. 103, page 15, Schlienkamp; Milchpruefung, Methoden der.
W., vol. 10, page 487; M. Sc., vol. 1864, page 852, Millon & Commaille; Milchuntersuchung, Ausfuehrliche.
W., vol. 10, page 487; A. Ph., vol. 114, page 227; A. Ph., vol. 115. page 26; F. Z., vol. 1863, page 445, Wittstein; Milchpruefung.
Analyst, vol. 10, page 12, Liebermann; Fat, Determination of.
Analyst, vol. 47, page 35; Fat, Determination of.
Analyst, vol. 10, page 55, A. C. Wilson; Fat, Apparatus for Eestimation of.
Chemical News, vol. 47, page 35; Fat, Determination of.
Chemical News, vol. 48, page 118; Fat, Determination of Volume of.
Chemical News, vol. 51, page 24; Fat, Determination of.
Chemical News, vol. 50, page 68; Fat, Determination of.
Chemical News, vol. 43, pages 101, 102, 173, 228, 230, 247; Fat, Estimation of.
J. Am. Ch. Soc., vol. 1, page 358, F. Soxhlet; Fat, Gravimetric Determination of.
J. Am. Ch, Soc., vol. 1, page 570, Feser; Fat, Optical Estimation of.
W., vol. 23, page 822; A., vol. 189, page 358; D. J., vol. 226, page 418; M. Sc., vol. 1877, page 94; Ch. Ct. bl., vol. 1878, page 9; Chemical News, vol. 1877, page 257, Lehman; Fettbestimmung in der Milch.

W., vol. 25, page 937; D. J., vol. 232, page 461; Ch. Ct. bl., vol. 1879 page 510, F. Soxhlet; Fett, Bestimmung der.
W., vol. 26, page 703, Friedlunder; Fettbestimmungen, Vergleichende mittelst des Feserschen Lactoscops.
W., vol. 26, page 703, Vieth; Fettbestimmungen, Vergleichende, mittelst des Fesersehen Lactoscops.
W., vol. 26, page 703, Soxhlet; Fettbestimmung.
W., vol. 27, page 836; Z. f. B., vol. 1881, page 110, Egger; Fettbestimmung.
W., vol. 27, page 836; A. d., Ph., vol., 219, page 34, Marpmann; Fettbestimmung.
M. Z., vol. 1882, page 149, A. Ott; Fettbestimmung, Araeo metrisch.
W., vol. 28, page 926; Z. F. B., vol. 18, page 1; Z. Ldw., V. v. B., vol. 1882, page 18, Emmerlich and Soxhlet; Fettbestimmung, Araeo metrische.
W., vol. 29, page 973; Z. a. Ch., vol. 1883, page 383, Liebermann and Wolf; Fettbestimmung.
W., vol. 30, page 1054; Z. a. Ch., vol. 1884, page 476, Liebermann; Fettbestimung.
W., vol. 7, page 476; J. f. pr. Ch., vol. 82, page 13; Rep. d. Ch. p., vol. 3, page 416, A. Muller; Fettbestimmung.
W., vol. 20, page 810; P. Ct. Bl., vol. 1874, page 72, Lace; Lactodensimeter.
Am. Ch., vol. 1875, page 24, Stoddard; Lactometer.
W., vol. 22, page 891; B. L. Ch., vol. 25, page 30, Tinker; Lactometer.
W., vol. 23, page 825; D. J., vol. 225, page 283, Hensner; Lactoskop.
W., vol. 20, page 825; Dt. Ind. zt., vol. 1877, page 266, Loebner; Laktoskop.
W., vol. 24, page 990; P. N. Bl., vol. 1878, page 75; Bay. Ind. u Gew. Bl., vol. 1878, page 20; D. J., vol. 200, page 80; Ind. Bl., vol. 1878, page 113; Ch. Ct. bl., vol. 1878, page 313, Feser; Laktoskop.
W., vol. 24, page 992; C. R., vol. 87, page 425; M. Sc., vol. 1878, page 1146; Chemical News, vol. 1878, page 173; Ch. Ct. Bl., vol. 1878, page 712, E. Marchand; Lactobutyrometer.
W., vol. 27, page 836; L. V. st., vol. 27, page 133, Portele; Laktoskop, Feser's.
W., vol. 27, page 836, Schmoeger; Laktobutyrometer.
W., vol. 28, page 923, G. Recknagel; Lactodensimeter.
W., vol. 30, page 1055; Am. J. Ph., vol. 13, page 244, Pile; Lactometer.
Analyst, vol. 2, page 70; Microscope and diseased milk.
W., vol. 18, page 612; A. Ch. Ph., vol. 25, page 382; D. J., vol. 205, page 65; P. Ct. Bl., vol. 1872, page 1236, Boussingault; Mikroskop, die Milch unter dem.
W., vol. 16, page 519; Chemical News, vol. 1870, page 5, Dancer; Mikroskopische Untersuchung der Milch.
Chemical News, vol. 51, page 193; Milk-sugar and Cane-sugar, Determination of Mixtures of.
Analyst, vol. 10, page 62, Stokes and Bodener; Milk-sugar and Cane-sugar, Determination of Mixtures of.

Analyst, vol. 10, page 81, W. Johnston; Apparatus for the Optical Estimation of Milk sugar.
Chemical News, vol. 57, page 118, Milk-sugar and Galactose.
Analyst, vol. 10, page 30, C. W. Stevens; Milk-sugar, Determination of.
A. d. Ch., vol. 1884; Ch. Ct. Bl., vol. 1884, page 499, Bignamini; Saccharose, Glukose und Laktose, Bestimmung von.
W., vol. 19, page 687; Z. a. Ch. vol. 9, page 285; Ch. Ct. Bl., vol. 1873, page 9, A. Muller; Milchanalyse, Beitraege zur.
W., vol. 24, page 993, Skalweit; Milchproben, Werth der.
W., vol. 24, page 993; M. Z., vol. 1878, page 425; D. J., vol. 230, page 80; Ch. Ct. bl., vol. 1878, page 767, Klenze; Milchproben, Werth der.
W., vol. 26, page 703; D. J., vol. 238 page 413, Mittelstrass; Optische Milchprobe.
W., vol. 28, page 925; B. D. Mullh., vol. 51, page 401, Zweifel und Peter; Optische Milchproben.
W., vol. 15, page 489; D. J., vol. 193, page 396; Chemical News, vol. 45, page 131, Heeren; Optische Milchprobe.
W., vol. 27, page 835, Heeren; Pioskop.
Analyst, vol. 10, page 22, Heische; Results, Calculation of.
W., vol. 23, page 595, Jaergensen; Refractometer, Milchpruefung mit.
Analyst, vol. 10, page 99, Adams; Sour Milk, Treatment of, for Analysis
Chemical News vol. 46, page 124, Scherff's Method.
W., vol. 30, page 1060, Pellet; Milchzucker, Bestimmung von.
Chemical News, vol. 38, page 255–91; Testing.
Chemical News, vol. 39, page 264; Total Solids, Determination of.
W., vol. 3, page 333; Ch. Ct. Bl., vol. 1851 page 579, W. Knop; Trockensubstanz, Bestimmung von.
W., vol. 26, page 703; B., 1880, 1910; Schmoeger; Trockensubstanzbestimmung.
W., vol. 27, page 835; B., 1881, 2121; Schmoeger; Trockensubstanzbestimmung.
W., vol. 20, page 920; Ch. Ct. bl., 1882, page 13; Janke; Trockensubtanz-bestimmung.
Chemical News, vol. 46, 244; Urine, Reaction of, in Milk.
W., vol. 14, page 593; A. d. Ph., vol. 132, page 220; Z. a. ch., 1868, page 384; Erdmann; Vogelsche Milchprobe, Untersuchung in.
W., vol. 4, page 419; D. J., vol. 148, page 374; P. Ct. bl., 1858, page 1377; C. Brunner; Wassergehalt der Milch, Bestimmung von.
W., vol. 5, page 443; J. b. K. L. G. Schule zu Ansbach, 1858, page 59; K. u. G. bl., 1859, page 601; Seichelt; Probe zur Ermittlung.
W., vol. 25, page 936; D. J., vol. 235, page 61; Petri and Muencke; Wassergehalt, Apparat zur Bestimmung.

COMPOSITION AND PROPERTIES.

Analyst, vol. 3, page 230, A. W. Blythe; Albuminoids, Fatty Metamorphosis of.
Analyst, vol. 9, page 198; Albuminoids in Human Milk.
Analyst, vol. 49, page 168; Albuminoids of.
Analyst, vol. 50, page 177; Albuminoids in Human Milk.
Chemical News, vol. 42, page 96; Albumen of Milk.

Chemical News, vol. 43' page 130; Albumen, Researches on.
Chemical News, vol. 37, page 22; Albumen and Serum in.
Chemical News, vol. 48, page 170; American Milk.
Analyst, vol. 4, page 4; Acid of.
Chemical News, vol. 36, page 257; Caseine and Fats in.
W., vol. 23, page 821; Carbo-hydrat, Vorkommen eines vom.
J. f. pr. Ch., vol. 15, page 348, H. Ritthausen; Milchzucker verschieden in der Milch.
Chemical News, vol. 47, page 237; Changes in, with Certain Medicines.
Chemical News, vol. 31, page 54; Chemistry of.
Chemical News, vol. 42, page 271; Chemical Constitution of.
Chemical News, vol. 43, page 235; Chemical Constitution of.
W., vol. 18, page 617; J. f. pr. Ch., vol. 6, page 1; Soxhlet; Chemie, Physiologischen zur.
W., vol. 20, page 806; Ch. Ct. bl., 1874, page 791; Hammersten; Coagulation der Milch.
Analyst, vol. 10, page 67, Vieth; Composition of Milk and Milk Products.
Chemical News, vol. 50, page 263; Composition and Method of Analysis.
Chemical News, vol. 50, page 280; Composition and Method of Analysis.
Chemical News, vol. 50, page 289; Composition and Method of Analysis.
Chemical News, vol. 50, page 301; Composition and Method of Analysis.
Chemical News, vol. 39, page, 240; Composition of Cows', of Different Breed.
W., vol. 2, page 277; J. f. pr. Ch., vol. 48, page 1; F. Crusius; Zusammensetzung, Veraenderung in, und in Nahrungswerth.
W., vol. 2, page 277; D. J., vol. 142, page 75; Rohde; Zusammensetzung, uber die verschiedene, bei öfterem Melken.
W., vol. 2, page 277; A. Ch. Ph., vol. 97, page 150; J. f. pr. Ch., vol. 98, page 24; P. Ct. bl., 1855, page 695; J. Lw. d. K. Han., 1855, page 417; Beredecke and Struchmann; Zusammensetzung, Aenderung in der normalen zu den verschiedenen Tageszeiten.
W., vol. 4, page 421; Bay. K. & G. bl., 1858, page 194; Feichtinger, Lintner and Rhien; Zusammensetung der Kuhmilch.
W., vol. 5, page 445, H. Ritthausen; Zusammensetzung, Untersuchungen ueber.
W., vol. 6, page 430; Arch. f. Path. Anat., vol. 17, page 417; F. Hoppe; Die Bestandtheile der Milch.
W., vol. 8, page 519; W. Lw. Ct. bl., vol. 2, page 19; Voelcker; Zusammensetzung und Natur der Milch.
W., vol. 19, page 687; J. d. Ph. Ch., vol. 17, page 337; A. Béchamp; Untersuchung ueber die Milch.
W., vol. 20, page 809; B., 1874, page 1463; Ch. Ct. bl., 1874, page 774; Selmi; Untersuchung ueber die Milch.
W., vol. 22, page 889; Arch. d. Ph., 1876, page 440; Ind. bl., 1876, page 457; E. Reichardt; Zusammensetzung, Verschiedenheit in der unverfaelschten Milch.

W., vol. 23, page 818; M. Z., vol. 1877, page 181; D. J., vol. 224, page 459, Fleischman; Die Bestandtheile der Milch.

W., vol. 24, page 989; Ch. Ct. bl., vol. 1878, page 588, Schreiner; Beschaffenheit der Milch.

W., vol. 25, page 933; B., vol. 1879, page 1490, Schischkoff; Zusammensetzung der Milch.

J. F. Inst., vol. 1882, page 269, Haines; Zusammensetzung der Milch.

W., vol. 21, page 905; Med. Ct. bl., vol. 1875, page 295; Ch. Ct. bl., vol. 1875, page 310, Schmidt; Zur Kentniss der Milch.

Analyst, vol. 9, page 34; Condensed.

Chemical News, vol. 49, page 167; Constitution of Milk.

W., vol. 17, page 670; C. R., vol. 72, page 123, Samson; Constitution der Butter-kuegelchen.

W., vol. 30, page 1050; Ph. Ct. bl., vol. 1884, page 316, Schroede; Contraction der Frischen Milch.

W., vol. 17, page 670; Ph. Trans., vol. 31, page 605; J. Ch. Soc., vol. 9, page 165; Ch. Ct. bl,, vol. 1871, page 325, Wanklyn; Condensirte Milch, Newnham's.

Chemical News, vol. 38, page 112; Cow-tree, Composition of.

Analyst, vol. 7, page 129; Analyst, vol. 7, page 164; Analyst, vol. 7, page 176; Fat, Specific Gravity and Solids not Fat, Relation between.

Chemical News, vol. 50, page 248; Formation of.

Chemical News, vol. 54, page 305; Fat in.

Chemical News, vol. 45, page 88; Fat in.

Chemical News, vol. 39, page 240; Fat, Influence of Food on Quantity of.

W., vol. 12, page 484; Dt. M. G. zt., vol. 1; Hess. Gew. bl., vol. 44, page 345; P. Ct. bl., vol. 1866, page 414, A. Vogel; Fettgehalt der Frischen und Abgerahmten Milch.

W., vol. 26, page 702; M. Z., vol. 1880, page 186, Janke; Fettgehalt, Schwankungen im, der Gesammtmilch.

W., vol. 28, page 924; Z. a. d. G. d. Vht., vol. 1882, page 195, Vieth; Fettgehalt, Aenderung beim Milchverkauf.

W., vol. 2, page 278; A. Ch. & Ph., vol. 98, page 124, Wicke; Fett und Wasser Gehalt der Ziegenmilch.

Chemical News, vol. 31, page 106; Decomposition of.

Analyst, vol. 9, page 196; Human Milk, Composition of.

J. A. C. Soc., vol. 6, page 252, Leeds; Human Milk, Composition of.

Chemical News, vol. 47, page 281; Milk Studies.

J. A. S. Ch. Ind., vol. 3, page 452, Urech; Milk Sugar, Action of sod. hydr. on.

W., vol. 30, page 1050, Heuppe; Mikro-organismen, Zersetzung der Milch durch.

Chemical News, vol. 33, page 261; Nitrogen and Albumen in.

Chemical News, vol. 35, page 11; Nitrogen, Determination of.

W., vol. 27, page 836; Arch. Ph., vol. 219, page 41, Arnold; Ozon in der Milch.

Chemical News, vol. 49, page 104; Physical Property of.

J. S. Ch. Ind., vol. 1, page 334, O. Loew; Preserved Milk, Changes in.

W., vol. 28, page 916, O. Loew; Preserved Milk, Changes in.

B., vol. 1882, page 1482, O. Loew; Preserved Milk, Changes in.
Chemical News, vol. 49, page 168; Rennet, Action of, on Milk.
Chemical News, vol. 45, page 58; Reaction of.
W., vol. 19, page 688; J. pr. Ch., vol. 7, page 137; Ch. Ct. bl., vol. 1873, page 339, Vogel; Milch zum Lakmusfarbstoff.
Analyst, vol. 5, page 151; Salt in.
W., vol. 5, page 445; C. R., vol. 47, page 1013; Rep. Ch. App., vol. 1, page 146; Ch. Ct. bl., vol. 1859, page 158, Tilhol and Joly; Schafmilch, Analysen Verschiedener Racen.
W., vol. 28, page 922; M. Z., vol. 1882, page 689, Schmidt; Schlempefuetterung, Einfluss auf die Beschaffenheit der Milch.
W., vol. 15, page 490; Chemical News, vol. 492, page 217, Cameron; Schweinemilch, Analysen der.
W., vol. 2, page 278; J. f. pr. Ch., vol. 68, page 224; Ph. Ct. bl., vol. 1856, page 649, Scheven; Schweinemilch, Zusammensetzung.
Analyst, vol. 4, page 11, Cameron; Solids in.
Chemical News, vol. 32, page 28; Solids in.
Chemical News, vol. 37, page 123; Solids in.
Chemical News, vol. 36, page 237; Souring of, During Thunderstorms.
Analyst, vol. 5, page 35, page 32, Dr. Muter; Sugar, Estimation of.
Analyst, vol. 9, page 84; Sugar in Milk.
Chemical News, vol. 37, page 221; Sugar, Lactic Fermentation of.
W., vol. 24, page 908; B., vol. 1878, page 154; Ch. Ct. bl., vol. 1878, page 198, Musso; Sulfaten u. Sulfocyanaten, Gegenwart in der Kuhmilch.
Analyst, vol. 7, page 129, Hehner; Specific Gravity of Milk.
W., vol. 29, page 972; F. G. d. Vht., vol. 1883, page 247, Vieth; Spec. Gewicht der Milch.
W., vol. 29, page 971; M. Z., vol. 1883, page 419, Recknagel; Spec. Gewicht der Milch.
W., vol. 17, page 670; L. V. St., vol. 14, page 194; Ch. Ct. bl., vol. 1871, page 805, Fleischmann; Studien ueber die Milch.
W., vol. 28, page 924; F. G. d. Vht., vol. 1882, page 191, Vieth; Trocken-substanz, Aenderung beim Aufbewahren.
W., vol. 28, page 927; Uffelmann; Verdaüung der Kuhmilch.
W., vol. 30, page 1050; M. Z., vol. 1884, page 493, Schrodt Andonard; Verfuetterung eingesaeuerter Ruebenschnitzel an Milchkuehe.
W., vol. 27, page 836; F. G. d. Vht., vol. 1881, page 63, Munke; Ziegenmilch, Fettgehalt.
W., vol. 22, page 389; A. Ct. Ph., (5) vol. 7, page 171; C. R., vol. 82, page 266; M. Sc., vol. 1876, page 277; Ind. bl., vol. 1876, page 236; B., vol. 1876, page 356; C. Ct. bl., vol. 1876, page 216, Tisserad; Einwvirkung der Kaelte auf Milch.

HANDLING OF MILCH.

W., vol. 27, page 828, Fleischmann, Sachtleben; Aufrahmsverfahren, Das Beckersche.
W., vol. 27, page 829; M. Z., vol. 1881, page 589, Arrium; Aufrahmssysteme.
W., vol. 37, page 833; M. Z., vol. 1881, page 177; V. Peter; Aufrahmung der Milch beim Transport.

Analyst, vol. 6, page 220; Centrifugal Machines.
W., vol. 25, page 940; Ind. bl., vol. 1879, page 100, Lefeld; Centrifuge zur Entrahmung der Milch.
W., vol. 10, page 406; D. J., vol. 174, page 149, Prandte; Concentration der Milch und beschleunigte Rahmerzeugung.
W., vol. 14, page 591; Bay. K. & G. bl., vol. 1868, page 571; P. Ct. bl., vol. 1868, page 1471, Werner; Condensirte Milch von Kempten.
W., vol. 16, page 515; D. J., vol. 198, page 168; Hess. G. bl., vol. 1870, page 305; P. N. bl., vol. 1870, page 357, Trommer; Condensirte Milch, Herstellung.
W., vol. 19, page 684; D. J., 210, page 61; Ch. Ct. bl., 1873, page 788, Trommer; Condensirte Milch, Bereitung.
W., vol. 19, page 687; Dt. Ind. zt., vol. 1873, page 440; P. Ct. bl., vol. 1873; page 1376, Gefall; Condensirte Milch, Darstellung.
W., vol. 20, page 813; D. J., vol. 211, page 150; P. Ct. bl., A. Ott; Condensirte Milch auf der Wiener Ausstellung.
W., vol. 22, page 891; D. J., vol. 222, page 539, Horsford; Condensirte Milch, Geschichte.
W., vol. 28, page 917; F. G. Vht., vol. 1882, page 137, Liebig; Condensirte Milch, Herstellung.
W., vol. 29, page 968, Sweetland; Condensirte Milch.
W., vol. 30, page 1058; M. Z., vol. 1884, page 281, Merz & Deutsch; Condensirte Milch.
W., vol. 30, page 1058; Pharm. J & Trans., vol. 1884, page 461, Meaden; Condensirte Milch.
W., vol. 19, page 687; B., vol. 1873, page 977, R. B. Stephens, Condensations-Apparat.
W., vol. 1, page 251; B. Soc. Enc., vol. 1885; D. J., vol. 138, page 142; P. Ct. bl., vol. 1855, pages 319, 1261, Mabru; Conservation der Milch.
W., vol. 17, page 671; Dt. Ind. zt., vol. 1871, page 248, G. Gahn; Conservirungsmittel.
W., vol. 24, page 996; D. J., vol. 229, page 199, Toninetti; Conservirung von Milch.
W., vol. 25, page 942; D. J., vol. 202, page 94, Voigt, Schulze; Conservirung von Milch.
W., vol. 27, page 828, Klebs; Conservirung von Milch.
W., vol. 28, page 913; M. Z., vol. 1882, page 321, Mayer; Conservirung von Milch.
W., vol. 28, page 914; B., vol. 1882, page 1259, Meissl; Conservirung von Milch.
W., vol. 28, page 918; Bdm. Ct. Bl., vol. 1882, page 789, Busse; Conservirung von Milch.
W., vol. 29, page 966, Roden; Conservirung von Milch.
W., vol. 29, page 966, Bertling; Conservirung von Milch.
W., vol. 29, page 966; Lw. V. St., vol. 28, page 321, Fleischmann and Morgen; Conservirte Milch.
W., vol. 29, page 968, Schrodt; Conservirung der Milch.
Analyst, vol. 2, page 18; Depreciation by serving Cream first.
Analyst, vol. 4, page 163, J. C. Bell; Difference in Can sold from between Beginning and End of Delivery.
W., vol. 28, page 916, Feser; Erwärmen der Milch, Apparat zum.

W., vol. 30, page 1861, Ahlborn; Erwärmen der Milch, Apparat zum.
Analyst, vol. 2, page 124; Keeping Natural Loss, through.
W., vol. 28, page 914; Dt. med. Wsch., vol. 1882, page 70, Bertling; Kochapparat.
W., vol. 28, page 919; M. Z., vol. 1882, page 177, Gaebel; Kuhler, der Lawrencische.
W., vol. 28, page 919; M. Z., vol. 1882, page 817, Fleischman and Schmoege; Kuehlung der Milch.
W., vol. 30, page 1058; Th. J. and trans., vol. 1884, page 582, Gibson, Kumiss.
W., vol. 30, page 1058; Chicago Drugg., vol. 1883, Vogeler; Kumiss.
W., vol. 30, page 1059; B. vol. 1884, page 313, Struve; Kephir.
Analyst, vol. 1, page 30; mixing old and new milk, effect of.
W., vol. 13, page 516; D. J., vol. 185, page 85; P. C. Bl., vol. 1867, page 1006; P. N. Bl., vol. 1867, page 260; Dt. Ind. zt., vol. 1867, page 416; M. Sc., vol. 1867, page 592, J. Liebig; Milch extract, Schweizerische, nach Bolley and v. Liebig.
W., vol. 14, page 592, Wagner; Milch Extract von Cham. and Landislaus.
W., vol. 20, page 81, Theil; milch extract.
M. Z., vol. 1878, page 353; D. J., vol. 229, page 345, Prandte; Milchtheiler.
W., vol. 28, page 919, De Laval; Milchschleuder.
W., vol. 28, page 919; D. J., vol. 244, page 373, Feser; Milchschleuder
W., vol. 28, page 920, Burmeister; Milchschleuder.
W., vol. 28, page 927, O. Lahrmann; Muttermilch, Herstellung Kuenstlicher.
W., vol. 30, page 1058, Gerhartz; Milchliqueur.
W., vol. 30, page 1860; F. G. Vht., vol. 1884, page 307, Merz; Milchzucker, Herstellung von.
W., vol. 29, page 966; M. Z., vol. 1883, page 329, Vieth; Pferdemilch, Condensirte.
Chemical News, vol. 35, page 94; Preserving, directions for.
J. Soc. Ch. Ind., vol. 2, page 358; D. J., vol. 47, page 376; Preserving.
J. Soc. Ch. Ind., vol. 4, page 543, F. Hueppe; Preservation of.
J. Soc. Ch. Ind., vol. 1, page 159, Gerber; Preserved and Condensed
W., vol. 18, page 615; B. vol. 1872, page 227, Wanklyn and Eassie; Præservirung der Milch.
Analyst, vol. 8, page 149; Samples taken at Railroad Station.
Chemical News, vol. 53, page 37; Sampling, Method of.
Chemical News, vol. 33, page 6; Sampling for Analysis.
Analyst, vol. 2, page 117, J. Shea; Sewage Farm Milk.
W., vol. 30, page 1058; M. Z., vol. 1884, page 164, Vieth; Stutenmilch, Condensirte.
Analyst, vol. 4, page 51, Estecourt; Temperature, Effect of low.
Analyst, vol. 4, page 117, Gatehouse; Tin cans, selling Milk from.
W., vol. 5, page 445; P. Ct. bl., vol. 1859, page 317; P. N. bl., vol. 1860, page 7, A. Fleck; Zink Nachtheile von Milchgefaessen aus.

Legal Cases.

Analyst, vol. 2, page 185; Adulterated, Action to Recover Fines.
Analyst, vol. 2, page 185; Adulterated, Paid for Selling.

Analyst, vol. 2, page 82 ; Adulteration in New York.
Analyst, vol. 2, page 28 ; Adulteration in New York.
Analyst, vol. 3, page 261 ; Adulterating and Stealing, Conviction for.
Analyst, vol. 5, page 152 ; No Adulteration if Substance Added is not Injurious to Health.
Analyst, vol. 6, pages 11, 32 ; Adulterations, Sent by Train.
Analyst, vol. 7, page 30 ; Adulterations, Convictions for, in New York.
Analyst, vol. 7, page 154 ; Adulterations and Parliament.
Analyst, vol. 8, page 123 ; Adulterations in New Jersey.
Analyst, vol. 8, pages 154, 156 ; Adulterations in New York.
Analyst, vol. 8, pages 128, 151, 185 ; Adulterations at Manchester.
Analyst, vol. 9, pages 9, 29, 31, 32, 52, 71, 112, 191, 210 ; Adulterations, Prosecutions for.
Analyst, vol. 9, page 132 ; Adulterated, Supplied to Work-house.
Analyst, vol. 9, page 190 ; Amending Food Act.
Analyst, vol. 5, pages 149, 153, 187, 212, 213 ; Analyst's Certificate as to Adulterations must use Words of Act.
Analyst, vol. 2, page 67 ; Diseased Milk, Action to Recover Price of.
Analyst, vol. 3, page 296 ; Farmers Convicted for Adulterating.
Analyst, vol. 5, page 11 ; Milkmen Selling Milk in Streets, bound to Serve Inspector.
Analyst, vol. 1, pages 33, 34, 35, 55, 56, 81, 117, 118, 136, 153, 168, 200, 215 ; Prosecutions for Selling Adulterated.
Analyst, vol. 2, pages 18, 30, 69, 85, 122, 137, 184, 185, 201, 202, 227; vol. 3, pages 245, 261, 296, 297, 298, 332, 333, 348, 350, 351, 367, 368, 386, 387 ; Prosecutions for Selling Adulterated.
Analyst, vol. 4, pages 13, 14, 34, 36, 55, 73, 74, 118, 119, 137, 155, 174, 194, 196, 212, 216, 235, 236, 233, 234; Prosecutions for Selling Adulterated.
Analyst, vol. 5, pages 151, 170, 188, 190, 191, 212, 226 ; Prosecutions for Selling Adulterated.
Analyst, vol. 6, pages 32, 11, 52, 72, 74, 152, 153, 155, 195, 212, 234 ; Prosecutions for Selling Adulterated.
Analyst, vol. 7, pages 71, 89, 92 ; Prosecutions for Selling Adulterated.
Analyst, vol. 7, pages 32, 186, 125 ; Prosecutions in Dublin, France, New Jersey.
Analyst, vol. 8, pages 10, 48, 51, 91, 110, 118, 128, 162, 181, 183, 185, 260 ; Prosecutions for Selling Adulterated.
Analyst, vol. 1, page 214 ; Skimmed, Conviction for Selling as new.
Analyst, vol. 1, page 203 ; Skimmed, Conviction for Selling as new.
Analyst, vol. 7, page 30 ; Skimming, Conviction for.
Analyst, vol. 7, page 45 ; Skimming, Conviction for, in New York.
Analyst, vol. 1, page 55 ; Summons for Refusing to Sell to Inspector.
Analyst, vol. 1, page 15 ; Walks, As to being Premises, Shops or Stores.
Analyst, vol. 1, page 81 ; Warranty, False.

MILK TRAFFIC.

Analyst, vol. 4, page 58; Anglo-Swiss Condensed Milk Co., Profits of.
Analyst, vol. 9, page 131 ; Brooklyn and New York, Inspection in.
Analyst, vol. 8, page 2 ; Control of.
W., vol. 25, page 938; Ind. bl., vol. 1879, page 155 ; D. J., vol. 232, page 381; Controle der Verkaufsmilch.

Analyst, vol. 4, page 159; Dairymen and Milk Analysis.
Analyst, vol. 8, page 243, Wigner; London, Supply of.
Analyst, vol. 3, page 242, Edge; Milkmen and Farmers.
Analyst, vol. 3, page 275; Milkmen and Analyst.
Analyst, vol. 4, page 214; Analyst, vol. 4, page 234; Analyst, vol. 4, page 238; Farmers and Milk-dealers.
W., vol. 27, page 834; Milch Verkaufsordnung in Darmstadt.
Analyst, vol. 5, page 188; Skimmed Milk, Profits of.

Standards.

Analyst, vol. 7, pages 34; and 37 Samples below the Society's Limit.
Analyst, vol. 8, page 245; Solids, Valuation of, Instead of a Limit.
Analyst, vol. 1, page 40, A. Hill; Standards.
Analyst, vol. 1, page 80; Standards, Correspondence as to.
Analyst, vol. 9. page 116, D. Muter; Standards and Analyses.
Chemical News, vol. 31, pages 266, 280; Solids, Minimum in.
W., vol. 20, page 810; Chemical News, 1874, 224; Horsley; Werthermittlung der Milch.

Unhealthy and Abnormal.

Analyst, vol. 1, page 47, Patterson; Abnormal Sample of New.
Chemical News, vol. 50, page 248; Blue Milk.
Chemical News, vol. 47, page 176; Blue, Observations on.
W., vol. 27, page 832; M. Z., 1881, page 28; Herter; Blauwerden der Milch.
W., vol. 29, page 975; J. d. l'Ag., 1883; Russet; Blauemilch.
Analyst, vol. 2, page 70; Diseased Milk.
Analyst, vol. 3, page 249, Heisch; Diseased Milk.
Analyst, vol. 3, page 251, Wigner; Diseased Milk.
Analyst, vol. 7, page 165; Diseased Cow, Milk from.
W., vol. 27, page 832, Nalline; Gelbe Milch.
W., vol. 28, page 922, Schmidt; Fahdenziehende Milch.
Chemical News, vol. 31, page 232; Health and Disease in.
Chemical News, vol. 32, page 312; Health and Disease in.
Chemical News, vol. 39, page 226; Health and Disease in.
W., vol. 25, page 935; B., 1879, page 1216; A. W. Blythe; Composition of, in Health and Disease.
Chemical News, vol. 40, page 226; Human, Composition of Abnormal.
W., vol. 27, page 833; Ind. bl., 1881, page 9; Airy; Krankheitstoffe, Uebertragung durch die Milch.
W., vol. 30, page 1058; M. Z., 1884, page 341; Fleischmann; Pasteurisirte Milch.
W., vol. 29, page 974; J. pr. Ch., vol. 27, page 249; Struve; Tuberculose, Uebertragung durch die Milch.
W., vol. 29, page 974; Arch. Hyg., 1883, page 121; May; Tuberculose, Uebertragung durch die Milch.

EXPLANATION OF ABBREVIATIONS.

Am. Ch.—American Chemist.
Am. J. Ph.—American Journal of Pharmacy.
A. Ch. Ph.—Annales de chimie et physique.
A.—Annalen der Chemie.
Arch. Ph.—Archiv der Pharmacie.
Arch. p. Anat.—Archiv der pathalogischen Anatomie.
Bay. K. & Gew. bl.—Bayrisches Kunst u. Gewerbeblatt.
Bay. Ind. & Gewb.—Bayrische Industrie u. Gewerbeblatter.
B.—Berichte der deutschen chemischen Gesellschaft.
Bull. Soc. d'En.—Bulletin de la Societé d'Encouragement.
B. S. Ch.—Bulletin de la Societé chimique de Paris.
B. Mulhause—Bulletin de Mulhause.
Biedermann Ct. bl. Ag. Ch.—Biedermann Central blatt fuer Agricultur u chemie.
Ch. Ind.—Chemische Industrieblatter.
Ch. Ct. bl.—Chemisches Centralblatt.
Ch. News—Chemical News.
Ch. Tech. Mth.—Chemische, Technische Mittheilungen.
Dt. Ill. Gew. zt.—Deutsche Illustrirte Gewerbezeitung.
Dt. Ind. zt.—Deutsche Industriezeitung.
Dt. med. Wochsch.—Deutsche medicinische Wochenschrift.
D. J.—Dingler's polytechnisches Journal.
F. G. Vht.—Fortschritte auf dem Gebiet der Viehaltung.
Fuerther Gew. zt.—Fuerther Gewerbezeitung.
F. T.—Fresenius Zeitschrift fuer analytische Chemie.
Hess. Gew. bl.—Hessische Gewerbeblaetter.
Ind. bl.—Industrieblaetter.
J. pr. Ch.—Journal fuer praktische Chemie.
J. Soc. Ch. Ind—Journal of the Society of Chemical Industrie.
J. Am. Ch. Soc.—Journal of the American Chemical Society.
J. Ph. & Ch.—Journal de Physique et de Chimie.

REPORT ON BUTTER AND ITS ADULTERATIONS.

By ELWYN WALLER, Ph. D., assisted by EDWARD W. MARTIN, WALTER MOELLER, Ph. B., and RUSSELL W. MOORE, A. B. M. Sc.

HON. J. K. BROWN, *State Dairy Commissioner:*

SIR—I have the honor to report the following with regard to butter and its adulterations:

Butter is the fatty portion of the milk of the cow. As manufactured it contains besides the butter fat, some salt which has been added to preserve it and bring out its flavor, and as constituents incidental to the mode of manufacture, water and curd.

In the accompanying table from Koenig's Nahrungsmittel, 2d ed, Berlin, 1883, vol. 1, are given the results of the examination of 123 butters of all kinds sold in different parts of Europe. Koenig expressly states, however, that the mean of these results do not represent a good butter. Many of the samples represent what is to be found in the markets of Europe, but many of them, though made from milk or cream, were badly made either through ignorance or dishonesty. He gives (p. 279) as the composition of good fresh butter the following: Water, 11·7 per cent; fat, 87·0 per cent; caseine, 0·5 per cent; milk sugar, 0·5 per cent; salts, 0·3 per cent.

Dr. James Bell reports (Report of Board of Inland Revenue, May 31, 1876, Jour. Roy. Agri. Soc., 13, 1877, p. 181), the examination of 117 butters obtained from farmers in various parts of Great Britain (Table 2). The object was to obtain a fair standard of composition from which to judge of the question of adulterations. A. H. Allen (Commerc. Organic Analysis, vol. 2, p. 204), remarks that Dr. Bell assumed that the farmers were all entirely innocent of any fraudulent intent in preparing their butter for the market, a position that is not invariably tenable. For the report made to the Lancet which led up to legislation on the subject of food adulteration in England forty-eight butters were examined.

Angell and Hebner (Butter Analysis, 2d ed., London, 1877, pp. 14 and 15) give the results of their examination of thirty butters such as are ordinarily encountered, besides three of exceptional composition. Wanklyn examined fifty different samples of butter supplied to the work-houses in London (Milk Analysis, London, 1874, p. 57).

Caldwell (Second Report to New York State Board of Health, p. 510), reported to the New York State Board of Health the result of the examination of some forty samples of butter, genuine and artificial, delivered to him by the inspectors.

TABLE I.

Analyses of Butter.

(Koenig, Nahrungsmittel, 2d ed, vol. 1, Berlin, 1882.)

Number.	REMARKS.	Date of investigation.	Per cent of water.	Per cent of fat.	Per cent of caseine.	Per cent of milk sugar.	Per cent of nonazotized substance.	Per cent of ash.	Per cent of fat, on dry substance.	Analyst.
1	Swedish butter from various places, and prepared by various methods.	1864	13.67	85.00	0.51	0.70	......	0.12	98.46	Alex. Müller and Eisenstuck. Landw. Versuchsstat, 1864, vol. VI, page 8.
2		1864	14.02	82.08	2.58		......	1.87	95.41	
3		1864	13.16	82.01	2.01		......	2.82	93.36	
4		1864	11.42	85.35	1.92		......	1.31	96.35	
5		1864	11.71	84.89	2.58		......	1.32	95.58	
6		1864	11.29	85.75	1.30		......	1.70	96.66	
7		1864	6.10	90-18	1.87		......	1.85	96.04	
8		1864	10.07	87.05	1.87		......	1.51	97.78	
9		1864	12.69	83.95	1.81		......	1.55	95.01	
10		1864	16.13	81.86	1.85		......	0.16	97.60	
11		1864	20.20	77.33	2.28		......	0.19	96.78	
12		1864	13.46	83.84	0.99		......	2.21	96.80	
13		1864	10.08	86.57	1.20		......	2.15	96.27	
14		1864	14.84	83.51	1.59		......	0.06	98.06	
15		1864	18.29	79.61	1.93		......	0.17	97.43	
16		1864	21.10	74.29	4.35		......	0.26	94.16	
17		1864	8.87	88.89	0.91		......	1.33	97.58	
18		1864	18.59	80.52	0.80		......	0.09	98.91	
19		1864	17.03	81.55	1.33		......	0.09	98.29	
20		1864	23.47	74.60	1.80		......	0.13	97.48	
21		1864	14.96	83.85	1.59		......	0.10	97.99	
22		1864	15.25	83.03	1.57		......	0.15	97.97	
23		1864	9.24	83.52	0.64	0.60	......	6.00	92.02	

24		1864	18.82	84.78	1.27		......		0.13	98.38	
25		1864	7.66	88.83	0.29		......		2.59	96.20	
26		1864	9.93	88.13	0.44	0.46	......		1.04	[illegible]	
27		1864	12.56	83.57	0.78	0.48	......		2.66	95.57	
28		1864	15.91	8.22	0.45	0.35	......		0.07	98.97	
29	Swedish butter from various places and prepared by various methods.	1864	15.30	8.89	0.75	0.88	......		0.18	97.86	Alexander Müller and Eisenstuck.
30		1864	8.18	.45	1.08	1.11	......		0.18	9.10	
31		1864	13.00	8.69	0.62	......	0.69			.89	
32		1864	14.00	84.90	0.54	......	0.56			98.72	
33		1864	11.44	86.08	0.54	0.70	......		1.24	97.20	
34	....	1867	13.82	84.78	1.27		......		0.13	98.38	
35	....	1867	12.56	83.57	0.78	0.43	......		2.66	95.57	
36	....	1867	15.91	83.22	0.45	0.35	......		0.07	98.97	
37	September Holstein...............	1867	10.25	86.88	0.52	0.49	......		1.86	96.80	Alexander Müller, Landw. Verstat. vol. 9, page 276.
38	Spring Sweden...	1867	11.45	83.82	1.63		......		3.60	94.09	
39	Summer Sweden..................	1867	948	87.00	3.60		......		3.13	96.11	
40	Cream butter.....................	1868	13.11	85.97	0.84		......		0.08	95.94	
41		1868	19.96	78.54	1.25		......		0.25	98.13	O. Lindt, Jahresber, Agri. chem., 68-69, p. 711.
42	Main fine........................	1872	11.68	86.95	0.19	0.85	......		1.43	.85	
43	Holstein medium fine............	1872	12.09	84.76	0.39	0.81	......		1.95	95.28	
44	Holstein quite fine...............	1872	10.35	6.96	0.26	0.82	......		4.83	9.00	
45	Holstein moderate...............	1872	10.09	85.50	0.28	0.69	......		2.24	8.10	
46	Holstein.........................	1872	12.64	84.10	0.58	0.86	......		.89	8.27	A. Emmerling, Landw. Wochenblc. Schles.-Holstein, '72 p. 49.
47	Holstein..	1872	14.42	8.91	0.50	1.07	......		1.78	96.88	
48	Holstein oily................ . ..	1872	10.81	86.43	0.32	0.75	......		1.85	96.91	
49	Holstein normal and gd........	1872	12.29	85.50	0.57	0.59	......		0.93	97.48	
50	Salted butters...	1873	5.50	9.96	0.50	......	......		3.02	96.25	R. Alberti, Jahresber agri. chem., 73-74 pp. 289-290.
51		1873	10.60	8.62	0.63	......	......		2.03	95.77	
52	Tea [illegible]ter..........................	1873	14.20	85.55	0.25	......	......		0.11	99.68	
53	Good market butter..............	1873	13.77	86.06	0.42	......	......		0.12	99.70	J. Moser, ib.
54	Bad market butter................	1873	17.08	82.60	0.72	......	......		0.20	.88	
55	From sweet cream	1873	12.06	84.00	0.75	0.64	......		2.45	95.52	
56		1873	12.05	84.43	0.51	0.47	......		1.48	8.00	
57	From sour cream..	1873	11.88	83.21	0.92	0.47	......		1.98	96.00	F. Dahl, ib.
58		1873	12.49	82.75	0.90	0.74	......		3.17	94.43	
59	Price 11 Sgr......	1873	28.75	63.95	3.25	0.61	......		3.11	94.56	
60	Price 12 Sgr.....................	1873	15.05	76.55	4.70	......	......		4.05	89.75	
61	Price 13 Sgr.....................	1873	12.72	88 60	2.60	......	......		3.70	90.11	N. Gräger, ib.
62		1873	8.16	8.54	......	......	6.30		1.12	95.78	
63	Fresh butter	1873	10.02	85.84	......	......	4.57		......	98.14	
64		1873	11.68	83.09	......	.. ..	5. 3		.. ...	94.84	E. Schulze, ib..
65	Boiled butter..................	1873	(0 07)	9.00	......	......	0.93		......	94 08	
66		1873	(0 04)	9.63	......	.. ..	0.33		......	(99 07)	

BUTTER — (*Continued*).

Number.	REMARKS.	Date of investigation.	Per cent of water.	Per cent of fat.	Per cent of caseine.	Per cent of milk sugar.	Per cent of other subst'ces not containing nitrogen.	Per cent of ash.	Per cent of fat in dry substance.	Analyst.
67	Boiled butter	1878	(0.06)	97.61	......	......	2.83	......	(97.67)	E. Schulze, ib.
68		1876	85.12	61.09	......	......	1.22	2.57	(94.16)	
69		1876	25.27	71.99	......	......	1.87	1.87	96.83	J. Konig and C. Brimmier. Landw ztg. f. West f. 1876, p. 3.
70		1876	27.55	69.82	......	......	1.07	1.56	96.37	
71	Ordinary butter from the Münster market.........................	1876	34.12	63.97	......	......	1.37	0.54	97.10	
72		1876	30.42	66.68	......	......	1.15	1.75	95.88	
73		1876	17.45	80.60	......	......	0.82	1.13	97.64	
74		1876	27.53	67.65	......	......	2.33	2.49	93.35	
75		1876	25.53	70.15	......	......	1.81	1.51	95.48	Salt.
76	From sweet cream	1876	13.12	83.92	0.62	......	0.68	0.14	96.59	1.23 — V. Storch. Milchztg., 1876, p. 1722.
77		1876	13.41	83.82	0.61	0.46	0.74	0.12	96.80	1.30
78		1876	10.45	85.40	0.54	0.32	0.58	0.16	95.37	2.92
79	From sour cream..	1876	17.09	80.01	0.87	0.18	0.71	0.15	96.50	1.17
80		1876	11.57	85.43	0.62	0.17	0.39	0.12	96.61	1.87
81		1876	9.60	86.77	0.61	0.23	0.62	0.15	95.98	2.25
82		1876	9.80	83.86	1.00	......	0.80	0.25	92.42	4.97
83	Sp. Gr. 0.927, melting point 22.5 C.	1859	18.11	80.22	1.59	......	......	0.08	97.96	Traces. — C. Karmrodt.
84	Sp. Gr. 0.921, melting point 19.9 C.	1859	16.59	81.07	1.24	......	......	0.10	97.19	1.00
85	Sp. Gr. 0.925, melting point 19.3 C.	1859	17.74	80.99	1.15	......	......	0.08	98.45	0.04
86	Sp. Gr. 0 920, melting point 19.2 C.	1859	10 58	86.34	1.19	......	......	0.14	96.56	0.14
87	Sp. Gr. 0 927, melting point 21.9 C.	1859	15.94	81.88	2.02	......	......	0.10	97.41	0.10
88	Alpine butter, melting pt. { 37.1 C.	1877	10.81	86.14	1.86		......	0.90	96.58	W. Eugling.
89	37.5 C.	1877	14.97	83.33	1.46		......	0.24	98.00	
90		1879	17.97	76.95	4.77		......	0.31	93.81	M. Schrodt and Ph. du Rol. Milchztg. 179, p. 558.
91	Obtained by churning milk.......	1879	16.03	81.17	2.69		......	0.10	96.67	
92		1879	15.33	81.87	2.65		......	0.15	96.69	
93		1879	14.65	82.76	2.51		......	0.08	96.97	
94	Obtained by churning cream	1879	13.78	83.80	2.28		......	0.14	97.19	
95		1879	13.79	84.32	1.75		......	0.14	97.81	

	Italian butters:						Lac. Acid.			
96	Lodi	1878	13 66	(84.95)	0.36	0.00	.....	0.48	98.39	
97	N	1878	19.78	77.18	1.89	0.47	0.12	0.56	8.21	
98	Milan	1878	14.78	8.87	0.60	0.44	0.10	0.20	98.42	
99	Codogno	1878	15.08	83.38	0.65	0.57	0.09	0.23	98.19	A. Menozzi.
100	Lodi	1878	15.47	8.34	0.58	0.81	0.14	0.16	8.59	
101	Mierei Maleo	1878	15 35	82.45	1.45	0.88	0.17	0.19	97.40	
102	Lodi	1878	14.53	8.98	0.53	0.71	0.14	0.11	8.26	
103	Lodi	1878	15.25	83.25	0.79	0.36	0.15	0.19	98.26	
104	Lodi	1878	14.82	83.08	0.91	0.53	......	0.13	97.53	
105	Lodi	1878	15.59	83.57	0.99	0.59	0.10	0.16	99.00	
106	Lodi	1878	15.72	82.44	0.93	0.56	0.22	0.14	97.82	
107	Lodi	1878	15.03	83.56	0.67	0.45	0.14	0.15	98.84	R. Just Lombard. (II.) XII 79, p. 294.
108	Lodi	1878	16.12	82.01	0.83	0.72	0.14	0.17	97.77	
109	Lodi	1878	15.56	83.02	0.56	0.69	0.04	0.14	98.82	
110	Lodi	1878	14.18	84.57	0.66	0.33	0.12	0.14	98.54	
111	Lodi	1878	14.43	*83.94	0.88	0.47	0.19	0.15	98.10	
112	Westphalian Swartz method	1878	13.42	85.57	0.92		......	0.09	98.83	
113	Westphalian Swartz method	1878	13.09	85.71	0.92		.. ..	0.28	98.62	J. Konig u. C. Krauch, Landw. Versuchsstat. Munster, 78, 80, p. 47.
114	Westphalian Swartz method	1878	12.92	85.71	0.95		......	0.42	98.43	
115	Same old method	1878	15.42	81.94	1.74		......	0.90	8.88	
116	Centrif. cream	1878	14.88	83.85	0.60	0.44	0.04	0.20	98.52	
117	Best Lombardy butter.	1878	15.08	8.37	0.65	0.57	0.09	0.23	98.17	
118	Best Lombardy butter.	1878	15.85	82.45	1.45	0.38	0.17	0.19	97.40	G. Cantoni. L. 'Industrie Laitière, 1878, No. 44.
119	Best Lombardy butter.	1878	14.53	8.98	0.53	0.71	0.14	0.11	98.26	
120	Best Lombardy butter.	1878	15.25	83.25	0.79	0.36	0.15	0.19	8.23	
121	Mixed low grade	1878	19.78	77.48	1.09	0.47	0.12	0.56	96.58	
122	Sweet cream	1880	13.65	8.62	0.61	0.56	0.04	1.52	8.84	M. Schmöger. Milchztg. 1880, p. 273.
123	Sour cream	1880	14.23	82.83	0.80	0.54	0.05	55	96.57	
	Minimum	......	5.50	76.87	0.19	0.85		0.08	89.75	
	Mm	......	8.12	85.25	4.74	1.16		5.65	99 70	
	rge	......	14.49	83.27	0.71	0.58		0.95	97.34	
	ARTIFICIAL BUTTER. (Nitrogenous substance).									
1		1877	12.01	82 03	0.74		......	5.22	93.28	A. Mott.
2		1877	11.25	815	0.57		......	1.03	98.20	T. Brown.
3		1877	10.01	87.91	0.23		......	1.85	97.69	J. Konig.
4		1878	7.77	91.56	0.40		......	0.27	99.27	J. Moser.*
	Mean	......	10.33	87.16	0.43		......	2.08	97.09	

*1st Bericht Versuchsstation Wien, 1878. Table XXIX.

TABLE II — Dr. BELL'S Table.

Results of the analysis of 117 samples of butter.

Number.	WHENCE OBTAINED.	Per cent of water.	Per cent of salt.	Per cent of curd.	Per cent of butter fat.	BUTTER FAT. Specific gravity at 100 degrees Fahr.	BUTTER FAT. Melting point, deg. Fahr.	Per cent of fixed fatty acids.
1	Surrey	4.15	....	....	......	.91349	86	
2	Surrey	6.80	3.27	.80	89.13	.91309	88	
3	Surrey	15.50	2.10	1.70	80.70	.91309	88	
4	Surrey	11.40	.76	.77	87.07	.91228	90	
5	Surrey	7.55	1.03	1.15	90.27	.91389	85	
6	Surrey	12.70	.80	.86	85.64	.91279	86	
7	Irish salt butter.........	11.67	2.20	.86	85.27	.91228	87.5	87.20
8	County Galway	11.79	3.39	.68	84.14	.91309	88.3	
9	County Galway	14.04	1.63	1.51	82.82	.91158	88.8	
10	County Galway	10.12	2.62	.70	86.56	.91299	88.2	
11	County Galway	4.91	1.54	.43	93.12	.91208	89	87.42
12	County Galway	11.78	2.11	.47	85.69	.91299	89	
13	County Galway	11.83	1.14	.80	86.23	.91269	89	86.60
14	Devonshire	13.22	1.34	.68	84.76	.91269	87.8	
15	Cornwall	16.99	2.65	1.36	79.00	.91239	88.7	
16	Cumberland	12.26	4.52	.94	82.28	.91289	87.5	
17	Cumberland	11.92	4.22	1.52	82.34	.91289	87.5	
18	Cumberland	12.96	3.80	.36	82.88	.91299	88	
19	Cumberland	9.72	2.82	.28	87.18	.91198	88	
20	Cumberland	8.18	3.14	.92	87.76	.91269	87	
21	Cumberland	12.84	2.78	.98	83.40	.91269	87.5	
22	Dorsetshire............	16.85	2.77	.11	80.27	.91188	87.8	
23	Dorsetshire............	16.37	3.22	.56	79.85	.91188	87.5	
24	Dorsetshire............	17.06	2.13	.88	79.93	.91198	87.3	
25	Dorsetshire............	17.03	2.25	.86	79.86	.91228	87.8	87.30
26	Dorsetshire............	18.37	1.63	.39	79.61	.91208	87.3	
27	Dorsetshire..	13.24	1.25	.40	85.11	.91218	87.5	
28	Cumberland	12.22	.61	.34	86.83	.91138	89.2	
29	Cumberland	13.02	.72	.61	85.65	.91128	90.2	
30	Cumberland	11.74	1.32	.42	86.52	.91168	90	
31	Cumberland	8.72	.58	.70	90.00	.91218	88.5	
32	Cumberland	9.55	4.17	.24	86.04	.91198	88.7	
33	Suffolk	14.41	3.10	.64	81.85	.91239	89	86.87
34	Suffolk	20.75	3.82	.61	74.82	.91158	92	87.30
35	Suffolk	14.26	3.82	.22	81.70	.91289	88.5	86.45
36	Suffolk	9.11	8.28	.40	82.21	.91279	88.5	86.00
37	Suffolk	11.52	3.92	.41	84.15	.91389	87.5	85.50
38	Suffolk	9.60	6.45	.82	83.13	.91228	89	
39	Devonshire	14.36	2.66	1.46	81.52	.91299	88.5	
40	Devonshire	15.52	4.08	1.54	78.86	.91178	89	87.40
41	Devonshire	17.56	2.98	1.14	78.32	.91239	89	
42	Devonshire	17.18	3.00	1.24	78.58	.91299	88.5	
43	Devonshire	16.28	3.32	1.56	78.84	.91279	88.5	
44	Devonshire	18.72	2.24	1.36	77.68	.91289	89	
45	Devonshire	16.42	2.80	1.60	79.18	.91228	88	86.87
46	Devonshire	13.62	3.00	.60	82.78	.91078	90	88.00
47	Suffolk	13.14	5.74	2.96	78.16	.91397	88	
48	County Londonderry....	19.40	3.70	.56	76.34	.91296	89	
49	County Londonderry ...	13.70	2.30	1.86	82.14	.91228	90	
50	County Londonderry...	15.94	2.40	2.68	78.98	.91106	91.5	
51	County Londonderry ..	18.52	4.84	2.16	74.48	.91191	90.5	
52	County Londonderry....	14.90	6.04	1.50	77.56	.91188	90.5	
53	County Londonderry....	14.98	3.74	1.14	80.14	.91097	91.5	
54	Kent	11.71	3.04	.76	84.49	.91314	88.5	
55	Kent....................	13.51	2.90	.70	82.89	.91309	88.5	
56	Kent....................	18.64	2.68	.79	77.89	.91050	93	88.60
57	Kent....................	17.60	2.60	.98	78.82	.91063	93	

TABLE II — (*Continued*).

No.	WHENCE OBTAINED.	Per cent of water.	Per cent of salt.	Per cent of curd.	Per cent of butter-fat.	BUTTER-FAT. Specific gravity at 100 degrees Fahr.	BUTTER-FAT. Melting point, Fahr.	Per cent of fixed fatty acids.
58	Surrey	13.55	2.49	.80	83.16	.91098	92	88.85
59	Surrey	14.16	....	....	......	.91223	89	
60	County Cork	13.63	.44	.62	85.31	.91106	89.5	87.72
61	County Cork	16.46	1.13	1.12	81.29	.91019	92.5	88.75
62	County Cork	13.57	.65	.84	84.94	.91140	89	87.50
63	County Cork	14.98	.68	.68	83.66	.90987	92.5	89.15
64	County Cork	15.84	.40	.69	83.57	.91062	91	
65	County Cork	14.64	.46	.82	84.08	.91041	91.5	
66	Carnarvonshire	11.41	3.08	.70	84.86	.91239	92	87.01
67	Carnarvonshire	10.43	2.46	.57	86.54	.91148	89	
68	Carnarvonshire	13.79	2.96	1.26	81.99	.91063	93	88.32
69	Carnarvonshire	11.05	7.71	.44	80.80	.91073	93	
70	Carnarvonshire	11.86	4.97	1.04	82.63	.91106	93	88.42
71	Carnarvonshire	16.24	9.20	.40	84.16	.91134	93	88.12
72	Normandy	11.71	8.60	.95	73.74	.91145	90	
73	Irish salt butter	16.89	8.56	1.23	73.32	.91148	89.5	
74	Wiltshire	11.59	1.49	.44	86 48	.91201	90.5	86.96
75	Wiltshire	13.21	1.74	.56	84.49	.91179	90.5	
76	Wiltshire	12.52	2.12	.79	84.57	.91146	90.5	87.35
77	Wiltshire	11.99	2.28	.99	84.79	.91182	89	
78	Wiltshire	12.57	1.58	.89	84.96	.91148	90.5	87.65
79	Cumberland	11.81	8.38	3.06	76.75	.91251	89	86.96
80	Cumberland	12.08	2.39	3.74	81.79	.91160	92	87.74
81	Cumberland	12.89	3.69	3.18	80.27	.91208	90	86.92
82	Cumberland	13.08	2.33	2.72	81.87	.91060	92	88.29
83	Cumberland	11.18	1.79	5.32	81.71	.91174	91.5	87.60
84	Cumberland	19.12	3.93	4.02	72.93	.91094	92.5	88.40
85	County Monaghan	13.39	6.68	1.62	78.31	.91042	92.5	
86	County Monaghan	15.60	6.51	.54	77.35	.91014	92	88.90
87	County Monaghan	13.59	15.08	1.36	69.97	90947	93.5	
88	County Monaghan	13.50	2.58	.55	83.37	.91104	91.5	
89	County Monaghan	14.55	5.86	1.31	78.28	.91030	93	
90	County Monaghan	12.43	3.55	.55	83.47	.91070	92	
91	County Londonderry	11.81	2.85	.70	84.64	.91085	92	88.62
92	County Londonderry	13.88	3.15	.75	82.22	.91147	90	87.66
93	County Londonderry	14.34	3.31	.78	81.57	.91188	90.5	
94	County Londonderry	12.57	4.32	.51	82.60	.91065	92	88.74
95	County Londonderry	13.56	2.29	.75	83.40	.91208	90.5	87.42
96	County Londonderry	11.56	2.82	.47	85.15	.91179	90.5	88.05
97	Dorsetshire	13.92	2.13	.52	83.43	.91058	94	88.65
98	Dorsetshire	8.88	4.50	.50	86.12	.91085	92.5	88.46
99	Dorsetshire	12 55	2.22	1.35	83.88	.91220	91.5	
100	Dorsetshire	12.81	1.78	.74	84.67	.91080	92.5	88.17
101	Staffordshire	10.61	1.11	.63	87.65	.91094	91.5	88.21
102	Staffordshire	12.87	1.56	.76	84.81	.91244	89.5	87.14
103	Staffordshire	12.84	1.67	.56	84.93	.91129	90	87.90
104	Staffordshire	13.11	1.66	.46	84.77	.91173	91.5	
105	Staffordshire	10.93	1.25	.62	87.20	.91190	89	87.30
106	Staffordshire	12.79	1.03	.66	85.52	.91011	93	
107	County Sligo	12.36	3.24	.87	85.53	.91011	93	
108	County Sligo	11.02	1.89	.87	82.22	.91176	91	
109	County Sligo	14.61	3.86	.85	80.68	.91091	90.5	88.46
110	County Sligo	14.12	2.28	1.06	82.54	.91280	90	
111	County Galway	13.78	.90	.85	84.47	.91241	91.5	86.79
112	County Galway	10.24	3.99	1.22	84.55	.91141	92.5	87.79
113	County Galway	11.75	3.33	1.93	82.99	.91151	93	87.51
114	County Galway	15.17	1.96	1.99	80.88	.91128	92.5	87.66
115	County Galway	14.37	3.21	1.89	80.53	.91937	94.5	89.90
116	County Galway	14.50	1.44	1.61	82.45	.90939	95	89.80
117	County Galway	15.70	1.54	1.49	81.27	.91178	92	

TABLE III.

Summary of results of analyses of numerous samples of butter.

ANALYST.	No. of samples.	WATER. Maximum.	WATER. Minimum.	WATER. Average.	FAT. Maximum.	FAT. Minimum.	FAT. Average.	CURD. Maximum.	CURD. Minimum.	CURD. Average.	SALTS. Maximum.	SALTS. Minimum.	SALTS. Average.	Remarks.
Konig	123	35.12	5.50	14.49	85.25	76.37	83.27	4.77	0.25	1.29	5.65	0.08	0.95	From European markets
Bell	117	20.75	4.15	14.2	......	......	......	4.02	......	1.2	15.08	0.5	....	Great Britain farmers.
Hassall	48	15.43	4.18	......	96.93	67.72	......	....	......	....	2.91	0.3	....	Fresh butters G't Brit'n
		28.6	8.48	......	96.93	67.72	......	....	......	....	8.24	1.53	....	Salt butters.
Angell and Hebner	30	16.	6 4	......	90 2	76.4	......	5.1	1.1	....	8.5	0.4	....	Great Britain.
Wanklyn	50	24.9	8.6	......	......	......	......	....	......	....	10.7	0.1	....	London workhouses.
Caldwell	26	30.75	10.45	......	......	......	......	....	......	....	......	......	....	
Ellis	12	10.5	4.9	......	89.7	80.8	......	4.9	1.1	....	6.2	.1	....	Toronto, regarded as unadulterated.
Larue	12	16.5	3.	......	86.9	79.14	......	5.5	1.5	....	3.60	0.4	....	
Cornwall	....	......	....	......	......	......	82–90	....	......	....	......	......	....	
Dätzsch	....	......	....	10–15	......	......	85–90	....	......	...	......	......	....	
Hoorn	....	......	..	12–14	......	......	......	....	......	....	......	......	....	
Fleischmann. fresh	....	......	..	18.	......	......	80.	....	......	....	......	......	2.	
Fleischmann. salt.	....	......	....	12.	......	......	83.5	....	......	....	......	......	6.5	
Blyth, 1st ed. p. 69	5	12.984	8.58	......	87.223	82.643	85.45	5.137	2.054	2.5	3.151	0.424	....	
Schacht	8	9.00	1.25	......	98.00	87.00	......	0.5	......	....	6.	0.57	....	

Exceptional samples.

ANALYST.	No. of samples.	WATER.	FAT.	CURD.	SALTS.	Remarks.
Hassall	....	......	......	......	28.6	
Hehner and Angell	....	24 and 42.3	......	6.9–7.8	......	
Caldwell	....	30.75	......	......	......	
Johanson	....	40–49	......	......	......	
J. B. [illegible], Montreal, five samples.	....	20	49.8	16.6	13.6	
		9	49.8	16.6	24.6	
		......	......	10 per cent and over.	......	

[illegible]ann and Hassall say it is possible to introduce 20-50 per cent H_2O. Conviction of fraud, 23.4 per cent H_2O. Analyst, IV, 16.

TABLE IV.

Limits given by different authorities.

AUTHORITY.	Water should not exceed.	Fat should not be less then.	Curd should not exceed.	Salts should not exceed.
	Per cent.	Per cent.	Per cent.	Per cent.
Wanklyn, fresh	18	......	......	......
Wanklyn, salted	12	......	......	6
Angell & Hehner	12	80	......	8
Caldwell	......	80	......	....
Soc. Pub. Anal.	......	80	......	....
Cornwall	......	80	......	8
Blyth	......	80	......	8
Hoorn	20	......	......	....
Dietzsch	20	......	......	....
Fleischman, fresh	18	80	6	
Fleischman, salted	12	83·5	......	4·5
Tollens	......	80	....	....
Bischoff	......	75	......	....
J. B. Edwards, rept	......	......	5	4
J. B. Edwards, butterine	10	......	......	10
Hassal, 428, fresh	12	......	4	4
Hassal, salted	12	......	4	8
Handb. Hyg., p. 255, Hilger, fr'h	12	86	2	....
Handb. Hyg., p. 255, Hilger, sal'd	......	......	......	3
Storage butter (Dauer butter)*.	16	......	......	8
Calvert	10	......	......	2·5
Chevreul	......	83	......	....
Duflos	......	80–83	......	....
Schacht	......	80	......	....

* Hassall, 430.

So far as results on genuine butters obtained in my own laboratory are concerned the figures have been:

Fat, 83 to 85 per cent.
Water, 8 to 10 per cent.
Curd, 1 to 3 per cent.
Salts, 3 to 5 per cent.

As may be seen, the minimum of butter fat permissible is usually taken as eighty per cent.

Many German analysts lump together the water, curd and salts under the term "buttermilk." Others again omit the salts from the "buttermilk," and report percentages of fat, buttermilk and salts.

The amount of mineral matter contained in the butter or remaining on incineration of the sample, is called by some "ash," by others "salts," and by others again "salt" (meaning thereby ordinary kitchen salt). The term "ash" probably describes it the most accurately. "Salts" is applied because the curd also contains a small proportion of mineral salts (phosphates, etc.), whereas "salt" is applied by many because the proportion of mineral salts obtained from the curd is so small as to practically constitute no greater impurity than is ordinarily encountered in ordinary kitchen salt. (Hassall Food, p. 434.) From the following results it will be seen that common salt constitutes about 93 to 97 per cent of the ash of butter and its substitutes.

	Number.	Ash.	Salt = Na. Cl.
Butter	1268	3·81	3·65
Butter	1337	2·88	2·68
Butter	1269	2·77	2·70
Butter	1314	6·17	5·87
Oleomargarine	1329	4·43	4·24
Oleomargarine	1399	7·63	7·30

A large proportion of salt is sometimes added to preserve the butter which has been so unskillfully made that an undue proportion of curd remains in it.

The organic non-fatty substance in butter, consists of the non-fatty portion of the milk from which it is obtained. It is chiefly caseine, and accordingly may figure in reports of butter analysis as "caseine" or "curd." Some chemists determine the amount of milk-sugar present or the product from the change of milk sugar in the process of souring, *i. e.*, lactic acid. A reference to Nos. 96 to 111, on Koenig's table shows that Menozzi found lactic acid in the twenty-two butters which he examined to the extent of 0·04 to 0·22 per cent, with an average of 0·12 per cent.

Cane sugar is sometimes added to butter to partially replace the salt as a preservative, but it is not generally liked by consumers, and such addition is infrequent.

Small amounts of salicylic acid, nitre or of borax are also used as preservatives. By many these are not regarded as legitimate additions.

Coloring matters, usually added in winter when cows are on dry feed, are in a similar position. Koenig says that coloring butter has become a necessary evil through a perverted taste in color on the part of consumers. The coloring does not add to the flavor, and should not more be expected by consumers than the coloring of wines, however harmless the coloring matter.

The rancidity of butter is the result of a separation of the butyric acid in the butter from its combination with glycerine. It is usually believed to be caused by the alteration of the caseine — a very unstable substance when in the moist state.

Hageman (Landw. Versuchsstat., 1882, vol. 28, page 201) attributes it to the action of the lactic acid formed from the milk sugar present. V. Lang asserts that the rancidity of butter is at first due to the formation of butyric acid from lactic acid under the influence of a ferment. At a later stage only the butyric acid is separated from its combination with glycerine in the butyrin of the fat. (Kunstbutterfabrikation.)

Numerous methods of restoring rancid butter have been tried. Thus far, though success has attended the effort to remove the rancidity, it has always been attended with the loss of the delicacy of flavor which would cause the butter to be rated of high-grade. The methods consist essentially in washing the butter with water alone, or with water containing minute amounts of alkali.

The amount of butyric acid necessary to impart a very rancid flavor and odor, is extremely small, and as butyric acid is soluble in water, or can be neutralized by alkali, the rancidity can be removed by the above means. On the other hand the difficulty of working water into a mass of butter to wash out the butyric acid and the working it out again if the butter is solid, usually makes the treatment "cost more than it comes to," and if the butter is melted and then washed, the "grain" is spoiled and the flavor is still further impaired.

Adulterations of Butter.

The definition of adulteration in the case of food or drink, which is usually accepted here and in England, is given by G. W. Wigner in the prize essay prepared for the competition instituted by the American National Board of Trade (Analyst vol. 6, p. 3), which has been incorporated into the State statute, chap. 407, Laws of 1881. It is as follows: Section 3. An article shall be deemed to be adulterated

(*b.*) In the case of food or drink.

1. If any substance or substances has or have been mixed with it so as to reduce or lower or injuriously affect its quality or strength.

2. If any inferior or cheaper substance or substances have been substituted wholly or in part for the article.

3. If any valuable constituent of the article has been wholly or in part abstracted.

4. If it be an imitation of, or be sold under the name of, another article.

5. If it consists wholly or in part of a deceased or decomposed, or putrid or rotten animal or vegetable substance, whether manufactured or not, or in the case of milk if it is the produce of a diseased animal.

6. If it be colored, or coated, or polished, or powdered, whereby damage is concealed, or it is made to appear better than it really is, or of greater value.

7. If it contained any added poisonous ingredient, or any ingredient which may render such article injurious to the health of a person consuming it, provided, that the State Board of Health may, with the approval of the Governor, from time to time declare certain articles or preparations to be exempt from the provisions of this act; and provided further, that the provisions of this act shall not apply to mixtures or compounds, recognized as ordinary articles of food, provided that the same are not injurious to health and that the articles are distinctly labeled as a mixture, stating the components of the mixture.

The following substances have been mentioned as adulterants of butter:

Excessive amounts of water, curd and salt.

The introduction of such large amounts of water as have been sometimes found (up to fifty per cent) is said to be only made possible by warming the butter until it is quite soft and then stirring in water until it hardens sufficiently to retain it.

Excessive amounts of any preservatives other than salt, as saltpeter, borax or salicylic acid. By some, any addition of these substances is regarded as an adulterant.

Vegetable matters for bulk as starch from grain or potatoes, potatoes dried and ground, carrot scrapings, syrup, carregeen mucilage.

Animal products for bulk. White cheese, gelatine, butyric acid.

Mineral matters for bulk and weight as chalk, soapstone, heavy spar gypsum, white lead, alum, water glass or sodium silicate and alkalies. The use of borax or alum is said to make the incorporation of larger proportions of water more easy. (Dietzsch, p. 206).

Fatty substances. Vegetable oils, as benne or sesame oil, cottonseed oil, mangosteen oil, rape-seed, cocoanut, palm, peanut, olive and other oils.

Animal fats, as beef, mutton and bone fat, lard and lard oil, etc.

Hurtful coloring matters as chrome yellow (chromate of lead) and coal tar colors, Victoria yellow (Saffron substitute). (Potassium Dinitrocresylate). Martius yellow (calcium or potassium Dinitronaphthylate), etc. Vegetable coloring matters ordinarily reckoned as harmless are not considered as adulterants by many. Such are annatto carrot juice, turmeric and marigold flowers. (Calendula off). The fruit of the winter cherry (Physalis alkekingi) has been mentioned, but Tollen's Handbuch off. Gesundheitswesens, vol. 1, p. 46, says that it is improbable.

DETERMINATION OF WATER AND OTHER NON-FATTY CONSTITUENTS.

Determination of water and other non-fatty constituents.

It has been noted by Husson that when much water is present in a butter if a knife is plunged into it droplets of water appear in the track of the knife. He also notes that the butter is crumbly when large amounts of water are present.

In taking a sample for the quantitative determination of the water, care must be taken to obtain what fairly represents the average constitution of the sample. If the sample is soft it may sometimes be previously mixed with a stout spatula. If hard, a fair proportion of the interior and exterior must enter into the portion taken. Since the water and fat do not readily mingle the mixture of the two can seldom be made quite uniform. The simplest method in theory proposed for the determination of the water, consists in melting a known amount in a graduated tube, and reading off the volume of water. Although the volume of water is slightly increased by the curd and salt present in it, the increase is umimportant in this relation, and the relative proportion of the water by weight can be calculated. (Hassall). The difficulty with this method is that the curd only partially separates from the fat, and the line of demarcation between the water and fat is rendered indistinct.

A method used by Heeren, Lefeldt, Birnbaum and others consists in melting the butter as before mentioned and forcing a more complete separation in a short time by placing the graduated tube in an iron cylinder attached to a rotating arm. The apparatus described and figured (first report, NewYork State Dairy Commissioner, p. 106), as in use for the determination of the volume per cent of cream in milk can be used, or a rougher device of the same character can be improvised. (v. Babo Landwirth correspblatt., Baden, 1863 p. 65. Martigny Die Milch, etc., 1, 1871-2, 195).

The use of some solvent for the fat as ether, or petroleum naphtha to facilitate the separation is a method adopted by some. Babo uses ether and the centrifugal apparatus already alluded to. Other chemists as Hoorn, Johanson, Caldwell (2d Report N. Y. State Board of Health, p. 526), and others use petroleum ether, and after thorough shaking allow the graduated tubes to stand, and read off.

An objection, though not a very serious one, to these methods is that the results are obtained in percentages by volume and not by weight, the usual form in which they are to be recorded or reported. On this account most analysts prefer the determination of water by weight.

Weighing the butter in a cylinder, adding about twice its weight of water, keeping the fat melted until clear, solidfying and cooling, and then running out the water and weighing the fat remaining is a method used by some. (Duflos and Hirsch, Die wichtigsten Lebensoeduerfnisse, etc). Griessmayer, Die Verfalschung der wichtigsten Nahrungs und Genussmittel, 1880. Elsner Die Praxis des Nahrungsmittel—Chemikers, 1880.

Dietzsch (Nahrungsmittel und Getraenke, 4th ed. Zurich, 1884, p. 214), accomplishes this in an ingenous manner by putting the mixture. of butter and warm water into a tube, corking it up, shaking and cooling in an inverted position so that the fat occupies the lower end of the tube and the water may be readily poured off.

These methods are properly a determination of the amount of fat and not of the water, though by them data can be obtained from which the approximate percentage of water may be calculated.

A practical objection to this method is that with many butters, as well as substitutes therefor, much of the curd holding moisture will remain suspended in the melted fat, rendering the complete separation of the water difficult or impossible. The water poured off contains the salt and a part only of the curd, so that with the most careful weighing it affords only an approximate estimation of the amount of fat present. Objections of this character are also presented by Cauldwell (Loc. cit).

Drying a weighed quantity of butter in a dish over a boiling water bath is another more accurate method (Blyth).

It requires some considerable time, during which there is a danger of some loss of the fat at the temperature required. The water being heavier sinks to the bottom of the melted fat and is protected by it against rapid evaporation. The addition of a small amount of absolute alcohol to cause the water and fat to commingle and allow water and alcohol to be driven out together, only partially remedies this difficulty since the curd acts as a sponge and obstinately retains the water.

The method which has been found the most satisfactory (Angell and Hehner, p. 12), consists in drying a weighed portion of the sample in a capacious platinum dish over a low flame, the mass being continually stirred with a thermometer which should not indicate over 105 to 110 degrees C. (220 to 230 degrees F.), throughout (Blyth, foods).

The operation is finished when bubbles of steam cease to escape through the fat. The method requires constant attention, but is more rapid and satisfactory than drying over the water bath. From seven to ten grammes is a good quantity to use for this test. It is well to heat cautiously for some little time until the curd contracts in bulk from loss of water before allowing the thermometer bulb to come in contact with the curd, otherwise it attaches itself very firmly and is removed with difficulty. The small amount of fat adhering to the thermometer bulb may be rinsed into the dish with a few drops of ether, and the ether driven off by a few minutes heating on the water bath.

The other non-fatty adulterants mentioned, except coloring matters which usually remain with the fat, are frequently neglected by analysts because of their infrequency (Allen 2, p. 202). These analysts content themselves with a determination of the fat alone, or of the water and the fat. If the fat is eighty per cent or over they regard the sample as unadulterated so far as non-fatty constituents are concerned.

Bischoff (Handb. off. Gesundheitswesens, 2.509) states that in none of the samples of butter examined by him in Berlin, since 1875 (some 800), has he found any of the mineral or vegetable adulterations mentioned, though accusations of that sort have frequently been made. The term "bosh butter" has been applied according to Hassall (p. 430), to butters adulterated with starch or substances containing it. According to a witness before the New York Legislative Committee on public health (p. 227), the name was derived from the name of the town Hertogenbosch, in Holland where large amounts of artificial butter are made.

To determine the character of the non-fatty adulterants other than water, Hilger (Handb. der Hygien. Nahrungsmittel, Leipzig, 1882, p. 256), recommends taking five to ten grammes of the butter, adding to it in a test tube about twice as much water (preferably containing alcohol) and warming for some time at the melting point of the fat. The fatty layer retains the foreign coloring matters, the aqueous layer below dissolves the salt, borax, alum, alkali, water glass, salicylic acid, etc., while the insoluble portions, as starchy material, heavy spar, chalk, etc., settle to the bottom.

Dietzsch (Loc. cit.) gives a detailed method of testing the solution and sediment obtained by a similar treatment. The sediment is filtered out, then the caseine is dissolved off by ammonia, leaving starch and insoluble mineral matters to be tested by the microscope by iodine for starch and by other well known chemical methods for chalk, gypsum, heavy spar (barium sulphat), soapstone, etc.

In the water solution tests may be made for the soluble adulterants. J. B. Edwards, of Montreal, states that borax to the extent of one or two per cent is not, in his estimation, an adulteration. By a large number of chemists, however, the presence of any borax is regarded as a hurtful addition, since it is believed by them that even small amounts in food interfere materially with the processes of digestion, the effect being naturally more marked where the digestive powers are weak.

To detect lead, copper or zinc compounds, the two latter usually being present as the result of using brass, copper or zinc vessels in the manufacture or storing of butter Dietzsch (p. 217), also Tollens (Handb. off. Ges. p. 492), recommends boiling fifty grammes of the butter for fifteen minutes with 100 to 120 cubic centimetres of water, and ten cubic centimetres pure hydrochloric or nitric acid, filtering through a wetted filter, and testing the acid solution for these metals with sulphuretted hydrogen, potassium ferrocyanide, etc.

The addition of gelatin to butter, presumably for the purpose of giving it more consistency in warm weather, is a matter which has come up but recently and has not yet been mentioned in the literature of the subject. The presence of the gelatin is very difficult of detection since in many of its properties it bears a close resemblance to the albuminoids of the curd. Besides this, our knowledge of gelatin and of the albuminoids in milk is as yet very incomplete.

A method for its detection was devised which seems to answer the requirements of the case tolerably well. The gelatin was extracted from fifty to seventy-five grammes of the butter by heating with water, the water solution partly cleared by filtration or straining through muslin, was then boiled with the addition of a drop of dilute acetic acid and a few drops of potassium ferrocyanide until the precipitate, which only forms after a time, turned blue and was in a state to filter readily. In the warm filtrate the gelatin was tested for by solution of tannic acid or chlorine gas. With most butters unadulterated with gelatin, only a slight turbidity was produced. With a few containing much buttermilk or with milk serum the turbidity was more dense, sometimes affording a flocculent precipitate. When gelatin was present a stringy p e tate of very different character and very decided was obtained.r Examinations .of the fat, so far as the experiments went, did not indicate that any reaction between the fat and the gelatin had taken place.

The addition has been chiefly suspected in the case of artificial butters, and with these the question comes up as to whether some gelatinoid substance may not be extracted from the membrane in the process of rendering the fat. The presence of some such substance is specified in some of the patents relating to the manufacture of oleomargarine and other fats. (Paris, Pat. No. 19,011, Jan. 1882, Huet. Eng. Pats. 5,249, 1881, and 134, 1882, Huet.)

Butyric acid is reported as an adulterant of butter by J. Zanni (Quelques essais sur des Beurres fondus, Constantinople, Fres. zeits. f. Anal. Chem. 23, 87). To detect it he recommends washing the butter fat and testing for and determining the butyric acid in the washings.

On account of the presumable infrequency of adulteration by organic non-fatty substances or mineral substances other than salt, it is a common practice among analysts to call all the mineral constituents of butter "salt," and the organic non-fatty substances "curd" or "caseine" (Allen 2, 204, etc).

The mineral constituents of butter, salt and mineral adulterants, are determined by incineration of weighed amounts of the material directly, or incinerating the residue after extracting the fat with ether or some similar solvent. Lead, if present, might be thus volatilized and escape detection, saltpetre also would be destroyed, but all other mineral adulterants enumerated will remain and can be detected by the well known methods of mineral analysis.

If a strong heat is applied before the removal of the water there is danger of mechanical loss by the boiling of the water beneath the layer of fat. After removing the water the fat can be burned off quietly. The salt often fuses about some of the partially burned carbon in the residue so as to protect it from oxidation, leaving the ash gray, or even black, and if the heat is continued until the ash is perfectly white, so much salt has been volatilized that the quantitative results are erroneous. If the ash is light gray, without long heating the results are sufficiently close. If black, the most accurate method is to dissolve all the soluble salts with water, and burn the carbonaceous residue, add the water solution, evaporate, ignite and weigh. Titration with standard silver nitrate solution will then show whether the mineral matter is entirely or only partially common salt. The fat may be determined approximately by the method of Dietzsch. The amount of fat is determined by melting the butter until the curd has settled, pouring off as much of the clear fat as possible into a weighed dish, dissolving off the remainder of the fat into the same dish with ether, petroleum spirit or carbon disulphide, evaporating off the solvent and weighing.

Some chemists use the solvent from the start without pouring off any portion of it. In the case of some butters where the curd separates with difficulty, this is the only satisfactory plan. The residue insoluble in the ether may be dried and weighed, giving curd and salt. If the amount of water is known the percentage of fat may be determined indirectly by difference, a plan recommended as accurate by Allen (2, p. 105), Hehner and Angell (p. 13, etc.).

Some prefer to mix the butter with sand or pounded glass and extract in a fat extraction apparatus as Soxhlets. Hilger (Handb. Hyg. Nahr. p. 256) Tollens, Soxhlet, Thom., etc.

The proportion of caseine or curd, or what is generally reckoned as such is usually obtained by difference. The approximate estimation of the fat by melting with water in a tube and pouring off the water has already been mentioned.

EXAMINATION OF THE FAT.

The most important point at the present day in the examination of butter is the determination of the character of the fat, and first, it may be stated that what is known in commerce or in daily life as a simple fat, consists almost invariably of a mixture of several different substances called "fats" by the chemists. In the chemical sense a fat results from a combination of a fatty acid with glycerine, water being eliminated at the time of the union. Such fats are called glycerides, and the names by which they are designated terminate in "in," *e. g.*, stearin is the glyceride of stearic acid, palmitine is the glyceride of palmitic acid, arachin, that of arachidic, etc., etc.

Glycerine has the property of uniting with one, two or three molecules of fatty acid affording mono-di or tri-glycerides designated according to the acid, *e. g.*, mono-stearin, distearin, tristearin, monopalmatin, dipalmatin, tripalmatin, etc., etc.

In almost all the fats occurring in nature the fats are tri-glycerides, tristearin, tripalmatin, triolein, etc., and in speaking of natural products the prefix "tri" is omitted, being understood.

The mode of combination of glycerine with fatty acids may perhaps be best understood from the following example :

Glycerine. Stearic acid. Tristearine, glyceryl tristearate. Water.

$$C_3H_5\left\{\begin{array}{l}HO\\HO\\HO\end{array}\right. + \begin{array}{l}H\,C_{18}H_{35}O_2\\H\,C_{18}H_{35}O_2\\H\,C_{18}H_{35}O_2\end{array} = C_3H_5\,......\left\{\begin{array}{l}C_{18}H_{35}O_2\\C_{18}H_{35}O_2\\C_{18}H_{35}O_2\end{array}\right. + \left.\right\} \;3H_2O$$

Glycerine. Stearic acid. Distarine, glyceryl distearate. Water.

$$C_3H_5\left\{\begin{array}{l}HO\\HO\\HO\end{array}\right. + \begin{array}{l}H\,C_{18}H_{35}O_2\\H\,C_{18}H_{35}O_2\end{array} = C_3H_5\,......\left\{\begin{array}{l}C_{35}H_{35}O_2\\C_{35}H_{35}O_2\\OH\end{array}\right. + \left.\right\} \;2H_2O$$

Glycerine. Stearic acid. Mono-stearine, glyceryle mono-starate Water.

$$C_3H_5\left\{\begin{array}{l}OH...\\OH.\\O\end{array}\right. + H\,C_{18}H_{35}O_2 = C_3H_5\,......\left\{\begin{array}{l}C_{18}H_{35}O_2\\OH\\OH\end{array}\right. + \left.\right\} \;H_2O$$

The formula of stearic acid may be also written $C_{18}H_{36}O_2$. This particular fatty acid has been selected for illustration as probably the most familiar one of those existing in butter and other commercial fats. It belongs to the best known series of fatty acids which are most common in these products; the series designated as the $C\ n\ H_2\ w\ O_2$, series where the number of hydrogen atoms is twice that of the carbon atoms, two atoms of oxygen always being present. The few members of this series may be given by way of illustration:

Formic acid, $C\,H_2\,O_2$.
Acetic acid, $C_2\,H_4\,O_2$.
Propionic acid, $C_3\,H_6\,O_2$.
Butyric acid, $C_4\,H_8\,O_2$.
Valerianic or valeric, $C_5\,H_8\,O_2$, etc. etc.

There are other series as the $C\,n\,H_{2n-2}\,O_2$ to which the oleic acid belongs etc. In the accompanying table are given the formulæ occurrence, etc., of the principal fatty acids which have been isolated and identified:

It will be noted that margaric acid is not recognized as a constituent of butter or any other natural fat. A substance in the condition of glyceride and the corresponding fatty acid was described in his researches on fats by Chevreul and by him named margarine and margaric acid. He supposed it to be an individual acid with the formula $C_{17} H_{34} O_2$, and that it was a frequent and important constituent uent of animal and other fats. It has since been proved that what he examined was a mixture of palmitic and stearic acids, $C_{16} H_{32} O_2$, and $C_{18} H_{36} O_2$, and that the true margaric acid was not a natural constituent of any of our common fats.

It was supposed by Chevreul to exist in butter constituting together with olein the chief substances in their fat. Hence the manufacture of the term "oleomargarine" as a trade name for the substitute for butter.

Chevreul's statement was that butter fat consisted of stearin, margerin and olein, with small quantities of butyrin, caproin, and caprin to which its odor is due. According to Heintz, it contains oleine, much palmitin, some stearin and very small quantities of glycerides, yielding myristic and butic (or arachidic)acids (Watts Dict. 1, 687). Until 1874 the only recorded analysis was that of Bromeis, which was given thus:

	Per cent.
Margarine	68
Butyroleine	30
Butyroleine, caproine and caprine	2

Later researches have proved this analysis to be erroneous in every detail. The margarine, as indicated above, has been shown to be a mixture of stearin and palmitin. That called margarine by Bromeis was also found to contain glycerides of butyric and capric acid. The butyrolein, as appears from the experiments of Gottlieb is nothing but triolein, while the butyrine, caproin, caprin, exists in much larger proportion in butter, as proven by the experiments of Angell and Hehner (Butter, its Analysis and Adulterations, London, 1874. See review of the same, Chemical News, vol. 30, page 174).

Blyth states that the general composition of butter appears to be as follows:

Glycerides:

	Per cent.
Oleine	42·21
Stearine and palmitine	50·00
Butyrine	7·69
Caproine, Capryline, Caprine (Rutin)	·10
	100·00

Fatty acids:

Oleic acid	40·40
Stearic and palmitic acids	47·75
Insol. acids	87·90

g
f
f

g
t
p
i
U
g

M
E
E

d
n
w
b
t
p
H
re

fo

O
S
B
C
C
C

Butyric acid	6·72
Caproic	......
Caprylic	......
Capric (Rutic)	......
Total acids calculating soluble as butyric..............	94·62

Muter obtained 40·4 per cent and 34·8 per cent oleic acid in genuine butter (Analyst, II, 73).

According to E. Wein (Sitzungsber, d. Phys. Med. Soc. Erlangen, XI, page 1664) butter contains:

	Glyceride.		Corresponding acid.
Butyrine	$C_3 H_5 (C_4 H_9 O_2)_3$	Butyric	$C_4 H_8 O_2$
Caproine	$C_3 H_5 (C_6 H_{11} O_2)_3$	Caproic	$C_6 H_{12} O_2$
Capryline	$C_3 H_5 (C_8 H_{15} O_2)_3$	Caprylic	$C_8 H_{16} O_2$
Caprine or Rutin	$C_3 H_5 (C_{10} H_{19} O_2)_3$	Capric	$C_{10} H_{20} O_2$
Myristine	$C_3 H_5 (C_{14} H_{29} O_2)_3$	Myristic	$C_{14} H_{28} O_2$
Palmitine	$C_3 H_5 (C_{16} H_{31} O_2)_3$	Palmitic	$C_{16} H_{32} O_2$
Stearine	$C_3 H_5 (C_{18} H_{35} O_2)_3$	Stearic	$C_{18} H_{36} O_2$
Arachin or Rutin	$C_3 H_5 (C_{20} H_{39} O_2)_3$	Arachidic	$C_{20} H_{40} O_2$
Olein	$C_3 H_5 (C_{18} H_{33} O_2)_3$	Oleic	$C_{18} H_{34} O_2$

and this list is the one generally accepted. The author also states that formic and acetic acids have been found in minute quantities. Reichardt (Arch. Pharm. [3] X, 339), includes laurine or laurostearine, $C_3 H_5 (C_{12} H_{23} O_2)_3$, in the list; so also Koenig, Lecithin $C_{42} H_{84} NPO_9$, is also a constituent of butter (Schaedler, page 641).

The elementary composition of butter as compared with other fats will be found in the following table:

Table VI.

Elementary composition of animal fats.

NAME OF FAT.	Melting point.	Solidification point.	Per cent carbon.	Per cent hydrogen.	Per cent oxygen.
	Degrees.	Degrees.			
Mutton fat.............. ...	41.0–52.5	24–43	76.61	12.03	11.36
Ox fat.....................	41.0–50.0	36	76.50	11.91	11.59
Pig fat	42.5–48.0	28	76.54	11.94	11.52
Dog fat (omentum).........	40.0	26	76.66	12.01	11.33
*Cat fat..............	38.0	Temp. of room.	76.56	11.90	11.44
Horse fat (comb)...........	Fluid at ordinar	y temperatures.	77.07	11.69	11.24
Human fat (kidney)........	41.0 Soft	at temp. of room.	76.44	11.94	11.62
Human fat (omentum)......	Fluid at temper	ature of room.	76.80	11.94	11.26
Butter (fresh)..............	37.0	....	75.63	11.87	11.50

*Extracted from tissues and intestines of a lean cat.

The fat was taken from the freshly killed animal, freed from blood and meat, dried, minced, melted and filtered, the remaining membrane exhausted with boiling ether and residue united with fat first obtained. The difference in fats taken from different portions of the body is very small. Kidney fat is generally the hardest and that from the omentum the softest.

The mean composition of mutton, ox and pig fat corresponds to the empirical formula:

$$C_{106} H_{99} O_{12}$$

	Calculated.	Found.
C	76·53	76·5
H	11·91	12·0
O	11·56	11·5
	100·00	100·0

Some careful experiments by Dr. J. Bell would seem to prove that the glycerides in butter are not simple but complex in character. He certainly shows by them that the ordinary butter-fat is not the same as an artificial mixture of the component glycerides. By treating such an artificial mixture with solvents he was able to separate out different proportions of the glycerides in the mixture, tributyrin, triolein, etc., whereas on treating butter-fat with the same solvents no such separation was effected.

The hypothesis is offered that the glycerides are compound in character, *e. g.*:

$$\left.\begin{array}{ll}\text{Butyric,} & C_4 H_9 O_2 \\ \text{Stearic,} & C_{18} H_{35} O_2 \\ \text{Oleic,} & C_{18} H_{33} O_2\end{array}\right\} C_3 H_5.$$

The point is theoretically interesting and calls for further investigation.

THE MANUFACTURE OF SUBSTITUTES FOR BUTTER.

Although the admixture of lard and fats of a similar character with butter was not altogether unknown previous to the experiments of Mège Mouriès, the systematic manufacture of substitutes for butter from materials derived in part or entirely from the fat of beeves or other domestic animals dates from the time of his invention. About 1866 or 1867 Mège Mouriès at the instance of the French government, instituted experiments with a view to the preparation of a cheap substitute for butter which might be suitable for use in the navy and by the poorer classes.

By 1870 the experimenter had a factory in operation at Poissy, near Paris, where the butter substitute was manufactured and sold under the name "oleomargarin." The Franco-Prussian war then intervened and it was not until April 1872, that the Council of Health of the Department of the Seine, accepting the favorable report of M. Felix Boudet, admitted the new article to the trade, carefully stipulating, however, that it was not to be sold as butter.

† Schulze and Reinecke Ann. Chem. Pharm., 142, 191.

The process was patented in England in 1869 and in the United States in 1873. It was described in many of the scientific and trade periodicals at the time. Among these may be mentioned the report of M. Boudet (Monit. Sci., an abstract of which is published in American Chemist, vol. 4, p. 370, Th. v. Gohren, Die Kunstbutterfabrikation. Fuehlings landwirthsch. Zeitung, vol. 25, Heft 1. V. Lang, Die Fabrikation Der Kunstbutter, Sparbutter und Butterine, Wien, Pest, Leipzig, 1878, etc.) From the latter, which gives a very complete account of Mège Mouriès process the following is abstracted:

For the purpose the lung and kidney fat from freshly-slaughtered beeves must be selected. The operations are:

1. Washing the fat.
2. Hashing.
3. Melting.
4. Crystallizing.
5. Pressing.
6. Churning.

The most scrupulous cleanliness is absolutely essential — even a small amount of fat if allowed to adhere to the apparatus and utensils used is liable to decompose in such a way as to spoil the succeeding batch of material worked up.

The fat must be kept quite cool while in transport from the slaughter house.

1. In washing, the fat selected for the purpose is placed in a tank and water at a temperature of 16 degrees to 18 degrees C. (65 to 69 F.) is sprayed over it until it is covered. The fat is then worked in this for about an hour when the water is run off. Water is sprayed over it again and it is again worked for an hour. This must be repeated until the water runs off perfectly clear. With proper care the second washing may suffice.

2. In "hashing" as it is termed, the endeavor is made to create as fine a state of the subdivision of the membrane containing the fat as possible, since the yield is materially affected by this circumstauce inasmuch as the temperature for melting is restricted to a certain limit.

3. For melting it is necessary to use steam heat which is carefully regulated. Since some of the finer portions of the membrane may mix with the fat, rendering it difficult to remove it without material loss, the pepsin extracted from the stomachs of freshly killed calves or pigs, or else the stomachs themselves washed and cut in pieces are introduced, together with a little caustic alkali or alkaline carbonate. The charge usually is:

Hashed raw fat,	1,000 kilos.	2,000 lbs.
Water	300 kilos.	600 lbs.
Sodium (or potassium):		
carbonate,	1 kilo	2 lbs.
stomachs,	2 kilos	4 lbs.

If caustic alkalies are used, one-half kilo (1 lb.) of soda or potash is sufficient. The water is first run into the melting tank and by means of the steam coil quickly brought up to 45 degrees C. (113 degrees Fahr.) The dissolved alkali is then added and the chopped stomachs thrown in. After keeping the whole at 45C. (113 deg. Fahr.,) for ten minutes the fat is run in from the hasher. The mass is stirred together, being kept as nearly as possible at 45C. (113 deg. Fahr.) The operation requires two to two and a half hours, at most three. After settling, the fat is run off from the top, through a fine hair sieve into the settling tanks. These tanks are provided with double walls into the space between which steam is admitted in sufficient proportion to keep the temperature at 45 C. In these the fat must stand until it is clear. The addition of about two per cent of salt aids the clearing operation by dissolving in the water suspended in the fat, and rendering it more dense. By such addition the fat is cleared in about two hours.

4. By cooling in a chamber which is rigidly kept at a temperature of 23 to 25 C. (73 deg. to 77 deg. Fahr.) a large proportion of the stearin and palmitin are crystallized out, the fat becoming apparently solid. The operation is complete in about twenty-four hours.

5. The fat is then put in bags and the bags alternating with iron plates which are placed in a vertical position are subjected to the action of an hydraulic press. The temperature of the pressing room is maintained at 25 C.

The oily product expressed constitutes the oleomargarine oil and in it the olein predominates. The hard fat remaining in the press consists chiefly of stearin and palmatin and is turned over to the candlemaker. The average yield is stated as:

Hard fat, stearine, palmatine 40 to 50 per cent (melting point 40 deg. to 50 deg. C.)

Oleomargarine oil 50 to 60 per cent (melting point 20 deg. to 22 deg. C.)

This oleomargarine oil without further treatment constitutes at present an article of commerce. When run into metallic boxes of a certain form previously heated to 100 deg. and allowed to cool down to 18 or 20 deg., it constitutes what has been known in France as "Marine butter."

6. To make the "oleomargarine butter" from this "oleomargarine oil," it is necessary to impart to it the color and aroma which it still lacks. For coloring a mixture of extracts of annatto and of turmeric is recommended, and directions are given for the preparation of each.

To give the aroma Lang recommends cumarin, the substance said to give the odor to new-mown hay and to various field plants which is found in larger quantities in the Tonka bean.

Fresh milk is also necessary to impart the proper aroma and flavor. To effect an intimate mixture of the flavoring and coloring matters with the fat is necessary that the latter should be minutely subdivided and mixed. This requires that the fat shall be emulsified. Mège Mouriès discovered that the udder of the cow contains a substance extractable by water which will emulsify a fat.

The churns for making the butter should be comparatively small as time is thereby saved. A capacity of 200 liters is the maximum though that of 150 to 160 liters is preferable. They consist of casks mounted on a rotating axis which passes diagonally through them. To prepare them for use paraffin heated to 130 deg. to 140 deg. C. is poured into them and they are rotated until the paraffin has cooled to about 60 deg. to 65 deg. C. when it is run out, leaving only enough to thoroughly saturate the wood.

The charge for each churn varies. For the best product the charge contains 40 to 50 per cent of fresh milk. The usual charge is:

Liquid oleomargarine. 50 kilos.
Fresh milk, 20 to 25 liters.
Aqueous extract of the udder, 20 to 25 liters.
Coloring matter, aroma, etc., in suitable quantities.

The product is still better if the oil is washed with warm water just previous to churning. After the churn has been in rotation for fifteen or twenty minutes, the contents are all in a state of emulsion and resemble cream. The rotation is kept up for two hours at as nearly 17 C. (63 deg. F.) as possible. At the end of that time the butter will be found to have separated in masses resembling those to be found when churning genuine cream. The product is treated essentially in the same way as ordinary butter. It is worked in water of 8 deg. to 10 deg. C. or colder, washed by passing through a machine which spreads it in a thin layer, while a stream of ice-cold water plays upon it, drained, salted and packed.

The yield is stated to be as follows:

One ox affords, 83 kilos of crude fat; 28 kilos caul fat, giving 18 kilos of artificial butter, besides 3 kilos scrap. Yielding 85 kilos stearo-palmitin and 16·5 kilos oleomargarine, 55 kilos crude tallow. Giving 26·5 kilos tallow, 18·5 kilos wet scrap.

The cumarine is thus prepared: Tonka beans are cut in small pieces and soaked for some time in strong alcohol. The extract contains cumarine, fat and brown coloring matter. It may be used in this state, but it is preferable to extract the pure cumarine by distilling the alcohol off with a carefully regulated heat, boiling the remaining syrup with water, filtering through several filter papers which absorb the fat and cooling the solution down, when the cumarine crystilizes out in needles. It is quite soluble in alcohol and hot water, very slightly soluble in cold water. For use it is dissolved in strong alcohol, and the solution added to cold water so long as the solution remains clear. Numerous substances are reported to have been prepared for this purpose by dealers in extracts and essences.

Lang's account of the American method of manufacturing oleomargarine oil and butter described a similar treatment of the fat, omitting, however, the use of stomachs of calves, etc. He also states that 100 liters of the oleomargarine oil are churned for one and one-half hours with twenty liters of sour milk, from which the curd has been filtered off. The account does not altogether agree with the description given by Mott of the process pursued by the United States Dairy Company, which is briefly as follows.

H. A. Mott, Artificial Butter, Proceedings Am. Chem. Soc., I, 156.

The fat is first sorted; all portions to which blood adheres being washed separately. It is first washed with tepid water, and then three times with cold water.

The fat is then passed through the "hasher," by which it is cut very fine, and finally forced through a fine sieve. The product is then placed in the "melting tank," in which it is heated by heating the water surrounding it, until the temperature of 122 or 124 degrees Fahrenheit is attained. Experiments have shown that this temperature cannot be passed and a sweet and odorless oil obtained.

In this process the fat must be continually stirred to maintain an even temperature. When the fat has melted, it is allowed to stand that the "scrap" may settle to the bottom. The fat is then skimmed and the clear oil is drawn off and set aside to cool in wooden cars. Should the scrap refuse to settle, vigorous stirring will usually cause it to do so. If not, throwing on some salt and stirring it will have the desired effect. This melting process occupies some two or three hours. The oil in the cars requires at least twelve to twenty-four hours to granulate, which should take place at a temperature of about 70 degrees Fahrenheit.

The solidified oil is now taken to the "press-room," which is kept at a temperature of about 89 to 90 degrees Fahrenheit. Here it is packed in cloths, and is then put under the press. The pressure should only be gradually increased, and is continued until the oil ceases to flow. The oil is caught in a suitable receptacle, while the cakes which remain consist of pure white stearine, which is ready for sale.

The oil is again cooled to 70 degrees Fahrenheit. One hundred pounds of the oil are then taken, and with fifteen to twenty pounds of sour milk, are placed in the churn. Two and a half to three ounces of solution of annatto, containing one-half to three-fourths ounce of bicarbonate of soda, may then be added, and the whole is agitated for about ten or fifteen minutes, when it is at once run into a tub containing pounded ice, the oil being kept in constant motion, until suffi-sufficiently cooled. By this process the grain is completely removed. After remaining in contact with the ice for two or three hours, it is then dumped on an inclined table and crumbled up so that the ice may melt out. Then about thirty pounds at a time are put into a churn with twenty to twenty-five pounds of churned sour milk, and the whole is churned for about fifteen minutes. By this last process the flavor and odor desired are imparted to the butter.

Working, draining and salting the product (three-quarters of one ounce of salt to the pound) complete the manufacture.

Another account of the process as pursued in Europe at a more recent date may here find a place.

The process consists (1) in washing the fat (2) crystallizing the fat (3) pressing the crystallized fat (4) churning with olive oil etc., and milk.

The fresh suet is first freed from all adhering tissue and is then thrown into large tubs where the blood is carefully washed off by means of cold water. It is then put through a meat hasher, where it is cut and torn up into a white mass which is delivered into a kettle, jacketed with warm water and supplied with a stirring apparatus. Here it is warmed up to 50 deg. C. with constant agitation for two hours. The stirring is then stopped, water is introduced through H. and the rendered fat is forced into a jacketed tub through a pipe. From here it is drawn into small trays and in twenty-four hours is cooled down to 27 deg. C., when it is wrapped in cloths and by hydraulic pressure the oleomargarine (a mixture of stearin, palmitin and olein) is squeezed out of it. The oleomargarine is then put into a churn together with milk, olive sesame or peanut oil and a little butter color. In ten to fifteen minutes the churning is completed. The churned mass is then cooled with ice and is worked in the same manner as butter to remove milk and water.

Artificial butter carefully prepared will keep for months without becoming rancid.

In rendering the fat Mége used gastric juice which he obtained from a sheep's stomach and converted it into pepsin by adding carbonate of soda and phosphate of calcium. He also thought that the excretions of the mammary glands were necessary to convert the oil into butter and he therefore added them to the contents of the churn; later experiments, however, have proved that those two precautions are unnecessary. His patent, however, involves the two points (1) the extraction of the oil from the fat at a low temperature; (2) converting that oil into butter by churning with milk.

Lang also mentions the mode of manufacture of Hamburg, as reported to be practiced in that city.

Rape-seed oil and dry potato starch, in the proportion of six pounds of the former to six ounces of the latter, are heated together for some time at a high temperature, about 300 C., (572 deg. F.) until a clear yellow oil is formed. This when mixed with beef tallow gives a product resembling the boiled butter, Schmalzbutter or Rindschmalz, suitable for use in cooking. Lang, however, states that by following the directions he was unable to obtain an edible product.

In the report of the Committee on Public Health (Investigation of Dairy Products, Albany, 1884), are to be found the accounts of the most recent methods of manufacturing oleomargarine and other butter substances.

The only materials used, according to the statements of the manufacturers, are oleomargarine oil, lard, cotton-seed and sesame oils and coloring matter (annatto). The oleomargarine oil is made from picked caul fat (fresh) which is hashed, washed with ice water, and rendered at 120 to 140 F. The fat is kept warm for some time and then pressed. The solid residue is called stearine, the liquid portion constitutes the oleomargarine oil. The manufacturers strenously object to having the name "tallow" applied to this part, as conveying a false impression.

The lard is prepared by rendering pure leaf lard at 140 F., and the product is then deodorized by long soaking in ice water containing salt. One witness stated that some other chemicals were used besides salt, but refused to tell what they were. Another stated that nitric acid was used which was denied by others. Sesame or benne oil is purchased. It is said to be an imported article made from the seed of an African plant. It softens the product, preventing it from becoming brittle and giving it a gloss. For this purpose it goes further than the same quantity of butter-fat. The annatto is rendered soluble by treating with alkali.

The proportion of materials taken varies somewhat with the season. For winter oleomargarine, oil and the prepared lard are taken in the proportion of five to one. To every two thousand pounds of this mixture are added thirty to forty pounds of sesame oil. The whole is then melted together and then churned for twenty to twenty-five minutes with about five hundred pounds of milk, the coloring matter being also added in suitable quantity. The oil is then run on ice, drained, salted, worked and packed. The formula given by another manufacturer is, one thousand pounds oleomargarine oil, five hundred pounds neutral lard, ten gallons sesame oil, melted together and churned with five to six hundred pounds of milk and eighty ounces coloring matter. This product it was calculated would contain at most five per cent of butter. The cost of oleomargarine butter to the manufacturer is stated to be thirteen to fourteen cents a pounds. It was also stated that there are twenty-eight or thirty factories in the country working under the Mége patent, two of which are in this State, the inference that these factories do not use lard in the manufacture being implied in the testimony.

Butterine consists, according to the evidence given, of lard and butter prepared by churning the deodorized lard with milk in proportion according to the quality demanded. It is also colored with annatto. The lower grades contain twenty per cent of butter, the higher up to fifty per cent.

"Suine" seems to be a term applied to the lower grades of butterine where the proportion of lard is high. There seems to be no sharp division between the two.

Our State laws are said to classify all substitutes for butter under the name "oleomargarine," and hence a certain amount of confusion results. It would be difficult, however, to foretell the various names under which these substitutes for butter may be placed on the market.

It is most strenuously insisted by the manufacturers that the fat from diseased animals, or even from those not freshly killed, cannot be deodorized and used for this purpose. The resources of modern chemistry, however, have frequently been drawn upon in just such cases, and there is no good reason for the assumption that such deodorization is impossible.

In table VII are given the results of some analyses of oleomargarine butter. It will be seen that in respect to proportions of fat, water, curd and salts, they do not differ materially from those of genuine butter. In the earlier days of the oleomargarine manufacture it was claimed that the product always contained less water and matters not fat than the genuine butter. That was true to only a slight extent and it is a matter of no more difficulty to incorporate twenty per cent or more of water with oleomargarine than with butter.

The list of substances mentioned in the patents relating to the manufacture of artificial butter when put together is of the most motley description and no one who knows what these names mean would imagine that he was reading over a list of substances to be used in the preparation of an article of food.

Numerous oils, chiefly vegetable, are used or purport to be used in the manufacture of these substitutes for butter. A number of these oils were examined as will be seen later on, in order, if possible, to afford data for their possible detection when mixed with other fats. A description of these oils and their sources may here find a place.

Cotton-seed Oil.

Cotton oil..........................	Kootin Beerson (Arabian).
Baumwollensaatoel..................	Rapase Tula (Bangalem).
Baumwollensamenoel...............	Hotten (Arabian).
Huile de coton........................	Rokapas Root (Hindostan).
Oleum Gossypii.....................	Watta (Java).
Poombeh (Persian).................	Rapen (Ceylon).

Fig. 201.

Olivenbaum. Olea europaea Linn.

Fig. 194.

Baumwolle. Gossypium herbaceum Linn

Cotton oil is obtained from the seeds of the various species of *Gossypium*, Linn, the most important of which are :

Gossypium herbaceum, Linn. found native in Asia and most extensively cultivated to-day.

Cotton Gossypium herbaceum, Linn. Branch with blossoms and fruit. 1. Crown spread out showing the stamens. 2. Calyx with embryos. 3. Embryos cross sections. 4. Embryos long sect. 5· Capsule with adhering leaves of calys. 6. Burst capsule with seeds. 7. Capsule without seeds. 8. Seeds surrounded with cotton. 9. Seeds without cotton. 10. Seeds long sect. 11. Seeds cross-cut showing the folded leaves of the bud. V. Schaedler.

The cotton seed grows wild in Asia, Africa and America, and can be advantageously cultivated in locations having an average temperature of 25 to 35 degrees C.

The seeds which yield the oil are separated from the fiber by means of the cotton gin. They contain about twenty-five per cent of oil, and until the beginning of this century were allowed to go to waste.

In making the oil the seeds are dried, the shell is removed and the oil extracted by means of pressure or solvents. The crude oil thus obtained is purified by coagulating the suspended material with boiling water or steam. The clear liquid is drawn off and heated with alkaline carbonates or caustic alkalis when the oil separates in three layers: the upper one consists of the refined oil, the middle one of the saponified fat and the lower is a dark colored lye containing the coloring matter.

Cotton seed oil is used to adulterate olive oil, also as a luminant, a lubricant, a liniment and in the manufacture of soap. The shells of the seeds serve as a fuel or are mixed with the oil cake and fed to cattle, they are also used in the manufacture of paper. The oil consists of a mixture of oleiu and palimtin.

The United States produces annually 600,000 tons of cotton seed, which yields 750,000 barrels of oil, 250,000 tons of oil cake, and 3,000 tons of shells used in the manufacture of paper.

Sesame Oil.

Benne oil...........................	Oleum Sesamé.
Jinejili oil..........................	Krishna tit (Hindostan).
Gingelly oil.........................	Yettoo-cheddie (Tamul).
Tilor Teel oil.......................	Noovooloo (Telingu).
Sesam oil	Schiteloo (Malabar.
Huile de Sesame...................	Bareek-til (Deccan).

Kunjed (Persian), Duhn-es-Simsim (Arabian). It is obtained from the *Sesamum indicum*, Linn., a plant which grows wild in Southern Asia, where it has been cultivated for ages, and is found to-day in most tropical and warm climates.

The seed contains about fifty per cent of the oil and are usually pressed twice, yielding a fine table oil which as such can fairly compete with olive oil, and an ordinary oil which is used as a burning oil and a lubricant, also in the manuacture of soaps and perfumes and in the adulteration of olive oil.

The oil extracted by solvents is only used to make soap. The oil is a mixture of stearin, olein, palmitin and myristin.

Sesam-Sesamum Indicum.

Branch with blossoms.

1. Blossom natural size. 2. Calyx. 3. Opened crown. 4 and 5. Stamens. 6. Germs showing hilum. 7, 8 and 9. Fruit opened cross-section. 10. Seed. 11. Seed, long section. V. Schaedler.

Ben oil.

Lorinja oil......................	Oleum Balanium.
Behen oel	Oleum Behenn.
Ben oel.........................	Mooringhy (Hindostan).
Huile de Ben....................	Merekoolu Ceylon.
Huile de Ben oele...............	Gaumurunga, Ceylon.
Sainga Saigut...................	(Bombay).

This oil is obtained from the seeds of *Moringa ptereygosperma*, Gaertner—*Geulandina Morringa*, Linn, and *Moringa aptera*, Gaertner, a tree thirty feet high found native in Egypt, Arabia, Syria and the East Indies and cultivated in the tropics of America.

The bark of the roots tastes and smells like horse radish, the leaves flowers and shells of the unripe fruit serve as vegetables for the poor classes, and the seeds after being ground and pressed yield an oil depends on the skill exercised in pressing.

The oil is a mixture of glycerides of stearic, palmitic, myristic, oleic and Behenic (75 per cent palmiti ; 25 per cent myristic acids) and is used as a salad oil, a liniment, a solvent for aromatic substances (oil of violets, jasmin, etc.,) in the manufacture of perfumes as a hair oil and for making violet soap.

Moringa pterygosperma.

1. Branch and blossoms. 2. Bud. 3. Blossom. 4. Blossoms after removing the leaves of the calix. 5–6. Anthers. 7. Embryo. 8. Long sect. of embryo. 9. Fruit with seed cross-cut. 10. Fruit section between two seeds. 11–12. Seed with three wings. 13. Seed cross-cut. 14–15. Seed without wing cross-cut.

Peanut oil.

Peanut oil......................	Huile de Pistache de terre.
Earthnut oil....................	Oleum arachidis.
Groundnut oil...................	Moeng Phullic (Hindostan).
Erdnuss oel	Katjang-tannah (Java).
Erdeichel oel...................	Cochang-gomug (Sumatra).
Arachis oel.....................	Mandobi (Brazilian).
Huile d' Arachide.	Amendoine (Brazilian).

Peanut, Arachis hypogaea.

1. Plant with blossom and fruit. 2. Blossom. 3–4. Anthers. 5. Germ. 6. Fruit natural size. 7. Fruit, longitudinal section. 8. Seeds, natural size longitud section showing the plumule. 9. Seed cross-cut.

Peanut oil is obtained from the fruit of *Arachis hypogaea*, Linn, which is cultivated in the tropics of America, Africa and Asia. The oil is found in the paronchymal cells of the cotyledons and is extracted by

Fig. 200.

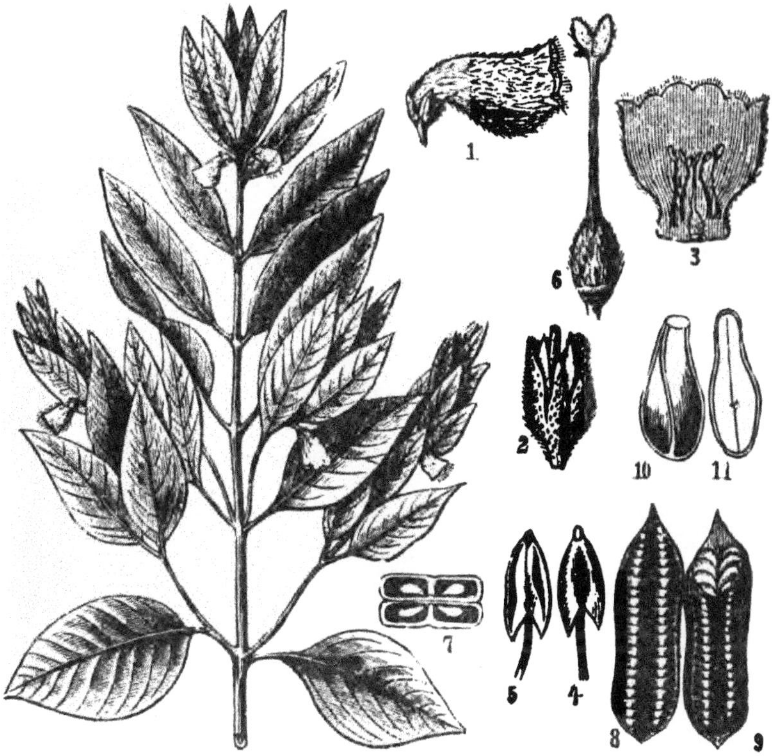

Sesam. Sesamum indicum.

Fig. 191.

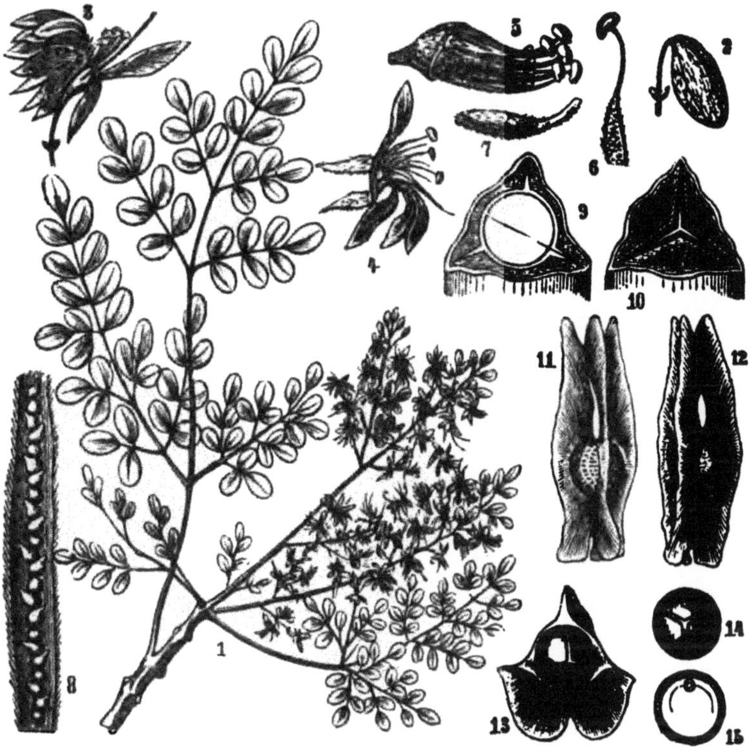

Oelmoringie. Moringa pterigosperma

Fig. 190.

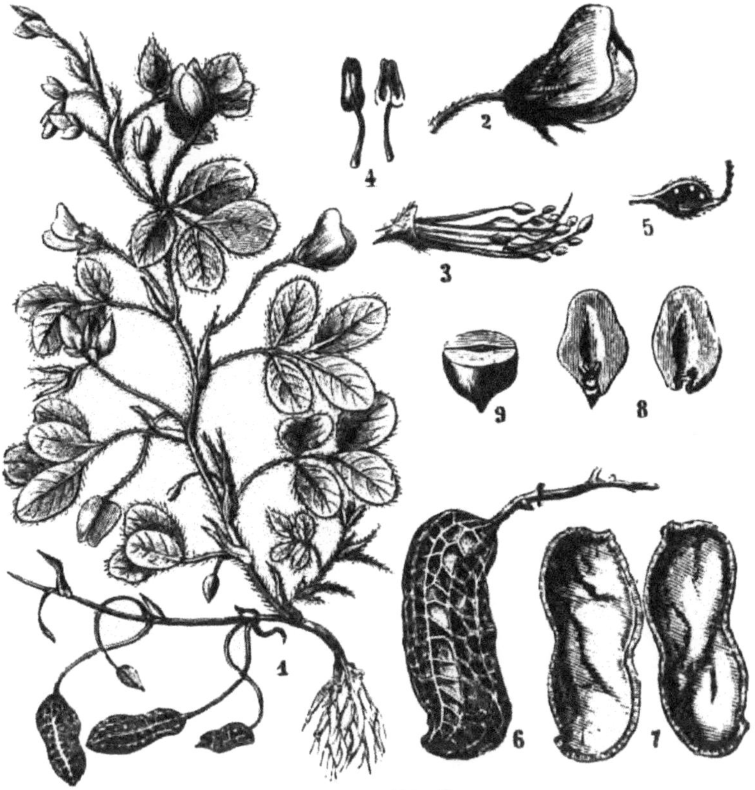

Erdnuss. ***Arachis hypogaea.***

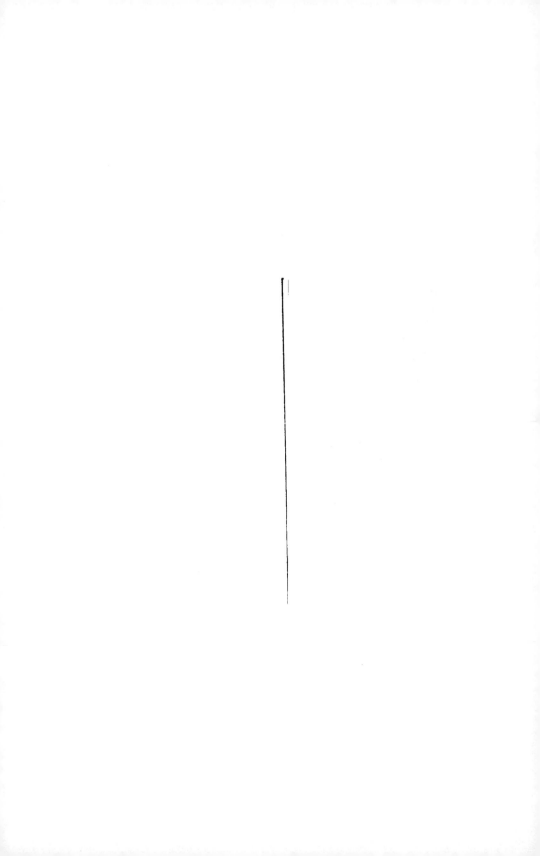

Fig. 197.

Schwarzer Senf. Brassica nigra.

pressure or by solvents. If obtained by pressure the operation comprises three steps: two cold and one warm pressing. The oil obtained by the first pressing is used as a fine table oil; the second pressing yields a second class table oil or a burning oil; the oil obtained by this third pressing is used in the manufacture of Marseilles soap. The residual oil cake is fed to cattle. Peanut oil is composed of the glycerides of the following acids:

Palmitic, $HC_{16} H_{31} O_{21}$, Hypogaeic $HC_{16} H_{29} O_2$ and Archidic, $HC_{20} H_{39} O_2$.

It is but slightly soluble in alcohol, but readily soluble in ether, chloroform and oil of turpentine.

The oil is quite stable and serves as an excellent luminant.

Mustard seed oil.

Schwarzsenf oel.
Huile de moutarde noire.
Oleum Sinapis nigri.

Brassica nigra, Koch; *Sinapis nigra*, Linn; from the seeds of which mustard oil is made, grows wild in Europe, and is cultivated in Europe Asia, Africa and America. The ethereal mustard oil, allylisothiocyanate, does not exist ready formed in the seeds but is produced by the decomposition of myronic acid; $C_{10} H_{18} H N S_2 O_{10}$ under the influence of myrosin is shown by the reaction:

$$\underset{\text{Potassium myronate.}}{C_{10} H_{18} K N S_2 O_{10}} = \underset{\text{sugar.}}{C_6 H_{12} O_6} + \underset{\text{bicarbonate of potassium.}}{K H S O_4} + \underset{\text{allylisocyanate.}}{\left.\begin{matrix} C N \\ C_3 H_5 \end{matrix}\right\} S}$$

Upon this decomposition depends the virtue of mustard plasters and the pungent taste of table mustard. The oil is used in the manufacture of soap and the oil cake for the preparation of mustard paper.

Black mustard, Brassica nigra.

1. Blossoms. 2. Blossom with petals removed. 3–4. Honigdrusen. 5–6. Pods. 7. Pods opened. 8. Pod inclosed cross-section. 9. Seed 10. Seed, cross-section.

Closely related to *Brassica nigra* is *B. juncea, Sinapis juncea*, Linn.; which grows wild in India and is cultivated in Russia. The oil is used as a table oil and the oil cake is ground up and comes into commerce as "Sarepta, mustard flour".

Both oils are obtained merely as products.

White mustard oil.

Weissen oel.
Huile de moutarde blanche.
Oleum Sinapis albi.

This oil is obtained from the seeds of *Sinapis alba*, Linn.—*Brassica alba Boiss*—a plant growing wild in southern Europe and cultivated in Germany, England and in India. The seeds when ground yield a yellow powder which on being heated with water does not evolve mustard oil since the seeds do not contain myronic acid. The glucoside, *Sin. albia*, $C_{30} H_{44} N_2 S_2 O_{16}$, is present in the seed, and this under the influence of myrosin which is also present, undergoes a decomposition similar to that of myronic acid.

The substance thus produced acrinyeisothio-cyanate can be obtained by distillation; it blisters the skin but little; white mustard oil like black mustard oil is a by-product and mixed with other oils is used as a burning oil and as a lubricant.

White mustard, Sinapis alba—Linn.

1. Flower. 2. The same with petals removed. 3. Ripe pod. 4. Burst pod. 5. Enlarged cross-section of pod. 6. Seed. 7. Seed cross-section.

Colza oil.

Coleseed oil	Huile de Colza.
Kohlsaat oel	Oleum Brassicae.
Colza oel	

Rape-seed oil.

Rape oil	Oleum Napi.
Rappsamen oel	Reps oel.
Raps oel	Sursov (Bombay).
Huile de naoltte	Sursul (Guzerat).

Rubsen seed oil.

Rubsen oil	Rata aba (Ceylon).
Rueboel	Rudaghov (Tamilien).
Reubsen oel	Aooloo (Tellingu).
Huile de rabette	Rhurdal (Arabian).
Rae Bubral (Bengal)	Sir schuf (Persian).
Oleum rapasum.	

Winter rubsen, Brassica Rapa oleofera biennsis.

1. Calyx with anthers. 2. Anthers. 3. Petals. 4. Root. 5. Young plant. 6. Young pod. 7. Ripe pod opened. 8–9–10. Enlarged seeds different views.

The several varieties of Brassica furnish colza, rape and rubsen oil.
Summer rubsen oil, from B. Rapa annua.
Winter rubsen oil, from B. Rapa biennis.
Summer rape oil, from B. Napus annua.
Winter rape oil, from B. Napus biennis.
Colza oil, from B. campestris.

The seeds of these various plants are difficult to distinguish from one another and the oil which they yield are very much alike in physical and chemical properties. They are of a dark brown color; leave a repulsive, scratchy taste due to the presence of some volatile sulphur compounds, are odorless when fresh but soon become rancid and obnoxious.

The presence of sulphur seems to be characteristic of the oils of all Crucifers, and may be detected by means of lead plaster. This dissolves in the oil, sulphite of lead is formed and the oil is colored black The oils are but sparingly soluble in alcohol; they absorb oxygen from the atmosphere, become rancid and thicken but do not dry. They consist of a mixture of the glycerides of stearic, brassic, $HC_{22}H_{41}O_2$ and oleic acids.

Fig. 196.

Winterrübsen. Brassica Rapa oleifera biennis.

They are used as lubricants and as luminants, but seldom in the manufacture of soaps.

Treated with starch or with sweet spirits of nitre the repulsive taste can be removed and the oil may thus be prepared for the table.

Gold of Pleasure; Camelina sativa.

1. Blossom enlarged. 2–3. Blossoms with leaves of calyx removed. 4. Germ enlarged, long sect. 5. Fruit. 6. Germ enlarged, cross-section. 7. Embryo with two cotyledons. 8–9. Seed, cross and long sections. 10. Seeds natural size. 11. Seed enlarged.

Cameline oil.

German Sesam oil..................	Huile de Cameline.
Leindotter oel......................	Huile de Camomille.
Dotter oel..........................	Huile de Sesame d' Allemagne.
Rape dotter oel.....................	Oleum Camelinae.
Deutsches Sesam oel...............	Oleum myagri.

The seed of *Myagrum sativum*, Linn.; *Camelina sativa* Cez, a plant found in Asia and Europe; contain about 25 to 30 per cent of an oil which dries up slowly in the air by oxidation, burns with an excellent flame and consists of the glycerides of oleic, and linoleic acids. The cold pressed oil is occasionally used as a table oil, but more generally as a luminant and especially in the manufacture of soaps.

The oil cake serves as a good food for cattle and fowl.

Cocoanut Oil.

Cocosoel...........................	Oleum cocois.
Cocosnussoel.......................	Klopper olie (Holland).
Kokoo butter.......................	Nasil Nasel (Hindostand).
Huile de Coco......................	Tynga (Tamul).
Beurre de Coco	Kalbri (Tellingu).

Is contained in the seeds of *Cocos nucifera*, Linn.—the ordinary cocoa palm. This palm, a native of the South Sea Islands and the Coco Islands west of Sumatra, is confined to the tropics of both hemispheres; it grows in forests to a height of from sixty to ninety feet and bears a fruit, the cocoanut, from its eighth to its one hundredth year. The meat of the cocoanut contains about seventy per cent of oil. This is boiled in water for some time then crushed in mortars and pressed. By this means a milky mass is obtained which is warmed in kettles and the oil which floats on top is skimmed off. The residue is fed to the cattle as is also the oil cake. The meat contains two oils of different consistency which can be separated from each other by pressing.

Cocoa palm, Cocosnucifera.

1. Bluethenkolben. 2. One female and two male blossoms. 3. One male blossom. 4. One male blossom enlarged showing the anther. 5. The female blossom with calyx removed. 6. The stem fruit. 7. Long section of 6 in., which a fibrous central layer surrounds

the stem shell, which is pierced with three holes. 8. Cross section of shell showing the whole seed. 9. The lower part of the shell showing the three holes. 10. The seed, cross section, the cavity of which is filled with milk. 11. The iower part of the seed, long section, showing the germ. 12. The germ.

When fresh the oil is white, melts at 22 degrees C., and is soluble in alcohol and ether. It contains of glycerides of the volatile fat acids, capric, caproic and caprylic and of the non-volatile fatty acids, lauric myristic and palmitic.

Cocoanut oil is used in the manufacture of soap and candles, as a substitute for codliver oil, and in the adulteration of olive oil.

Palm Oil.

Palmfett	Huile de Palme.
Palm butter	Oleum Palmæ.
Palmoel...........	Monteca delceroro (Spanish).
Beurre de Palme..................	Thiothio (Antilles).
	Caiane (Brazil).

Elaeis guineensis, Jacq., and *Elais melano cocca,* Gaert., are natives of Africa, and are cultivated to-day in South America. They bear a stony fruit from which two varieties of oil are obtained, palm oil from the meat, and palm nut oil from the the stone. Palm oil is extracted from the meat by pressure or by boiling it with water. When freshly prepared it is of a dark yellow color, becoming light-colored and rancid by exposure to the atmosphere. It may be readily bleached by the action of light, air and heat, and also by stronger oxidizing agents such as chlorine gas, oxygen and strong acids. The coloring matter does not exist in suspension but in solution in the fat, and cannot be extracted therefrom by means of water and alcohol.

The oil consists of a mixture of olein and palmitin also contains large quantities of free fat-acids and glycerine. It is used in the manufacture of soap and candles.

Oil palm — Elais guineensis.

1, 2 and 3. Male blossom. 2 and 3. Same magnified. 4. Anthers. 5. Stem fruit, showing three hilums. 6. Shell, showing of. 5. Showing the three holes. 7. Long section of 6 showing the seed. 8. Seed, cross section.

Cacao Butter.

Oil of Theobroma..................	Beurre de Cacao
Cacao butter.........................	Oleum Cacao
Cacao oel..............................	

Is obtained as a by-product in the manufacture of chocolate.

Thebroma Cacao, Linn., and the allied varieties, are trees eighteen to thirty-six feet in height, are natives of West Indies, grow wild in Central America and are cultivated in South America. The seeds or beans contain about fifty per cent of this oil. They are shelled and

Fig. 219

Oelpalme. Elaeis guineénsis.

Fig. 214.

Cacaobaum. Theobroma Cacao.

roasted, warmed up to 75 degrees C. and pressed. The oil thus obtained is filtered through a dry filter and the oil cake is worked up into chocolate.

This oil solidifies at 21 degrees C., can be kept for a long time without becoming rancid; it is soluble in either chloroform, oil of turpentine and absolute alcohol. The glycerides of the acids palmitic, oleic, stearic and arachidic enter into its composition. It also contains about fifteen per cent of the alkaloid Theobromine, $C_8 H_8 N_4 O_2$, three to four per cent cacao red and eight per cent humic acid.

Cacao tree, Theobroma Cacao.

1. A closed blossom. 2. An opened blossom. 3. A petal. 4. Sepals and petals. 5. Germ with opened crown and petals. 6 and 7. Seeds. 8. Seeds long section.

Cacao butter is used in the manufacture of soaps, perfumery and cosmetics; in medicine it is used as a suppository, etc.

Bone fat.

Knochenfett.................................... Graisse Pos.
Suif Po's.. Petit suif.

Bone fat is prepared from the bones of all animals either by boiling them in an open kettle with water, or by subjecting them to the action of steam. The steaming process is carried on in upright cylinders, digesters having a false bottom. Into these the bones are placed and are there steamed for several hours under a pressure of from three to six atmospheres. In both instances the fat must be purified by remelting it with common salt. The fat obtained from fresh bones is used as wagon grease, and that obtained from older bones is converted into soap.

In many establishments the fat is removed from the bones by solvents, and for this purpose many inventions have been patented. This method possesses the advantage of securing a better quality of glue, bone black and superphosphates.

Seltsam's apparatus for extracting bone fat.

A. ketttle. B. pump. C. reservoir. D. conduit for the solvents. E. pipe through which the vapors are carried off. F. condenser. C. H. connection between kettle A. and the distilling apparatus. J. K. conduit between the distilling apparatus and condenser.

Tests on the Fat of Butter.

Specific gravity.

When oleomargarine was first introduced, and experiments began to be made as to methods of distinguishing the imitation from the genuine article, it was noted that a sufficient difference between the specific gravities of the two existed on which to base at least a suspicion as to the character of the fat. It was also soon found that the temperature at which the determination was made was of the greatest importance, requiring more precaution than usual with many other substances. A reason for this may be found in the fact that the most convenient temperature at which we can work with butter and with similar fats are not very far removed from the melting point of those substances, and it is well known that all substances at or near a change of state (solid to liquid, etc.) exhibit anomalies in their rates of expansion. By way of illustration of this point as applied to fats, see experiments of Duffy (Jour. Lon. Chem. Soc. 5, 197). A coefficient of expansion (ratio of expansion for degree of temperature) for all oils is given by Stillwell (Am. Chemist 1, 407) as 0·00063 for 1 deg. C., (equivalent to about 0·00035 for 1 degree Fahr.).

The coefficient of expansion for butter fat between 100 and 112 degrees (212–234 degrees) Fahr. is recorded by Wigner (Analyst 4, 184) as 0·078 per degree C. or 0·0434 per degree Fahr. That for lard and butterine is nearly the same (0·042 per degree Fahr.).

Some of the methods used for determining the specific gravity of butter and other fats may here be described:

First, For the solid fats at ordinary temperatures 60 Fahr. or 15 C. are naturally the temperatures usually preferred.

Casamajor (Jour. Am. Chem. Soc. 3, 83) proposed to drop a globule of the melted fat into alcohol of 53·7 per cent (Sp. Gr. 0·926) and alcohol of 59·2 per cent (Sp. Gr. ·915) the liquid being kept at 15 C. In a mixture of equal volumes of the two, the butter will sink and the oleomargarine will float while both are warm and liquid. Afterward the butter will become solid, while the oleomargarine may still remain liquid. The butter will rise to the top of the alcohol, the result of expansion on solidifying.

In the lighter liquid the oleomargarine sinks while the butter will be held in equilibrio. In the heavier, the oleomargarine will be in equilibrio while the butter floats. By making various mixtures of known proportions of the pure liquids any specific gravity between these points may be obtained in which a fat consisting of a mixture of butter and oleomargarine will be in equilibrio.

Hager's method (Analyst 4, 206) is the same in character but is intended to be more accurate. Globules of the fat are formed by dropping the melted fat into alcohol, and by mixing alcohol, water, and glycerine mixtures are prepared in which the globules will remain suspended. The specific gravity of this mixture is then taken. The experiments are made at 15 to 16 degrees C.

Blyth weights a small test tube with mercury, weighs it, introduces some of the fat, and weighs the combination in air and then in water at the conventional point (Analyst 5, 76).

Many analysts have preferred to take the specific gravity of the melted fat as being in some respects more convenient.

The pycknometer specific gravity bottle has been extensively used for the determination at 100 F. It consists simply of a small bottle of one to one and one half ounces capacity, with stopper usually carrying a delicate thermometer and a side capillary tube by which the excess of liquid introduced can escape. It is weighed empty, then full of distilled water at the required temperature, filled with liquid fat at a temperature somewhat below 100 degrees and then immersed in water at or a little below 100 degrees until the thermometer indicates that the fat has reached 100 degrees. The fat escaping from the end of the capillary tube as the result of expansion is carefully wiped off from time to time. When the contents of the pycknometer are at 100 degrees it is taken out of the water, dried cooled and weighed. The ratio of the weights of equal volumes of fat and water the specific gravity may then be ascertained. This method, though susceptible of the highest accuracy, is nevertheless very trying to the patience of the operator.

The Mohr-Westphal balance, apparatus consisting of a modified scale beam from one arm of which depends an elongated weighted glass bulb carrying a thermometer, while the arm itself is graduated for the reception of weights of different sizes and values, has been used by some. (Estcourt, Chem. News 34, 254; Bell, Chem. News, 38, 267, v. table 9).

The glass bulb is suspended in the melted fat and by use of the weight the balance is brought into equilibrium. Reading off the weights and their position gives the specific gravity.

Wigner (Analyst 1, 145), also Estcourt (Chem. News, 35, 10) proposed the use of glass bubbles or beads of known weight and volume. The bubbles are thrown into the melted fat. Those which are lighter than the fat rise to the surface; those heavier sink, while those of the same specific gravity as the fat are in equilibrio, and if marked with specific gravity to which they correspond, the determination is easily made. It is, however, a matter of some difficulty to get such bubbles exactly adjusted for use at 100 degrees F. or 100 degrees C. (212 F.) See table X.

Table IX.

Wigner's Bubbles.

The following results obtained by Mr. Wigner are given as an illustration of the reliability of the method adopted. The clean melted butter-fat was weighed according to Dr. Muter's suggestion at 100 degrees F. and 135 degrees F.:

Specific gravity at 100 degrees F. compared with water at 60 degrees F.	907·2
Corresponding to "actual density" at 100 degrees F. compared with water at 100 degrees F.	912·1
Specific gravity at 135 degrees F. compared with water at 60 degrees F.	895·2
Corresponding to actual density at 135 degrees F. compared with water at 135 degrees F.	906·7
Bubbles had specific gravity of:	
A. and B.	889·0
C. and D.	888·0
E.	896·0

TABLE X.

Number.	Percentage of butter.	Actual density	Temperature at which the bubbles sank. Degrees Fahr.				
			A.	B.	C.	D.	E.
1...	....	905·3	127	126	129	129	114
2....	16	906·2	131	131	132	132	117
3....	33	907·1	136	136	137	137	122
4....	50	908·6	139	139	141	140	124
5....	66	910·8	141	141	142	143	128
6....	83	911·2	145	145	146	145	132
7....	100	912·1	146	147	149	148	135

TABLE 8.

Estcourt Westphal, 92 *degrees C.*—208 *degrees F.*

Beef fat	860·0	
Mutton fat	860·6	
Lard, home rendered	862·8	
Butter, M	870·0	
Butter, B. B	870·7	Cal.
Equal weight of mutton, fat and butter	865·6	865·0
Equal weight of beef and butter	865·7	865·8
Equal weight of lard and butter	865·8	866·4
Dutch butterine	865·2	

It is more convenient in many respects to take the specific gravity at the boiling point of water or a little below. Leune and Harburet (Monit. Sci. Apr. 1881) used for this purpose an areometer, or specific gravity spindle, which is floated in the butter, placed in a cylinder standing over a boiling water bath. It is thus easy to obtain comparative results at the same temperature although in the arrangement described the temperature of the liquid standing in the air over boiling water is not quite up to 212 degrees Fahr. It is not easy to obtain from our instrument-makers here a spindle accurately adjusted for use at such a high temperature, although an instrument apparently of this kind is described under the name "Margarimeter" (Les Mondes No. 5, 1881). It is arbitrarily graduated like the lactometre, the zero mark corresponding to the gravity of pure butter fat, the 100 mark to that of beef fat. The intermediate degrees are supposed to be read off in percentages of foreign (beef) fat. Gaebel (Milch Zeitung, 1882, p. 437) reports that it is useless, one degree of temperature making a difference of five per cent in the readings.

The Sprengel tube (Journ. Lond. Chem. Soc. 26, 577) is a convenient method of taking the specific gravities of fat in the liquid state,

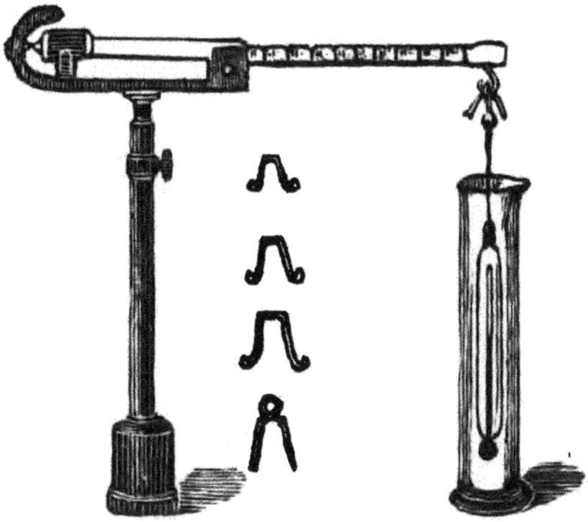

and especially so when the temperature at which the determination is made is taken 100 degrees C. (212 degrees F.). It consists of a small glass tube bent in U form, the ends being drawn out to capillary tube. In effect it is a pycknometer in tube form instead of the ordinary glass form, and is used in essentially the same manner.

The table of the specific gravities of butter fat and some other fat at the different temperatures of observation is given. The difference between butter and oleomargarine, etc., is in the second place of decimals counting water as one, so that the accuracy required in these determinations is greater than in some other determination which may be required of a chemist or physicists. Riche (Jour. Pharm. Chem. 1881) says that the difference of specific gravity between butter fat and its possible substitutes is too limited to make it of any practical value. Muter (Analyst 7, 93) calls attention to the fact that a product from cotton-seed oil has the same specific gravity at 100 degrees Fahr. (37.8 degrees C.) as butter. E. W. Jones (Analyst 4, 39) notes that the age of the sample materially effects the specific gravity.

Some analysts discard the specific gravity test altogether in consequence, while others use it only as a rough guide to the probable character of the fat under examination since butter fat is usually 0·910 or over at 100 degrees Fahr. while oleomargarine is usually 0·904 or thereabout at the same temperature as may be seen from the following results obtained in the laboratory.

Butter fat at 100 degrees Fahr. specific gravity	0·911
Oleomargarine sold as such, specific gravity............	0·9044
Oleomargarine sold as butter, specific gravity...........	0·904
Oleomargarine sold as butter, specific gravity...........	0·9048

TABLE XI.

Animal fats and oils; specific gravity taken at 15 *degrees C, unless otherwise stated; where given the temperature is expressed in certigrade degrees:*

	Name of fat.	Specific gravity.	Authority.
Temperature not given.	Lard	·861	König.
	Lard	·938–·940	Schaedler.
	Lard oil....................	·9175	Stillwell.
	Lard oil....................	9165–·9200	Schaedler.
	Lard oil....................	·915	Allen.
	Beef suet.................	·915–·9200	Schaedler.
Temperature not given.	Beef suet.................	·860	König.
	Beef suet.................	·9137	Stillwell.
	Tallow oil..................	·616	Allen.
	Tallow elain	·9011	Stillwell.
	Mutton suet...............	·913	Schaedler.
Temperature not given.	Horse fat..................	·861	König.
	Neats foot oil...............	·914–·916	Allen.
	Neats foot oil...............	·915–·916	Schaedler.
	Neats foot oil...............	·9142	Stillwell.
	Bone oil....................	·914–·916	Allen.

Vegetable fats and oils.

Temperature.	Name of fat.	Specific gravity.	Authority.
	Olive oil, light greenish yellow,	·9144	Stillwell.
	Olive oil, dark green.........	·9145	Stillwell.
	Olive oil, virgin very light yellow..	·9163	Stillwell.
	Olive oil, virgin dark clear yellow..................	·9169	Stillwell.
23° ...	Olive oil, Provence...... ...	·912–·914	Dietrich.
23° ...	Olive oil, green.............	·909–·915	Dietrich.
	Olive oil, green.............	·9173	Valenta.
	Olive oil, yellow	·9149	Valenta.
37°-38° ...	Olive oil...	·915	
	Olive oil...................	·914–·917	Allen.
	Olive oil..........	·9177	Schaedler.
	Cotton-seed oil	·922–·930	Allen.
	Cotton-seed oil, raw.........	·9224	Stillwell.
	Cotton-seed oil, refined yellow,	·9230	Stillwell.
	Cotton-seed oil, salad........	·9231	Stillwell.
	Cotton-seed oil, white winter.	·9288	Stillwell.
	Cotton-seed oil	·9228	Valenta.
	Cotton-seed oil	·917–·921	Dietrich.
	Cotton-seed oil, crude.......	·9306	Schaedler.
	Cotton-seed oil, refined......	·9264	Schaedler.
	Sesame oil	·923–·924	Allen.
23° ...	Sesame oil	·919	Dietrich.
	Sesame oil	·9213	Valenta.
	Sesame oil	·9235	Schaedler.
	Rape-seed oil, winter white...	·9144	Stillwell.
	Rape-seed oil, dark yellow...	·9168	Stillwell.
37°–38°....	Rape-seed oil...............	·915–·913	
38°–39°....	Rape-seed oil, refined........	·913–·911	
23°....	Rape-seed oil...............	·910	Dietrich.
	Rape-seed oil and colza.......	·914–·917	Allen.
	Rape-seed oil summer Brassica Rapa	·9139	Schaedler.
	Rape-seed oil, winter Brassica Rapa..................	·9154	Schaedler.
	Rape-seed oil, summer Brassica Napus	·9147	Schaedler.
	Rape-seed oil, winter Brassica Napus..................	·9184	Schaedler.
	Earthnut oil................	·9154	Stillwell.
	Earthnut oil................	·9193	Valenta.
23°.....	Earthnut oil................	·917–·918	Dietrich.
	Earthnut oil........... . .	·916–·920	Allen.
	Earthnut oil, 1st quality fresh.	·918	Schaedler.
	Earthnut oil, 2d quality old...	·9202	Schaedler.
	Almond oil (sweet)..........	·9186	Valenta.
	Almond oil..................	·917–·920	Allen.
	Almond oil..................	·919	Schaedler.
	Mustard oil.	914–·920	Allen.

Temperature.	Names of fat.	Specific gravity.	Authority.
	Mustard oil black	·917	Schaedler.
	Mustard oil white	·9145	Schaedler.
	Sunflower oil	·924–·926	Allen.
23°	Sunflower oil	·920	Dietrich.
	Sunflower oil	·9260	Schaedler.
	Niger seed oil	·926–·928	Allen.
	Apricot seed oil	·9191	Valenta.
	Apricot seed oil	·915	Schaedler.
	Illipe oil	·9175	Valenta.
	Illipe butter	·972	Schaedler.
	Pumpkin seed oil	·9241	Valenta.
	Pumpkin seed oil	·9231	Schaedler.
	Ruell oil	·9248	Valenta.
31–33°	Dotter oil	·924	
	Dotter oil	·9228	Schaedler.
30–31°	Hemp oil	·930–·928	
	Hemp oil	·925–·931	Allen.
	Hemp oil	·9276	Schaedler.
	Linseed oil	·930–·935	Allen.
	Linseed oil, raw	·9399	Stillwell.
	Linseed oil, boiled	·9411	Stillwell.
19–30°	Linseed oil, pure	·931–·929	
	Linseed oil	·9350	Schaedler.
	Poppy oil	·924–·937	Allen.
	Poppy oil	·9245	Stillwell.
32–33°	Poppy oil, and walnut	·926–·924	
	Walnut oil	·925–·926	Allen.
	Walnut oil, cold pressed	·9250	Schaedler.
	Walnut oil, hot pressed	·9268	Schaedler.
	Castor oil, cold drawn	·9667	Stillwell.
23°	Castor oil, pure	·964	Dietrich.
	Castor oil	·960–·964	Allen.
	Castor oil	·9667	Schaedler.
	Croton oil	·942–·943	Allen.
	Croton oil, fresh	·942	Schaedler.
	Croton oil, old	·955	Schaedler.
100° C. with water at 15°	Cacao butter	·857	Allen.
	Cacao butter	·8900–·9000	Schaedler.
	Palm oil	·9046	Stillwell.
	Palm oil	·945	Schaedler.
100° C. with water at 15°	Palm oil	·857	Allen.
100° C. with water at 15°	Palm seed oil	·866	Allen.
	Palm seed oil	·952	Schaedler.
100° C. with water at 15°	Cocoanut oil	·868	Allen.

Temperature.	Name of fat.	Specific gravity.	Authority.
	Cocoanut oil	·9250	Stillwell.
	Nutmeg butter	·990	Schaedler.
	Shea butter	·953	Schaedler.
100° C. with water at 15°	Japanese wax...............	·873	Allen.
	Japanese wax..............	·970	Schaedler

Marine Oils.

Temperature.	Name of fat.	Specific gravity.	Authority.
	Sperm oil..................	·875–·883	Allen.
	Sperm oil, bleached winter...	·8813	Stillwell.
	Sperm oil, natural winter	·8815	Stillwell.
	Sperm oil..................	·910	Schaedler
	Bottle nose oil, or black fish..	·876–·880	Allen.
	Bottle nose oil, or black fish..	·9180	Schaedler.
	Shark-liver oil.......... ...	·865–·867	Allen.
	Shark-liver oil..............	·870–·880	Schaedler.
	African fish oil..............	·867	Allen.
	Sea elephant oil.............	·9199	Stillwell.
	Cod-liver oil, pure	·9270	Stillwell.
	Cod-liver oil, yellow..........	·923	Schaedler.
	Cod-liver oil, dark brown ...	·929	Schaedler.
	Cod oil (Tanner's)	·9205	Stillwell.
	Cod oil (Labrador)..........	·9237	Stillwell.
	Cod oil (Straits')............	·9290	Stillwell.
	Cod oil (Banks').............	·9320	Stillwell.
	Cod oil	·923–·930	Allen.
	Seal oil, natural	·9246	Stillwell.
	Seal oil, racked.............	·9286	Stillwell.
	Seal oil....................	·924–·929	Allen.
	Seal oil, clear oil....	·9165	Schaedler.
	Seal oil, brown oil...........	·9170	Schaedler.
	Sea calves' oil	·9155	Schaedler.
	Whale oil, natural winter....	·9254	Stillwell.
	Whale oil, bleached winter...	·9258	Stillwell.
33°–34°....	South Sea train oil..........	·924–·922	
	Whale oil..........	·920–·930	Allen.
	Menhaden oil, dark..........	·9292	Stillwell.
	Mendaden oil, light	·9325	Stillwell.
	Menhaden oil	·929–·932	Allen.
	Porgy oil	·9332	Stillwell.
	Porpoise oil	·920–·923	Allen.
	Porpoise oil	·9220	Schaedler.

Appendix to Table XI — Fatty acids from oils.

Source of fatty acids.	Specific gravity.
Pure olive oil	·8444
Pure olive oil	·8429
Earth nut oil	·8475
Pure rape oil	·8439
Colza oil	·8464
Niger seed oil	·8562
Linseed oil	·8599
Train oil	·8597
Palm oil	·8389
Cotton-seed oil	·8494

The above specific gravities were all taken at 100 degrees C. and compared with water at the same temperature by L. Archbutt.

Melting Points.

One of the methods first tried for determining the genuineness of a butter fat was the fixing of a melting point. It was long ago observed that butter fat becomes liquid at lower temperatures than the fats usually used as adulterants or substitutes. What point, however should be regarded as the melting point is a matter of more difficulty than may appear at first sight. The melting point of butter is variously stated by different authorities; the range given is from 19 degrees C., (66 degrees F.) to 40 degrees C, (104 degrees F.)

Two obvious reasons exist for these discrepencies:

First — The methods of determining the melting points are very different and hence wide differences are made to appear and

Second — That butter as well as other fats is a mixture of various fats which melt at different temperatures, so that instead of passing into the fluid state in a sharp and decided manner, it passes only gradually into that state with a steady rise of temperature, becoming from a consistent solid softer and softer until it is finally liquid. Other reasons as the specific heat of the constituent fats and various other properties no doubt have had much influence.

Some have observed the point at which the fat becomes transparent. Hassall applied this method in test tubes inserted in water the temperature of which was gradually raised. The point observed was called by him the "point of clearance."

TABLE XII — HASSALL.

Points of fusion and of clearance of other fats.

	Points of fusion.	Points of clearance.
Beef kidney	46·5	47·5
Beef caul	45·7	46·7
Mutton kidney	48·6	49·5
Mutton caul	46·0	47·0
Veal kidney	38·7	39·4
Veal caul	41·1	42·2
Lamb kidney	48·4	49·5
Lamb kidney	51 6	52·9
Lamb caul	48·5	48·7
Lamb caul	46·2	47·5
Pig kidney	47·7	50·0
Pig caul	47·4	49·8
Lard.... Home	43·7	46·7
Lard.... Rendered	42·6	45·4
Lard, Irish	44·6	47·9
Beef dripping	43·5	45·8
Beef dripping, sirloin	45·5	46·7
Mutton dripping, loin	48·2	50·1
Mutton dripping, leg	42·3	43·3
Pig dripping	43·5	44·7

Table XII — Hassall — (*Continued*).

Points of Fusion and of Clearance of Butter.			I. Beef-kidney Series.			II. Mutton-kidney Series.		III. Pig-kidney Series.		IV. Lamb-kidney Series.	
	Point of fusion.	Point of clearance.	Percentage of foreign fat.	Point of fusion.	Point of clearance.	Point of fusion.	Point of clearance.	Point of fusion.	Point of clearance.	Point of fusion.	Point of cleara'e.
1...........	34·0	35·7	10......	35·9	37·1	37·8	40·0	36·4	37·5	37·3	38·6
2...........	34·2	35·6	20......	37·8	40·0	40·1	42·1	38·2	39·3	39·0	40·3
3...........	33·4	35·3	30......	39·8	41·7	41·9	43·3	40·2	41·3	40·9	42·1
4...........	32·8	35·1	40......	41·4	43·0	43·1	44·2	42·1	43·5	42·5	44·0
5...........	33·6	35·2	50......	42·7	44·1	44·8	45·8	43·6	44·9	44·2	45·6
6...........	33·8	36·3	60......	44·0	45·0	45·8	47·0	44·5	46·5	45·6	46·6
7...........	34·3	35·4	70......	45·0	46·2	47·0	48·2	45·2	47·5	46·3	47·5
			80......	45·6	47·0	47·5	48·7	46·1	48·4	47·0	48·2
Mean........	33·7	35·5	90......	46·1	47·4	48·1	49·2	47·0	49·2	47·7	49·0
			100......	46·5	47·5	48·6	49·5	47·7	50·0	48·4	49·5

TABLE XII — HASSALL — (*Continued*).

V. First Mutton Dripping Series.			VI. Second Mutton Dripping Series.		VII. Pig Dripping Series.		VIII. Mixtures of Stearin and Olein.		
Percentage of foreign fat.	Point of fusion.	Point of clearance.	Point of fusion.	Point of clearance.	Point of fusion.	Point of clearance.	Point of fusion.	Point of clearance.	Appearance.
10	36·8	38·7	35·3	35·7	35·0	36·0	*47·5	50·2	Half liquid.
20	38·6	40·5	36·3	37·4	35·9	36·8	52·1	54·1	Buttery.
30	40·5	43·0	37·0	38·0	37·3	38·0	55·0	56·5	Lardaceous.
40	42·2	44·5	37·7	39·2	38·0	39·1	56·5	57·3	Consistence of beef fat.
50	43·6	45·9	38·7	40·1	39·0	40·1	57·5	58·3	Consistence of mutton fat
60	45·0	47·2	39·4	41·0	39·9	41·0	58·4	59·1	Very hard.
70	46·0	48·2	40·9	41·9	40·7	42·0	59·0	59·7	Very hard.
80	46·7	49·0	40·9	42·5	41·6	42·7	59·7	60·4	Very hard.
90	47·5	49·7	41·5	43·1	42·5	43·8	60·4	61·3	Very hard.
100	48·4	50·1	42·3	43·3	43·5	44·7	61·5	62·6	Very hard.

*Per cent stearin the same as per cent of foreign fat.

The test was applied in a similar manner by Pohl (Polyt. Centr., 1855, 165), except that he smeared the fat on the bulb of a theremometer, and noted the point at which the mercury in the bulb was clearly visible. Roster (Gaz. Chim., 10, 13) applies this method in especially constructed apparatus.

The point of obscuration of the graduations on the thermometer was also used by Dr. Campbell Brown (Butter Anal., Hehner and Angell, page 23), as an aid in fixing the melting point. Bergman (Kunst u. Gewerbbl. Bayern, 1867, Jan.) noted the point of becoming clear when the fat was confined in a capillary tube also Duffy (Jour. Lond. Chem. Soc., 5, 197), and many others both before and since. It has been remarked by Wimmel (Pogg. Anal., 133, 121), that butter and most other fats are liquid at a temperature at which they are still turbid, although for Japanese wax the reverse was the case. The transparency is attributed by Hehner and Angell (Butter Anal., page 21), to the action of the olein present in dissolving the stearin, etc., its solvent power being greatly increased by the rise in temperature.

The method recommended by Dragendorff (Pflanzen Analyse, page 13), consists in placing the fat on the surface of mercury in juxtaposition with a thermometer bulb, and noting the temperature when the fat fuses so as to spread over the surface in a film. Redwood (Pharm. Jour. Trans. [3] 6, 1009) previously described this method.

Redwood (Analyst, vol. 1, page 51) also modifies this method by placing the fat just between the thermometer bulb and a clean dry glass rod and noting the point at which the fat fuses sufficiently to raise by capillary attraction between the two.

TABLE XIII.

Dr. C. Brown (H. & A. p. 26).	When heated.		On cooling.			Solid at.
	Softens.	Melts.	Obscures reading.	Stem indistinct.	Stem invisible.	
	Degs.	Degs.	Degs.	Degs.	Degs.	Degs.
Butter from town-fed cows	20.5	24.4	28.3	24.4	22.8	16.7
Irish butter	28.9	31.7	22.7	21.7	20.5	16.7
Irish butter, best	20.5	26.6	25.5	23.3	21.6	18.9
Irish butter, low	24.4	31.7	27.7	25.0	23.3	20.5
Cornish butter	22.2	26.6	26.6	25.5	22.2	14.4
Canadian butter	23.3	32.2	21.6	20.5	20.0	18.8
Kiel butter	28.8	32.2	28.3	22.2	21.6	21.1
Suspected butter	21.1	35.5	41.1	28.8	24.4	22.7
Lard	28.8	35.5	....	....	35.5	29.4
Lard	26.1	30.5	26.6	26.1	24.4	20.0
Lard	30.5	35.5	26.6	26.1	25.5	23.3
Palm oil	27.2	33.8	....	31.1	26.6	20.5
Butter with 20 per cent of lard	27.7	35.5	30.0	27.7	24.4	21.6
Butter with 20 per cent of tallow	31.1	37.2	26.1	25.0	24.4	22.7
Butter with 50 per cent of dripping	27.7	33.8	33.3	27.2	23.8	21.6

Another series of methods of determining melting points consist in noting the point at which the fat becomes soft enough to detach itself from the walls of a tube in which it has been introduced and allowed to harden. Kellner (Landw. Versuchsstat xxiii 45), uses a capillary tube with a club-shaped top which is closed by melting the glass together. While this end is still hot, the cold end is immersed in the molten fat, and some of the fat is drawn into the tube by the contraction of the air remaining in the bulb. When about an inch of fat has been thus drawn up, the tube is immersed in cold water, which solidifies the fat leaving the imprisoned air still somewhat rarified by further cooling. Then on immersing this tube in cold water with a thermometer and gradually raising the temperature, the point at which the fat is still further drawn up in the tube may be noted.

Another method (Buis. Ann. Ch. and Pharm., xliv 152) involves the use of a capillary tube open at both ends. Into one end is drawn about an inch of the melted fat which is solidified as before by immersion in cold water, and then on immersing the tube (fat end down) in water so far that the top of the fat is below the level of the water, and raising the temperature, the point may be noted at which the fat softens sufficiently to allow the water to force it up the tube. The method is described by Tripe (Chem. News xxxi, 205) and results obtained with it are given by Heisch (Chem. News, xxxi, 227. See Table XI. A correspondent of the Chemical News, (xxxi, 228) points out that this is more properly a determination of the softening point than of the melting point.

A method somewhat similar in principle is described by Cross and Bevan. (Jour. Lond. Chem. Soc. 40, 111) . A cup is attached to the bulb of a thermometer and into this cup some of the molten fat is run. While it is still soft a platinum wire attached to a small bulb is introduced. When the fat is hard the whole is immersed in mercury which is slowly heated. When the fat melts it releases the bulb which rises to the surface and at that moment the reading of the thermometer is taken.

Hehner and Angell (Butter Analysis p. 22) have determined what they call the "sinking point" by noting the temperature at which a small glass bulb placed on the top of the butter will sink into it. The bulb is made as nearly as possible one cubic centimeter in bulk and weighted with mercury until it weighs about 3·4 grammes.

Table XIV.

Heisch's table capillary tubes. Melting points.

	Table No. 1.		Table No. 2.	
	Degrees C.	Degrees F.	Degrees C.	Degrees F.
Number 1	36·3	97·5	35·6	96·0
Number 2	31·3	88·5	31·1	88·0
Number 3	31·1	88·0	31·7	89·0
Number 4	30·9	87·5	30·6	87·0
Number 5	32·2	90·0	32·2	90·0
Number 6	33·0	91·5	32·8	91·0
Number 7	32·2	90·0	33·3	92·0
Lard	38·8	102	39·5	103·0
Suet	44·3	112	43·8	111·0

Table XV.

Sinking point with genuine butter.

No.		Centigrade.
1.	Butter from town-fed cows	35·3
2.	Farm butter from Hertfordshire	36·1
3.	Farm butter from St. Lawrence	34·8
4.	Farm butter from Chale	36·3
5.	Farm butter from Newport	35·3
6.	Butter from Ventnor	35·3
7.	Butter from Ventnor	35·5
8.	Farm butter from Niton	34·8
9.	Salt butter from Ventnor	35·9
10.	Salt butter from Ventnor	34·8
11.	Fresh butter from Ventnor	35·1
12.	Farm butter from Chale	35·6
13.	Farm butter from Chale	36·3
14.	Rancid butter	35·8
15.	Farm butter from Chale	36·3
16.	Farm butter from Ventnor	35·5
17.	Farm butter from Ventnor	34·3
18.	Sussex butter	35·5
19.	Normandy butter	35·7
20.	Butter from Ventnor	36·1
21.	Butter from Ventnor	35·5
22.	Jersey butter	35·3
23.	Butter from Guildford	35·7
24.	Butter from Guildford	35·3
	Average of the 24 samples varying from 34·3 degrees to 36·3 degrees	35·5

TABLE XVI.

Sinking points of fats other than butter.

No.		Degrees.
1.	Tallow from candle	53·3
2.	Butterin, patented	31·3
3.	Ox fat, from	48·3
4.	Ox fat, to	53·0
5.	Mutton fat, from	50·1
6.	Mutton fat, to	51·6
7.	Lard, from	42·1
8.	Lard, to	45·3
9.	Dripping	42·7
10.	Dripping from beef	43·8
11.	Dripping from beef	44·5
12.	Dripping from veal	47·7
13.	Dripping, mixed	42·6
14.	Cocoa butter	34·9
15.	Palm oil	39·2
16.	Stearin	62·8

Sinking points of mixture.

No.		Found.	Cal., deg.
1...	66·7 per cent butter and 33·3 per cent tallow..	43·1	42·08
2...	73·0 per cent butter and 27·0 per cent mut. fat	42·3	40·2
3...	10·0 per cent butter and 90·0 per cent mut. fat	48·8	49·6
4...	85·0 per cent butter and 15·0 per cent ox fat..	38·1	38·1
5...	69·8 per cent butter and 30·2 per cent ox fat..	39·5	39·8

Adulterated butters.

No.		Centigrade.
1.....	Butter from Ventnor, 1s 2d per pound	35·9
2.....	Butter from London, 1s 2d per pound	42·7
3....	Kiel butter, 1s per pound	38·4

Sinking point of fatty acids from butter.

1....	41·8
2....	42·1
3....	40·5
4....	41·1

A method for the determination of melting points by electrical connection has been devised. A mass of mercury in a porcelain dish is connected with one pole of a battery through a small alarm bell. The

Fig. 10.

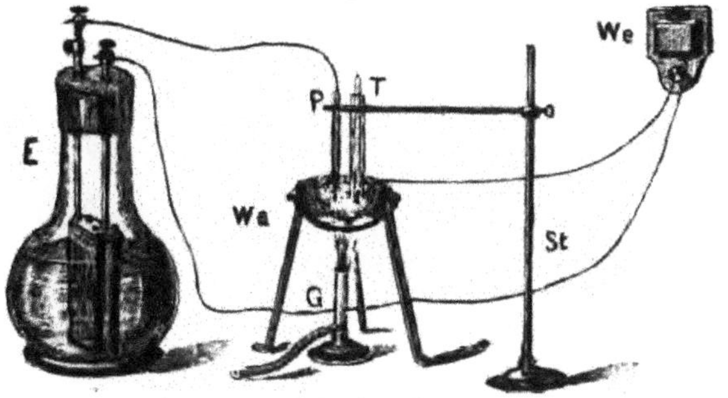

Apparat zur Schmelzpunktbestimmung.

connection with the other pole of the battery is made with a stout platinum wire, the end of which is protected by a mass of the fat, the melting point of which is to be determined and this is then plunged into the mercury. The bulb of a sensitive thermometer is also immersed in the mercury. On slowly raising the temperature of the mercury the bell rings when the fat melts off so as to allow the circuit to close.

The discrepancies in the case of other fats are as great as those in the case of butter fat.

It is evident that the method has much to do with the results obtained, and consequently that no statement of the temperature of melting for a fat is of important value unless some intimation of the mode of its determination is given.

Wimmel (Loc. cit.) has shown that fat heated considerably above its melting point and then allowed to cool, if tested at once will show a different melting point from another portion of the same fat heated to a lower point. Some of his results are:

	Degrees, C.
Lard two hours after being melted showed melting point of..	33·0
Lard two days after being melted showed melting point of...	42·0
Butter two hours after being melted showed melting point of.	25·0
Butter two days after being melted showed melting point of.	31·5

The same observation has also been made by other observers. Wimmel attributes this to a slowness on the part of a fat to solidify completely. In the case of some fats he recognizes the existence of two melting points. Some other experimenters also find that there are two melting points (Heisch Chem. News, 31, 227 Redwood), while Duffy (Jour. Lond. Chem. Soc., 5, 197) recognizes three.

The age of the sample also seems to have an influence upon the melting point of a butter. Dupré (Analyst, 4, 43) notes especially that the change is irregular with different samples, apparently according to no fixed law.

Repeated melting and solidifying the sample also has a marked effect on butter as well as other fats (Dupré, Analyst 1, 91).

The point of solidification of fats being not quite so variable as the melting point has been adopted by some as of more service. The French chemists have pretty generally given it the preference. It is, however, liable to some variability depending on the treatment of the sample, mode of testing, etc., as may be seen from an inspection of the table herewith given. At the point of solidification there is a rise in temperature sometimes of only a fraction of a degree, in other cases of several degrees.

In consideration of the facts set forth or implied in the forgoing account, and further in consideration of the fact that it is not very difficult to prepare mixtures of fat which will show essentially the same melting or solidifying point as genuine butter fat, it is not surprising that this method of testing butter for its genuineness which seemed at first to promise to be an aid has been practically discarded by almost all analysts.

Schaedler, page 100.

A mixture of		Fuses at.	Solidifies at.	Character of solidified mass.
Stearic acid. $C_{18}H_{35}$ O HO.	Palmitic acid. $C_{16}H_{31}$ O. HO.			
		Degs. C.	Degs. C.	
100 parts....	0 parts...	69·2	....	
90 parts....	10 parts...	67·2	62·5	
80 parts....	20 parts...	65·3	60·3	
70 parts....	30 parts...	62·9	59·3	
60 parts....	40 parts...	60·3	56·5	
50 parts....	50 parts...	56·6	55·0	
40 parts....	60 parts...	56·3	54·5	
35 parts....	65 parts...	55·6	54·3	
32·5 parts..	67·5 parts.	55·2	54·0	
30 parts....	70 parts...	55·1	54·0	
20 parts....	80 parts...	57·5	53·8	
10 parts....	90 parts...	60·1	54·5	
0 parts....	100 parts...	62·0	....	

TABLE XVII (*Schaedler*, p. 53).

NAME OF OILS AND OTHER FATS.	Freezing points.	Melting points.	Solidifying points.
Cotton-seed oil..................	—2	....	....
Oil of ben	+0	....	....
Butter, fresh....................	....	31·0–31·5	19–20
Butter, old	....	32·5	24
Cocoa butter	....	33·5–34	20·5
Cocoanut oil	....	24·5	20·0–20·5
Peanut oil	—3–4	....	....
Mutton fat, fresh	....	46	36
Mutton fat, old	....	49·5	39·5
Japan wax........................	....	53·5–54·5	40·5–41
Linseed oil	—27–27·5	....	....
Corn oil	—10	....	....
Almond oil	—2–3	....	....
Muscat butter....................	....	43·5–44	33
Oleic acid	—6	....	....
Olive oil........................	+2–+4	....	....
Palm oil fresh, soft	....	30	21
Palm oil fresh, hard	....	36	24
Palm oil, old....................	....	41	37

TABLE XVII — (*Continued*).

NAME OF OILS AND OTHER FATS.	Freezing points.	Melting points.	Solidifying points.
Rape-seed oil	—2–3	....	....
Beef tallow, fresh	....	43	33
Beef tallow, old	....	42·5	34
Lard	....	32–33	26
Mustard oil (sinap. alb.)	—16	....	....
Mustard oil (sinap. nigr.)	—18	....	....
Sesame oil	— 5	....	....
Wax, yellow	....	62·5	....
Wax, white	....	64	....
Spermaceti	—17–18	....	....

TABLE XVIII.

Table of melting and solidifying points according to various authorities.

AUTHORITY.	MELTING.		SOLIDIFYING.	
	Degrees C.	Degrees F.	Degrees C.	Degrees F.
Dubois and Pade	26·4	79·5	23·8	75
Schubarth	40	10·4	....	....
Chevreul	32–26·5	89·5–80	....	....
Wimmel	19–32·5	66–90·5	19·5–25·5	67–78
Dietzsch	36	97	23	73·5
Elsner	31·5	89	20	68
Fleischmann	36	97	23	73·5
Hassall	32·8–34·9	91–95	....	....
Husson	31–32·5 28	88–90·5 82·5	....	....
Bell	29·5–35	85–95	....	....
Hehner & Angell	35 5 34·3–36·5	96	14·4–21·1	58–70
Moser	33–37	91·4–98·6	....	....
Heisch	30·6–36·3	87–97·5	....	....
Redwood	27–29	80·5–84·5	....	....
Brown	24·4–32·2	....	....	....
Blythe	35·8	96·5	....	....
Lang	22	71·5	....	....
Hilger	30·8–35	87·5–95	....	....
Hanssen	31–38	88–100·5	....	....
Reichardt	33–37	91·4–98·6	....	....

THE WASH PROCESS.

As has been said, when oleomargarine was first manufactured nothing was satisfactorily known of trustworthy methods for distinguishing it from butter.

Relying upon what could be gleaned from the reports of former investigators but little difference was to be found by any known analytical methods, and the advocates of the new product confidentially asserted that there was no difference. In 1876, however, Messrs. Hehner and Angell, public analysts in Great Britain, undertook a series of experiments for the purpose of discovering if possible some method of distinguishing the imitation from the true, and possibly determining their proportions when mixed.

The results of these experiments have in one way or another formed the basis for essentially all of the methods of examining butter-fat now in common use. The only quantitative analysis of butter then on record was that of Bromeis years before which has been already quoted. The percentage of butyrin, caproin and caprin given by Bromeis was 2, implying the presence of at most 1·74 per cent of butyric acid, assuming the caprin and caproin to exist as mere traces. For the purpose of the research, butter-fat, free from salts, curd and water was taken.

The first experiments made show that the amount of butyrin, caproin, etc., was nearly four times as great as stated by Bromeis, also that the amount of volatile fatty acids in butter-fat is very constant, and almost independent of the variety of cow, feed, time of year, etc.

Attempts to distill off the volatile fatty acids were however not very successful so far as constancy of results was concerned. From butter fat were obtained amounts of fat acid equivalent to a, 6·52 b, 6·146, c, 7·48, d, 5·094, e, 4·796, f, 7·452, g, 7·259 and h, 6·26 per cent of butyric acid; b and c were from the same butter-fat; also d, e, f and g.

To obtain the volatile fatty acids the butter-fat was saponified with caustic potash, which afforded soaps of the various fatty acids and glycerin then by addition of excess of sulphuric acid, which took away the potash from its combination with the fatty acids, and subjecting the mixture to distillation, the volatile fatty acids came over with the steam and were condensed with it.

It seemed impossible to distil off the acids completely, the distillate retaining a marked odor of butyric acid, and continuing to be acid in its action on test paper even on pushing the operation to the point where the sulphuric acid was so much concentrated as to attack the glycerin remaining in the retort. This method of analysis was therefore abandoned and an indirect method resorted to. The line of reasoning in was effect this : Since butter contains approximately nearly 7 per cent of butyric acid, that implies the presence of about 8 per cent of tributyrin. The remainder of the butter consists chiefly of olein, stearin and palmitin. In those glycerides the fatty acids of which are not volatile they constitute on an average about 95·5 per cent. A mixture of eight parts or more of butyrin with sufficient stearin, palmatin and olein to make it up to 100 parts, should therefore yield something under 90 per cent of non-volatile insoluble fatty acids and the insoluble fatty acids of butter should necessarily be considerably less than the insoluble fatty acids of such fats as lard, tallow, etc., which consist almost exclusively of glycerides of the insoluble fatty acids. The correctness of this view was established by experiment, it being found by Hehner and Angell that the insoluble fatty acids of butter constituted 85·4 to 86·2 per cent of the fat. By more perfect methods of analysis it has been found by those gentlemen that the range is from 86·6 to 87·5 per cent, though it may be in rare cases as high as 88·5 per cent. The authors first published a book on the subject in April 1874, describing the process, and in 1877 were awarded a prize for "A trustworthy method of butter analysis," by the Leipzig Pharmaceutical Association,(Chem. News, 36, 248).

The prize was announced to be offered in November, 1876, two and one half years after the appearance of Hehner and Angell's book (Analyst 1, 180).

The question of priority of the idea of this process has been mildly disputed by Dr. Muter (Analyst 1, 122 and 147). A second edition of the "Butter Analysis" by Hehner and Angell was published in 1877.

The method there given consists in dissolving three or four grammes of the fat in about 50 c. c. of alcohol by the aid of heat and stirring, in a rather deep porcelain dish. One or two grammes of pure caustic potash were then dropped in and the mixture so kept warm, but not boiling, over a water bath until the addition of water fails to produce any turbidity, an indication that saponification is complete. The dish is then heated on the water bath until all odor of alcohol has disappeared. Water is added and the dish is heated to dissolve the soap when hydrochloric acid is added in sufficient amount to render the solution strongly acid and thus liberate the fatty acids. The heating is continued until the fatty acids have melted to a clear oil, and the acid liquid below is nearly clear. The contents of the dish are then filtered through a filter paper previously dried and weighed and then wetted and partly filled with boiling water. The filter, (five inches in diameter should be of the best quality, close in texture and well fitted to the funnel. The entire contents of the dish must be rinsed onto the filter with boiling water and great care must be taken that all of the fatty acids are removed from the dish. The fatty acids which will not readily run through the wet paper are then washed with boiling water until the filtrate has a bulk of at least 750 c. c. and a few c. c. of the wash water then tested with delicate litmus solution shows no acidity. The fatty acids are then solidified by plunging the funnel into cold water dried in the air bath for two hours and weighed, dried for two and a half hours more and weighed again if necessary, dried until the weight is practically constant.

Hehner and Angell give the following which may be taken as an instance of the determination of the insoluble fatty acids in genuine butter-fat:

Beaker and butter-fat	39·3226
Beaker	36·0395
Butter-fat taken	3·2831

Tube and filter	15·4281
Tube empty	14·8451
Filter	0·5830
Beaker empty	20·9967
Beaker and filter	21·5797
Beaker, filter and fatty acid (2 hours)	24·4642
Beaker, filter and fatty acid (2½ hours)	24·4500
Beaker, filter and fatty acid (3 hours)	24·4505
Fatty acids	2·8703

Fatty acids, 87·42 per cent.

Tests to determine the change by heating:

Beaker and fatty acids after two hours' drying	24·4642
Beaker and fatty acids after two and a half hours' drying.	24·4500
Beaker and fatty acids after three hours' drying	24·4505
Beaker and fatty acids after three and a half hours' drying	24·4504
Beaker and fatty acids after four and a half hours' drying	24·4504
Beaker and fatty acids after six hours' drying	24·4517
Beaker and fatty acids after seven hours' drying	24·4556
Beaker and fatty acids after eleven hours' drying	24·4553
Beaker and fatty acids after seventeen hours' drying	24·4526

Dr. Muter (Analyst 1, 7) examined this process and expressed his approval of it, although in some matters of detail, he found, in his opinion, that its accuracy might be improved. He also proposed that inasmuch as the soluble fatty acids are washed out they should be determined by trituration. The principal points of diffence consisted in saponifying in a flask with potash and acidifying and washing in the same flask, thus avoiding any transfer of the fatty acid. The fatty acids are washed in the flask by means of alternate treatment with hot and cold water, a long tube being inserted in the cork of the flask to act as return cooler and thus prevent any loss of butyric acid by volatilization during the process. He also dries the fatty acids at 105 degrees C., instead of 100 degrees.

The acidity of the wash waters, their contents in sulphuric acid, and of potassium sulphate (after neutralization) furnish data from which the proportion of soluble fatty acids could be calculated.

Some experiments in the laboratory have been made in the course of this investigation to determine whether it is possible to wash fat or fatty acids completely from beakers and flasks with boiling water. The results confirm Dr. Muter's assertions. From four beakers and two flasks containing fat or fatty acids the fatty matter was removed as thoroughly as possible by a jet of boiling water from a wash bottle using 300 to 500 c. c. of water. The amounts of adhering fat then determined by drying, weighing, washing out with ether, drying and weighing again. The amount of adhering fat was found to be:

	Beakers.	Flasks.
A	0·0052 gm.	0·0098 gm.
B	0·0027 gm.	0·0085 gm.
C	0·0010 gm.	......
D	0·0080 gm.	......

The amounts retained are material and irregular.

Dupré (Analyst, 1, 87) also examined the process, after first confirming by several experiments the point made by Messrs. Hehner and Angell that the proportion of butyric and other soluble acids in butter fat was considerably greater than as stated by Bromeis. He also showed that if it is intended to determine the soluble acids, it is necessary to saponify in a closed vessel to avoid loss of butyric acid in the form of butyric ether, but that after saponification has once taken place, the alcohol may be boiled off without loss.

His method consists in saponifying the butter fat in alcoholic soda solution of known strength in a closed flask, adding sulphuric acid of known strength in sufficient amount to render the solution distinctly acid. The washing is performed in the manner recommended by Muter the wash waters being passed through a filter and any particles of insoluble acid caught on the filter are finally returned to the flask by passing ether through the filter. The acidity of the wash waters then represents the amount of acid used which has not been neutralized by the soda used in saponification plus the amount of fatty acid washed in the butter fat. The soluble acids consisting chiefly of butyric are reckoned as all butyric acid, while the insoluble fatty acids are dried and weighed. In drying them it is often an economy in time to add to them a little absolute alcohol which will cause the small amount of water present to mingle with the fat and the alcohol and water are expelled together. Dupre does not accept the assertion of Muter that the sum of the soluble and insoluble fatty acids should amount to about ninety-four per cent in every trustworty analysis of a fat. (Analyst, 1, 114.)

An increase in the age of the sample of butter seems to cause an increase in the amount of soluble fatty acids (Wanklyn, Stevenson, Analyst, 1, 14).

Dr. Muter also has confirmed the assertion of Hehner and Angell that the rancidity of the sample does not materially affect the process, also that the acidity of rancid samples is very small, ordinarily 0·2 per cent butyric acid or less. He fixed the limit at 89·5 per cent insoluble acids with five per cent of soluble.

Messrs. Hehner and Angell criticize Dr. Muter's criticism and disapprove of any attempt to determine soluble acids (Butter Analysis, page 70) as an unnecessary elaboration, which cannot be carried out with accuracy where much work of the kind is to be done.

The process was critically examined in England by Bell, Turner and others besides Muter and Dupré. Dr. Bell tested fifty-one of the one hundred and seventeen samples examined for Somerset House by this process (Pharmaceutical Jour., July 1876). The extreme of insoluble fatty acids recorded by him is 89·9 per cent. Fleischman and Vieth fixed the limit at 89·73 per cent (Fres. Zts., 17, 289). Kretschmar (Bericht D. Chem. Gesell. 10, page 2091) regards ninety per cent as the proper limit. John (Arch. Pharm., 9, 339), finds over eighty-nine per cent in genuine butters; so also De la Source (Report. Pharm. 38, 446) in case of cows fed on oil-cake.

Milne (Analyst 4, 40) tested the process and finds that the determination of soluble fatty acids is of great value and importance. The maximum of insoluble fatty acids found was 89·73 degrees.

Abroad the process was examined by Heintz (Fres. Zts. 17, 160). He asserts that lauric acid, if present, is only imperfectly washed out, and that the amount of wash water must always be a definite amount. Sachsse (Fres. Zts. 17, 151) found the process was applicable to very rancid butter. Bachmeyer (Dingler Polyt. Jour. 226, p. 103) finds the process satisfactory. E. W. Jones (Analyst 2, 19 and 37) made experiments on washing the insoluble acids with larger amounts of water than usual and found that the insoluble acids were to all appearances slowly dissolved by washing. Dr. Harland (Analyst 8, 163) testified that two hours drying of the fatty acids was insufficient, twelve to fourteen hours being required.

The results of work in the laboratory in connection with this investigation go to show that the method of washing makes much difference with regard to this point. If the insoluble acids are incompletely washed the acids which are less soluble or volatile may volatilize slowly, and hence a long time required to bring them to constant weight. From two to three grammes require six or seven installments of 100 to 125 c. c. of water each, the operation being performed in a flask as recommended by Muter. Hanssen (Inaug. Dissert. Erlangen, 1883) found some of the soluble acids remaining with the washed insoluble acids. His results have not been either disproved or confirmed by other experimenters.

Blyth (Analyst 2, 112) suggests a modification of the method of washing, using an inverted flask, but that does not seem so convenient as the "stop-cock flask" or large separating funnel proposed by Corfield (Chem. News 37, pp. 7 and 30).

That the process is unable to detect cocoanut oil, which contains much lauric acid, is admitted by Hehner and Angell in their original paper. The peculiar odor of cocoanut oil has been, however, up to the present time, a bar to its use as an adulterant.

A method of deodorizing cocoanut and other oils of the kind, has been patented in Germany (Jeserich and Meinert, Wagner's Jahresbericht 1882, p. 932), and it remains to be seen how far such addition to butter may be made without detection by this method of analysis.

A priori it seems very probable that the extreme slowness with which the soluble acids are extracted by washing, and the low figures which would be obtained for the sum of soluble and insoluble fatty acids, would serve to indicate the presence of such an adulterant. Cornwall (8th Report of N. J. State Board of Health, p. 195) regards the possibility of adulteration by cocoanut oil as a strong objection to the wash process and in this respect is followed by Moore (Jour. Am. Chem. Soc. 7, 188).

Reichardt says (Arch. der Pharm. 13 Heft. 2) that the limits are too high. Kuleschoff (Bull. Soc. Chim. 29, 17) on the other hand obtained a maximum in his experiments of 88·57 degrees, and concludes that they are too low. Riche (Jour. Pharm. Chem., Sept. 1880) finds the figures given to be correct.

Fleischmann (Fres. Zts. 17, 287) approved the principle, but finds the maximum percentage of insoluble fatty acids should be set at 90 per cent.

The method was adopted in the Paris Municipal Laboratory (Moniteur Scientifique, 3, 11, 393) and is still in use there (Analyst 10, 198), the standard taken being 87·5 per cent insoluble fatty acids, the solidifying point of which should not be over 38 degrees C.

The process is highly commended by Dr. Bell in his address on the occasion of the health exposition in London in 1884, but he says with truth that in working by it, it might be possible for 40 per cent of adulterations to escape detection in special cases, and hence we cannot as yet regard butter analysis as by any means perfected.

On the other hand Ambuhl (Rep. Anal. Chem. 1, 171) speaks of the process as antiquated and pronounces in favor of the Reichert process to be described.

Results by the process on fats and oils other than butter and the fat of farm animals are given by Dietzell and Kressner (Fres. Zts. 18, 83). The point made by them is that the vegetable oils as a rule yield very nearly the same results as beef fat, tallow, etc., about 95 per cent or over of insoluble fatty acids.

A method has been proposed by J. West Knights (Analyst 5, 155) which is in effect a modification of this process. After saponifying, as usual, the insoluble fatty acids are precipitated as barium salts, and from this the fatty acids are separated by addition of an acid. Experiments with it did not impress one with favor toward it either on the score of convenience or of accuracy. It has met with but little approval (Caldwell, 2d Report to N. Y. State Board of Health, p. 524).

The method used in this laboratory has been essentially that of Dupré, the fatty acids having been washed, as has been said with six or seven or even more installments of hot water (about 100 c. c. each time), rinsing off between each with about 25 c. c. of cold water. For the purposes of this investigation separate determinations were made on the acidity of each washing.

The Wash Process Results Obtained in Laboratory.

Examination of Fats by the "Wash Process."

	Insoluble.	Soluble.
1268 Butter:		
1358a Butter	88·119	5·319
1359a Butter	88·5	5·67
1359b Butter	87·161	6·114
1359b Butter	87·164	6·116
1384 Butter	88·140	5·514
1507 } Butter two years old	88·28	4·99
1507 } Butter two years old	88·35	4·93
1547 Holstein	89·13	4·61
Doubful:		
1337	89·91	5·32
1468	90·52	4·305
1469	91·24	4·839
1470	90·39	4·484
Oleomargarine:		
1303	94·85	0·38
1314	95·34	0·35

1314	95·15	0·12
1316	95·88	0·39
1317	95·41	0·05
1318	95·18	0·37
1329	96·14	0·35
1330	95·36	0·86
1331	95·66	0·52
1333	96·32	0·24
1334	96.60	0·47
1336	96·05	0·42
1338	95·80	0.16
1339	95·63	0·14
1341	94·74	0·56
1397	95·87	0·71
1398	96·11	0·38
1399	95·91	0·19
1441	96·28	0·37
1533	94·65	0·42
1661	95·60	0·28
1662	96·01	....
1662	96·30	0·58
1367, Beef fat	95·961	0·275
	95·792	0·155
1368, Mutton fat	95·662	0·114
1369, Lard	95·668	0·035

Results of Analysis of Commercial Samples of Oleomargarine.

Number.	Fat.	Water.	Curd.	Salts.
1269	......	9·21	......	2·77
1303	85·96	8·87	0·72	4·45
1314	82·72	10·05	1·09	6·14
1316	86·34	9·31	0·74	3·61
1317	84·68	9·04	0·80	5·48
1318	84·93	10·21	0·29	4·57
1329	84·39	9·89	1·29	4·43
1330	85·92	9·29	0·91	3·88
1331	83·56	10·79	1·56	4·09
1333	84·97	9·73	1·10	4·20
1334	86·59	8·68	1·19	3·54
1336	81·18	10·72	1·87	6·62
1338	81·66	10·41	1·62	6·31
1339	74·81	11·67	1·75	11·77
1341	86·28	10·10	0·95	2·67
1397	86·24	7·78	0·31	5·67
1398	85·86	9·91	0·35	3·88
1399	80·70	10·01	1·66	7·63
1441	85·36	9·57	0·66	4·41

Some abstracts from Hehner and Angell's work on this subject are herewith given.

TABLE XX.

(Results obtained by Hehner & Angell, p. 25.)

Insoluble Acids in Genuine Butter.

Per cent.	Per cent.	Per cent.	Per cent.	Per cent.
87·9	87·1	87·2	87·5	87·8
87·1	87·3	86·9	87·6	86·9
87·2	86·8	87·2	86·8	87·8
87·5	86·8	87·3	87·1	87·5
87·8	87·2	86·8	87·2	88·2
87·0	87·3	87·3	86·9	86·1
87·1	87·3	87·0	87·5	87·3

Average, 87·24 per cent.

Samples of Normandy Butter, genuineness doubtful.

Per cent.	Per cent.	Per cent.	Per cent.
89·1	89·4	88·6	88·9
89·6	88·7	88·6	88·7

TABLE XXI — H. & A., p. 19.

Adulterated Butter.

Number.	Water, per cent.	Salt, per cent.	Curd. per cent.	Insoluble fatty acids, per cent.	Microscopic appearances.
1....	15·6	8·5	0·41	89·33	Crystalline.
2....	16·2	2·9	2·57	93·87	Very crystalline.
3....	16·5	6·9	1·1	89·9	Spheroid crystal.
4....	22·19	5·68	1·02	87·51	Spheroid crystal.
5....	20·94	5·6	1·8	89·45	Spheroid crystal.
6....	6·6	2·6	5·08	92·2	Spheroid crystal.
7....	8·9	8·4	1·2	93·2	Spheroid crystal.
8....	13·68	2·2	2·79	93·3	Spheroid crystal.
9....	13·1	3·5	3·4	86·6	Crystalline.
10....	9·95	4·08	3·8	92·53	Very crystalline.
11....	13·90	2·91	2·32	91·9	Very crystalline.
12....	19·6	8·0	1·96	87·4	Very crystalline.
13....	4·08	2·33	7·39	91·79	Very crystalline.

TABLE XXI — (*Continued*).

Number.	Water, per cent.	Salt, per cent.	Curd, per cent.	Insoluble fatty acids, per cent.	Microscopic appearance.
14....	12·0	·9	1·29	92·1	Many small crystals.
15....	5·9	3·7	2·7	91·9	Crystalline.
16....	6·9	3·85	2·5	93·3	No distinct crystals, whole mass depolarized light.
17....	11·67	3·22	1·31	91·04	Spheroid crystal.
18....	5·1	1·5	2·9	93·9	Not recorded.
19....	16·9	9·2	1·29	88·0	Crystalline.

TABLE XXII — H. & A., p. 72.

Showing the relative proportions of insoluble and soluble acids.

Insoluble acids.	Soluble acids corresponding.
86 per cent........................	8·88 per cent.
87 per cent........................	7·97 per cent.
88 per cent........................	7·06 per cent.
89 per cent........................	6·15 per cent.
89·5 per cent........................	5·69 per cent.

TABLE XXIII — H. & A., p. 76.

Table 1 — Results of analysis of animal fats by Bell.

DESCRIPTION OF SAMPLE.	Specific gravity at 100 degrees Fahr.	Percentage of fixed fatty acids.
Mutton suet........................	902·83	95·86
Beef suet........................	903·72	95·91
Fine lard........................	903·84	96·20
Dripping (commercial)................	904·56	94·67
Mutton dripping (genuine)............	903·97	95·48

TABLE XXIV.

Results on fats other than butter, obtained by the process of Hehner & Angell.

Name of fat.	Per cent insol. acids.	Authority.
Palm oil........................	95·6	Dietzell & Kressner.
Rapeseed oil........................	95·0	Dietzell & Kressner.
Rapeseed oil........................	95·14	Benseman.
Poppy oil........................	95·38	Dietzell & Kressner.

TABLE XXIII — (*Continued*).

Name of fat.	Per cent insol. acids.	Authority.
Benne oil	95·60	Dietzell & Kressner.
Benne oil	95·86	Benseman.
Olive oil	94·03	
Olive oil	95·43	Benseman.
Almond oil	94·02	
Fat from cow's milk	87·77	Benseman.
Cottonseed oil	95·75	Benseman.
Peanut oil	95·86	Benseman.
Cacao fat from Maracaibo beans ...	94·59	Benseman.
From Caracas bean	95·31	Benseman.
From Trinidad bean	95·65	Benseman.
From Porte Plata	95·44	Benseman.
From Machala Quayaquila	95·24	Benseman.

Dietzell and Kressner, Fresenius Zeitschrift für analytische Chemie, vol. 18, page 83.

Benseman, Repertorium der analytischen Chemie, vol. 11, page 165.

——— ———, Archiv der Pharmacie, vol. 9, 1878, page 134.

The following results were obtained in the laboratory:

The first nine of these samples with one exception were from Mr. J. B. Dunham, of Almoral, Delaware county, Iowa, and in every case were accompanied by affidavits stating that the sample was to his own knowledge genuine butter obtained from cream of cow's milk without any admixture of foreign fat, coloring or salt. The sample starred was from Schock and Bolender, Orangeville, Illinois, and was accompanied by an affidavit to the same effect, but stating that the sample was colored. They will be called for reference affidavit samples.

In the following tables the analyses bracketed are on the same sample; those numbered are on different samples from different sources.

The following fifteen samples are of known purity, being made directly from cream either in the laboratory or by the inspector.

TABLE

Number	NAME.	Amount taken in grms.	KOH used to saponify expressed in grms. of NaOH.	H_2SO_4 used expressed in grms of NaOH.	Cb. cm. of $\frac{n}{10}$ NaOH used to neutralize each successive wash of 100 cb. cm. hot and 25 cb. cm. of cold water. 1.	2.	3.	4.	5.	6.	7.	8.	9.	10.	11.	Per cent of soluble acids calculated at butyric.	Per cent of insoluble acids.
1691..	Creamery August, 1885.....	1.0220	.1960	.2800	27.15	.45	.20	.20	.15	.10	..	..	..	..	..	5.91	87.56
		1.9255	.3920	.5200	43.60	1.30	.40	.40	.30	.15	.10	..	..	..	..	6.20	87 88
1692..	Dairy August, 1885.........	1.0085	.1960	.2800	26.70	.35	.25	.15	....	....	..	..	..	..	..	5.39	88.76
		2.6610	.4900	.5800	31.15	2.05	.65	.65	.25	.30	.15	..	..	..	..	5.89	88.62
1711..	*Gathered cream August 1885	1.1115	.2940	.4000	34.80	.45	.25	.15	.15	.09	..	..	..	..	..	6.76	88.47
		2.3193	.3920	.6000	70.45	1.39	.44	40	.20	.15	..	..	..	..	..	6.67	88.61
1726..	Dairy October, 1885........	1.2870	.2548	.3040	28.21	.70	.30	.10	....	....	..	..	..	..	..	5.81	88.07
		2.6035	.4900	.5800	36.20	1.00	.70	.40	.10	....	..	..	..	..	..	6.18	88.82
1727..	Creamery October, 1885....	1.0195	.1960	.2800	25.30	.50	.30	.20	.10	....	..	..	..	..	..	4.77	88.31
		1.0225	.21993	.2900	21.5	.80	.30	.10	....	....	..	..	..	..	..	4.96	88.21
1757..	Dairy November, 1885......	1.8355	.34407	.4000	22.87	1.20	.31	.31	.27	.13	.07	..	..	..	..	5.35	87.97
		3.3205	.68079	.7200	38.20	2.30	.80	.53	.35	.23	.12	.23	.07	..	..	5.42	87.71
1758..	Creamery November, 1885.	1.3270	.25805	.3200	21.1	.46	.35	.31	.17	.07	..	..	..	..	..	4.59	89.22
		2.3600	.516082	.6200	30.0	1.20	.42	.38	.23	.15	.07	..	..	..	..	4.49	88.93
1785..	Creamery December, 1885..	2.2465	.45125	.50896	28.70	.31	.20	.10	....	....	..	..	..	..	..	5.49	89.33
		2.2485	.45125	.50896	26.00	.93	.36	.31	.10	....	..	..	..	..	..	5.57	89.58
1786..	Dairy December, 1885......	2.9350	.60162	.68183	32.00	1.14	.47	.20	.10	....	..	..	..	..	..	5.08	90.27
		2.064	.40108	.44467	21.80	.58	.31	.20	.10	....	..	..	..	..	..	5.21	90.21
E 3158	Native Upland...	2.362	.79282	.82821	25.50	.83	.42	.20	.10	....	..	..	.	..	..	6.80	87.69
		2.003	.66520	.70974	25.60	.62	.42	.20	.10	....	..	..	..	..	..	6.97	87.81
E 3158	Same with gelatin..........	2.1205	.79282	.82821	21.4	.83	.42	.30	.10	....	..	..	..	..	..	6.98	88.13
		2.300	.79282	.82821	25.4	1.14	.38	.10	....	....	..	..	..	..	..	7.18	88.17
3159	Native Upland.............	2.5100	.79282	.82821	25.1	1.04	.52	.31	.10	....	..	..	..	..	..	6.44	88.61
		2.5075	.79282	.82821	26.5	.52	.20	.10	....	....	..	..	..	..	..	6.52	8871
E 3159	Same with gelatin..........	2.3185	.79282	.82821	24.3	.73	.52	.10	....	....	..	..	..	..	..	6.40	88.69
		2.3475	.79282	.82821	25.0	.94	.62	.05	....	....	..	..	..	..	..	6.55	88.85

E 3160	Native lowland	2.0590	1520	.48782	287	.78	.42	.10	....	....	..	..	..	..	..	6.21	88.66
		1.8530	.277776	.354924	27.4	.68	.42	.10	....	....	..	..	..	..	..	6.14	88.80
	Same with gelatin	1.7145	6650	.414078	23.9	.73	.31	.10	....	....	..	..	..	..	..	5.97	88.63
		2.5110	1540	.59154	32.45	1.04	.42	.10	....	....	..	..	..	..	..	5.70	88.83
E 3161	Dutch lowland	0490	408	.44467	23.40	.73	.42	.10	....	....	..	..	..	..	..	5.95	89.70
		2160	408	4467	22.86	1.04	.86	.05	....	....	..	..	..	..	..	5.97	89.22
	Same with gelatin	2.1355	.421134	5065	34.06	.94	.47	.10	....	....	..	..	..	..	..	6.13	88.89
		1.6345	9216	4078	28.70	.84	.42	.10	....	....	..	..	..	..	..	6.04	89.11
E 3162	Lowland pasture	1.9365	4060	4592	24.34	.42	.20	.10	....	....	..	..	..	..	..	5.32	88.23
		2.5970	5850	5455	29.54	.47	.68	.86	.10	....	..	..	..	..	..	5.88	88.49
	Same with gelatin	2.1700	4340	5065	34.8	.68	.47	.86	.05	....	..	..	..	..	..	5.22	88.62
		2160	4080	4891	25.5	.42	.20	.10	....	....	..	..	..	..	..	5.50	88.52
1549..	aNve butter	1.8330	3326	4000	44.1	.65	.40	.10	....	....	..	..	..	..	..	5.76	87.09
		1.8860	851590	.4800	42.6	1.10	.50	.10	....	....	..	..	..	..	..	5.84	87.30
1548..	Ayrshire butter	1.2465	284	3800	43.0	1.80	2.05	.40	.05	....	..	..	..	..	..	7.25	87.65
		6260	5653	6400	51.1	1.80	1.00	.30	.05	....	..	..	..	..	..	7.18	87.58
1550..	Devon butter	1.0061	2000	2292	12.28	.78	.25	.22	.18	.09	.05	..	..	..	..	6.86	.....
		5340	41340	.4807	25.90	1.18	.44	.81	.18	.18	.13	.13	.09	.13	.13	6.72	*86.88
		8918	2610	2292	11.51	.61	.22	.17	.26	.09	.05	..	..	..	..	6.86	*86.79
1546..	Jersey butter	1.6012	45180	6270	51.70	1.94	.48	.28	.28	.14	.14	.14	.28	..	..	6.37	*87.89
		1.6402	.41840	4807	26.07	1.14	.44	.40	.18	.18	.18	.13	.13	.13	.13	6.61	*87.22
1547..	Holstein butter	0490	3928	8200	36.4	.40	.20	.15	.05	....	..	..	..	..	..	5.06	89.3
		2.3915	4254	6000	48.8	.40	.80	.15	....	....	..	..	..	..	..	5.11	89.00
1542..	Reid butter	2.5565	38197	.6000	42.35	.70	.50	2.30	.05	....	..	..	..	..	..	6.53	88.48
		1.8345	3590	.4800	44.2	1.10	.30	.30	.10	....	..	..	..	..	..	6.66	88.52
1507..	Butter	1.8120	3326	.4068	31.8	.60	.42	.23	.07	....	..	..	..	..	..	6.18	86.31
		2.4870	455	.5331	33.96	1.10	.73	.50	.26	.23	.08	..	..	..	..	5.94	86.49
1752	Oleomargarine made in laboratory	1.6930	......	......	......	....	....	....	....	....	..	..	..	..	..	....	94.62
		2.3830	......	......	......	....	....	....	....	....	..	..	..	..	..	...	94.63

Animal fats and oils.

1557..	Beeswax	2.0240	9828	.3600	32 5	.50	....	....	....	....	..	..	..	..	..	.18	99.90
		1.2830	9888	.5200	295	.70	.10	....	....	....	..	..	..	..	..	.18	99.87
1551..	Rose wax	3.5400	6648	.8000	41.7	.75	.60	.40	.10	....	..	..	..	..	..	3.53	90.00
		3.2255	6648	.8000	40.0	.95	.80	.30	.10	....	..	..	..	..	..	3.42	90.37
1636..	Horse oil	3235	6082	.6000	22.9	.31	.31	.23	.15	.07	..	..	..	..	..	1.12	93.18
		1.6285	4005	.4800	20.7	.46	.31	.31	.15	.07	..	..	..	..	..	1.28	92.84
1637..	Dog oil	5550	3898	.4000	23.9	.46	.31	.19	.07	....	..	..	..	..	..	2.17	92.73
		2.0250	6657	.5200	25.7	.42	.28	.07	....	....	..	..	..	..	..	1.69	92.57
1727(a)	Elephant fat (Jumbo)	2.4810	.544764	. 7699	34.45	.63	.15	.07	....	....	..	..	..	..	..	.24	95.40
		1.7815	.430057	. 090	11.79	.80	.31	.23	.23	.87	..	.	.	..	..	.10	95.21
1789..	Bears' fat	2.0670	......	......	......	....	....	....	....	....	..	..	..	..	..	....	94.87
		1.9515	......	......	......	....	....	....	....	....	..	..	..	..	..	....	95..0

TABLE XXIV — (*Continued*).

Vegetable fats and oils.

Number.	NAME.	Amount taken in grms.	KOH used to saponify, expressed in grms., of Na OH.	H_2SO_4 used, expressed in grms. of Na. OH.	Cb. cm of $\frac{N}{10}$ Na OH used to neutralize each successive wash of 100 cb. em. hot and 25 cb. em. of cold water. 1.	2.	3.	4.	5.	6.	7.	8.	9.	10.	11.	12.	Per cent of soluble acids calculated at butyric.	Per cent of insoluble acids.
1575..	Olive oil..................	2.1015	.4101888	.6000	47.75	.40	.15	.10	....	....	..	..	..	..	..	..	.501	95.09
		2.0880	.390656	.4200	32.50	.45	.15	.10	....	....	..	..	..	..	..	..	.343	94.82
1561..	Linseed oil (1)	2.4785	.468787	.7000	61.15	.70	.15	.05	....	....	..	..	..	..	..	..	.51	94.06
		2.6615	.468787	.6000	35.05	.45	.25	.10	....	....	..	..	..	..	..	..	.67	94.37
......	Lin-seed oil (2)	2.0095	.390656	.4800	23.25	.20	.30	.10	....	....	..	..	..	..	..	..	.66	94.35
		2.0740	.390656	.5200	33.10	.70	.10	....	....	....	..	..	..	..	..	..	.66	94.37
1572..	Sesame or Benne oil (1),..	2.0840	.3807657	.4800	25.70	.27	.19	.07	....	....	..	..	..	..	..	..	.61	95.42
		2.0640	.3807657	.4800	25.40	.36	.19	.07	....	....	..	..	..	..	..	..	.54	95.38
1638..	Sesame oil (2)	2.2010	.49999	.6200	27.50	.50	.35	.15	....	....	..	..	..	..	..	..	.73	95.68
		1.0080	.21993	.3800	40.25	.30	.30	.05	....	....	..	..	..	..	..	..	.76	95.49
1576..	Cotton oil	2.3630	.435154	.5600	33.80	.33	.31	.19	.15	.15	.07	..	..	.	..	..	1.52	94.46
		2.6195	.471418	.6000	35.70	.31	.23	.15	.23	.15	.07	..	..	..	..	..	1.55	95.04
1759..	Peanut oil..................	2.0755	.390656	.5200	33.9	.35	.10	....	....	....	..	..	..	..	..	..	.35	94.94
		2.0190	.390656	.5200	33.2	.40	.10	....	....	....	..	..	..	..	..	..	.59	95.09
1558..	Bayberry tallow..........	.9995	.19994	.2410	15.1	.35	.10	....	....	....	..	..	..	..	..	..	1.70	89.95
		2.5835	.539836	.6000	19.8	.60	.10	....	....	....	..	..	..	..	..	..	1.62	89.93
1545..	Cocoanut oil..............	2.2195	.44832	.6000	41.80	1.05	1.15	.90	.60	.50	.45	.45	.45	.30	.40	..	6.12	80.69
		......	....	......	.55 .50	.35	.30	.40	.40	.30	.30	.30	.50	.35	.30	..		
		......	....	......	.20 .15	.20	.20	....	....	....	..	..	..	..	..	..		
		2.4940	.585984	.7200	40.35	1.65	1.05	.80	.70	.75	.55	.60	.50	.25	.55	..	7.00	80.87
		......	....	......	.65 .55	.50	.45	.40	.35	.20	.40	.40	.50	.55	.30	..		
		......	....	......	.15 .30	....	....	....	....	....	..	..	..	..	..	..		
1563..	Mustard-seed oil	2.5370	.4883196	.6400	39.60	.30	.25	.10	....	....	..	..	..	..	..	..	.81	95 66
		3.1120	.6055168	.7600	38.80	.50	.20	.10	....	....	..	..	..	..	..	..	1.05	95.18

Marine oils.

1753..	*Porpoise jaw oil, skimmed and strained.	1.9795	.430057	.5200	63.45	3.73	.90	.50	.40	.27	.23	.15	.08	..	..	..	21.11	86.35
		2.5950	.559080	.6400	74.3	4.16	2.40	1.10	.70	.35	.27	.15	.03	..	..	..	21.77	68.48
1748..	Porpoise jaw oil, skimmed and strained.	1.5150	.352800	.4400	48.70	2.08	.40	.40	.15	....	..	..	..	..	..	..	17.38	.....
		2.6725	.58800	.6800	64.55	7.50	1.70	.55	.25	.10	..	..	..	..	..	..	16.98	71.93
		4.3675	.878976	1.0400	98.50	9.90	5.85	.80	.10	....	..	..	..	..	..	..	......	72.18
1754..	Porpoise jaw oil, not skimmed or strained.	2.3050	.5160824	.6000	19.30	.80	.19	.07	....	....	..	..	..	..	..	..	.00	96.51
		2.4370	.5160824	.6000	20.55	.37	.19	.07	....	....	..	..	..	..	..	..	.06	96.49
1755..	Black-fish body oil.......	1.7325	.3870696	.4800	61.14	3.80	.54	.31	.19	.07	..	..	..	..	..	..	21.74	66.55
		2.0590	.430057	.5240	67.84	4.16	1.00	.54	.46	.23	.23	.15	.08	..	..	..	21.84	66.02
1756..	Black-fish body oil.......	1.8955	.3870696	.4800	27.74	.38	.38	.27	.23	.38	.23	.07	..	..	..	..	2.49	93.17
		2.2270	.430057	.5240	26.56	.38	.38	.38	.46	.27	.35	.31	.31	.11	.07	..	2.43	92.97
1580..	Porpoise body oil.........	2.4795	.453268	.6000	45.28	.90	.54	.54	.54	.46	.27	.35	.19	.07	..	..	3.48	89.80
		2.7710	.507680	.6400	40.77	.90	.46	.38	.54	.23	.19	.15	.07	..	..	..	3.37	90.09
1577..	Sea elephant oil..........	2.4410	.4351549	.5600	43.81	.90	.46	.46	.38	.38	.70	.46	.38	.23 .23	.31 .19	.12 .23	6.40	87.27
		2.2930	.4170235	.5600	46.28	.69	.42	.27	.27	.15	.19	.19	.15	.23	.23 .23	.12 .38	5.84	87.16
1579..	Seal oil..................	2.8315	.546918	.6800	41.5	.80	.50	.35	.10	....	..	..	..	..	.	..	5.78	90.56
		3.7140	.72271	.8400	48.35	.90	.50	.35	.10	....	..	..	..	..	..	..	5.87	90.80
1582..	Shark liver oil...........	3.6170	.7031808	.8400	48.3	.65	.35	.10	....	....	..	..	.	..	..	..	3.69	93.95
		3.9515	.761779	.6800	41.7	.65	.25	.10	....	....	..	..	..	..	..	..	3.75	93.89
1583..	Menhaden oil	1.4955	.286718	.3600	19.56	.46	.35	1.15	.50	.39	.31	.38	.7	.35 .9	.15 .39	.12 .55	4.69	85.62
		2.1665	.430057	.5200	22.25	.55	.50	.46	.38	.31	.31	.39	.92	.31 .31	.15 .35	.12 .55	4.36	85.79

In the analysis of the marine oils much difficulty was experienced in obtaining constant results. It was necessary to use wax to collect the insoluble fatty acids, and on drying the acids the presence of a substance was observed which attached itself to the sides of the flask and was evidently of a non-fatty character. The oils are of unknown composition and contain unsaponifiable matter. Further investigation of this subject would doubtless lead to interesting results.

THE REICHERT PROCESS.

The publication of the results of Hehner and Angell's investigations as to the constitution of butter fat naturally stimulated a search for other methods of analysis. Recognizing the fact that the soluble acids of butter were also the volatile ones, attempts were made in this direction, viz.: to determine the volatile acids by distillation, and in this way to get data from which safe conclusions as to the character of the fat could be drawn. (Lechartier, Perkins, Analyst 4, 142). The path in that direction seemed to be blocked by the fact already recorded by Hehner and Angell, that however far the distillation was carried, some butyric acid still remained in the retort.

Reichert conceived the idea that it was unnecessary to attempt to distill all the fatty acids over, but that by taking a definite amount of the fat, every time, saponifying with the same quantities of alcohol, evaporating, diluting and acidifying with the same amounts of water and acid and finally taking the same amount of water and acid and finally taking the same amount of distillate, a sufficient uniformity of results might be obtained on which to base an opinion.

The result was a process which, though arbitrary in every detail is practicable and counts among its advocates many skilled and careful chemists. The process (Fres. Zts, 18, page 68) consists in weighing out 2·5 grammes of fat in a flask of 150 c. c. capacity (Erlenmeyer's form preferred), saponifying by one gramme of caustic potash in 20 c. c. of 80 per cent alcohol, evaporating off the alcohol, dissolving it in 50 c. c. of water, and adding 20 c. c. of dilute sulphuric acid (1 to 10). The mixture is then distilled, platinum spirals or pumice stone being introduced to prevent explosive ebullition, the distillate being passed through a filter as it drops into the receiving flask. The first portions of the distillate are returned to the original flask and the distillation is then kept up until 50 c. c. have passed over. The acidity of this distillate is then determined by means of a tenth normal solution of caustic soda, and the results are expressed in the number of c. c. of this dilute alkaline solution used. Butter fat was found to require 13 to 15 c. c. when thus treated, while all other fats showed much less, oleomargarine and beef fat, etc., requiring under one c. c.

It is necessary to run through a blank test with the chemicals intended to be used as they frequently contain impurities which affect the result. This constant quantity being found, the proper deductions can be made as long as the stock lasts.

Another point not very strenuously insisted upon by Reichert, though regarded as essential by some who have examined the process, is that the alcohol should be completely driven off after saponification (Cornwall, 8th Report, N. J. State Board of Health, page 195).

The following are the results upon which the method and conclusions of Reichert are based.

Table XXV.

Name of sample.	cb. cm. $\frac{n}{10}$ Na OH.
Butter	14·5
Butter	14·05
Butter	13·75
Butter	14·3
Bntter	14·0
Butter	14·4
Butter	14·0
Butter	13·25
Butter	13·8
Butter	14·95
Butter	14·20
Butter	13·00
Butter	13·40
Commercial butter	10·50
Cocoanut oil	3·70
Oleomargarine	·95
Lard	·30
Kidney fat	·25
Rapeseed oil	·25

The process was subjected to further trial in tho hands of its author by applying it to various mixtures of butter with foreign fats. In the first case lard was used, with the following results:

	Per cent.	Per cent.	Per cent.	Per cent.	Per cent.	Per cent.
Butter	0	20	40	60	80	100
Lard	100	80	60	40	20	0
c. c. $\frac{n}{10}$ Na OH	.3	3.1	5.9	8.5	11.5	14.3

In the second case a mixture of the following composition was used to mix with the butter:

	Per cent.
Lard	50
Kidney-fat	30
Rapeseed oil	20
	100

The results obtained were as follows:

	Pr ct.	Pr ct.	Pr ct.	Pr ct.	Pr. ct.	Pr ct.	Pr ct.	Pr ct.	Pr ct.	Pr ct.	Pr ct.
Butter	0	10	20	30	40	50	60	70	80	90	100
Mixture	100	90	80	70	60	50	40	30	20	10	0
c. c. $\frac{n}{10}$ Na OH..	.2	1.65	3.15	4.6	5.95	7.3	8.9	10.15	11.55	13.0	14.5
c. c. $\frac{n}{10}$ Na OH..	.2	1.65	3.15	4.5	6.10	7.45	8.75	10.20	11.60	13.2	14.5
Average	.2	1.65	3.15	4.55	6.02	7.37	8.82	10.17	11.58	13.1	14.5

The formula given by Reichert for the calculation of the percentage of Butter fat present in a mixture is $B=(7.30\pm0\cdot24)(n-0\cdot30)$ in which B=percentage of butter present and n=number of c. c. of tenth normal soda used.

He allows a minimum of 12·5 c. c. for the poorest butter.

In connection with the Reichert process the observation of Duclaux (Ann. Chim. Phys. (5) 2, p. 333) is pertinent. He found that if a liquid containing any fatty acid be distilled, each successive fraction contains an amount of acid practically constant in proportion to the total acid present for the same fraction, but varying according to the nature of the acid employed. Thus if one-eleventh of a solution of butyric acid is distilled off, it will contain 16·4 per cent of all the acid present. Another point brought out by Duclaux in this investigation was that contrary to expectation the more volatile acids, when free from water, were slower in coming over. Thus only 5·9 per cent of the total acetic acid present makes its appearance in the first eleventh of the distillate as against 16·4 per cent for butyric acid as quoted above. Yet pure acetic acid boils at 119 degrees C., while pure butyric acid boils at 163 degrees C., isobutyric acid boiling at 154 degrees C. Taking the figures given by Duclaux as a basis, the distillate in the Reichert process should contain 83 to 89 per cent of the whole amount of butyric acid present. It is, however, possible that the presence of the oily fat acids may have some modifying influence. Duclaux found that the presence of glycerine had some influence on the proportions of acid obtainable in the different fractions, and that alcohol made the first fractions much weaker in acids, a reminder of the importance of completely removing the alcohol from the soap after saponification has been effected. Twelve c. c. of the soda solution by this process would indicate 4·224 per cent of butyric acid, 14 c. c. represents 4·928 per cent.

The method has experienced little or no modification of importance since it was first proposed. The proposition of Meissl (Dingl. Polyt. Jour. 233, p. 329) is almost the only one. This consists in taking double the amount of butter and reagents proposed by Reichert and distilling off 110 c. c. instead of 100 c. c., which would correspond with the 50 c. c. of distillate by Reichert. The alcohol he used is also somewhat weaker (70 per cent instead of 80 per cent). His results are comparable with those of Reichert. The lowest figure usually obtained with genuine butter he gives as 24 c. c. (12 of Reichert).

Munier in testing the method modified it by using much less alcohol in the saponification, and adding phosphoric instead of sulphuric acid (Fres. Zts. 21, p. 394) besides apparently using a different method of removing the alcohol. He calls these modifications unessential, but in this respect some other chemists do not agree with him. (Reichard, Arch. Pharm. 222, 393 ; Sendtner, Archiv. fur Hygiene, 1883, p. 137 ; Wagner's Jahresber. 29, 979 ; Cornwall, 8th Report, N. J. State Board of Health, 195.) His results ranged from the very low figure 9·2 c. c. for December butter to 12·4 for the warmer months. His conclusion naturally is that Reichert's figure is too high.

On the other hand Beckursts finds them if anything too low (Pharm. Central H. 1883, 557). Ambuhl (Repert. Anal. 1, 171, Wagner's Jahresber. 27, p. 839) also obtains figures somewhat higher than Reichert.

Reichard's figures (Arch. Pharm. 222, 93) range from 13·8 to 14·7 c. c., with an average of 14·16.

Sendtner (Loc. cit.) gives as the lowest permissible figure for genuine butter 24 (12 of Reichert), Meissl's process being used. Further examination would be necessary if 23 degrees or 23·5 degrees c. c. were obtained, as the fat might be an exceptional kind of butter fat.

Besides the chemists already mentioned Medicus and Scherer (Fres. Zts. 19, p. 159), Hanssen (Inaug. Diss. Erlangen 1882), Elsner (Praxis der Nahrungsmittel chemikers, 1880), Caldwell (Second Report, N. Y. State Board of Health), A. H. Allen (Analyst 10, 103), McCay (Chem. News 50, p. 151) and others have expressed their approval after testing it.

The experiments of Medicus and Scherer indicate that there is some tendency in melted butter fat to "stratification," the lower layers showing a little less by the Reichert process than the upper.

Hehner objects to the process on the ground that it serves only to estimate a part of the substance present (Analyst 10, 105). Its arbitrary character constitutes a serious objection to it in the estimation of many.

Bischoff (Eulenberg's Handb. off. Gesundeitsw. 2, p. 509) asserts that with neither Reichert's or Koettstorfor's process has been able to obtain sufficiently concordant results to take them into court.

In experimenting with the process on butters and other fats and oils we have obtained figures confirming those of Reichert. But we have pushed it further in order to test the value of a distillation process. After taking off the first 50 c. c. of a distillate and titrating, 50 c. c. more water was added to the contents of the flask, and another 50 c. c. distilled off, titrated, another 50 c. c. added and again distilled, and so on until the acidity of the distillate was too small to be of importance.

According to Duclaux's experiments, if the liquid in the flask consisted only of butyric acid and water, the third or fourth 50 c. c. would bring over practically all of the butyric acid, of which the first distillate would contain about 86 per cent. In practice there was in the retort volatile acids other than butyric, the glycerine resulting from the decomposition of the fat, the non-volatile fatty acid and potassium sulphate besides free sulphuric acid, and these substances, one or all, modified the results. Usually five to nine distillations were necessary before the distillate was practically free from acidity, and the first distillate contained about 80 per cent of all the acid obtained, reckoning it all as butyric acid. Besides this a certain amount of solid fatty acid made its appearance in the condenser in most cases. Thus far we have been unable, for lack of opportunity, to determine the character of this fatty acid, but hope soon to do so. The results agreed very well with those obtained by the washing process.

Medicus and Scherer obtained the following results on various oils:

Name of oil.	cb.cm $\frac{n}{10}$ Na OH used.
Olive oil	·3
Palm oil	·5
Benne oil	·35
Rapeseed oil (desulphuretted)	·4
Rapeseed oil (pure)	·3
Lard	·2

The following results were obtained in the laboratory:

TABLE XXVI.

Samples of known purity.

	b. cm. $\frac{n}{10}$ Na OH used for each 50 cb. cm. distillate.									Per cent volatile fatty acids calculated to butyric acid.	Percent of total acidity in first distillate.
	1.	2.	3.	4.	5.	6.	7.	8.	9.		
Creamery, August, 1885	13.4	1.4	.55	.35	.15	......	......	......	......	5.52	84.54
Dairy, August, 1885	13.6	2.13	.87	.70	.53	.42	.15	......	......	6.41	78.90
Gathered cream, August, 1885 (colored)	13.8	2.00	.65	.55	.45	.2	.25	.35	.20	6.67	72.24
Dairy, October, 1885	13.7	1.10	.6	.4	.4	.4	.2	......	......	6.48	78.88
Creamery, October, 1885	13.2	1.8	.55	.45	.45	.2	......	......	......	5.84	78.80
Dairy, November, 1885	15.5	1.5	.77	.69	.38	.38	.19	.07	......	6.87	79.25
Creamery, November, 1885	12.9	1.9	.69	.67	.38	.28	.19	.07	......	6.00	75.85
Creamery, December, 1885	13.3	2.4	.99	.49	.39	.29	.09	......	......	6.30	74.04
Dairy, December, 1885	12.2	1.6	.49	.39	.19	.09	......	......	......	5.29	81.25
Butter, native upland	14.9	1.6	.7	.35	.4	.35	.3	.1	......	6.58	79.6
Same with gelatin	14.3	1.8	1.7	.5	.4	.4	.2	.1	......	6.83	78.7
Butter, native upland	14.4	2.5	.45	.5	.15	.1	......	......	......	6.87	79.6
Same with gelatin	15.05	1.9	.65	.5	.35	.1	......	......	......	6.53	81.1
Butter, native upland	15.0	1.2	.6	.5	.4	.07	......	......	......	6.17	84.9
Same with gelatin	14.3	2.05	.5	.35	.2	.35	.1	......	......	6.28	80.1
Butter, Dutch lowland	12.45	1.8	.65	.4	.2	.2	.1	......	......	5.56	78.8
Same with gelatin	13.0	1.8	.55	.35	.3	.25	.3	.1	......	5.86	78.1
Butter, lowland pasture	13.85	1.65	.6	.3	.3	.25	.15	.1	......	5.88	79.9
Same with gelatin	13.1	1.85	.55	.4	.3	.3	.1	......	......	5.74	78.9
Butter, Jersey	14.15	1.65	.55	.35	......	......	......	......	......	....	....
Butter, native	14.65	2.05	.55	.45	......	......	......	......	......	....	....
Butter, Ayrshire	15.20	1.65	.55	.45	......	......	......	......	......	....	....
Butter, Jersey (sour cream)	14.80	1.90	......	......	......	......	......	......	......	....	....
Butter, Devon	16.30	1.49	.50	.30	......	......	......	......	......	....	....

Commercial Samples.

New York market	18.9	1.6									
Oleomargarine (1)	1.4	.30	.15	.10	.05					0.70	60.00
Oleomargarine (2)	.7										
Oleomargarine (3)	1.09										
Oleomargarine (4)	1.30										
Oleomargarine (5)	1.29										
Oleomargarine (6)	1.29										
Margarine (7)	1.20										
Butter (Western)	12.85	1.4	.4	.35							

c. c. $\frac{n}{10}$ Na OH required for each 50 c. c. of distillate.

Suspected butter (1), 1468	10.5	.9									
Suspected butter (2), 1469	10.0	.75	.30								
Suspected butter (3), 1470	10.7	1.0	.55								
Suspected butter (4), 1662	.8	.2									
Suspected butter (5), 1661	.25	.2									

Vegetable fats and oils.

Muskat butter	2.0	1.6	1.15								
Cocoa butter (1)	.2	.1	.1	.05							
Cocoa butter (2)	1.6										
Cocoanut oil (1)	3.6	1.9	1.25	1.1	1.0	.07					
Cocoanut oil (2)	3.7										
Cocoanut oil (3) washed	2.7										
Palm oil (1)	.45	.15	.15	.15							
Palm oil (2)	.8										
Bayberry tallow	.35	.3	.15								
Olive oil (1)	.2										
Olive oil (2)	.2										
Cottonseed oil (1)	.25	.1									
Cottonseed oil (2)	.3										
Linseed oil (English)	.25	.1									

TABLE XXVI—(*Continued*).

Vegetable fats and oils.

	c. c. $\frac{n}{10}$ Na. OH required for each 50 c. c. of distillate.							Per cent volatile fatty acids calculated to butyric acid.	Per cent of total acidity in first distillate.
	1.	2.	3.	4.	5.	6.	7.		
Mustard seed oil (1) (American)	.2	.1	......	......	......	......	......	....	....
Mustard seed oil (2)	1.00	......	......	......	......	......	......	....	....
Sweet almond oil (1)	.1	.15	......	......	......	......	......	....	....
Sweet almond oil (2)	.3	......	......	......	......	......	......	....	....
Castor oil	.3	.05	......	......	......	......	......	....	....
Benne oil (1)	.15	.10	......	......	......	......	......	....	....
Benne oil (2)	.6	......	......	......	......	......	......	....	....
Olive oil (1565)	.2	.05	......	......	......	......	......	....	....
Cotton-seed oil (6)	.6	.25	.15	......	......	......	......	....	....
[illegible]in oil	.4	.3	.2	......	......	......	......	....	....
Poppy oil	.5	......	......	......	......	......	......	....	....
Linseed oil	.2	......	......	......	......	......	......	....	....
Rape seed oil	.3	......	......	......	......	......	......	....	....
Pean oil	.4	......	......	......	......	......	......	....	....

Animal fats and oils.

	1.	2.	3.	4.	5.	6.	7.		
Beef suet	1.00	.4	......	......	......	......	......	....	....
Mutton suet	1.42	.35	.35	.2	......	......	......	....	....
Lard	1.02	.4	.2	.2	......	......	......	....	....
Stearin	0.5	......	......	......	......	......	......	....	....
Beeswax	.6	.25	.2	.2	.1	......	......	....	....
[illegible]e wax	1.3	.35	.2	......	......	......	......	....	....
Lard oil	3.25	1.15	.75	.45	.25	......	......	....	....
Horse oil	.75	.25	......	......	......	......	......	....	....
	1.00	......	......	......	......	......	......	....	....
Dog oil	.9	.5	.3	.2	......	......	......	....	....
	1.9	......	......	......	......	......	......	....	....
Olein	1.3	.6	.4	.2	......	......	......	....	....

Marine oils.

Porpoise jaw oil (1) skimmed and strained	47.60	1.95	.4	.2	.05	......	......	17.09	94.9
	47.95	1.00	.5	.15	......	......	......	17.28	94.6
Porpoise jaw oil (2) skimmed and strained	55.88	2.28	.79	.2	.1	......	......	20.901	95.4
	56.18	2.77	.6	.8	.1	......	......	21.05	93.55
Porpoise jaw oil not skimmed or strained	2.02	.7	.5	.8	.8	.1	......	1.44	70.9
	2.15	1.05	.45	.25	.1	......	......	1.41	58.79
Black-fish jaw oil	63.95	1.8	.5	.8	.1	......	......	23.46	97.4
	65.88	2.35	.8	.8	.1	......	......	24.15	95.6
Black-fish body oil	5.6	.6	.35	.2	.1	......	......	2.41	....
	5.6	.45	.2	.1	......	......	......	2.23	....
	5.25	.75	.4	.25	.1	......	......	2.37	....
Cod liver oil	2.1	.8	.4	.8	.2	.1	......	1.86	58.8
Menhaden oil (1)	4.2	2.0	1.2	.7	.45	.08	......	....	....
Menhaden oil (2)	2.4	1.85	.75	.45	.3	.2	......	....	....
Sea elephant oil	8.6	1.6	1.0	.4	.25	......	......	....	....
Porpoise oil	8.55	.7	.4	.2	......	......	......	....	....
Seal oil	1.85	1.0	.65	.8	.25	......	......	....	....
Menhaden (3)	.85	.2	.15	......	......	......	......	....	....
Menhaden (4)	.35	.2	.15	......	......	......	......	....	....
Menhaden (5)	.8	.4	.25	......	......	......	......	....	....
Spermaceti	.7	.4	.2	.15	......	......	......	....	....

Variations of process.

	c. c. $\frac{n}{10}$ Na OH required for each distillate of 50 c. c.		
	1.	2.	3.
One gramme of KOH dissolved in 20 c. c. of 80 per cent alcohol and distilled fractions of 50 c. c	.2	.1	.05
One gramme of KOH dissolved in 20 c. c. of 80 per cent of alcohol, the alcohol driven off, 5 c. c. of water added, also 20 c. c. dilute H_2SO_4 and distilled	.15	.10	.05
The same repeated	.20	.05	....

Variations of process — (Continued).

	Amount taken.	c. c. of water used.	Volume of fraction.	c. c. $\frac{n}{10}$ Na OH used for each distillate of 50 c. c.				
				1.	2.	3.	4.	5.
Orange county butter.....	2.425	1.50	.50	10.1	3.8	.95	.4	.....
Orange county butter.....	2.500	1.50	.50	9.85	3.55	2.45	.75	.5
Orange county butter.....	2.500	.50	.50	14.05	1.05	.45	.4	.3
Orange county butter.....	2.500	.50	.50	13.80	1.35	.45	.4	.3
Orange county butter.....	2.500	.25	.25	12.55	1.35	.40	.3	.5

THE KOETTSTORFER PROCESS.

Inasmuch as the experiments of Hehner and Angell had demonstrated that butter fat contained a large proportion of the fatty acids that have a low combining weight, it occurred to Koettstorfer that the combined fatty acids of butter fat ought consequently to require more alkali to combine with them than the combined fatty acids of most other mixtures of glycerides which might be used as a substitute for butter. The result was the process known by his name (Fres. Zts. 18, page 199).

For instance, the combining weight of butyric acid is 88, that of stearic acid is 284. Each of these amounts will neutralize 56 parts of caustic potash. One hundred parts of butyric acid will therefore neutralize about three times as much potash as one hundred parts of stearic acid, and a mixture of the two will neutralize the more potash, the larger the proportion of butyric acid. As palmitin, stearin, and olein which predominate among the glycerides of insoluble fatty acids of butter fat have nearly the same combining weights, and on the other hand butyrin is the predominating glyceride of the soluble acids present therein, a prior conclusion would favor the correctness of Koettstorfer's supposition.

The process is described (Loc. cit.) as follows :

Weigh out one to two g ammes of the pure fat in a tall beaker of about 70 c. c., capacity, and r25 c. c. of standardized alcoholic potash solution and warm it on the water bath with stirring until the fat is saponified. Cover with a watch glass and allow it to stand hot some fifteen minutes longer. Stir for a moment, add a few drops of phenolphthaleine and determine the excess of potash present by titration with standardized solution of half normal hydrochloric acid. The amount of potash used less that neutralized by the hydrochloric acid gives the amount neutralized by the acids of the fat.

Koettstorfer remarks that the alkalinity of an alcoholic solution of potash is slightly diminished by heating in the air in consequence of oxidation of the alcohol, and resulting formation of acid, hence the strength of the alkaline solution should be determined under the same conditions as are produced in the actual test. The use of sulphuric acid instead of hydrochloric is objectionable as the potassium sulphate which is formed in titration being insoluble in the alcohol used obscures the final point of neutralization.

The results are expressed in the number of milligrammes of caustic potash required by one gramme of the fat. The figure for butter ranges between 221 and 232; for beef fat, oleomargarine, etc., it is 195 to 197.

Allen (Analyst 4, page 162) has proposed that the figures should be divided into 56·1 the combining number of cautics potash, thus giving the combining number of the fats present. Expressed in this way the figure for butter would be about 247·1, and for beef fat, lard, etc., about 286.

The suggestion, however, has not been very generally adopted.

Koettstorfer applied his process to a number of samples of butter and other fats and obtained the following results:

TABLE XXVII.

Name and source of sample.	No. m. g K OH per gm. required.
Butter from neighborhood of Fiume:	
1	227·3
2	227·2
3	225·7
4	232·5
5	226·1
6	222·2
7	225·5
8	221·8
9	229·6
10	231·3
Melted butter from known source and purity:	
September	221·5
October	224·4
November	223·3
Beef tallow melted in laboratory	196·5
Commercial suet	196·8
Pig kidney fat melted in laboratory	195·8
Pig fat from unsmoked bacon	195·7
Common lard (probably American)	195·4
Mutton tallow melted in laboratory	197·0
Olive oil	191·8
Rapeseed oil	178·7
*Melted butter (commercial)	226·7
*Melted butter (commercial)	214·1
First quality Vienna oleomargarine from maker Sarg	195·8
Butter from Halle	228·0
Butter from Carinthia	224·2
Butter from Milan	229·0
Commercial sample, Fiume	227·5
Commercial sample, Fiume	229·5
30. Commercial sample, Fiume	231·3
Melted butter from Carinthia	224·2
Butter from Krain	233·0

*Often adulterated with lard.

Dr. Koettstorfer's assistant, Untchij, obtained the following results on commercial samples of known purity obtained in Fiume :

1	231·1
2	224·0
3	227·6
4	228·7

The following tests were made after the butter had been allowed to stand for six weeks in the laboratory in order to note the effect of rancidity :

	Fresh.	Rancid.
Sample 9	229·6	228·1
Sample 10	231·3	229·9

Finest Vienna butter from Moll, price 100 florins for 100 Kg.	221·7
Ordinary butter from same source, price 68 florins for 100 Kg	200·7
Commercial sample, Fiume	224·2
Commercial sample, Fiume	197·3
Commercial sample, Fiume	197·7
Butter from Milan	192·5

Examination of a sample which separated into a solid and a fluid portion :

Solid portion	221·8
Fluid portion	225·0

The following figures are calculated :

Stearin	188·8
Olein	190.0
Palmitin	208·0

A modification is proposed by Becker (Corr'bl'tt ver. Anal. Chem. 2, 357), to avoid the difficulty attendant upon keeping the alcoholic solution of potash. A standard aqueous solution of potash is kept for use, and alcohol is added only at the time the test is made.

The process was examined by Wigner, (Analyst 4, 182) who gives it a qualified approval. He finds that the carbonic acid of the laboratory inevitable in the proximity of the flame of lamps may interfere with the results. The potash used may act on the glass and affect the results. An excess is always necessary. The process he finds useless with old samples of butter or other fats which have been repeatedly heated and cooled. It is however in his estimation a good auxiliary test.

A modified form of the method apparently worked out independent of Koettstorfer was proposed by Perkins. (Analyst 4, p. 142). It consists in saponifying, acidifying with oxalic acid and distilling off the volatile fatty acids which are titrated separately. The insoluble fatty acids after washing are dissolved in alcohol, and their acidity determined as above described.

Methods depending on the same principle have been proposed for other examinations of fats by Schepper and Geitel (Dingler, 245, p. 295) Hausman (ib. 240, p. 62) Groeger (ib. 244, p. 303 and 246 p. 286) and Zulkowsky (Ber. D. Chem. Ges. 16, p. 1140 and 1315).

The test has been and is still used in the Paris Municipal Laboratory (Monit. Sci. [3] 11, p. 393 and Analyst 10, p. 198) as a preliminary test. The limit there taken is 221; below that figure further examination is required.

As regards the working of this process, although apparently so simple in theory and mode of execution, it has been found to yield very variable results unless great care is exercised. The weakening of the potash by its action on the glass can apparently be avoided by moderate care in selecting the glass ware. The danger from this source appears to be slight so far as our experiments have indicated. The danger of absorption of carbonic acid from the air, and that of loss of butyric acid as butyric ether during saponfication appears to fall into the same category. They seem to be avoidable where only moderate care is used.

The danger of incomplete saponification, through the use of insufficient amounts of potash, or giving it too little time, is probably one important reason for irregularity of results. The most constant results have been obtained by us on evaporating the alcohol entirely off, dissolving and then titrating. Another, probably the most fruitful source of the anomalies presented by this process, is the point alluded to by Koettstorfer himself: the possible oxidation of the alcohol under the conditions produced, to which might be added the possible oxidation of the fat or oil itself. We have observed that certain oils and fats, especially the marine oils, afford results of the most variable character.

It does not seem unreasonable to suppose that if the alcohol contains fat or a soap of the fat readily oxidizable, that both will experience oxidation, in the case of the alcohol in larger proportions than under other circumstances. This, however, is for the present only an hypothesis; we hope to be able to make some experiments which may throw light on the point.

A peculiar point has been noticed in connection with this process that when the titration is finished in an alcoholic solution, dilution with alcohol does not disturb the neutrality of the solution whereas, on diluting with water, the liquid invariably shows an alkaline tinge. It probably arises from a partial decomposition of the soap into an acid soap and free alkali, the change which ordinarily takes place in washing with soap.

The following results are by A. H. Allen. and E. Valenta as indicated by initial:

Koettstorfer process.

Name of oil.	Mg. KOH per grm.	Fatty acids.	
Olive oil	191·7	203	V.
Olive oil	191–196	....	A.
Cotton-seed oil	195·0	203·9	V.
Cotton-seed oil	191–196·5	....	A.
Palm oil	202–202·5	206·5–207·3	V.
Palm-nut oil	270–275	....	A.
Palm-nut oil	247·6	265·8	V.
Benne oil	190	199·3	V.
Sweet almond oil	194·7–196·1	....	V.
Linseed oil	189–195	....	A.
Rape-seed oil	177	....	V.
Rape-seed oil	175–179	....	A.
Cocoanut oil	257·3–268·4	270·7–275·5	V.
Cocoanut oil	270–275	....	A.
Peanut oil	191·3	....	V.
Apricot-seed oil	192·9	....	V.
Bank cress oil	174	....	V.
Pumpkin-seed oil	189·5	....	V.
Oil-cake oil	188·6	....	V.
Rull oil	186·0	....	V.
Castor oil (1st pressure)	181·0	....	V.
Castor oil (2d pressure)	181·5	....	V.
Castor oil	176–178	....	A.
American bone oil	190·9	....	V.
Goose grease (melted in lab.)	192·6	....	V.
Lard	195·5–196·6	....	V.
Lard	195·4–195·8	....	A.
Cod liver oil	213·2	....	V.
Tripalmitin	208·8	....	A.
Tristearin	189·1	....	A.
Triolein	190·4	....	A.
Tributyrin	557·3	....	A.
Drippings	196·5–197	....	A.
Tallow	196·8	....	A.
Horse grease	199·4	....	A.
Lard oil	191–196	....	A.
Niger-seed oil	189–191	....	A.
Whale oil	190–191	....	A.
Seal oil	191–196	....	A.
Cod oil	182–187	....	A.
Herring oil	186–187·5	....	A.
Sperm oil	130–134·4	....	A.
Dugong oil	184·2	....	A.
Shark oil	84·5	....	A.

The following results were obtained in the laboratory and are all duplicate determinations on the same samples which were examined by the wash and Reichevl process :

Butters.

	Mg. KOH per gm.
Creamery, August, 1885	223·9
Gathered cream, August, 1885 (colored)	221·2
Dairy, August, 1885	224·0
Dairy, October, 1885	230.1
Creamery, October, 1885	224·6
Dairy, November, 1885	221·8
Creamery, November, 1885	220·7
Creamery, December, 1885	224·7
Dairy, December, 1885	221·4
Jersey Butter (sweet cream)	230·15
Jersey Butter (sour cream)	231·0
Holstein butter (sour cream)	221·9
Ayrshire butter (sweet cream)	224·75
Native butter (sweet cream)	225·2
Devon butter (sour cream)	222·1
Butter, native upland	228·9
Same with gelatin	228·4
Butter, native upland	220·6
Same with gelatin	221·0
Butter, native lowland	232·0
Same with gelatin	231·3
Butter, Dutch lowland	227·0
Same with gelatin	226·1
Butter, lowland pasture	225·1
Same with gelatin	225·1
Rancid butter	223·0
Same sample deodorized	219·45
Insoluble fatty acids of butter	214·25

Oleomargarine, etc.

Oleomargarine	188·65
Oleomargarine made by Dr. Clark	191.6
Butter color	183·55
Butter color (color extracted)	188·6

Animal fats and oils.

Mutton suet	203·25
Beef suet	199·2
Lard	195·85
Bear's fat	199·6
Dog oil	193·9
Horse oil	191·05
Lard oil	206·0
Olein	189·25
Oleic acid	192·4

	Mg. KOH per grm.
"Oleinic" acid	175·8
Butyric acid	625·75
Stearin	196·7
Japanese wax	212·95
Beeswax	7·0

Vegetable fats and oils.

Linseed oil	177·9
Linseed oil	195·2
Corn oil	183·2
Cotton-seed oil (1)	162·0
Cotton-seed oil (2), summer	190·75
Cotton-seed oil (3), summer	180·3
Cotton-seed oil (4), winter	193·05
Cotton-seed oil (5)	191·2
Benne oil (1)	185·3
Benne oil (2)	199·45
Benne oil (3)	192·4
Rape-seed oil (1)	163·95
Rape-seed oil (2)	170·45
Rape-seed oil (3)	183
Olive oil (1)	183·4
Olive oil (2)	183·95
Olive oil (3)	185·2
Mustard-seed oil (1), English yellow	167·65
Mustard-seed oil (2), refined	164·55
Mustard-seed oil (3)	173·9
Sweet almond oil (1)	184·95
Sweet almond oil (2)	187·9
Castor	175·7
Palm (1)	195·75
Palm (2)	196·6
Cocoa butter (1)	188·85
Cocoa butter (2)	199·8
Muskat butter	167·65
Cocoanut oil (1)	249·7
Cocoanut oil (2)	250·3
Sample 2 washed (3)	246·2
Bayberry tallow	203·3
Poppy oil	192·8

Marine fats and oils.

Porpoise jaw oil	143·9
Porpoise jaw oil (skimmed, etc.)	253·7
Porpoise jaw oil (skimmed, etc.)	272·3
Black-fish jaw oil	197·3
Black-fish jaw oil (skimmed, etc.)	290.0
Sea elephant oil	200.5
Seal oil	202.5
Porpoise body oil	195.95

	Mg. KOH per grm.
Shark liver oil	44·95
Menhaden oil (1)	183·95
Menhaden oil (2)	191·55
Spermaceti	12·0
Cod liver oil	194·7

Results of variations of treatment of cottonseed oil of which an exact gramme was taken :

Treatment of solution.	Mg. KOH per grm.
Heated until clear	155·
Evaporated to dryness	189·
Evaporated to dryness and diluted with water	160·7
Evaporated to dryness	189·3
Evaporated to dryness	193·3
Different sample :	
Heated until clear	178·
Heated until clear	171·8
Heated until clear	175·6
Heated until clear and diluted with water	168·3
Saponified for five minutes	184·2
Evaporated to dryness	191·7
Evaporated to dryness	194·1
Heated until clear	149·9
Saponified for five minutes	189·4
Evaporated to dryness	194·0

Determination of free fatty acids.

Name of fat.	Grms. of Na OH required to neutralize 100 grms of fat	Calculated to butyric acid. Per cent.
Porpoise oil	·119	·2618
Gathered cream butter	·141	·31
Jersey butter	·038	·08
Holstein	·089	·196
Ayrshire	·069	·152
Native	·072	·1584
Devon	·029	·0638
Elephant fat	·328	7·216
(Stearic acid)		23·29

The Hübl method.

It was observed by Baron Hübl that almost all the fats so far as our present knowledge of them extends, consist of glycerides of members of three groups of fatty acids viz :

1. "The acetic," of which butyric and stearic acid may be taken as examples. $C_n H_{2n} O_2$.

2. "The acrylic," of which oleic acid is the most important. $C_n H_{2n-2} O_2$.

3. "The tetrolic" the principal member of which is linoleic acid. $C_n H_{2n-4} O_2$.

The behavior of these series towards halogens (iodine, bromine or chlorine) forming what are called addition products is as follows:

1. The acetic series unites with none.
2. The acrylic series with two atoms, *e. g.* $C_n H_{2n-2} I_2 O_2$.
3. The tetrolic series, with four atoms. $C_n H_{2n-4} I_2 O_2$.

The proportion of the different glycerides in the fats and oils of commerce varies within comparatively narrow limits, and if we determine the amount of iodine for instance, that a fat will take up and form the above compounds, that amount should be tolerably constant for other samples from the same source and nature. Fats containing only glycerides of the acetic series should absorb no iodine, while those containing glycerides of the acrylic and tetrolic series should absorb iodine in proportion to the amount of those glycerides which they contain. Hübl preferred to use iodine for the purpose and worked out a process by which its combination with the fatty acids could be easily regulated. The process is as follows (Dingler Polyt. Jour. 253, page 281): Twenty-five gramme of iodine are dissolved in 500 c. c. of strong alcohol (95 per cent), 30 grammes of mercuric chloride are also dissolved in 500 c. c. of alcohol of the same strength, and the solutions are mixed and allowed to stand for some time. The solution is standardized by a solution containing 24 grammes of hypo-sulphite of sodium per liter. Then the fat is weighed out in a flask in proportions depending upon its character:

For drying oils, 0·2 to 0·3 grammes.
For non-drying oils, 0·3 to 0·4 grammes.
For fixed fats, 0·8 to 1·0 grammes.

It is then dissolved in 10 c. c. of pure chloroform and 20 c. c. of the iodine reagent run in, mixed by agitation and the mixture allowed to stand. If it decolorizes in a short time more of the reagent must be added so as to keep the iodine always in excess, and the color of the mixture should indicate such excess even after one and one-half to two hours standing; 10 to 15 c. c. of potassium iodine solution (10 per cent) is then added and 150 c. c. of water, and the amount of iodine unabsorbed by the fat is determined by the solution of sodium hyposulphite with the aid of starch paste in a manner well known in all laboratories. By deducting from the total amount of iodine used the amount found to remain unabsorbed, it is found how much iodine was taken up by the fat. The results are expressed in the number of grammes of iodine absorbed by 100 grammes of the fat. This, Hübl calls the iodine number for that fat.

The theoretical considerations of Hübl are well sustained by the results of experiment, *e. g.*, pure oleic acid should theoretically have an iodine number of 90·07. The mean of four experiments showed 90·1.

The process was examined some months since by Mr. Moore (Chem. News, 41, page 172; Am. Chem. Jour., 6, page 416), and Mr. Moeller (School of Mines Quarterly, 6, page 276). The results obtained agree essentially with those of Hübl. They, together with other more recent determinations are given in the table.

The process has as yet received less attention apparently than it merits; it is however, comparatively new.

For mixtures of fats or oils differing greatly in their "iodine numbers," the process is of great service as an adjunct in the examination; where the iodine numbers come nearer together its value is somewhat lessened, though in conjunction with other tests, it may throw much light on the composition of the fat. For purposes of calculation in the ease of mixtures of two fats Hübl gives the formula

$$X = 100 \frac{I-2n}{m-n}$$

In which X is the percentage of one of the fats present; m the iodine number of x; n the iodine number of the other fat; I the iodine number of the mixture.

Morawski and Demski (Dingl., 258, page 39) applied the process to the fatty acids separated from the fat by a process essentially that of Hehner and Angell. The solution in chloriform they regard as unnecessary.

The following results were obtained by Hübl:

TABLE XXIX.

Name and origin of fat.	Iodine No.	Fatty Acids. Melt at, deg.	Fatty Acids. Solidify at, deg.
Linseed oils.			
Oil fifteen years old	156	17	13·5
Upper Austria, I	157	....	....
Silesia	158	....	....
Upper Austria, II	159	17·5	13·5
Hungarian commercial	160	16·5	13·0
Hemp oil.			
Hungarian commercial	143	....	....
Walnut oils.			
Pressed in Vienna	142	20·0	16·0
Obtained from Bavaria	144	....	....
Poppy oils.			
From Gounelle, Marseilles	135	19·0	16·0
Obtained from Germany, I	136	....	....
Obtained from Germany, II	137	22·0	17·0
Pumpkin-seed oil.			
Crude oil of Hungarian origin	121	28·0	24·5
Sesame oils.			
From Gounelle, Marseilles	105	24·5	21·0
Vienna commercial, I	106	....	....
Vienna commercial, II	107	27·0	23·0
Vienna commercial, III	108	26·5	23·0
Peanut oils.			
From J. Stettner, Trieste	101	28·5	24·7
Vienna commercial	105	27·0	23·0
Obtained from Hungary	133	20·0	15·7

Cotton-seed oils.

Name and origin of fat.	Iodine No.	Fatty Acids. Melt at, deg.	Fatty Acids. Solidify at, deg.
Obtained from Marseilles	105	38·0	34·0
Obtained from J. Stettner	106	35·0	30·0
Obtained from Hull	108	32·0	27·5

Rape-seed oils.

Vienna commercial, I	97·0	....	....
Vienna commercial, II	98·0	....	....
From Silesia	98·1	21.5	14·0
Oil fifteen years old	100	....	....
Vienna commercial, III	100	....	....
Crude Hungarian	100	20·5	12·0
Crude Hungarian	104	18·5	10·5
Crude Hungarian	105	....	....
Crude Hungarian horse radish	105	....	....

Apricot-seed oils.

From J. Stettner, Trieste	102	....	....
From the Imperial Court Apothecary:			
Fresh pressed	99·8	14·0	5·0
Pressed in Vienna	99·1	....	....

Almond oils.

From the Imperial Court Apothecary:			
From sweet Bari almonds	97·5	....	....
From sweet Avola almonds	99·0	....	....
From bitter Candia almonds	98·9	....	....

Castor oils.

From Italian seeds	84·0	....	....
Vienna commercial, I	84·3	....	....
Vienna commercial, II	84·4	13·0	2·5
Obtained from Italy, colorless	84·6	....	....
Obtained from Italy, yellowish	84·7	....	....

Olive oils.

Salad oil, Nice	81·6	26·5	20·5
Salad oil, Leghorn	81·8	26·5	20·2
Salad oil, south of France	81·8	26·0	20·5
Manufacturing Dalmatian, I	81·8	....	....
Manufacturing Dalmatian, II	82·1	6·5	21·8
Salad oil from J. Stettner	82·2	26·5	22·0
Salad oil, Lucca	82·2	....	....
Manufacturing Durazzo	82·7	25·5	20·5
Salad oil, Bari	82·7	....	....
Manufacturing Dalmatian, III	82·7	....	....
Manufacturing Jaffa	83·7	26·5	22·0

Name and origin of fat.	Iodine No.	Fatty Acids. Melt at, deg.	Fatty Acids. Solidify at, deg.
Salad oil, Dalmatian	83·9	26·0	21·8
Manufacturing Dalmatian, IV	84·0	26·0	22·0
Manufacturing Candia	84·5	24·0	20·0
Manufacturing Valona	84·5	....	....
Olive-seed oil.			
Obtained from Italy	81·8	....	....
Bone oils.			
Neatsfoot oils	66·0	....	....
Vienna commercial	70·0	30·0	28·0
Lard.			
Melted in laboratory	57·6	....	....
Vienna commercial	60·0	....	....
Artificial butter.			
From Sarg in Liesing	55·3	42·0	39·8
Palm oil.			
Vienna commercial	52·4	....	....
From candle-works, Brunn	50·4	....	....
Bayberry oil.			
Obtained from Trieste	49·0	27·0	22·0
Tallow.			
Press tallow	16·6	52·0	51·5
Vienna commercial	39·2	....	....
Melted in laboratory	40·0	45·0	43·0
Rognolato tallow	18·8	48·0	47·0
Suint	36·	....	....
Cocoa butter from J. Stettner	34·0	52·0	51·0
Muskat butter	31·0	42·5	40·0
Butter.			
One year old (very rancid)	30·9	....	....
Vienna market butter I	32·7	....	....
Vienna market butter, II	35·1	....	....
Silesian butter I	31·9	38·0	35·8
Silesian butter II	32·4	....	....
Silesian butter III	29·4	....	....
From the Vienna Dairy	31·7	....	....
Very hard butter	26·0	....	....
Cocoanut fat, fresh	8·9	....	....
Japanese wax	4·2	....	....

TABLE XXX.

Morawski and Demski (loc. cit.) give the following results of the Hübl method on fatty acids from different fats:

Fatty acids from	Iodine number.	
Rape-seed oil	96·3	99·02
Peanut oil	95·5	96·90
Sesame oil	108·9	111·4
Cotton-seed oil	110·9	111·4
Linseed oil	155·2	155·9
Hemp oil	122·2	125·2
Castor oil	86·6	88·3
Beef tallow	25·9	32·8
Cocoanut oil	8·39	8·79
Palm nut oil refined	3·6	4·7
Bone oil refined	55·7	57·3
Palm nut oil crude	12·07	
Bone oil crude	57·40	
Olive oil	86·10	
Palm nut oil (not given by Hübl)	13·4	13·6

The following results were obtained in the laboratory:

Ayrshire butter		34·7
Jersey butter (sweet cream)		36·7
Jersey butter (sour cream)		30·5
Native butter		30·5
Devon butter sour cream		37·0
Rancid butter		40·5
Suspected butter (1)		40·7
Suspected butter (2)		41·08
Suspected butter (3)		38·
Suspected butter (4)		53·75
1691, creamery, August 1885		34·8
1692, dairy, August, 1885		32·4
1711, gathered cream, August 1885 (colored)		36·7
1726, dairy, October, 1885		40·3
1727, creamery, October, 1885		43·8
1757, dairy, November, 1885	31·1	30·9
1755, creamery, November, 1885	37·5	34·0
1785, creamery, December, 1885		35·9
1786, dairy, December, 1885		40·6
Suspected butter		45·1
Suspected butter		45·2
3161, native upland butter		40·4
3161, gel. Same with gelatine		40·9
3162, native upland butter		37·0
3162, gel. Same with gelatine		39·1
3159, native lowland		42·4
3159, gel. Same with gelatine		41·1
3158, Dutch lowland		35·2
3158, gel. Same with gelatine		34·0
1752, oleomargarine made by Dr. Clark		51·5
3160, gel. Dutch lowland		38·8

Fatty acids from	Iodine number.
3160, same with gelatine	37·8
1699, oleomargarine (1)	50·9
1697, oleomargarine (2)	51·0
1698, oleomargarine (3)	54·6
1702, oleomargarine (4)	53·4
1700, oleomargarine (5)	54·9
Butter color, with color	95·0
Same, without color	96·0
Butter color, Richardson	96·3
Butter color, Danish	92·7
Linseed oil (1)	175·7
Linseed oil (2)	155·2
Poppy oil	134·0
Corn oil	119·2
Cottonseed oil (1)	107·9
Cottonseed oil (2)	108·7
Benne oil (1)	107·2
Benne oil (2)	102·7
Sweet almond oil (1)	101·7
Sweet almond oil (2)	98·1
Rapeseed oil (1)	99·4
Rapeseed oil (2)	103·6
English mustard oil (1)	94·6
American mustard oil (2)	85·5
Mustard oil	96·0
Peanut oil	87·4
Castor oil	84·6
Olive oil (a)	81·3
Olive oil (b)	102·9
Olive oil (1)	83·0
Olive oil (2)	80·9
Palm oil	48·6
Palm oil	50·3
Cocoa butter	34·4
Muskat butter	31·6
Cocoanut oil	6·8
Cocoanut oil	8·9
Beeswax	5·3
Japanese wax	5·61
Bayberry tallow	1·38
Mutton fat	57·3
Lard	55.0
Lard oil	47·2
Beef suet	38·4
Oleic acid	86·2
"Oleinic" acid	85·7
Butyric acid	0.0
Olein	82·3
Commercial stearin	1·7

Fatty acids from	Iodine number.
Marine oils :	
Shark liver oil	268·2
Menhaden oil	170·8
Porpoise jaw oil (1)	49·6
Porpoise jaw oil (2)	30·9
Porpoise jaw oil (different treatment)	76·8
Blackfish jaw oil	32·8
Blackfish body oil	99·5
Porpoise oil	131·2
Seal oil	103·4
Sea elephant oil	88·5
Cod liver oil	91·7

A number of simple and rapid processes for the examination of the butter fat have been recommended which have more or less value. They serve chiefly to distinguish between butter pure and simple and oleomargarine pure and simple, and are of little or no use when mixtures are in question. The following descriptive examination of these various processes is intended to present a clear idea of their character and mode of procedure.

According to Lechartier (Biedermann's Centralblatt, vol. 6, 1877, 146; Annales Agronomiques, vol. 1, 1875, 456), pure butter when melted becomes immediately a clear oil while the artificial article takes a longer time to completely liquefy, and forms at first a turbid emulsion. In the case, however, of poorly worked butters containing a large amount of curd the heating can be continued for some time before a clear oil is obtained, which fact naturally impairs the value of this test. In this connection the behavior of the fatty acids on decomposing the soap in the Reichert process before described may be mentioned. When the contents of the flask are heated the fatty acids of butter soon become clear while those of oleomargarine remain turbid for a long time. A practised observer can judge with tolerable certainty of the character of the sample under examination before obtaining the final data of the process.

A. v. Bastelaer (Chem. Centralblatt, 1882, 731) judges of the character of the sample by the odor it evolves when heated to 100 to 120 degrees to drive off the water in the course of proximate analysis. A tallowy odor indicates the presence of oleomargarine.

Donny (Payen Précis theorique et pratique des substances alimentaires; Fresenius' Zeitschrift, 3, 1864, 513) gives a simple test according to which the sample of butter (which it is not necessary to free from the water, curd and salt) is heated in a test tube. Pure butter foams somewhat and the whole turns brown. Artificial butter foams and spirts violently and the casein separates in clots.

De Smedt extracts the fat from the butter with ether, allows the ether to evaporate spontaneously, and gently heats the residue noting the odor. A tallowy smell is observed in the event of oleomargarine being present.

A process was suggested by Hager (Pharm. Centralhalle, 18, 413) which consists in dipping a wick into the melted fat. The wick is lighted and blown out and the odor observed. A pure butter gives little or no odor, while a tallowy smell denotes oleomargarine. The test is practically the same as those before given, and the value of all these tests is rendered very doubtful by the fact that all fats when heated to decomposition yield vapors of acreoline which smell the same in all cases. That part of the fat volatilized which has suffered only partial decomposition is what is observed and is at best a very uncertain quantity. Add to this source of error the fact that old samples of butter have naturally a decidedly tallowy taste and smell and it will be seen that the odor in any case is a very uncertain test.

Dietzsch (Die wichtigsten Nahrungsmittel und Getrænke, 1879) mentions that if a piece of blue litmus paper be placed on a sample of butter and exposed for several hours to the action of sun-light it will turn red if the sample is genuine butter, but in the case of oleomargarine no such change will be observed. This change is due to the readiness with which butyrin is decomposed by which butyric acid is liberated, thus accounting for the change of the litmus. Oleomargarine, however, contains a small amount of butyrin, and it is not impossible for the same change in the litmus to occur when it is tested.

Blyth (Foods, their Composition and Analysis, London, 1882, p. 290) has experimented on the different patterns which various fats assume, when melted and dropped upon water the temperature of which is sufficiently low to secure their solidification. The temperature is of importance, as well as the height from which the fat falls. In the case of butter, when the fat was melted and kept in the air both at a temperature ranging from 40 to 80 degrees and dropped from a clear glass rod on to water of a temperature of 10 to 15 degrees, a characteristic film the pattern of which resembled a pelargonium was obtained. The best temperature for butter is 55 degrees, the water being at 10 degrees. Glass plates are chemically cleansed first by treatment with alcoholic soda and subsequently by washing with ether. These plates are dipped in pure water and a thin film of water thus obtained. The patterns of butter thus obtained was of such extreme tenuity that no photographs could be taken. In the case of oleomargarine the pattern is less distinct, but comparative results may be obtained which may be of value. The pattern of oleomargarine resembles closely that of tallow.

A test similar in character is given by Tomlinson, who drew attention to the peculiar cohesive figures of various fats when dropped upon water (Phil. Mag., 1861 and 1862). An experiment was tried by Blyth, who took for the purpose pure butter fat, butters adulterated with five and ten per cent of lard, respectively, and pure lard. These four samples were dropped upon water of a temperature of 44 degrees. In the case of butter fat the drop spread itself out immediately into a thin film, become agitated by a rapid circular motion and threw off small droplets of fat. The motion gradually ceased, the drop extended, became irregular in outline, crenated at the edges and contraction finally took place. The butter drops containing lard were extremely slow in flattening, were agitated by a gyratory motion and threw off no droplets of fat and ultimately broke up with extreme slowness. The drop of pure lard exhibited no such phenomena and showed no changes up to the moment of solidification.

The degree of solubility of butter and other fats in various solvents has been made the basis of several tests.

According to Horsley (Chemical News, 4, 1861, 230; Fresenius Zeitschrift, 2, 1863, 100) and Ballard (Chemical News, 4, 1861, 283, 222; Fresenius Zeitschrift, 2, 1863, 100), the pure filtered fat is treated with ether at 18·5 degrees; if it is butter it will dissolve readily, while lard, tallow and other butter-substitutes do not dissolve so readily; they are moreover precipitated from this solution by the addition of methyl alcohol, while butter remains in solution. What is essentially the same test somewhat elaborated is given by Hoorn (Maandblatt, 1870, 16; Chem. Cenbralbl., 2, 1871, 149; Fresenius Zeitschrift, 2, 1863,100) who dissolves the butter fat in petroleum ether of the specific gravity of ·69 expels the solvent and then dissolves one gramme of the residue in 7 c. c. of ether. The flask containing the solution is then corked and allowed to stand for an hour in water of 10 to 15 degrees. Pure butter fat will under these circumstances remain entirely in solution while foreign animal fats will be deposited if present in larger quantities than 10 per cent. The manner in which various fats are affected by ether is also described by J. C. Brown (Chemical News, 1873, 39), who notes the amount of stearine deposited from the ethereal solution of the fat under certain circumstances. Butter is distinguished by the fact that no stearine is deposited. Hebner and Angell have examined this process and pronounce it any thing but satisfactory.

Dubois and Padé (Bull. Soc. Chim. XLIII, 207, XLIV, 187 and 602), have made a series of researches on butter and other fats, and recommend strongly the test of degree of solubility in 100 grms. of alcohol of the insoluble fatty acids, using the point of solidification of the fatty acids as a means of control. A saturated solution of the fatty acids is made in absolute alcohol at 12 degrees C, small portions of the fatty acids being added from time to time until the solvent refuses to take up more. Then a weighed quantity is taken out, evaporated to dryness, and the amount held in solution determined. They give a table of results as follows:

TABLE XXXII.

Solubility of fatty acids (insoluble) in 100 grms. of absolute alcohol.

Quantity of butter in per cent.	BUTTER				
	with oleomargarine.	with beef fat.	with veal fat.	with pork fat.	with mutton fat.
0	6·07	7·57	17·55	13·86	6·13
10	6·64	8·47	18·08	14·46	6·45
20	7·37	9·53	18·72	15·22	6·94
30	8·33	10.76	19·50	16·13	7·66
40	9·56	12·28	20·44	17·25	8·66
50	11·14	14·10	21·54	18·63	10·05

TABLE XXXII — (*Continued*).

Quantity of butter in per cent.	Butter				
	with oleomargarine.	with beef fat.	with veal fat.	with pork fat.	with mutton fat.
60	13·20	16·27	22·84	20·29	11·59
70	15·90	18·S8	24·36	22·24	14·56
80	19·46	22·04	26·13	24·53	18·20
90	24·20	25·88	28·30	27·40	23·30
100	30·59	30·59	30·59	30·59	30·59

The points of solidification of the fatty acids with such mixtures is as follows:

Degrees given in centigrade.

Percentage of butter.	with oleomargarine.	with beef fat.	with mutton fat.	with veal fat.	with pork fat.
0	45·6	44·2	49·4	42·7	42·
20	44·	43·3	47·1	41·8	41·2
40	42·4	42·2	44·7	40·8	40·4
60	40·8	40·7	42·3	39·8	39·5
80	39·2	39·2	40·	38·7	38·5
100	37·5	37·5	37·5	37·5	37·5

A difficulty with this would be to determine the proper degree of adulteration without some additional means of determining which fat had been added. For instance suppose 100 grms. of absolute alcohol dissolve 17 grms. of the fatty acids from a fat under examination.

This may mean 75 per cent butter fat, 25 per cent oleomargarine.

Or, 65 per cent butter fat 35 per cent beef fat.

Or, 100 per cent veal fat.

Or, 40 per cent butter fat, 60 per cent pork fat.

Or, 77 per cent butter fat, 23 per cent mutton fat, and if a third fat such as cotton-seed oil were present, not a violent supposition, the proportion of butter fat present would be easier guessed at than determined.

A test similar in character to the foregoing ones but rather more extended is given by Hager (Pharm. Centralhalle, 18, 43) who distills a portion of the filtered butter fat with a double volume of a mixture of sulphuric acid and alcohol. The butyric ether thus formed if the sample in question is genuine butter is easily distinguished by its odor. So strong and so characteristic is this odor that the process is rendered uncertain on this very account since oleomargarine always contains a small amount of butyrin owing to the fact of its having been churned with milk. Butyric ether is thus formed and the same phenomena are observed as in the case of genuine butter.

A similar test to this one was in use among butter dealers in New York city about a year ago. The fat was saponified with an excess of caustic potash or soda and evaporated down with alcohol. Butyric ether is then looked for as in the former case.

Gatehouse (Chemical News, 32, 1875 296) proposes a method depending on the fact that potassium stearate is insoluble in water when it is produced at a temperature of 200—216 degrees. The pure fat is accordingly saponified by heating to that point with solid potassium h drate and the soap after cooling is boiled with successive portions of distilled water amounting to 200 c. c. In the case of butter an almost clear liquid is thus obtained or at most opalescent, while if lard or any other fat containing much stearin has been added a very opaque liquid is obtained.

Husson (Le lait, la creme et le beurre, 1878), uses a mixture of glycerin with alcohol and ether as a means of distinguishing between butter and foreign fats. The butter is treated with ten parts of glycerin and is then digested with an equal mixture of alcohol and ether. Pure butter and oleomargarine will show no deposit at the juncture of the liquids, while lard or beef suet will show a decided deposit. These deposits alter in character on cooling and will alter still further if tincture of iodine be added. On addition of this substance oleomargarine shows very characteristic deposits under the microscope.

Filsinger (Pharm. Centralhalle, 19, 260; Fresenius Zeitschrift, 19 1880, 236), takes five grammes of the pure fat with seven c. c. of ether and the same quantity with fifteen c. c. of a mixture of four parts ether and one part alcohol. These mixtures are kept for twelve hours at a temperature of 18—19 degrees. A turbidity will indicate the presence of foreign fat while butter will remain clear.

Husson (loc. cit.) uses castor oil as a solvent, oleomargarine and suet requiring the highest temperature for complete solution. On cooling, butter and oleomargarine act in the same manner but on the addition of ninety per cent alcohol further differences in solubility are obtained. Crook (Analyst, 1879, IIII) treats one gramme of the pure filtered fat with 2 1-2 c. c., of a solution of carbolic acid (10 acid 1 H_2O), shakes and allows to stand. Pure butter will dissolve thoroughly If foreign fats, however, are present, two solutions of different densities are obtained. In the case of beef fat the lower solution will amount to 49·7 per cent of the entire volume, with lard to 49·6 per cent and with mutton 44 per cent. The results of Crook have been confirmed by Lenz (Fresenius Zeitschrift, 1880, 370).

Valenta (Dingler's Polytechuisches Journal, 252, 296) uses glacial acetic acid as a solvent but does not observe butter, though giving results on various substitutes for butter. The fat in question is dissolved in glacial acetic acid (density 1.0562) with the aid of heat and the temperature at which it becomes turbid is noted. The difference in temperature as shown by various oils can in some cases serve as a means of differentiation.

Dietzsch (Die wichtigsten Nahrungsmittel 1879) treats the pure fat with two volumes of strong sulphuric acid; butter will not develop much heat and the color changes first to a reddish and then to a brown translucent mass. In the case of foreign fats, however, the mixture becomes very black and hot. The butyric ether test is also given by Dietzsch but the application is limited to artificial butters made from rapeseed oil, tallow and lard only. Taylor (Analyst, 10, 13), notes that strong sulphuric acid added to pure butter produces at first a whitish yellow color which after ten minutes changes to a brick red. Oleomargarine made of beef fat changes at first to a clear amber and after twenty minutes to a deep crimson. The author claims that the color is not due to the action of the sulphuric acid on the annatto or other artificial coloring matter since annatto gives a dark bluish green with sulphuric acid.

Griessmayer (Die Verfalschung der wichtigsten Nahrungsmittel, 1879) uses sulphuric acid to test for the presence of rapeseed oil, lard and tallow and applies the butyric ether test to the residue.

Zanni (Pharm. Centralhalle, 24, 435; Fresenius Zeitschrift, 23, 89), gives a number of tests as follows: In order to distinguish vegetable oils the fat is treated with peroxide of hydrogen rubbed vigorously between the hands when the presence of sun-flower and cotton-seed oil is denoted by the smell.

Another way given is to melt the fat and to allow it to cool at 18 to 25 degrees, when the vegetable oil will separate out on the surface from which it can be pipetted and examined. Adulterations of animal origin are detected by heating to 130 degrees, allowing to cool for a day and noting the odor on stirring vigorously with a glass rod.

Zanni also judges of the character of the sample by the amount of ash yielded. One gramme of unsalted butter he claims will yield 12 m. g. of ash, and the same amount of oleomargarine 25 m. g. If the butter containing water and curd is taken even if unsalted the amount of the ash will in every case depend upon the amount of casein in the sample and not on the character of the fat. The following figures obtained in the laboratory show the amount of ash yielded by the pure filtered fat:

	Per cent.		Per cent.
Butter, 1	·00	Oleomargarine, 1	·006
Butter, 2	·005	Oleomargarine, 2	·010
Butter, 3	·00	Oleomargarine, 3	·042

The figures obtained so far as they go confirm the opinion of Zanni, but the quantities of ash obtained were so minute as to be of little or no practical use.

Each of the preceding tests has had its advocates; on the whole, however, they are uncertain and unsatisfactory and demand such an amount of skill and practice as to render them unserviceable and cumbersome. Add to this the fact that no fixed and uncontrovertible data can be obtained by their means and no percentages reckoned except the roughest approximations, and a correct idea of their value may be obtained.

The fact that the halogen elements behave differently toward various fatty acids has been a well-known fact for a number of years, and various processes have been devised based upon this fact with a view of differentiating between the various oils and fats. The data obtained by these processes were the various amounts of the halogen element absorbed by each oil or fat, and figures varying very widely for various fats and oils were thus obtained.

Cailletet, about the year 1857, proposed a process as follows: To the sample under examination add a 5 per cent solution of caustic potash, agitate thoroughly, and then add an excess of a solution of alcohol and bromine of known strength. To determine the amount of bromine absorbed, add a 2 per cent solution of turpentine and alcohol whose value in terms of bromine is known until the brown color disappears. The absorption for pure oils being known the amount of admixture can be readily calculated.

This process is at best a very crude and primitive one. The solution of bromine is unstable and is even liable to hourly change. The temperature also has a marked influence on the amounts of bromine absorbed.

Snodgrass and Mills (Journal of Society of Chemical Industry, 2, 435) modified this process first by using a weak aqueous solution of bromine, agitating the sample with the bromine and potassic iodide with starch as an indicator until a permanent blue color was obtained. Constant results were thus obtained, but the process was not recommended by the authors on account of difficulties in manipulation. They made a further modification by using a solution of bromine in bisulphide of carbon. The dried and filtered oil is dissolved in bisulphide to a solution of ten per cent or less, a definite volume of this solution is placed in a stoppered bottle and brought to 100 c. c. by addition of more bisulphide. A decinormal solution of bromine in bisulphide is then added from a burette until a color permanent for fifteen minutes is obtained; a blank test is made at the same time and brought to the same tint as the one made on the oil. The number of c. c. thus used is brought in as a correction and the amount of bromine absorbed can be readily calculated. If desired, an excess of bromine may be added and estimated volumetrically. The authors, however, have compared their manner of estimation with this latter manner and are satisfied that they agree well. They obtained the following results for butter and similar fats:

	Absorption. Per cent.
Butter from fresh cream	27·93
Butter, commercial	24·49
Butterine (Scotch ?)	36·32
Butterine (French)	39·71
Beef fat	35·01
Lard	37·29

For various vegetable oils :

Cocoanut oil	5·70
Palm oil	34·79
Cottonseed oil	49·97
Olive oil	54·00
Castor oil	58·34
Rapeseed oil	69·43
Linseed oil	76·09
Linseed oil, boiled	102·36
Stearic acid	0·00

A. H. Allen (1881, Journ. Soc. Chem. Industry, 2, 435) instead of free bromine used a solution of sodic hypobromite and hydrochloric acid. Mills and Snodgrass criticise this variation by stating that it involves the action of nascent as well as free bromine. Allen, however, considers the danger of variation from this source as slight.

This method, in any of its forms, has never come into use for the purpose of discriminating between butter and its substitutes, nor is it probable that it ever will, since the Hubl process before described combines all of its advantages, gives the same class of data and is much more simple, perfect and elegant.

David (Comptes Rendus, 94, 1477) for the estimation of glycerin proceeds as follows: One hundred grammes of fat are melted and sixty-five grammes of barium hydrate added. The mass is then heated until most of the water of crystallization has escaped, when 80 c. c. of 95 per cent alcohol are gradually added with continued stirring. Saponification then suddenly takes place, and in order to make it complete, the mass is stirred and heated until it becomes quite dry. A liter of water is now added and the whole boiled for an hour, when the solution containing the glycerin is poured off. The barium soap is rinsed with water several times and the washing added to the solution first obtained. This will contain a small excess of barium hydrate, and in order to remove it dilute sulphuric acid is added to barely acid reaction. The liquid is then boiled down to one-half its bulk, and a few grains of baric carbonate added to precipitate the slight excess of sulphuric acid. The sulphate of barium is filtered off and the filtrate concentrated to 50 c. c., when it is brought into a pycknometer and its specific gravity taken at 15 degrees C., with the help of the tables calculated by Lenz (Fresenius Zeitschrift, 19, 301), from which the per cent of glycerin can be reckoned. The advantage of this process lies in the fact that no loss of glycerin is occasioned by evaporation to dryness.

Liebschuetz (Jour. Am. Chem. Society, 7–134) objects to the method on the ground that the glycerin thus obtained is impure, as it contains compound salts of glycerin and barium, thus rendering the process liable to error. He makes use of the process in a very much modified form for the estimation of the glycerin in butter and oleomargarine. For this purpose ten grammes are taken and saponified with twenty grammes barium hydrate in the same manner as given by David. The essential modification is the purification of the glycerin after it had been obtained according to David's directions. It is for this purpose treated with strong alcohol by which a large quantity of salts are thrown down. These are filtered off, and the filtrate, after expulsion of the alcohol on the water bath, is dried in the air bath above 100 degrees C. to constant weight. It is then burnt and the loss gives the amount of pure glycerin. Under these circumstances an ash is always obtained which in the case of pure butter amounts to five per cent of the weight of the impure glycerin, while oleomargarine yields only from three-tenths to six-tenths per cent. After the deduction of this ash the following results are obtained:

	Per cent.
Butter	3·75
Oleomargarine	7·00

Determinations of glycerin even by the most approved methods are liable to error, and are untrustworthy both on account of the volatility of the glycerin and because as v. d. Becke has shown (Fresenius Zeitschrift, 19, 291) the same fat will yield different amounts of glycerin with different saponification agents. He made in this connection a large number of experiments which clearly demonstrate this point. They are as follows:

	Saponification with oxide of lead, per cent of glycerin.	Caustic potash, per cent of glycerin.	Lime, per cent of glycerin.
Butter	7·98	10·59	7·99
Cocoa butter	·23	5·99	2·19
Tallow	·13	7·84	2·43
Lard	6·60	8·27	9·27
Olive oil	3·76	6·41	....
Rape-seed oil	4·20	4·58	....
Linseed oil	4·40	6·20	....
Cocoa butter and butter	8·05	......	....
Cocoa butter and tallow	·09	......	....

Wanklyn and Fox (Analyst, 9–73) observing that when butter is saponified with alcoholic potash a strong smell of butyric ether is given off devised a method by which, by restricting the action of the potash, more than half of the butyric acid present in butter is converted into butyric ether. They proceed as follows: Five grammes of pure fat are placed in a retort of 200 cb. cm. capacity with 100 cb. cm. of alcohol (sp. gr. ·838) and ·5 gramme solid caustic potash. The distillate is collected in a stoppered bottle containing 40 cb. cm. normal caustic potash. When the distillation is completed the bottle is well shaken in order to effect a complete combination between the butyric ether and the alkali. The excess of alkali is then estimated volumetrically with phenol phtalein as an indicator, and the amount absorbed by the butyric ether calculated. The per cent of butyric acid in the sample can then be calculated.

The following results were obtained by the authors:

	I. Per cent.	II. Per cent.	III. Per cent.		
Butter, I	3·20	3·46	......	$C_4H_8O_2$	Mean of insoluble fatty acids, 87.80 per cent.
Butter, II	2·96	2·96	3·17	$C_4H_8O_2$	
Butter, III	3·17	......	.....	$C_4H_8O_2$	
Butter, IV	3·00	2·85	......	$C_4H_8O_2$	
Butter, V	3·40	......	......	$C_4H_8O_2$	
Butter, VI	3·26	3·13	3·40	$C_4H_8O_2$	

Three other samples of supposed butter gave.

	I. Per cent.	II. Per cent.	III. Per cent.		Insoluble fatty acids. Per cent.
Butter, I....	2·86	3·15	2·97	$C_4H_8O_2$	87·86
Butter, II....	None	......	......	....	95·17
Butter, III....	3·20	......	......	....	88·60

No. II. evidently contained no butter fat and several samples of cheap butter were examined with the same result. The fatty acids in the latter were a little over 91 per cent.

In the same way cocoanut oil and other fats and oils, some of which gave below 95 per cent of insoluble acids, were examined and yielded no butyric ether.

This process has been criticized by Allen who thinks the process liable to a grave source of error in the possible variation of the amount of potash used, the excess depending upon the nature of the fat under examination. A mixture containing only 20 per cent of butter would give a very considerable excess of potash which would seriously affect the result.

Hehner also disapproves of the process and states that when a great excess of alkali is used, as little as three-tenths per cent of butyric ether is formed. Since the amount of butyric ether obtained is a function of the quantity of the alkali the process becomes merely a qualitative test indicating the presence of butter in a mixture of fats.

It may also be added that whether the other volatile acids will yield distillates of their ethers does not appear to have been noticed by the authors of this process.

The process has never been used by the authors of any subsequent papers on butter analysis and does not appear to have met with much favor.

The Coloring of Butter.

Butter from grass-fed cows has a bright yellow color, due to the presence of a coloring matter which has been called lactochrome, the composition and nature of which is very little known.

When the cows are put on dry feed, or the pasture suffers from drought, the intensity of the color of the butter diminishes perceptibly and in mid-winter the butter in its natural state is almost entirely destitute of color.

Feeding the cattle with carrots and some other colored substances will often cause the butter to show a stronger color. Dairymen generally, however, assert that their trade is seriously injured by the absence of color from their butter and instead of coloring it through the medium of the cow, they add the color directly to their wares. The way is thus opened for the addition of hurtful as well as harmless ingredients to a substance found upon every table in the country.

It is claimed that the butter colorings put upon the market, and sold in large quantities to dairymen as well as manufacturers of oleomargarine are all of a harmless character, but the list of those recorded as found by different analysts does not include harmless substances alone.

Chrome yellow (lead chromate) and coal tar colors such as Victoria yellow and Martius yellow cannot be included in that category.

The coloring matters said to have been at one time or another used in butter are given in the descriptions of adulterants. It only remains to describe their properties and sources.

Chrome yellow, or lead chromate, is well know as a pigment. Its dangerous character arises chiefly from the lead present, although the chromic acid with which it is combined is by no means a harmless constituent. Its adaptability for this purpose is not great, which constitutes one principal safeguard against its use. The pigment is not dissolved by the fat, and on treatment with water or alcohol will readily separate out as a yellow powder.

Victoria yellow, golden yellow, saffron substitute, potassium dinitrocresylate ($K\ C_7\ H_6\ O\ [NO_2]_2$) is one of the coal tar colors. There are two or three different methods of preparing it, the simplest of which is by the action of nitric acid upon cresol, commonly known as cresylic acid, which is closely allied in constitution and properties to phenol or carbolic acid. The product bears the same relation to the cresol that picric (or carbazotic) acid does to phenol or that nitro-glycerine does to glycerine. In its effects on the human system it is similar to picric acid. Combined with some base usually potassium, it is used in dyeing, readily imparting a fine yellow color to silk or wool, for which it has a natural affinity.

Tollens (Hndb. off. Ges. 1, page 486), also Koenig (Nahrungsmittel, 2), and Hilger (Nahrungsmittel) mention it as an adulterant, and Bischoff (Hnb. off. Ges. 2, page 508), speaks of it as having frequently been found by him in samples at Berlin. He finds that it has frequently been incorporated with the butter in the form of powder, and by the aid of a hand magnifying glass he has been able to detect it, and on picking the particles out with a needle the necessary tests may be applied.

Martius yellow, calcium or potassium dinitronaphthylate, $K\ C_{10}\ H\ O_5\ (N\ O_2)_2$, or $Ca\ (C_{10}\ H_5\ O_3 \mid N\ O_2]_2)_2$, is also one of the coal tar colors, being made from naphthalene, a compound allied to anthracene, the source of alizarin the artificial coloring matter found naturally in madder. It is similar in constitution to the preceding and is very similar in its action as a dye. This compound was used in New York city about a year ago in coloring mustard (Analyst, 9, page 166) and some other articles of food. We have not yet encountered it in butter though it is said to be used.

The three following substances used in coloring butter are officinal:

Saffron consists of the stigmas of the crocus sativus. It is a native of Greece, but is cultivated in various parts of Europe for medicinal purposes, and in the United States as a garden flower. The English saffron is said to be the best for medical purposes, but it is now unobtainable and Spanish saffron is now considered the best. It has a sweetish odor and pungent taste. The coloring principle called polychroite is said to constitute 42 per cent of the entire material as found in the market. Its medicinal properties are due to an essential oil which is present, when the article is good, in the proportion of about 10 per cent. Its effects administered internally are stimulant and anti-spasmodic. It has been used as an emmenagogue in Europe.

Doses of two to three drachms are said to have produced fatal results. It is seldom prescribed as a remedy, its chief use being to color drugs and tinctures.

For use as a butter color, Lang (Kunstbutterfabrikation) says that it is too expensive, and this fact probably prevents its use here to any great extent.

Turmeric (Terra merita) is the root of curcuma longa, a plant indigenous to and cultivated in various parts of Southern Asia. The China turmeric is usually regarded as the best. It has a peculiar odor and a somewhat peppery, bitter taste. It contains a yellow volatile oil, some resin and the coloring matter (curcumin) which constitutes about one-tenth of its weight. In medicine it is used as a stimulant aromatic, being similar in its action to ginger. It is used in curry powder, also in dyeing as a constituent in what are called by the dyers "sour browns" on woven goods, and as a stain for wood, leather and paper.

To prepare the color for use with butter, Lang recommends to pulverize the root, cover it with alcohol of 40 per cent, allow it to stand for two days and then pour it off. Put on fresh alcohol, allow it to stand and pour off as before, repeating the process so long as any color is extracted. Alcohol of that strength leaves the most of the resin and oil behind. The greater part of the alcohol should then be distilled off, and the extract remaining kept in the dark protected from the influence of alkalies.

Carrot, the root of the daucus carota, is too well known as a garden vegetable to require description. The plant is said by some botanists to have been imported from Europe, but most of them believe it to be indigenous to this country. The root contains sugar, starch, gluten, malic acid, etc., besides the two coloring matters carotin, reddish yellow and hydro-carotin. In medicine the grated root is used as an external application to cancerous ulcers and ulcers which follow fevers. The carotin may be obtained from the expressed juice by treating with weak sulphuric acid and weak tincture of galls, extracting the hydro-carotin from the moist coagulum with strong alcohol, drying, exhausting with carbon disulphide and extracting with absolute alcohol. It is too unstable for service as a dye. A much simpler method of extracting the color for use in butter is used by some farmers. Cream is allowed to stand over night in contact with grated carrots, the mixture being kept cool in a pan of water. In the morning, the cream which has taken up a great deal of the color is poured into the churn and the operation is conducted as usual. The manufacturers of butter color keep their method secret, but the proceeding is probably essentially the same, some oil being used instead of the cream.

The following materials are not officinal:

Fustic is the yellow wood of the morus tinctoria, a tree growing in the West Indies and South America. It contains two coloring matters morin and moritannic acid, the latter resembling tannic acid in properties. It is extensively used in dyeing but apparently infrequently in coloring butter.

Marigold, the flower of calendula officinalis, is a well-known garden flower in this and other temperate climates. The odor, which is somewhat unpleasant, disappears on drying. The taste is bitter and harsh.

The coloring matter is known as calendulin. The material was at one time officinal, and was used as an internal remedy in cases of low fever, scrofula, jaundice, etc. Dr. Livesey states that it is beneficial as an external application in cases of lacerated wounds, etc. The material appears to be frequently used in Germany for coloring butter. As regards its use for that purpose in this country but little is known.

Annatto (anatto, annato, arnotta, orleana, roucou) is the pulp surrounding the seeds of the bixa orellana which grows in several places in South America. The method of preparing it consists simply in bruising the fruit, mixing with water, and straining from seeds, etc. The solid material carrying the coloring matter subsides when the water is poured off and the residue is made into cakes and dried. The Brazilian annatto is usually treated in this way. The best aunatto known as "French," because from French Guiana, has usually been allowed to ferment after bruising the seeds and mixing with water; by this means some of the useless material is destroyed or removed. The odor of the best material is unpleasant, the taste harsh and bitter. It contains two coloring matters, bixin red and orellin yellow, the former being the most desirable. As a dye annatto gives only fugitive colors. It is, however, used in combination with other dyes.

To prepare it for use in coloring butter, Lang recommends to boil one kilogramme of the best annatto for half an hour with five liters of water containing 100 grammes of crystalized carbonate of soda, cool and add a liter of strong alcohol, stir, let stand and filter. The extract must be kept from the light which bleaches it. For coloring oleomargarine he advises taking equal parts of this extract and that of turmeric described under that head, and using 60 to 100 grammes of the mixture to every 300 kilos of oleomargarine (0·01 to 0·03 per cent).

Besides the coloring matters enumerated it is said that almost any kind of scarlet blossoms and even red autumn leaves can be used to produce a butter color by suitable treatment.

To test for the nature of the color which may have been added, most chemists dissolve the fat in alcohol of only moderate strength and make tests upon that solution, Hilger (Loc. cit), Dietzsch (Loc. cit).

Tests upon this solution would show as follows:

Addition of *ammonia*, brown color, turmeric.

Hydrochloric acid, decolorization with formation of a yellow precipitate, slowly becoming crystalline. Victoria yellow or Martius yellow, filtered off and converted into the ammonia or potash, compound these substances, deflagrate on ignition.

Sugar solution followed by hydrochloric acid, red coloration, saffron.

Silver nitrate or *ferrous chloride*, solution turns black eventually, showing a grayish black precipitate, marigold.

Nitric or *citric acid*, greenish solution, saffron.

Evaporation to dryness and addition of *concentrated sulphuric acid*, blue coloration, saffron, greenish blue, annatto. If annatto is present dirty green flocks separate on diluting with a little water; *purple*, carrot.

Hilger (Loc. cit.) asserts that marigold and carrot color cannot be detected in butter.

If, as is sometimes the case, two or more colors are mixed, the difficulty of detecting them is usually much increased.

Mr. Martin (Analyst, 10) has devised a method of separating some of the coloring matters from artificially colored butters, which is of great service in detecting such additions as annato, etc. The fat is dissolved in carbon disulphide and this solution is shaken with dilute potassium hydrate. The coloring matter passes into the aqueous solution which may be drawn off and further examined. Carrot color cannot be detected by this means.

Mr. Moore has found that if a solution of the fat colored with carotin in carbon disulphide is shaken with alcohol, no color passes into the alcohol, but an addition of a drop of dilute ferric chloride, agitating and allowing to stand a short time, the alcohol takes up the color probably in consequence of some change in the constitution of the carotin. If only carrot color is present the carbon disulphide becomes colorless.

Comparatively few of the butter colors in the market were examined. Those which were tested were found to contain annatto and carrot color. The solvent was usually cotton-seed oil.

Optical Methods.

A careful scrutiny of the literature on this subject shows that little has been done, and that this little is so superficial in its nature as to be of but slight use to the investigator.

It must be remembered that the examination of complex compounds such as butter and its adulterations is attended with great difficulty for several reasons.

The chemical composition of many of the fats is, with the exception of butter, alike; and many of the optical tests are based upon the chemical composition, as for instance such variations in the refractive indices as may occur.

The experiments that have been made heretofore by many of the so-called optical methods are too limited in number to be of much practical value. That is, too few specimens have been examined and consequently the data on the subject are not sufficient in number to allow of absolute certainty in judging of the purity of the sample. It is well to remark that it is absolutely necessary to use all of these tests which are given later on samples of known purity. In other words they should be made comparative.

The methods may be divided into two classes:

I. The microscope.

The examination of butter by means of the microscope in order to determine whether the sample was adulterated or not seems to be the oldest method and is based upon two distinguishing characteristics:

(*a*) The presence in artificial butter of crystals of stearin, palmitin, etc., while in natural butter no such crystals could be detected.

(*b*) That natural butter under the microscope exhibits the presence of large numbers of globules of fat while in artificial butter only a few globules are seen, but in their place pear-shaped masses of fat very dissimilar to the butter globules both in appearance and size.

1. The presence of crystals in butter is indicative of adulterations but cannot be taken as conclusive evidence. Hassall in his work, Food Adulterations, etc., London, 1876, p. 430, writes: "Although in fresh butter no crystals are found they appear on longer preservation; moreover, they are found in greater numbers in cream."

Also in Dictionnaire des Alterations et Falsifications, 5th ed., p. 154, Chevallier and Baudrimont described and discussed the same facts.

An article by Angell (Analyst 6, p. 3) calls attention to the fact that butter made from scalded cream has sometimes a crystalline structure. All who have examined butter with the idea of determining its purity from the presence or absence of crystals are of the opinion that the method is unreliable.

During the past year some two hundred specimens of butter, oleomargarine and butterine were examined and although all of the samples of oleomargarine or butterine showed more or less of the crystalline structure and ninety per cent of the butters examined did not, still the uncertainty is too great to allow reliance to be placed on this test.

As in Hassall's experiments, it was found the longer the butter was kept, particularly if exposed to variations of temperature that the crystalline structure of that part in contact with the sides of the containing vessel was marked.

The method of examination for the presence of crystals is as follows: Take a small quantity of the sample on the point of a pen-knife, place it on the slide, smooth it gently with the back of the knife so as to get a rather thin layer, then place the cover on this thin layer and with the aid of a pencil, gently tap and press the cover down. In this way a thin layer is obtained without destroying the appearance. A one-quarter objective with the B eye-piece will be found to give sufficient enlargement to detect the presence of crystals. It is well to examine the specimen with a good 1-10 objective also.

2. To differentiate between natural and artificial butter by microscopic examination the specimen should be prepared by the method just given. Examine with the one-quarter or 1-10 objective. Three hundred to five hundred diameters will be found a suitable enlargement. (Fig 1, Plate V, represents natural butter.) It will be seen to consist of a vast number of globules of fat. In this sample crystals of fat cannot be seen. The square tabular crystals are those of salt and show the upper surface only of the cubical prism. A very large number of samples of butter have been examined in this way. All of them exhibited about the same appearance as in Fig. 1.

The samples examined were those whose genuineness was tested by the various methods of analysis. These appear in another part of this report.

Fig. 2, Plate V, represents the appearance of the sample of oleomargarine made by Dr. Clark, sample No. 1752.

The background is seen to consist of arborescent tracings, only a few globules appear and the characteristic pear-shaped masses of fat are seen.

Now, although this method of testing might be used with considerable certainty to determine whether the sample under examination was genuine or not, if a sample consisting of part butter and part foreign fat of some kind was examined the method becomes at once unreliable. The number of examinations of butter of known purity by the above method were 156. All looked similar to Fig. 1. About 100 samples of artificial butter were examined and Fig. 2 is similar to all of them, but when mixtures were examined the pear-shaped bodies were not always so apparent.

Plate V.

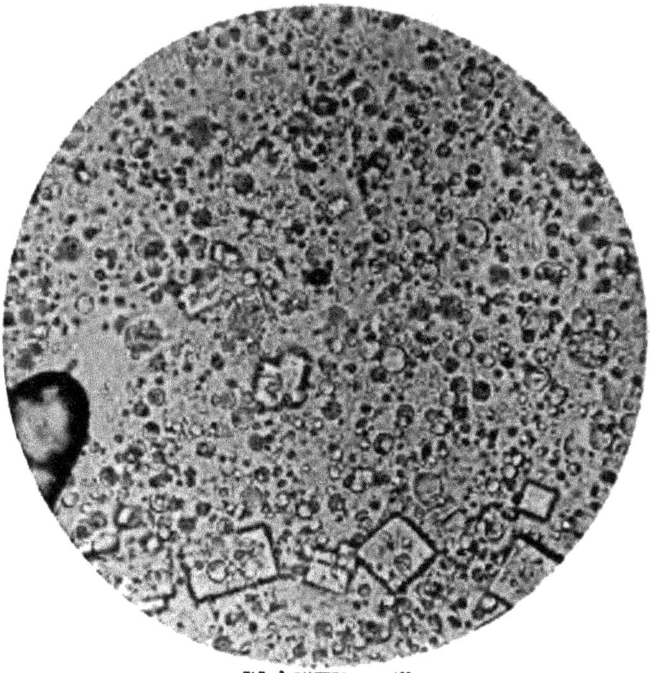

FIG. 1 BUTTER X 400.

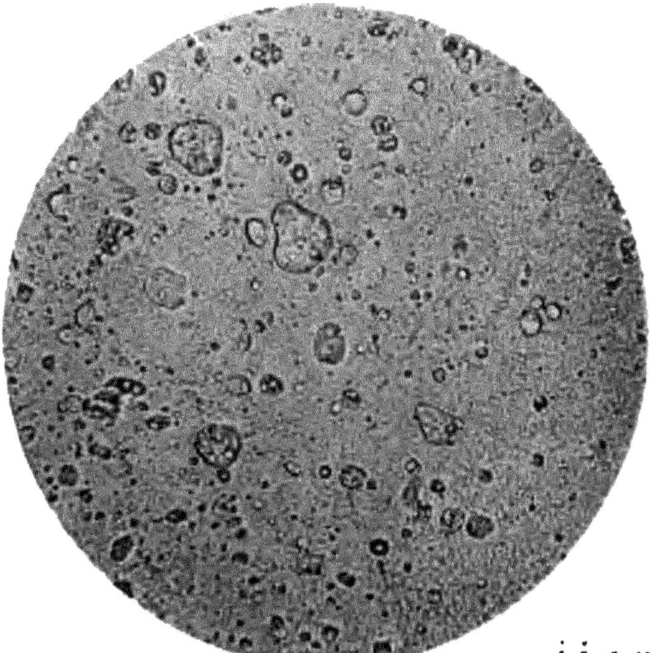

FIG. 2 OLEOMARGARINE. X 400.

We conclude that this method is one that may be used more as an indicator than as an absolute test. If the sample under examination presented the appearance of Fig. 2 there would be little doubt of the fact that it was not genuine.

Prof. G. LeChartier (Chem. News, vol. 26, page 912) states that fresh genuine butter which has not been melted appears under the microscope to be composed of ovoid granules and contains no crystals. M. P. Jalliard (Dingl. Jour., 226, page 325) places a portion of the suspected sample between two suitable slips of glass and examines it with a microscope. If the product is pure, fatty globules alone are seen, but if adulterated, arborescent crystals are seen between the globules.

The appearance of the crystals of fat formed by some arbitrary method of treatment was tried at the time of the hearing of the case *McGeoch, Everingham & Co.* v. *Fowler Bros.*, June 6 to August 21, 1883, at Chicago, and the experts testified that they could detect the difference between beef tallow and lard. This subject has been carefully gone into and a long and careful examination of many samples of lard and tallow undertaken. The method adopted was as follows: The crystals to be examined were prepared by Dr. Belfield's method, ten grains of the fat were dissolved in about one drachm of Squibb's ether, allowed to stand in an uncorked test tube at the ordinary temperature of sixty to seventy degrees Fahrenheit, for twenty-four hours and the crystals formed were put on glass slips and examined, enlarging to about three hundred diameters. When pure lard or pure beef fat was treated in this way no difficulty was experienced in determining the nature of the samples, but on mixing the lard and beef fat the difficulty increased in detecting the presence of one or the other or both of the fats, and when but twenty per cent of either was present the difference could not be detected with certainty.

Fig. 2, Plate VI, shows the crystalline form of lard multiplied by four hundred; fig. 4 the crystalline form of beef multiplied by four hundred. Now when we come to mix butter with lard and beef fat, the peculiar modifications caused by the admixture of butter fat makes this method perfectly unreliable for detecting the presence of either in butter.

Fig. 5, Plate VI, shows the appearance of oleomargarine butter treated by this method and magnified by 400. Fig. 3 shows the appearance of lard and butter treated by this method and magnified by 400.

The examination of the crystals of butter and other fats by means of the polariscope and microscope has owing to the labor of Prof. Taylor of the Agricultural Department been thoroughly investigated and promises to be of some use.

During the past year many samples of butter, oleomargarine, lard, beef tallow, mutton tallow, etc., were examined by this method. The results are, however, not as satisfactory as could be desired.

As long ago as 1856 Hassall speaks of this method but declares it to be unreliable (Chem. News, p. 1008).

E. Mylius finds that butter if examined under polarized light can be distinguished from other melted fats as the latter display a crystalline structure.

A writer in the Chemical News, does not consider that specific gravity affords a trustworthy means of deciding whether a sample of butter is genuine or sophisticated. Examination by polarized light supplies a much better characteristic. Mylius, who first called attention to this method, considers it of only limited applicability. The author is of a different opinion, as it is very rare for a butter to be sent for analysis after it has been melted. Pure butter, when examined with a magnifying power of two to three hundred diameters, appears as a conglomerate of round and roundish drops of different sizes, interspersed here and there with characteristic salt-crystals. All melted fats after congealing take a crystalline structure. On examining a factitious butter we find not the above-mentioned globular drops, but more or less perfectly developed crystals, readily detected by the experienced eye, especially with an oblique illumination. All doubt is at once removed on examination by polarized light. The crystals come out very distinctly, and if the upper nicol is slowly turned everything non-crystalline becomes gradually darker, whilst everything of a crystalline nature becomes lighter. The author finds, further, that different fats, like different minerals, produce characteristic differences in the polarization colors. He announces the early publication of a series of plates showing the characteristic forms and colors of each fat, whether raw, melted or crystalized from glycerin. Mutton tallow always gives a blue tone, and the contrasts when the nicols are exactly crossed are sharper than in case of any other fat, except, perhaps, cacao-butter. The latter differs most characteristically from all other fats, and the play of color from the deepest red to the brightest green does not admit of description. The fat of oxen displays merely green and white luminous effects. Small semi-lunar and vermicular bodies of a bright green appear by common light. Hog's lard displays many colors, especially red and blue, yellow which is very conspicuous in cacao-butter, being wanting. These optical reactions are available for the detection of foreign fats fraudulently added to chocolate or cocoa.

The following is the article of Dr. Taylor received by us some time since.

BUTTER AND FATS.

[Abstract of paper read by Dr. Thomas Taylor before the American Microscopical Society August, 1885, at Cleveland, Ohio.]

"Since 1876, when my first paper was published on butter and fats in the *New York Microscopical Quarterly Journal*, I have devoted a good deal of time to the investigation of this subject, principally with a view of finding a method by which I could, by the aid of the microscope, detect butter from butter substitutes. As a result of many experiments I find that a person experienced in the use of the microscope may distinguish the fats of various animals and of vegetables by following the methods herein described.

"The experimenter should first procure a specimen of common lard. This is composed mostly of crystalline starry forms which represent the solid fat of the lard. Real lard is composed of these and the oil common to lard. In very hot weather, when the thermometer is up in the nineties the crystals dissolve in the oil, and perfect crystals cannot then be obtained unless cooled slowly to about seventy degrees Fahr.

"Place a drop of sweet oil on a glass slide 3 x 1 inches, with the point of a needle. Place a small portion of the lard in the oil and mix them

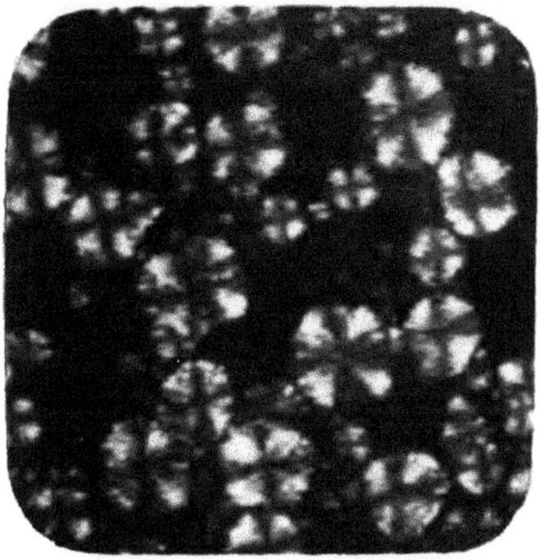

FIG 1 FAT CRYSTALS FROM BUTTER
Polarized Light X 400.

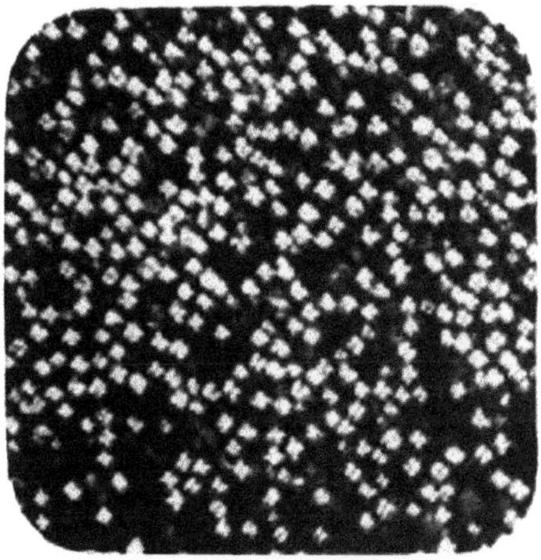

FIG. 2 FAT CRYSTALS FROM LARD
Polarized Light X 400

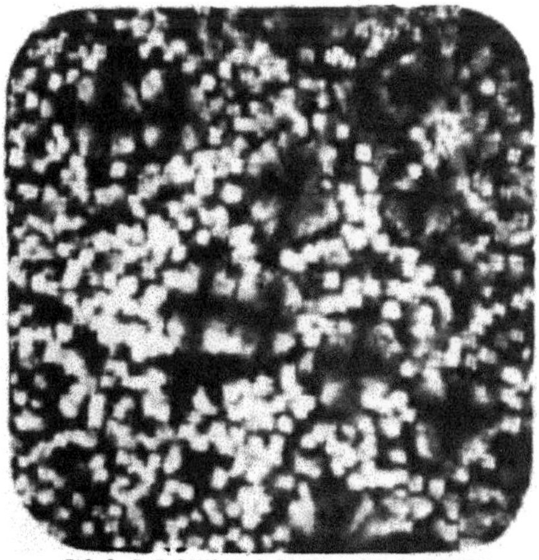

FIG 3 FAT CRYSTALS FROM LARD & BUTTER
Polarized Light X 400

FIG. 4 FAT CRYSTALS FROM BEEF.
Polarized Light. X 400.

FIG. 5 FAT CRYSTALS FROM OLEOMA
Polarized Light. X 400.

FIG. 6 COCOA-NUT OIL & BUTTER.

together. Place a microscopic glass disc over the lard and oil mixture and press gently. If held up to the light white granules will be seen if the temperature is not over eighty degrees Fahr.; these are fatty crystals. Under a low power of the microscope it will be observed that these crystals have stellar forms with dark centres and spines radiating from them. See figure 7.

"To procure normal crystals of beef-kidney fat, render a piece of this fat in an iron pan, without water. Strain, and add sufficient sweet oil to bring the fat to the consistency of butter. Cool slowly for a period of from twelve to twenty-four hours. Mount in oil as directed in the case of lard. The crystals in this case present quite a different appearance from those seen in lard. See Fig. 8. View them by polarized light, with and without selenite plate. The beef crystals, to be seen to advantage, require a power of at least five hundred diameters, being very small, although they appear very interesting objects with a power as low as eighty.

"When it is desired to examine the crystals of butter, boil about an ounce of pure, newly-made butter in a test tube or iron spoon for a period of several seconds; allow it to cool as directed in the case of beef and lard; place a few grains of it on a slip of glass; pour over it a few drops of alcohol (or better, with alcohol nine parts, carbolic acid one part); separate the crystals with a pin, and view them with a pocket lens; they will appear like the eggs of insects, Fig. 1. Place a second portion of the same butter on a glass slide 3x1 inches; combine it with a drop of sweet oil by means of a pin, reducing the butter to granules; cover with a thick disc of glass, and view first with plain transmitted light, when crystals like Fig. 2 will be seen. Second, by polarized light. In this case place the polarizer low down and turn this prism round until its face angle crosses the face angle of the analyzing prism above. Under these conditions a dark ground is produced, and the butter crystals, which are globular in form, are seen in bold relief. The butter globular crystals will now exhibit a well-defined black cross representing that known as St. Andrew's. See Fig. 3. Figure 4 represents a crystal of butter showing divisions produced in prismatic colors when the selenite plate is used with polarized light. If old butter or a poor oily butter is used in this experiment, the secondary crystals of butter are generally shown. These crystals are of rosette form, much smaller than that of the globular, and exhibit no cross. See Fig. 6.

"The globular crystals of butter, when kept for a month or more, seem to bud like a vegetable spore, and frequently every round crystal will show projecting from each a smaller crystal. See Fig. 5. The globular forms generally vary from fifteen ten-thousandths of an inch to the one-hundredth of an inch in diameter. These forms are never seen in pure beef or lard fats. Care should be observed not to press the crystals flat, especially the globular crystals, as the cross is not seen when severely pressed.

"Butter crystals vary slightly from each other in size and in some other slight particulars, such as color. A butter received from Tennessee, made from milk of Holstein and native breed, shows on its crystals indentations, a condition represented in no other butter yet observed. The butter crystals seen in the butter made at Mr. Frank Ward's dairy, from milk of Alderney cows, of Washington, also differ in some particulars from all others examined, being darker in color, spines longer, and of larger size. Specimens intended for permanent use should be mounted with a varnish ring, to prevent the cover from pressing on the crystals, and to prevent the movement of the cover used to protect them."

Dr. Taylor has examined quite a number of other fats, vegetable and animal, and finds thus far, that animals and vegetables of distinctly different genera, and even species, yield fats which give typical fatty crystals characteristic of the animals and plants which yield them, and he is confident that his new discoveries will prove highly useful to microscopists and chemists when investigating adulterated substances used as food or in medicinal preparations. Many scientific men have urged him to continue his investigations, the result thus far being highly appreciated.

A large number of examinations were made by this method and photographs taken of the results. These photographs were made in order that the exact appearance might be produced, and the error that might arise frrom an imperfect or careless drawing obviated. All were magnified 250 diameters.

Figure 4 gives the appearance of beef tallow.

Figure 2 gives the appearance of lard.

Figure 1 gives the appearance of butter.

Figure 5 gives the appearance of oleomargarine.

Figure 3 gives the appearance of lard and butter.

The appearance of the crystals of these fats viewed by polarized light are different when separated, yet so far as the so-called St. Andrew's cross is concerned the oleomargarine exhibits the appearance of butter

The lard and butter can, however, be distinguished with ease. It must be remembered that these results are from a long and patient examination of each and every sample.

The method will, we think, prove of some value in time. We as yet, however, fail to observe any characteristic difference between butter and oleomargarine, meaning by this latter term a mixture of beef tallow or other oleomargarine oil, cotton-seed and sesame oil.

Figure 6 shows the appearance of butter mixed with cocoanut oil multipled by 250.

The examination of butter, fats and oils by means of the microscope as given, is attended with this draw-back, viz.: The drawings produced either by means of the camera lucida or from memory are necessarily inaccurate. It is of great importance in these researches to have some method of exactly fixing the results so that they may be referred to from time to time as occasion requires.

Now the only method which will fulfil these conditions is photography.

At this time the introduction of dry plates places this art within reach of all. Any one who can make a photograph can make a simple form as is shown in Plate VII.

We have used this for some years past, and when not more than 500 diameters are required it answers every purpose.

The artotypes illustrating this report were made from negatives taken with this apparatus.

Plate VII

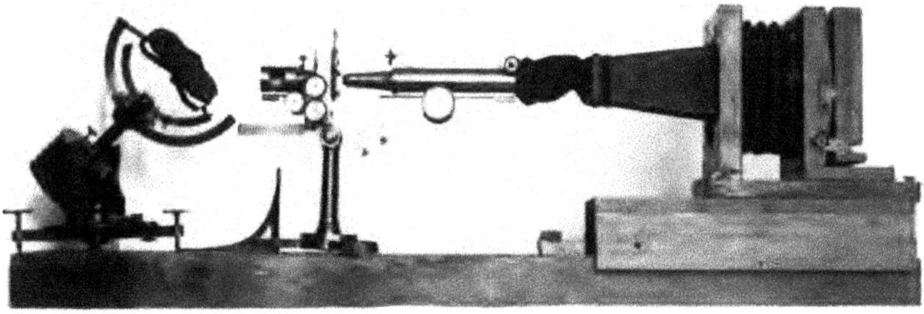

A B C D E F

FIG. 1. ARRANGEMENT FOR PHOTOGRAPHING FROM MICROSCOPE.

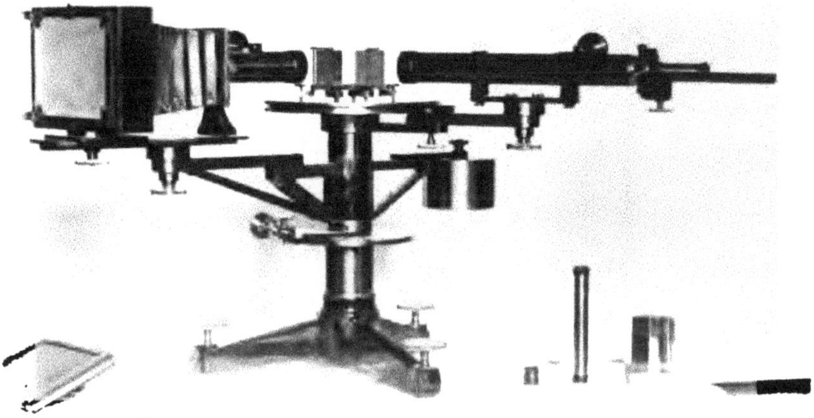

FIG. 2 SPECTROSCOPE, FOR PHOTOGRAPHING ABSORPTION BANDS.

We cannot urge too strongly upon all investigators to fix their results in this way. Of course some experience is required, but a few trials and a little patience will be found to to amply repay the experimenter.

The use of the spectroscope for the examination of oils and fats seems to have received but little attention.

In the Chemical News (1869), Dr. Sorlby makes some investigation with the micro-spectroscope but the limited manner of experiments made and the practically arbitrary scale used makes the results, as far as butter goes, almost useless to the investigator.

His scale was an interference spectrum with dark bands which divided the visible spectrum into twelve portions of equal optical value, and is so adjusted that the sodium line D is $3\frac{1}{2}$, then A equals $\frac{3}{4}$, B $1\frac{1}{2}$, C $2\frac{3}{8}$, D $3\frac{1}{2}$, E 5 11-15, b 6 3-16, F $7\frac{1}{2}$, G 10 5-8.

Butter dissolved in carbon bisulphide gave bands at 6 to $6\frac{1}{2}$.

The following is from Dingler's Polytechnisches Journal, vol. 198, p. 529, by J. Mueller, Spectrum Analysis of Fat Oils.

Frank, no circumpolarization in fatty oils.

Mueller and Steinheil's spectroscope was used.

The scale in the spectrum was determined by the following data:

Red lithium line, 134·0.

Yellow sodium line, 150·5.

Green thallium line, 165·0.

Blue strontium line, 200·0.

The red end of the spectrum came to 126 approximately.

Olive oil gave three absorption bands: One, very dark, situated in the red, from 132 to 137; the second appeared as a faint shadow, from 144 to 147; the third somewhat more marked, and extended from 164 to 167. The entire spectrum extended to 176.

Sesame oil gave a faint absorption band in the red from 133 to 135.

In the following figure the upper portion represents the spectrum of olive oil, the lower one the spectrum of sesame oil. Above the olive oil spectrum is the scale, and below the sesame oil spectrum is indicated the position of the brilliant lines of lithium, sodium and thallium. (Li. Va. Tl.)

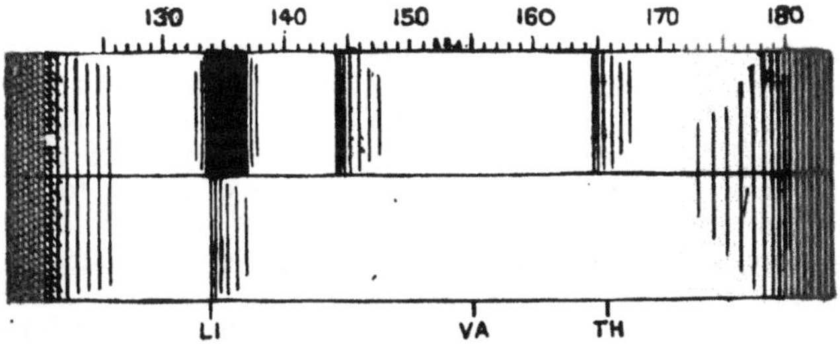

In a tube $2\frac{1}{2}$ centimeters in height the absorption bands of olive oil were less marked; the faint line in the red near the orange almost disappeared; the band in the green was fainter; the heavy line in the red was no fainter but narrower, but only on one side. It extended from 132 to 135.

A mixture of one-half sesame and one-half olive oil, and one consisting of one-half olive and one-half cotton-seed oil, in tubes 5 centimeters in length produced the same spectrum reaction as pure olive oil in a tube 2½ centimeters long.

Linseed oil produced the same absorption bands as olive oil, also elder oil. In the latter case, owing to its deep color, the blue was absorbed up to 164 so that the absorption band in the green could no longer be seen isolated.

The fact that the various oils produced the same absorption bands points to a common cause. On examination it was found that the absorption bands of olive oil and a solution of chlorophyll are identical and hence the above described absorption bands are due to the presence of chlorophyll.

The use of the spectroscope was thought to be worthy of attention and spectroscope arranged for photography was procured. The appearance of this instrument is shown in Plate VII, Fig. 2.

The second sized tube, about two inches in length, was found to be most convenient. All of the results were photographed.

It has long been known that rancid oils and fats give no absorption bands that can be seen, while all vegetable oils give definite absorption bands due to the presence of chlorophyll.

Besides the bands that can be seen it was thought probable that in the ultra violet region of the spectrum, bands might be present, which although not visible, could be reproduced and fixed by the aid of photography. A large number of experiments with this object in view have been made.

The oils experimented on have been sesame, cotton seed and olive; the fats, beef and mutton, suet, lard and butter.

Blyth's "pattern process" was examined and a large number of experiments tried, but this method is too comparative in its nature to be of practical value, for although the different fats and oils have different patterns, so much care is necessary to obtain concordant results that the process cannot be regarded with favor. Moreover when mixtures of the fats and oils are examined in this way, the variations in the patterns are not of such a character as to conclusively point out the kind of fat or oil. The following requirements are necessary:

1. A chemically clean dish.

2. Absolutely pure water. The water is poured into the dish. A casserole about four inches in diameter will be found to be a convenient vessel. The fat or oil is then made fluid by heat and a drop allowed to fall upon the surface of the water from a height of about four inches. The water should be at a temperature of about forty-five F.

It must be remembered that in each series of tests, all of the fats and oils must be treated in the same way. If, for instance, a sample of butter is being examined to determine whether it is adulterated with lard, etc., we must first test a sample of pure butter under exactly the same conditions as the suspected sample is tested, *i. e.*, the temperature of the water and of the sample must be the same, so that if the water has a temperature of forty-five degrees Fahrenheit and the fat a temperature of one hundred degrees Fahrenheit in one case, then in all of the tests the water and fat must be of the same temperatures.

Also the height above the surface of the water on which the fat or oil is dropped must be exactly the same in all cases. By paying strict attention to these details the process may be of some use, but as before stated, the results are hardly practical.

MANUFACTURE OF NEUFCHATEL CHEESE.

By Edward W. Martin and Russell W. Moore.

1st. Foreign Method.
2d. American Method.
A. *Pure.* B. *With lard.*

Foreign Method.

The milk, warm from the cow, is put in earthen pots. These pots are then placed in boxes, covered with blankets, and allowed to stand for forty-eight hours. The soft curd, thus formed, is transferred to closely woven baskets and allowed to drain for twelve hours; and next to bags of cotton cloth, and then kneaded at short intervals for twelve hours. The curd is now moulded into small cheeses of cylindrical form, each weighing about five ounces, and sprinkled with salt in the proportion of one-half pound of salt to one hundred pounds of cheese. After forty-eight hours they are placed on lattice work, covered with straw, allowing at least three inches between cheeses, and remain in this position for forty-eight hours; then turned and allowed to remain for another forty-eight hours. They are next placed on end for five days. This position is then reversed and they are allowed to remain for five days longer. The cheeses are now removed to an airy room, placed on racks until they become covered with a blue mold. This mold is removed and the cheeses are placed in a room warmed to seventy-five degrees Fahrenheit, and here remain until minute red spots appear. The cheese should now be soft and smooth and *not* "grainy."

American Method.

A. *Pure.*

The milk is run into vats, warmed and the caseine coagulated in the usual manner. The resulting curd is slightly pressed and is then formed into cheeses, weighing about five ounces each. These cheeses are now rubbed with salt, allowed to dry for a few hours, wrapped in paper and tin foil and finally packed in boxes marked "Neufchatel Cheese.

The "Cream Neufchatel Cheese" is made in a similar manner, but a certain proportion of cream is added to the milk before souring. About one hundred pounds of milk are required to make eighteen pounds of cheese.

B. *With lard.*

The milk is skimmed, a portion of it run through the emulsifier and mixed, or emulsified with lard, in the proportion of one and a half pounds of lard to one hundred pounds of skimmed milk. This compound is then run into vats containing a certain proportion of skimmed milk nearly at the souring point, and the whole well mixed. The milk is now coagulated with rennet, the curd pressed and made up into cheeses weighing about five ounces each, rubbed with salt, dried for a few hours, covered with paper and tin foil and packed in boxes and marked "Neufchatel Cheese." Sometimes a small amount of coloring matter is added to impart a richer appearance to the cheese.

METHOD OF ANALYSIS.

A. *Determination of water.*
B. *Determination of fat.*
C. *Determination of sugar and caseine.*
D. *Determination of lactic acid.*
E. *Determination of ash.*
F. *Determination of salt and ash.*
G. *Determination of nature of fat.*
H. *Determination of coloring matter.*

A. *Determination of water.*

Weigh out in a platinum milk dish two or three grammes of the cheese, place in the air bath at a temperature of 100 degrees C. until the curd is nearly dry. Weigh a small glass rod. Add to contents of dish 5 c. c. of 95 per cent alcohol. Stir carefully with the rod, allowing alcohol to evaporate at temperature of 90 degrees C. Then place in the air bath at temperature of 100 to 105 degrees C., dry to constant weight. The loss in weight equals *water.*

B. *Determination of fat.*

Extract the fat as in Waller's method of milk analysis, using petroleum ether and not ethyl ether, since lactic acid is soluble in ethyl ether.

C. *Determination of sugar and caseine.*

Indirectly as seen by the following formula :

$$(100-[A+B]-[D+E]=C.$$

In which A. per cent of water ; B. per cent of fat; D. per cent of acid ; E. per cent of ash ; C. per cent of caseine and sugar.

D. *Determination of acidity (lactic acid).*

Ten grammes of the sample are well shaken with 100 c. c. water ; 50 c. c. filtered off and titrated with $\frac{n}{10}$ NaHO, the acidity is calculated to lactic acid ($C_3H_6O_3$). 1 c. c. $\frac{n}{10}$ NaHO=0·00889 grammes lactic acid.

E. *Determination of ash.*

Incinerate at very low red heat, residue from B. and calculate the per cent of ash.

F. *Determination of Salt (Na Cl) in ash.*

Dissolve the ash in hot water. Add a few drops of $Fe_2(SO_4)_3$ and then precipitate with NH_4HO. The solution is filtered and the excess of NH_4HO is boiled out. The precipitate contains phosphates which might interfere. The solution is titrated with $\frac{n}{10}$ $AgNO_3$ and the per cent of NaCl calculated from this.

G. Determination of Character of Fat.

Any of the methods may be used, as Reichert's, remembering to extract the fat with petroleum ether. Otherwise the fat will certainly retain lactic acid.

H. Determination of Coloring. Matter.

Neutralize lactic acid; extract the fat by agitation with water; filter off fat and dry. Treat fat with bisulphide of carbon and alkaline water (Martin's method), Analyst, vol. 10. Alkaline water will take up coloring matter.

The following table shows the result of the examinations of samples of the different brands of Neufchatel cheese submitted to me for examination:

Number of sample and analysis.	Per cent of water.	Per cent of fat.	Per cent of caseine and sugar.	Per cent of ash.	Salt (NaCl) in ash.	Acidity calculated to lactic.
E. 3509	56.08	23·34	16·67	2·49	1·44	1·42
E. 3510	54·18	19·35	22·35	2·38	1·14	1·74
E. 3511	61·48	13·20	21·02	2·97	1·84	1·33
E. 3512	71·43	1·88	22·44	3·13	1·94	1·12
E. 3513	37·97	53·43	6·66	1·30	0·80	0·74
E. 3514	64·50	18·81	13·55	2·04	1·01	1·10
E. 3515	57·83	21·00	17·00	2·85	1·90	1·32
E. 3517	46·06	29·43	20·50	2·45	1·45	1·60
E. 3518	32·17	54·96	9·93	1·52	0·91	1·42
X............	54·96	1·42	38·71	4·91		

Samples E. 3513, E. 3517 and E. 3518 are "Neufchatel Cream Cheese."

The fat contained in the cheese was examined by Reichert's method. The following table shows the results:

Sample and analysis number.	c. c. $\frac{n}{10}$ Na. HO required to neutralize each fifty c. c. of distillate.								Per ct. volatile acids as butyric.
	1st 50.	2d 50.	3d 50.	4th 50.	5th 50.	6th 50.	7th 50.	8th 50	
E. 3509......	13.96	2.08	0.69	0.40	0.30	0.30	0.10	....	6.26
E. 3510......	13.17	1.98	0.94	0.54	0.30	0.40	0.30	0.10	6.28
E. 3511......	13.07	0.99	0.54	0.35	0.25	0.20	0.10	....	5.45
E. 3512......	11.39	1.78	0.59	0.49	0.30	0.20	0.10	....	5.22
E. 3513......	12.77	1.18	0.50	0.40	0.24	0.10	......	....	5.34
E. 3514......	13.66	.69	0.79	0.30	0.20	0.10	......	....	5.53
E. 3515......	13.55	1.68	0.79	0.50	0.40	0.20	0.10	....	6.00
E. 3517......	13.27	2.28	0.59	0.30	0.10	......	......	....	5.85
E. 3518......	12.97	1.68	0.59	0.59	0.24	0.20	0.10	....	5.76
X..........	12.4	1.9	......	......	......	......	......	....	....

Sample X is hard skim cheese made from milk that had been skimmed by the De Laval centrifugal separator and is here given for the purpose of comparison.

From the results obtained I should conclude that all the samples were made from milk alone and that foreign fats had not been added in any case.

Samples No. E. 3509, E. 3515 appear to have been made from whole milk.

Samples No. E. 3510, E. 3511, E. 3514, appear to have been made from partly skimmed milk.

Sample No. E. 3512 appears to have been made from skimmed milk, perhaps from a centrifugal separator.

Samples No. E. 3513, E. 3518 and E. 3517 appear to have been made from milk to which varying amounts of cream had been added.

BIBLIOGRAPHY.

Butter.

Analyst. VI. 10, 29, 32, 73, 74, 93, 153, 172, 195.
Wagner Jahr. 1856, C. Habich 279, Dingl. Polyt. Journ. 142–53. Polyt. Centralbl. 1857, 115.
Ancient samples of. Analyst, V. 17. Chem. News, 41, 46.
Irish Bog. Chem. News, 42, 47.
Yield of. Wagner Jahr., 1858, Moser Arensteins Allg. l. & forstw. Zeitung, 1858, 609.
Butter and Cheese. Wagner Jahr., 1858, 422.
Production of Butter and Cheese in the United States. Wagner Jahr., R. Doehn, 1871, 665. Deut. Industriezeit. 1871, 406.
Copper in Butter. Analyst II, 36.
Butter Coolers. Wagner Jahr., 1858. Dingl. Polyt. Journ., 150, 160. Polyt. Centralbl. 1858, 1246.
Centrifugal or Turbine Churn. G. M. de Stjerns waerd, Wagner Jahr., 1857, 336 fr. Bull. Soc. d'Encorm'nt 1857, 268. Polyt. Centralbl. 1858, 127.
Churning, Apparatus for. Petit. Wagner Jahr., 1857, 335. Genie, Ind., Dec. 1857, 328. Polyt. Centralbl., 1858, 202.
Butter Colors. Wagner Jahr., E. Schmitt, 1883, 981. Journ. des Sciences Med., Lille, V. 343.
Aroma in Butter. Chem. News, 37, 31 and 50.
Goats' Butter. C. Jehn, Wagner Jahr., 1883, 980. Arch. de Pharm., 221, 362.
Comparison of Butter with other Fats. J. Brown, Chem. News, 1873, 39.
Emulsification of Butter. A. Schischkoff, Wagner Jahr., 1874, 815. Ber., 1874, 486.
Formation of Butter on Churning. Storch, Wagner Jahr., 1881, 829. Milchzt., 1881, 589.
Preparation of Butter. Wagner Jahr., C. Fraas, 1857, 334; Die Natur d. Landw., 1857, 353; E. H. V. Baumhauer, Wagner Jahr., 1870, 520; Arch. Nederland, IV. 869, Dingler 195, 181; Chem. Centralbl., 1870, 943.
Preparation of Butter by Electricity. A. C. Tichenor, Wagner Jahr., 1884, 1061. D. R. P., 27, 795.
Preserved Butter. W. Fleischman, Wagner Jahr., 1883, Mlch. Zeit., 1883, p. 241, 2711.
Preservation of Butter. Bischof, Wagner Jahr., 26, 711.
Roll Butter from Normandy. Analyst, IV. 155.
Salting of Butter. H. Martini, Wagner Jahr., 1873, 690; Wochbl. f. Land, 1873, 202; Dingl. Polyt. 210, 478; Polyt. Centralbl., 1873, 1373.

Theory of Churning. Fr. Knapp, Wagner Jahr., 1869, 490; Dingl. Polyt. Jour., 192, 491; Polyt. Centralbl., 1869, 1387.
Rancidity of Butter. W. Hageman, Wagner Jahr., 1883, 975; Landw. Versuchss. 28, 201.
Cause of the Rancidity of Butter. Analyst 2, 358.
Rancid Butter. Chem. News, XLVI. 150.
Adulteration of Butter. E. Meissl Dingl., 232, 229.
Adulteration of Butter in America. Wagner Jahr., 1881, 838. Industriebl., 1881, 44, 153.
Adulteration in New York, Report on. Analyst VII. 94.
Adulteration of Butter. Chem. News, XLV. 141.
Adulteration and Parliament. Analyst VII. 154.
Law as to Adulteration in the United States. Analyst VIII. 139.

PROSECUTIONS, ETC., FOR SELLING ADULTERATED BUTTER AND BUTTERINE.

Prosecutions for Adulteration. Analyst IV. 12, 13, 34, 36, 73, 138, 155, 156, 157, 175, 194, 196, 212, 215; Analyst VIII. 48, 50, 90, 149; Analyst IX. 9, 52.
Prosecutions for Selling Adulterated and short weighted Butter. Analyst I. 34,55, 117, 118, 153, 154, 169; Analyst II. 50, 69, 82, 83, 84, 185, 201, 227; Analyst III. 243, 278, 298, 333, 348, 350.
Prosecutions for Adulterated Butter. Analyst V. 86, 132, 170.
Butterine and Declaring it According to the Act. Analyst II. 82.
Butterine, Prosecutions for Selling. Analyst VIII. 49, 162, 163.
Butterine, Prosecutions for Selling for Butter. Analyst I. 174' III. 314, IX. 29, 31, 51, 71.
Butter and Butterine Prosecutions. Analyst VII. 71, 104, 106, 186, 206.
Prosecution for Refusing to Sell Butter to Inspector. Analyst, I. 118, 154.
Conviction for Selling Fictitious Butter. Analyst, III. 314.
Butter Adulterated with Soapstone. Analyst, VI. 14.
Butter Adulterated with Tallow. Analyst, VI. 225.
Butterine Sold for Butter. Analyst, VI, 52, 196, 212, 235; V. 193, 229.
Butter Prosecutions at Cheltenham. A. H. Church, Analyst, II. 55. Mr. Horsley and Correspondence. Anyalst, II. 59.
Patent for Converting Fat into Butter. Analyst, I. 195.

ARTIFICIAL BUTTER.

Mège-Mouries, Wagner Jahr., 1872, 617; Monit. Sci., 1872, 740; Chemical News, 1872, 106; F. Boudet, 1872, 617; Monit. Sci., 1872, 741; E. Jacobsen Indus. bl., 1876, 405; Dt. Ind. Ztg., 1876, 425; E. Jacobsen, Wagner, 1876, 898; R. Godeffroy, Arch. d. Pharm., 10, 146; Ind. bl., 1877, 126; Dingl. Poly., Jour. 224, 204; Chem. Centr. bl., 1877, 319; D't Ind. ztg., 1877, 156; Jaroslouski, Wagner, Jahr., 1880, 711.
Preparation of Artificial Butter. A. Hilger, Jahr. ud. Fortschr d. Agri. Chem., 1880, 487. F. Boegel, Industrie bl., 1881, 258.
Manufacture of Artificial Butter. H. A. Mott, American Chemist, VII. 233; Monit. Sci., 1877, 1082; List. Wagner Jahr., 1880, 951

Remarks on the Manufacture of Artificial Butter. Towaillon, Uhland's Maschinen Construction, 1877, 233.
Value of Artificial Butter as Food. A. Mayer, Landw. Versuchss, 29, 215; Wagner, Jahr., 1883, 980.
On Butterine. A. Angell, Analyst, VI. 3.
Preparation of Butterine. Analyst, V. 105, 106; Wagner, 1881, 840, B. Hoffman.
The American Dairyman on Butter and Butterine. Analyst, III. 317.
Use of Butter in Butterine Manufacture. Analyst, VIII, 113.
Preparation of Oleomargarine. Industrie bl., 1877, No. 24, 219; No. 26, 233; No. 27, 244; No. 28, 250.
Manufacture of Fictitious Butter in the United States. Chemical News, 42, 165.
Butter and margarine. Chemical News, 44, 81.
Butter Substitutes. Analyst, VI. 167. A. Angell, Analyst, VIII. 110.

Butter Analysis.

J. Muter, Analyst, I, 39; W. Fox, Analyst, IX. 73; J. A. Wanklyn, Analyst, IX. 73; H. Leffman, Analyst, IX. 199; Chemical News, 37, 7 and 20; 39, 124; 49, 59; 51, 36; 1878, p. 7, Corfield.
Analysis of Butter-fat. F. P. Perkins, Analyst, IV. 142.
Butter Analysis in New York. Analyst, III. 317.
Butter Analysis in Somerset, Analyst, VIII. 51.
Butter Analysis in Scotland. Analyst, I. 11 and 105.
Improvements in Butter Analysis. A. Angell, Chemical News, 32, 69.
Analyses of Samples of Genuine Butter. E. Reichardt, Wagner, Jahr., 1884, 1062; Arch d' Pharm., 222, p. 93.
Composition and Analysis of Butter. A. Dupré, Analyst, I, 87, 114. A. Dupré, Arch. Pharm., XII. 76; Chem. Centr., 1878, 623. E. W. T. Jones, Analyst, II. 19 and 37.
New Method of Analysis. Dr. Koettstorfer, Zeits. Anal. Chem., 18, 199, 431; Analyst, 1879, p. 199; Arch. Pharm., 1879, p. 74; Ber., 1878, 1133; Chem. Centralhalle, 1870, 558; Dingl. Polyt. Journ., 232, 286.
Relation between Koettstorfer's and Old Processes. Analyst, IV. 182.
Koettstorfer's method of Butter Analysis. R. W. Moore, Chemical News, 50, p. 268.
Expressions of Results of Koettstorfer's Process. Analyst, IV. 162.
Results by Koettstorfer's Process. Analyst, IV. 162; A. H. Allen.
Results by Koettstorfer's Process. Analyst, IV. 192; G. W. Wigner.
New method of Butter Analyses. Dr. Milne, Analyst, IV. 40.
Testing of Butter. O. Hehner, zts. f. Anal. Chem., 1877, 145; Dingler, Polyt. Journ., 225, 404; Wagner Jahr., 1877, 827.
Hehner's Test for Butter. W. Heintz, zts. Anal. Chem., 1878, p. 160; Arch. Pharm., IX. p. 541; Chem. Centrbl., 1878, 624; Wagner Jahr. 24, 996. Kretschmar, Ber., 1878, 2091; Dingler, Polyt. Journ., 227, 591; Bull. Soc. Chim., 1878, 173; Chem. Centrbl., 1878, 104. Kuleschoff, Bull. Soc., Chim., 1878, 371; Wagner, Jahr,, 24, 999; Ber., 1878, 534; Chem. Centrbl., 1878, 344. Fleischman & Vieth, zts. Anal. Chem., 1878, 287; Wagner Jahr., 24, 999; Arch de Pharm., 1878, 257; Chem. Centrbl., 1878, 664. Jehn, Arch.

Pharm., 1878, 128; Wagner, Jahr., 24, 1000; Chem. Centrbl., 1878 234. Skalweit, Jahrb. d. Unt. f. Lbsult, 1878, 22. Sachssee, zts. Anal. Chem., 1878, 192. Wagner Jahrb., 24, 1000; Arch. Pharm., 1878, 543.

Butter Test, Reichert. Zts. Anal. Chem., 18, 68; Wagner, Jahr., 25 944; Ind. Blatt., 1879, 158; Chem. Centrbl., 1879, 49; Bull. Soc. Chem., 1879, 377.

Examination of Butter by Reichert. H. Beckert's Pharm., Centrbl., 1883, 557; Wagner, Jahr., 1883, 978.

Notes on Reichert's method of Butter Analyses. R. W. Moore, Chemical News, p. 151.

A Method of Testing Butter. E. Reichardt, Arch. d. Pharm., 1877, 339; Dingl. Journ., 225, 213.

Simple Method of Testing Butter. Mylius, Ind. Bl., 1879, 558; Wagner, Jahr., 25, 950; Chem. Centrbl., 1879, 558.

Butter Testing. Meissl, Dingl. Polyt. Jour., 233, 229; Wagner, Jahr. 25, 947; Chem. Centr., 1879, 586.

R. Sendtner, Examination of Butter. Ac. Meissl, Arch. f. Hygiene, 1883, 579; Wagner, Jahr., 1883, 979.

Detection of Oleomargarine. P. Casamajor, Chem. News, 44, 309; Wagner, Jahr., 1881, 839; Chem. News, XLI, 77.

Detection of Spurious Butter. Chem. News, 46, 281.

Detection of Adulteration, 40, 312.

Detection of Adulteration, 48, 170.

Detection of Fatty Matters introduced into Butter. Chem. News, 37, 163.

Detection of Adulteration of Butter with Animal Fats. E. Jaillard Les Mondes, 1877, Aug. No. 14; Wagner, Jahr., 1877, 829; Dingler Polyt. Journ., 226, 325. J. W. Gatehouse, Rapid Method of, Chem. News, No. 839, p. 297.

Examination of Adulteration of Butter. E. Valenta, Beilage d. ztsft. f. Landw., Gewrb., 1883, 9.

Examination of Butter for Tallow. O. Kunstman, Pharm. Centrh., 1875, 9; Wagner, Jahr., 1875, 909; Polyt. Centrbl., 1875, 392.

Examination of Butter. Cook, zts. Anal. Chem. 1879, 111; Wagner Jahr., 26, 712. Lenz, zts. Anal. Chem., 1880, 370; Wagner, 26, 712. Donny, Wagner Jahr., 26, 711; Chem. News, 47, 35; Chem. News, 47, 49; Chemical News, 47, 52. Hoorn, Polyt. Centr., 1871, 1188, Wagner, Jahr., 1871, 671; Dt. Indzt, 1871, 327; Chem. Centrbl., 1871, 49. Kressner, zts. Anal. Chem., 18, 83; Wagner Jahr., 25, 951; Chem. Centrbl., 1879, 83; Chem. News, 50, 68; Chem. News, 50, 192; Chem. News, 50, 213; Chem. News, 42, 306. Medicus & Scherer, zts. Anal. Chem. 1880, 159; Wagner, Jahr., 26, 712. J. Munier, zts. Anal. Chem., 1882, 394; Wagner Jahr., 1882, 928.

Contribution to Butter Examination. Koenig, Ind. bl., 1879, p. 455; Wagner Jahr.,25, 947; Dingl. Poly. Journ., 232, 286; Chem. Centrbl., 1879, 127. Magnier la Source.

Examination of Butter. Repert. Pharm., 38, 46; Wagner Jahr., 1882, 929. M. Leibschutz. Am. Chem. Soc., IV. 544; Analyst, VII. 134.

Distinguishing Between Butter and Oleomargarine. Analyst, VI. 54. A. Mayer, Fehling's Landw. Ztg., 31, 92.

New Method. W. G. Crook, Analyst, IV. III.
Foreign Fats in Butter. Chemical News, 50, 268; 46, 34.
Butter Testing. J. Horsiey, E. Ballard, Zts. and Chem., 1863, p. 99; Chem. News, 42, 307. Th. Taylor, Sci. American, 1878, 374; Dingl. Polyt. Jour., 220, 93; L'Industrie Laitiere, 1881; Milchzt'g, 1882, 27; Wagner Jahr., 1882, 930.
Butter Testing Apparatus. Butter Tester, W. Osten, Wagner, Jahr., 1882, 930; Ger. Pat., 19,078 and 20,695. D. Gaebel, Wagner, 1882, 930; Milchzt'g, 1882, 437.
Escourt's Butter Apparatus. Chemical News, 38, 267. J. G. Bell, Chem. Centrbl., 1879, 586; Wagner Jahr., 25, 947.
Fatty Acids of Butter. Chemical News, 42, 72. C. Schmidt, Bull d' Soc. Ind. d' Nord., 1884, 87; Wagner Jahr., 1884, 1062.
Insoluble Fatty Acids in Butter Fat. Analyst, V. 155.
Time of Drying. Analyst, VIII. 163.
Butter, Melting Points, Det. of, and Other Fats. T. Redwood, Analyst, I. 51.
Melting Point of Genuine and Artificial Butter. J. Moser, Stummin's Ingenieur, III. 97; Wagner Jahr., 1875, 909; Dingl. Polyl. Jour., 216, 288; Polyt. Centrbl., 1875, 392.
Oil Cocoanut, Relation of, to Various Methods of Butter Analysis. R. W. Moore, Jour. Am. Chem. Soc., VII. 188.
Polarized Light, Examination of Butter by. F. Fischer, Chemical News, 39, 358.
Ratio of Expansion by Heat of Butter. Analyst, II. 183.
Salicylic Acid, Detection of, in Milk and Butter. R. Portele, Landw. Versuchss, 27, 148; Wagner Jahr., 1881, 830.
Specific Gravity of Butter. G. Ambuhl, Schw. Wochshft. and Pharm., 1881, 7; Wagner Jahr., 1881, 839. A. W. Blyth, Analyst, V. 76.
Influence of Decomposition on Specific Gravity. Analyst, IV. 39, 93.
Spectroscope, Use in Detection of Adulteration. H. E. Sorby, Chem. News, 20, 316; Wagner Jahr., 1870; Dingl. Polyt. Jour., 198, 345.
Water in Butter. E. Johanson, Pharm. Ztsshft & Russland, 1881, 399; Wagner Jahr., 1881, 839.
Somerset House Standard for Water in Butter. Analyst, IV. 14 and 18.

Bibliography of Fats and Oils.

Chemical News, 31, 227, 228, 238, 205; 37, 173; 43, 101, 111, 170, 228, 230, 247.

Fatty Acids.

Conversion of Liquid Fat Acids into Solid Products. Dingl. Polyt. Jour., 254, 264.
Determination of Undecomposed Fat in Mixtures of Fatty Acids. Chemical News, 47, 72.
Determination of Fatty Acids in Oils. Analyst, IX. 125; Monit. Sci., XIV. 205; Analyst, VI. 100.
Easy and Rapid Method of Manipulating. A. W. Blyth, Analyst, II. 112.
Filter Washing of. Analyst, IV. 52.
Estimation of, in Oils. G. Richter, Seifen Sied. Zts. XVII. 198; Jour. Soc. Chem. Ind., II. 480.

Estimate of Neutral Fat in Mixtures of. M. Gröger, Dingl. Polyt. Jour., 246, 286; Journ. S. Chem. Ind., I. 508.
Halogen. Derivatives of. Chem. News, XXXV. 185.
New Fat Acid in nut of Calif Baytree. Shellman and O'Neill, Am. Chem. Jour., July, 1882; Jour. Soc. Chem. Ind., I. 370.
Nitrogen Radicals, Introduction of, into Compounds of the Fatty Series. Chem. News, 37, 172.
Nitro-compounds of Fatty Series. Chemical News, 38, 39.
Fatty Acids and their Alkaline Salts. Chemical News, 31, 269.
Suggested Application of, to Photography. Chemical News, 33, 126.
Olein Acid, Conversion of, to Palmitic Acid on a Manufacturing scale. W. Lant. Carpenter, Jour. Soc. Chem. Inds., II. 98.
Olein Acid Oxidation by $K_2 Mn_2 O_7$ in Alkaline Solution. A. Saytzeff. Jour. Prakt., 31, 541; Analyst, VII. 207.
Oleic Acid, Separation from Stearic. J. David, Comptes Rendus, 81, 416; Analyst, I. 360.
Superheated Steam, Acids Produced by distilling with. A. Cahours and E. Demarcay, Comptes Rendus, 1879, No. 6; Analyst, I. 403.
Adulteration, Fat sold for Lard. J. Muter, Analyst, VII. 93.
Adulteration of Lard. A. Seiffert, Schw. Woch. Pharm., 20, 287; Jour. Soc. Chem. Ind., I. 370.
Adulteration of Lard, Detection of Tallow. W. Rodiger, Chem. Zeit., 1882, 118; Journ. Soc. Chem. Ind., I. 162.
Adulteration of Almond Oil. Jour. Soc. Chem. Ind., II. 188.
Adulteration of Fatty Oils. G. Richter, Seifen sied Zts., 18, 210; Jour. Soc. Chem. Ind., II, 481. Detection of, Chemical News, 32, 140. G. Richter, Sefen sied. Zts. 16, 188; 18, 199; Jour. Soc. Chem. Ind., II. 384.
Detecting Cotton-seed Oil in Olive Oil. A. Buchheiter, Drag. Ztg., Jour. Soc. Chem. Ind., I. 32.
Reaction for Distinguishing Cotton and Olive Oil. M. Secchini Chem. Gaz., 1882, 61; Jour. Soc. Chem. Ind., I. 237.
Apparatus for Estimation of Fatty Bodies. F. Tschaplowitz, Zts. and Chem., 18, 492; Jour. Soc. Chem. Ind., I. 386.
Analysis of Fats. Chemical News, 47, 71; 49, 214. W. G. Crook, Analyst, IV. 111. F. P. Perkins, Analyst, 142.
Composition of Human Fat. Chemical News, 47, 10.
Composition of Fatty Matter of Wool. Chemical News, 31, 26.
Chinese Tree Fat and Grease of. Seifen sied Ztg., 35, 415; Jour. Soc. Chem. Ind., III. 181.
Constitution of Natural Fats. Chemical News, 31, 250.
Composition and mode of Action of Turkey Red Oils. A. Muller Jacobs, Dingl. polyt. journ., 251, 499; Journ. Soc. Chem. Ind., 3, 412.
Chemistry and Analytical Examination of fixed oils. A. Hallen, Journ. Soc. Chem. Ind., 2, 49.
Decomposition of neutral fat bodies. Chemical News, 31, 250.
Cotton-seed Analyses. Analyst, VI. 216.
Cotton-seed and Palm nut oils. Seifensied Ztg., 33, 390; Journ. Soc. Chem. Ind., 3, 181.
Cooling of fat. Chem. News, 37, 131.
Digestion and assimilation of fat in human body. H. C. Bartlett, Analyst, I, 212.

Estimation quantitative of oils and fats. Mills and Akelt, Analyst, Ind. 6, 245. Mills and Akelt, Journ. Soc. Chem. Ind., III, 366. Mills and Snodgrass, Journ. Soc. Chem. Ind. II, 435.

Estimation of fat in palm nut meal. V. v. Wilau, Journ. Soc. Chem. Ind., IV. 511.

Examination of Tallow, etc., especially for acidity. W. H. Dening, Jour, Soc. Chem. Ind., III. 540, III. 1643.

Examination of fat. Analyst, VIII. 154; Chem. News, 51, 172. R. Benseman, Repert, Anal. Chem., IV. 165; Journ. Soc. Chem. Ind., IV. 435.

Examining Fatty oils for mineral oils. Geissler, Seifen Sied, Zts., 16, 188. Jour. Soc. Chem. Ind., 2, 385.

Examination of Fixed Oils. Further notes on. W. Lant Carpenter, Jour. Soc. Chem. Ind., 2, 101.

Examination of fats. Hübl, Dingl. Polyt. Journ., 253, 281; Chem. News, 48, 87, 150.

Experiments on Fatty matters. Chem. News, 51, 202.

Extractive Apparatus for fats. T. O. C. Sloane, Jour. Am. Chem. Soc. 4, 250; Analyst, 8, 35.

Extraction and treatment of fats. Dingl. Polyt. Jour., 253, 415.

Figures assumed by various fats. A. Wynter, Blyth, Analyst, 6, 157.

Grapestone oil, production of. J. v. Jobst., Dingl., Polyt. Jour. 155, 450.

Lubricating oils, Investigations on. D. Lamansky, Dingl. Polyt. Jour., 248, 29; Jour. Soc. Chem. Ind., 2, 417.

Industry, Contributions to the fat. Chem. News, 39, 135.

Lard oil. H. Ohlo, Arch. d. Pharm. 142, 35; Wagner Jahrsb., 1857 340; Polyt. Centrbl. 1858, 223; Chem. Centrbl., 1857, 880. Bailly Technologiste, July, 528; Wagner Jahresb., 1857; Polyt. Centrbl. 1857, 1260.

Lard oil. Preparation of. C. Puscher, Wagner Jahr., 1855, 256.

Mechanical equivalent of fat in cattle being fattened. Chem. News, 41, 134.

Metals, action of certain on oils. M. Livache, Comptes Rendus Jan. Jour. Soc. Chem. Ind., 2, 349.

Mustard oil, hedge. E. Valenta, Dingl. Polyt. Jour. 247, 1; Journ. Soc. Chem. Ind., 2, 181.

Melting point, apparatus for detecting. Cross and Bevan, Hand. Chem. 43.

Melting point, determining. A. Terreil, Bull. Soc. Chem., 31, 155.

Melting point of fats. Chem. News, 44, 79; Chem. News, 32, 27.

Melting point of fats, determining. Chem. News, 31, 216. Kretschmar, Zts. Anal. Chem., 21, 399; Journ. Soc. Chem. Ind. 1,508.

Solidification of fatty oils. W. Lant Carpenter, Jour. Soc. Chem. Ind., III. 367.

Modification of animal fat, peculiar. H. C. Bartlett, Analyst, I. 175.

New form of fat. Analyst, 8, 65.

Oxidation of Turkey Red Oil in Dyeing. F. Schatz, Dingl. Polyt. Journ., 250, 9; Journ. Soc. Chem. Ind., 3, 106.

Physical Properties of Fats. Chem. News, 39, 265.

Ratio of Expansion by Heat of Different Fats. Analyst, 4, 183.

Resin Oil, Occurrence of Caproic Oil in. W. Kelbe and C. Warth, Chem. Ztg., 6, 221; Journ. Soc. Chem. Ind., 1, 143.

Saponification of Fat. Analyst, 5, 141.
Solubility of Iodine in Fatty Oils. W. Greenl. Archiv. de. Pharm., 223, 431.
Solvent Action of Acetic Acid on Fats and Oils. E. Valenta, Dingl. Polyt. Jour., 252, 296; Jour. Soc. Chem. Ind., 3, 524.
Specific Gravities of Fats. G. W. Wigner, Analyst, 1, 35.
———. Estimation of simple method. G. W. Wigner, Analyst, 1, 145.
———. Analyst, 4, 206.
Spectrum Analysis for Recognizing Oils. J. Mueller, Pogg. Ann., 198, 529; Wagner Jahr., 1871, 681; Chem. Centrbl., 1871, 717; Dingl. Polyt. Jour., 198, 529; Polyt. Centrbl., 1871, 316.
Synthesis of Neutral Fats. Chem. News, 49, 249.
Test for Sperm Oil. Jour. Soc. Chem. Ind., 1, 203.
Testing the Quality of Turkey Red Oil. A. Mueller Jacobs, Dingl. Polyt. Jour., 253, 473; Jour. Soc. Chem. Ind., 4, 115.
Testing Petroleum for Adulteration with Solar Oils. G. Heppe, Jour. Soc. Chem. Ind., 3, 232.
Testing by Volumetric. Analyst, 8, 121.
Stability of Hypobromite Solution and its use for Titration of Oils, Notes on. A. H. Allen, Jour. Soc. Chem. Ind., 3, 265; Analyst, 6, 99.
Stability of Fat Salts in Presence of Water. Chem. News, 31, 185.
Viscosity of Mineral Oils. R. Krause, Chem. Ztg., V. 7, 263. Dingl. Polyt. Jour., 219, 165. Jour. Soc. Chem. Ind., 2, 385.
Wax, Vegetable. M. Buchner, Repert Anat. Chem., 9, 137.
Wax, Bee's, Analysis by Koettstorfer Process. Hübl. Dingler Polyt., 249, 338; Jour. Soc. Chem. Ind., 2, 418.
Wax, Carnanba, Chemical Composition of. H. Sturcke, Ann. Chem. Pharm., 223, 283; Jour. Soc. Chem. Ind., 3, 448.
Wax, Bee's, Analysis and Composition of. O. Hehner, Dingl. Polyt. Jour., vol. 251; Analyst, 1883, 16.
Wool Grease, Decomposition of. Dingler, Polyt. Jour., 255, 88.

BUTTER.

To the Hon. J. K. Brown, *New York State Dairy Commissioner:*

Sir — I have the honor to submit the following report:

Butter.

This important article of food is, as is generally known, composed of, for the most part, the fat of the milk of mammalia.

That from the milk of the cow has a fine pasty consistency at ordinary temperatures. Its color varies from a lard-white to a beautiful golden yellow. The best quality has a fine delicate flavor peculiar to itself, which is easily lost or destroyed by careless management of both the article itself and the animal that produces it.

The word butter comes directly from the German or Bavarian "buttern" or "butteln," meaning to shake backwards and forwards.

Its etymology is, according to Stormonth, as follows: Bav., buttern or butteln; L., butyrum; Gr., boutyron.

According to the Imperial Dictionary: A. S., buter; Ger., butter; L., butyrum; Gr., boutyron, from bous, ox; tyros, cheese.

The ancient history of butter is involved in some cloudiness, but doubtless it was known at a very early period to the ancient civilized nations.

Dr. Edward Smith says: "It may be inferred that it was unknown to the ancient Greeks as no reference was made to it by Homer or Aristotle," but it is a well-known historical fact that the ancient Greeks and Romans used it as ointment in their baths, and furthermore, Roscoe and Scholemmer* state that Herodotus and Hippocrates, in the fifth century, B. C., both describe butter (*βουτρον*) as made by the Scythians by violently agitating mare's milk.

The manufacture of butter is also described by Stet. He says it was produced by agitating the milk of domestic animals and that the best butter is obtained from the fattest milk, as that of the sheep, although it can be made from goat's milk.

Recent developements relating to "Bog butter" may be cited in proof of the antiquity of butter. This is a substance which has been found in the bogs of Ireland, Scotland and occasionally of the Faroe Islands.

It has been considered to be a mineral resin of vegetable origin like bitumen, asphalt, amber, petroleum and ozokerite, and Mr. John Plant,† who denies that this is butter, calls it butyrellite, but the analysis of G. W. Wigner and Prof. Church‡ shows it to be genuine

* Roscoe and Scholemmer, vol. iii, part ii, page 364.
† Chemical News, vol. xli, page 205.
‡ Analyst, vol. v, page 17, 1880.

butter. The superstitious natives accounted for its presence by the agency of fairies, but the less fanciful considered it the work of smugglers, or that it was deposited there to preserve and ripen. A complete history of bog butter is contained in a paper by Dr. Wilde (In Proc. Roy. Irish Acad., vol. vi, page 369).

Its chemical analysis is given by Prof. E. Davy, in *Proc. Roy. Dublin Soc.*, 1826; by Williamson, in *Ann. Ch. Pharm.*, 1845, vol. liv, page 125; and by Brazier in *Chem. Gaz.*, 1852, page 375.

Wigner and Smith minutely describe the physical, chemical and microscopical characters of the sample they examined, and showed it to be genuine butter. They believed the specimen to be not less than one thousand years old and probably older. The analysis of the sample showed:

Moisture, 1·40 per cent.
Curd, fibre, and other matters insoluble in ether, 3·98 per cent.
Ash, ·32 per cent.
Fatty matters, 94·30 per cent.
The ash contained chlorine, ·033; chloride of sodium, ·054.
The analysis of the fat showed:
Volatile fatty acids calculated as butyric, ·06 per cent.
Soluble fatty acids not volatile, ·42 per cent.
Insoluble fixed fatty acids, 99.48 per cent.
Glycerin, minute traces.

They state: "It is interesting to observe how complete has been the decomposition of the original glycerides of butter, both the resulting glycerin and soluble fatty acids set free, having been almost entirely removed by the action of water only at a low temperature. Time has been an important factor in the change."

The microscopical examination of curd revealed considerable quantities of vegetable matter, fragments of wood, fungoid growth, mycelium, muscular tissue and some hairs.

The same chemists describe a specimen of butter taken from an Egyptian tomb which they considered two thousand and five hundred years old. It had been melted and sealed in a vase, and was in a comparatively good state of preservation. Their examination demonstrated it to have been made from the fat of milk.

The vessels in which bog butter has been found are described as rough wooden square, oblong and cylindrical boxes or kegs, and also as a hollow tree trunk. One of the latter kind is preserved in the museum of the Royal Irish Academy. The Edinburgh museum also contains specimens of these vessels.

Butter was used as an article of food to a considerable extent by the Hebrews. It is said to have been introduced to the knowledge of the Greeks by the Scythians, Thracians and Phrygians and to the Romans by the Germans.

It came into general use in England in the early part of the fifteenth century.* It seems to be a concomitant of civilization; for its early history shows it to have been used as an article of food only by the most civilized of ancient nations, and even at the present day it is very little used by barbarians.

Although Pliny † relates that butter was much valued by bar-

---

* Smith's Foods, p.
† Bell's Anal. and Adult. of Foods, p. 43 *et sequent.*

barian nations among whom its use served to distinguish the rich from the poor, and adds that it was prepared by agitating cow's, goat's and ewe's milk in long vessels with narrow necks.

The alchemists gave the name butter to many articles both vegetable and mineral which resembled butter in consistency, viz:

First. Vegetable butters: palm, cocoa, cocoanut, nutmeg and shea.

Shea butter, *i. e.*, tree butter, is used as a substitute for butter by the Africans and is made from the seeds of a fruit which resembles an olive. The butter is extracted by boiling the dried seeds. It is said to keep a long time. It is harder and whiter than cow's butter, and is said to have a more agreeable taste.

Second. Chemical butters; as butter of arsenic, antimony, zinc, bismuth, tin and rock butter. All except the latter are sublimated chlorides of the respective metals.

Rock butter is a greasy exudation from alum rocks, found in various parts of Europe; it is an iron alum.

Butter has been made from the milk of the reindeer, mares, camels, buffaloes, asses, goats, sheep and others. Ewes were extensively kept for milking at one time in Great Britain. But the butter from most of these animals has an unpleasant flavor characteristic of the animal, and at present butter is almost exclusively made from the milk of the cow.

Gheo* is a clarified butter made from the milk of the buffalo, and much used by the natives of India. It is manufactured by boiling the milk and after it has cooled adding sour milk called dhye, to coagulate it. It is then churned and hot water added. After standing a few days it becomes rancid when it is clarified by being boiled with dhye and salt, or betel leaf, and it is packed in sealed pots. It contains more casein than ordinary butter and has a disagreeable taste.

The amount of butter-fat in milk varies with the breed, age, time after calving, season, part of milk taken, frequency of milking, food, treatment, hygienic surroundings, etc., of the cow.

It is a well established fact that a cow will transmit her milking qualities to her offspring, and that butter, cheese and milk cows can be bred to augment these propensities the same as horses can be bred for speed.

As a general rule the milk of cattle which yield large quantities is deficient in the solid ingredients, especially fat, and under the same rule these are found among the larger and coarser built cows, and the butter producers among the smaller and closer built.

The favorite English breeds are the Ayrshire, Norfolk, Suffolk, Durham and Jerseys, for butter, and the Shorthorns, Herefords and North Devons for large quantities of milk and beef. The Germans prefer the Swiss, Allgauer and Dutch.

A cow is in the best condition for butter production from the fourth to the seventh year of her age, and the milk secreted from the end of the first to the end of the fourth month after parturition contains the greatest quantity and the best quality (other things being equal) of butter.

Temperature exercises considerable influence upon the condition of milk. Heat increases the amount of fat while cold increases the sugar and caseous elements, hence a cow will yield more butter in

*Smith's Food, p. 186.

winter if kept in a warm, comfortable place than if exposed. Dairymen well know that the last of a milking, known as strippings, contains much more butter fat than any other part, the first or "fore milk" containing the least. Hassall found the amount of cream in eight samples of strippings to be 141.5 per cent, while the fore milk contained but 61.5.

I found a sample of milk, about one-half pint taken from seven cows about one-half an hour after they had been milked, contained 7.37 per cent of pure dry fat, and a sample from a can into which the whole milk had been poured contained but 3.64 per cent of dry fat.

The great amount of fat in the last milking has generally been accounted for upon the principle that the fat obeyed the laws of gravity in the udder the same as it did out of it.

But Blyth shows this to be a fallacy as the same phenomenon occurs with human milk where the breast occupies nearly a horizontal position. The increase of fat in the last milk is probably due to the stimulation of the secretory cells by the act of milking. The secretion of fat being the most effected.

Morning's milk contains a larger proportion of fat than afternoon's, and that from cows milked but once a day more than either.*

Food probably exerts more influence upon both the quantity and quality of the fat of milk than any other cause. Much research and discussion have been given to this subject and hardly an agricultural journal is issued that does not contain more than one article upon the food question. This at once demonstrates its importance; and, indeed, it is the question which largely determines the profits and loss of the dairy farmer.

I do not propose to lay down a fixed regimen for cattle but to present a brief resume of the governing principles and physiological laws which have been adduced by experimental research.

It was formerly supposed that the fat of milk was taken by the gland out of the blood by a sort of filtering process, but this is found to be an error. There is strong evidence to show that it is manufactured by the epithelial cells which line the cavities of the gland. These cells contain a protein substance called protoplasm,† and it is believed that the fat, as well as all the solids, of milk is the product of the metamorphoses which take place in this substance.

If the fat was merely an extraction from the blood we would naturally expect the amount of fat in the milk to be increased by supplying the animal with food rich in fat, but on the contrary, it is found by experiment that the milk fat is diminished by fatty food and greatly increased by protein.

Moreover, it is an established fact that fat is manufactured in the animal body out of the proteids and carbohydrates which contain no fat in themselves, and not merely stored up from the fat taken in as food. The fats taken in are also probably broken up and reformed.

That fat is formed out of the proteids and carbohydrates was demonstrated as far back as 1843, by M. Persoz,‡ and later by M. Boussingault,§ who experimented on geese, ducks and pigs. They first

---

* Hassall's Foods and Adulterations, etc.
† See Yeo's Manual of Physiology.
‡ Annales de Chimie, tome XIV, p. 408.
§ French Chemistry of Agriculture, 1854.

ascertained the amount of fat contained in the bodies of the animals, then carefully determined the amount of fat consumed with their food for a period extending from thirty-five days to eight months. They then killed the animals and determined the amount of fat throughout the body.

The quantity of fat gained by the animals during the experiments was always much in excess of that taken with the food. The well-known experiments of MM. Dumas and Milne Edwards,* on the production of wax from honey by bees also goes to establish this fact.

Foster† says that Lawes and Gilbert have shown by direct analysis that for every 100 parts of fat in the food of a fattening pig, 472 parts were stored up as fat during the fattening period.

Hoffmann‡ determined the percentage of fat in a given weight of maggots' eggs. He then allowed an equal quantity to develop in defibrinated blood, and when they were matured found that their bodies contained a little more than nine times the amount of fat contained in the food and eggs from which they were developed.

We once produced so much fat in a turkey in a few weeks by the process of stuffing him with corn-meal as to render him unfit to eat.

The following list of experiments prove that the fat of milk is derived from some other source than that taken with the food:

Foster fed a bitch upon meat for a certain perion and ascertained that her milk during the experimental period contained more fat than she could possibly have taken in her food, and as she gained in weight the fat could not have been derived from that stored up in her body.

Liebig‖ ascertained that a cow nourished with 15 kilos (33 pounds) of potatoes and 7·5 kilos (16½ pounds) of hay per day, consumed in six days 756 grains (1 pound and 10½ ounces) of fat and yielded in her milk 3,116 grains (6 pounds and 8 ounces) of butter and 747·56 grains (about 1 pound 10 ounces) in her excreta.

Dr. Lyon Playfair, in a five days' experiment with a cow determined by analysis that she ingested on the second day ·486 pound of fat with her food, and yielded ·969 pound of butter.

On the third day she ate ·542 pound of fat and produced ·9 pound of butter. On the fourth day her food contained ·364 pound of fat and her butter weighed 1·36 pounds. She took ·29 pound fat the fifth day and gave 1·203 pounds butter. Therefore she produced in her milk for the five days 2·747 pounds more than she received.

The following experiments§ are interesting as showing the remarkable influence the proteids or nitrogenized food have upon the fat of milk:

Weiske fed a goat on a certain definite quantity of potatoes and straw and obtained 739 grammes (1·625 pounds) of milk per day. He then added a little fibrine (a highly nitrogenous substance) to the same amount of potatoes and straw and obtained 1,054 grammes (2·318 pounds) of milk.

He also fed a goat for a time on 1,500 grammes (3·3 pounds) of

* Annales de Chimie, tome XIV, page 400.
† Foster's Physiol, Reichert, 1885, page 505.
‡ Therapeutic Gazette, Nov. 16, 1885, page 733.
‖ Blyth's Foods, Composition and Analysis, page 268.
§ Mostly drawn from Blyth.

potstoes and 375 grammes (·825 pound) of chopped straw, and her daily yield of milk was 739 grammes (1·635 pounds) which contained 19·96 grammes of fat (·296 pounds). He then added 25 grammes of flesh-meat which increased the daily yield of milk to 1,054 grammes, and the fat to 33·21 grammes. He next substituted for the flesh-meat 250 grammes of bran and 125 grammes of oil which decreased the yield of milk to 588 grammes and the fat to 29·74 grammes. The oil was then dropped and in its place was given 85 grammes of stearic acid when the milk diminished to 506·2 grammes and the fat to 22·30 grammes.

Fleischmann and P. Vieth carried on a series of experiments, in reference to the influence of food on the quantity of milk and its fat, with 119 cows of the Red German Mecklenburg breed which extended over a period of one year. They also determined the morning and evening variations.

The experiments were divided into five periods of 65, 70, 61, 92 and 77 days respectively. During the second and third periods, including 131 days, 3·2 pounds of flesh-meat were added to the diet which caused an increase of the quantity of morning's milk of 1·73 pounds, of its fat ·05 pound; of the evening's milk, of 2·40 pounds, and its fat ·05 pounds.

The diet during the first period of sixty-five days and to which was added the flesh-meat during the second and third periods consisted of: 5·28 pounds of clover hay; 5·28 pounds of meadow hay; 15·84 pounds of oat and barley straw, chopped and mixed; 1·92 pounds of long oat straw; 2·2 pounds of wheat bran; 2·2 pounds of cocoanut cake.

During the last two periods the flesh-meat was withdrawn and the morning's milk fell off 3·62 pounds, and the evening's 4·25 pounds.

The following table gives a comprehensive view of these experiments:

PERIOD.	Food.	MILK PER COW.		FAT PER COW.	
		Morn'g.	Even'g.	Morn'g.	Even'g.
		lbs.	lbs.	lbs.	lbs.
1st. Jan'y 1 to March 5, 65 days	Chopped: 5.28 lbs. clover hay; 5.28 lbs. meadow hay; 15.84 lbs. oat and barley straw; 1.92 lbs. long oat straw; 2.2 lbs. wheat bran; 2.2 lbs. cocoanut cake	7.70	7.28	.26	.25
2d. March 5 to May 15, 70 days	Same with the addition of 3.2 lbs. of flesh-meat	8.84	8.49	.28	.27
3d. May 15 to July 15, 61 days	Pasturage on commons and flesh-meat	9.43	9.68	.31	.30
4th. July 15 to October 15, 92 days	Pasturage on clover grass	7.39	7.19	.24	.24
5th. Oct. 15 to Dec. 31, 77 days	9.1 lbs. clover hay; 8.55 lbs. meadow hay; 13.18 lbs. oat straw; 1.10 lbs. cocoanut cake; 1.10 lbs. rye meal	5.81	5.43	.20	.19
Average per day		15.630		.51	

While these experiments demonstrate that the fat of milk may be increased by the classes of food rich in starch and sugar and especially in albumen, yet it must not be inferred that either one or both can be used to the exclusion of fat. Physiologists have shown that the different classes of alimentary principles depend upon each other for properly carrying out their nutritive functions in the animal economy. And furthermore, that the withdrawal of either class will produce, in course of time fatal results.

Small quantities of fat aid in the digestion of the other principles and is an essential element for the proper constitution of the blood. The experiments with the bees, previously referred to, failed when they restricted the food to sugar, but when they substituted honey, which contains a little fat in the form of wax, the bees produced a goodly yield of wax.

The fact must not be lost sight of that the nervous system presides over the functions of all the animal organs and that the latter through the nervous system are in close sympathy with each other, so that the perfect action of one depends largely on the integrity of the whole.

Consequently a cow will produce the most and best butter in the long run when all the functions are in the best working order, so that, all things considered, the question of food resolves itself practically into this formula:

The diet containing the largest amount of nitrogenous, starchy and saccharine elements upon which a cow will thrive and maintain a good condition will be the most profitable one for the dairy farmer. As cows have their idiosyncracies as well as the higher animals, this diet will vary with individual cows. Practice has shown that the best diet for fattening cattle is also the best for butter, and also that a change of diet is greatly beneficial. In this latter fact probably lies the secret of much of the evil effect from the milk of stall-fed cows. They are generally kept upon the same food the year around, and even if it is in a healthy condition it becomes distasteful and fails to excite a copious flow of the digestive fluids, and consequently is imperfectly digested which leads finally to disease. I have recently had occasion to witness this state of affairs in a horse. A cow should be given more fatty foods in winter than in summer as fat is the great heat producing substance, and she should also be kept fatter in the winter because the accumulation of a thick layer of fat just underneath the skin has the mechanical effect of retaining the body heat.

Cream.

When milk is allowed to stand the most of the fat being lighter than the other ingredients, rises to the top bringing with it some of all the constituents of the milk, and which constitutes cream.

The first cream that rises contains fat freer from the other constituents of the milk than that which rises later.

We have often observed that the lower portion of the layer of cream in the cream gauge (which is equivalent to the deep "setting" process) after twenty to thirty hours loses its creamy appearance and resembles that of curd so that after the caseine has coagulated it is difficult to make out the line of junction between cream and curd.

This is evidence that some portion of the albuminous substances rise independently of the fat; and also that deep skimming will produce a cream less rich in fat.

There are three methods in vogue for obtaining cream from milk, viz.: shallow setting, deep setting and separating by centrifugal force.

We should say here that great diversity of opinion prevails upon each and every step in the process of making butter, beginning with the milk; and that I do not propose to enter into any discussion upon them or interpose any opinion upon a mooted point; but to present a general view of the subject and state what the weight of authority, at least, approves as the best methods.

It is generally conceded that as good results are obtained by the deep setting system as by the shallow, and it has the following advantages: First. There is a saving of utensils, space occupied and in cleaning. Second. Less waste by adhering to the vessel. Third. The milk is less exposed to the air and will consequently keep longer. Fourth. Temperature more easily regulated, etc.

The great desideratum in raising cream is to get all the fat of the milk to the surface. This has never yet been accomplished as is evidenced by the small quantity of fat remaining in the skimmed milk even after the most perfect method of separating has been employed.

Temperature is an important factor in raising cream.

When the caseine coagulates, or when it "curdles," as it is colloquially termed, the coagulum holds the fat in its meshes and prevents any further rise. So the temperature that will keep the milk sweet the longest without interfering with the other conditions is the best. This is generally said to be 48 to 56 degrees Fahrenheit, the lowest preferred.

M. Tissaraud,* the French inspector general of agriculture, performed experiments which tend to show that a larger quantity and better quality of cream is produced by a temperature of 32 degrees Fahr., though he is opposed in this conclusion by other experimenters.†

Probably the low temperature, aside from keeping the milk sweet, aids the rise of cream by changing the relative density of the constituents of milk in favor of the fat.

The time required for cream to rise varies from twenty to forty-eight hours, and probably the fat will continue to rise to some extent as long as the milk remains fluid.

The following experiments by Mr. W. A. Henry‡ shows the results of the deep setting system and the effects of temperature on the same. He placed an equal quantity of mixed milk in two Cooley tanks. For the first experiment one tank was kept at a temperature of 40 degrees Fahr., the other at 45 degrees Fahr.; for the second, one was kept at 45 degrees and the other at 50 degrees; and for the third, one was kept at 45 degrees and the other at 55 degrees. In each experiment the milk was skimmed after setting eleven hours and butter made from it. The experiments are embodied in the subjoined table arranged by Mr. Henry.

* Lancet, April 20, 1876.
† Biedermann's Centralbl., Jahrg., 1877.
‡ Second Annual Report of the Wisconsin Agricultural Experiment Station, 1885, p. 17.

TABLE.

Temperature at which milk was set.	Amount of milk set.		Amount of butter churn'd.		Per cent lost by high temperature
Degrees Fahrenheit.	lbs.	ozs.	lbs.	ozs.	
40..........................	101	14	4	8	....
45..........................	101	14	4	6	2.7
40..........................	109	6	4	12½	....
45..........................	109	6	4	9½	4.5
40..........................	95	4	4	8½	....
45..........................	95	4	4	5	4.8
40..........................	94	8	4	1½	....
45..........................	94	8	3	14	5.3
40..........................	87	8	4	0	....
45..........................	87	8	3	14	3.1
45..........................	92	12	4	5	....
50..........................	92	12	4	0	7.2
45..........................	88	4	3	0	....
50..........................	88	4	4	10	9.3
45..........................	96	4	4	9	....
50..........................	96	4	4	2½	8.9
45..........................	77	8	3	6½	....
50..........................	77	8	3	1	10.0
45..........................	76	0	3	4	....
50..........................	76	0	3	1	5.7
45..........................	67	0	3	4	....
50..........................	67	0	3	2	3.8
45..........................	70	0	3	5	....
55..........................	70	0	2	5	30.0
45..........................	73	0	2	15	....
55..........................	73	0	2	4	23.4
45..........................	68	0	3	3	....
55..........................	68	0	2	11	15.6
45..........................	80	4	3	8	....
55..........................	80	4	2	14½	16.9
45..........................	88	0	3	14	....
55..........................	88	0	2	15½	23.3
45..........................	73	8	3	14	....
55..........................	73	8	3	4	16.1

The same experimenter had occasion to test the effect of allowing milk to stand out of doors after being drawn for a time before setting. For this purpose he carefully mixed the milk, divided it into equal parts and put it into Cooley cans. Half of the cans were at once placed in ice water while the other half was allowed to stand out in the open air for a certain time before being placed in the ice water beside the other. The following table shows the result:

Number.	Amount of milk set.	Temperatu'e of the atmosphere.	Time lot II stood in open air.	Temperatu'e of milk when set.	Amount of butter churned.		Per cent of loss from delaying the setting.
	lbs.	Deg Fahr.	min.	Deg. Fahr.	lbs.	ozs.	
I.........	52	....	....	91	2	12	....
II	52	19	20	86	2	$9\frac{1}{2}$	5
I.........	41	....	....	91	2	5	....
II........	41	30	25	80	2	$1\frac{1}{2}$	9.4
I.........	51	....	....	93	2	13	....
II	51	28	25	86	2	11	4.4
I.........	51	....	....	91	2	$6\frac{1}{2}$	....
II	51	28	30	78	2	$4\frac{1}{2}$	5.2
I.........	50	....	....	93	3	1	....
II	50	....	30	82	2	13	8.1

The per cent of loss by high temperature is explicitly shown in the table.

The amount of fat remaining in the skimmed milk from the tank set at 40 degrees Fahrenheit was determined by analysis, and ranged from ·31 to ·44 per cent. The amount of fat in the full milk was about 4·5 per cent.

Mr. Henry concludes that this is as good a result as is practicable by any system except the centrifugal separator, and that all the cream obtainable will arise in eleven hours if set in deep cans with water containing plenty of ice around them.

He rightly claims that the early skimming has the advantage of yielding sweet skimmed milk for feeding. He also accounts for the dissatisfaction with the deep method expressed by some farmers by their failure to keep the temperature down especially where no ice or running water is used. As butter was lost by cooling the milk as is shown in the table, Mr. Henry wished to ascertain if there would be a gain by warming it up before setting the can in the ice water. He therefore heated one-half to either 110 degrees or 120 degrees Fahrenheit and the other half was placed in the ice water at once. He tabulated the results in a table as follows:

Number.	Amount of milk set.	Temperature of milk when set.	Amount of butter churned.		Loss by heating the milk.
	lbs.	degrees Fah.	lbs.	ozs.	
I	41	91	2	5	....
II	41	110	2	2½	6·7
I	46	90	2	5	....
II	46	110	2	3½	4·0
I	42	90	2	9	....
II	42	110	2	5½	8·5
I	46	91	3	0½	....
II	46	120	3	10½	12·3
I	41½	89	2	7	....
II	41½	120	2	7	0·
I	46	90	2	10½	....
II	46	120	2	7	8·2

Though tin is the most commonly employed, various vessels are used to "set" milk in; as glass, earthen, wood and zinc. Milk is said to keep sweet four or five hours longer in zinc vessels than in wooden or earthen, but it is unsafe as lactic acid and zinc form poisonous compounds.

The "ice water system,"* invented by Swartz, (a Swede) which has superseded all others in northern Europe consists of setting deep cans in cold or ice water of from 40 degrees to 50 degrees Fahrenheit.

This temperature is maintained by adding ice or snow to the water when necessary. All systems of setting at low temperature are modifications of Swartz.

In the Cooley method the whole can, which is covered with a close fitting lid is submerged in the water. It possesses the advantage of protecting the cream from external pollution.

In the "Harden plan" ice is used instead of water. He claims that by this method less ice is required and a purer, harder cream is obtained.

The method, a modification of Cooley's, pursued in many butter factories in America, is as follows: The milk as soon as possible after coming from the cow is "set" in round tin pails or pans, from eighteen to twenty inches deep and eight to ten inches in diameter. These are placed in pure running water at a temperature of from 48 degrees to 56 degrees Fahrenheit. The more uniform the temperature is kept the better. The water around the pail is kept a little above the surface of the milk.

Milk is said to be kept sweet for thirty-six hours in the hottest weather when treated in this manner.

*Anal and Adult. of Foods, by James Bell, part II, page 40, 1883.

In order to obtain all the cream some skim at the end of twelve hours then again at a later time. As might be expected the choicest butter is made from the cream which rises during the first twelve hours.

The most economical skimmer is found to be a non-perforated one as the perforated allow much cream to fall back into the pail with the curd. When quality is sought care should be taken not to skim too deep.

Frommer's method for setting consists in adding a little soda to the milk (in summer, 1½ per cent; in winter, 1 per cent). The milk is thus kept sweet until skimmed, and the disadvantage of churning "sour cream" avoided.

The following analyses show the composition of cream. König compiles the results of thirty-six analyses of cream made by different chemists, as follows (Blyth, page 276):

	Minimum.	Maximum.	Mean.
Milk fat	8·17	70·20	25·72
Albuminoids	2·20	7·40	3·70
Sugar	·74	4·57	3·54
Ash	·14	3·49	·63
Water	22·83	83·23	66·41

Analyses of Cream, Hassall.

Water	62.12	61·50	63·24	49·10	43·04	45·82
Fat	30·64	32·22	31·42	42·82	44·76	43·33
Caseine	5·83	5·14	2·70	5·20	7·40	6·38
Sugar of milk	1·27	0·74	2·36	2·46	4·45	2·92
Ash	0·14	0·40	0·28	0·42	0·35	0·50

Hassall says: "The first three were purchased of milkmen, while the other three were obtained direct from the dairy, and of their genuineness and quality no doubt could be entertained.

"The difference in amount of fat in the two cases is, as will be seen, very considerable, and appears to us, that since, as in the case of milk, as a standard for comparison is necessary it would not be pressing too hard upon the vendors to insist that cream should not contain less than thirty-five per cent of fatty matter.

The following tables were prepared by Dr. P. Veith, F. C. S., being the monthly averages of five hundred and thirty samples of cream analyzed in the laboratory of the Aylesbury Dairy Company, in 1883.

The samples were partly those received from farmers and partly separated by centrifugal force by the company (The Analyst, April, 1884):

Monthly average of cream analysis.

1883.	Total solids.	Fat.	Solids, not fat.
January	39·8	32·8	7·0
February	41·6	34·7	6·9
March	39·8	32·8	7·0
April	41·7	34·9	6·8
May	44·6	38·1	6·5
June	46·8	40·5	6·3
July	44·2	37·6	6·6

1883.	Total solids.	Fat.	Solids, not fat.
August	47·4	41·1	6·3
September	42·9	36·2	6·7
October	40·4	33·4	7·0
November	39·2	32·1	7·1
December	38·9	31·8	7·1
Yearly average	42·3	35·5	6·8

1884.	Total solids.	Fat.
January	39·6	32·6
February	40·6	33·6
March	41·1	34·2
April	40·6	33·7
May	43·1	36·4
June	43·7	37·0
July	45·4	39·0
August	44·3	37·7
September	43·3	36·3
October	42·4	35·6
November	40·6	33·6
December	40·8	33·9
Yearly average	42·1	35·3

The same chemist (The Analyst, April, 1884, page 61):

No.	Water. Per cent.	Total solids. Per cent.	Fat. Per cent.
1	59·1	40·9	34·1
2	52·0	48·0	41·0
3	40·1	53·9	55·2
4	66·2	33·8	25·5
5	68·6	31·4	24·0
6	39·4	60·6	56·8
7	52·9	47·1	40·4
8	36·6	63·4	58·8
9	73·9	26·1	18·1
10	57·1	42·9	35·9
11	34·5	65·5	60·9
12	52·9	47·1	40·4
13	36·6	63·4	58·8
14	69·3	30·7	22·8
15	63·4	36·6	29·5
16	54·3	45·7	37·5
17	61·4	38·6	31·2
18	42·2	57·8	52·3
19	61·0	39·0	32·2
20	53·3	46·7	40·1
21	62·8	37·2	29·9
22	53·2	46·8	40·0
Average	59·54	45·87	39·34

Six analyses by Wanklyn are as follows:

Water	72·20	71·2	66·36	60·17	53·62	50·00
Fat	19·00	14·1	18·87	33·02	38·17	49·90
Caseine, S. and A.	8·80	14·7	14·77	6·81	8·21	6·10

Analyses made by Edward W. Martin (First Annual Report New York State Dairy Commissioner, 1885, page 56):

	No. 1.	No. 2.	No. 3.
Water	50·02	52·21	65·75
Fat	41·80	41·16	26·60
Sugar	2·80	3·11	2·00
Caseine	5·06	3·40	4·21
Salts	0·31	0·12	0·44

No. 1 was cream raised in the usual way and draining off the skimmed milk from underneath. No. 2, was obtained by the "milk separator," which separates by centrifugal force.

Dr. Voelcker gives the following composition of cream:

Fat	33·43	25·40
Caseine	2·62	7·61
Sugar	1·56	7·61
Salts	0·72	2·19
Water	61·67	64·80

Clotted, clauted or Devonshire cream is a name given to cream that is obtained in the following manner:

The milk is put in large shallow pans which contain a little water, to prevent the milk from adhering to the pan, and the whole brought up to a temperature of about 180 degrees Fahrenheit, and kept there for twelve or more hours. The temperature is always kept below the boiling point. A thin layer or scum of albuminous and fatty matters forms upon the surface which serves to seal the cream beneath and prevents fermentation. This cream will keep sweet for many days in the hottest weather, if the seal remains unbroked, but if not lactic fermentation rapidly ensues. This treatment is said to cause more fat to rise, and in less time than the ordinary; also that its conversion into butter is much easier and the quality better. But there is a diversity of opinion as to its keeping quality; some say it keeps better than butter made in the ordinary way, and others that on account of its containing more caseine it is more liable to become rancid.

The following is the composition of clotted cream, according to Blyth:

	Per cent.
Milk fat	65·011
Caseine	3·530
Albumen	·521
Galactine	·050
Lactochrome	Undetermined
Milk sugar	1·723
Water	28·675
Ash	·490

This cream is often eaten on bread as butter, and as it possesses the advantage of keeping longer it would serve a good purpose as nourishment for the sick, especially in hot weather. Apropos, cream is growing in popularity as a nourishment for the sick.

The method of separating cream from milk by centrifugal force was introduced about ten years ago and is largely superseding the setting system in Europe.

The instrument is called a milk separator or milk centrifuges or cream separator.

The machine is based upon the difference in the specific gravity of the fat and the rest of the milk. The fat being the lightest goes to the centre and the remainder of the milk to the periphery.

A number of these machines have been devised with the object of more perfectly and quickly removing the cream from the milk. These are great advantages when large quantities of milk are dealt with; for instance, more cream is obtained, the skimmed milk is sweet and the many risks attending the keeping of milk, as decomposition and many accident familiar to dairymen, are avoided.

The whole milk is introduced into a cylinder revolving at the rate of 1,600 to 4,000 revolutions per minute. After a short time a division takes place in the milk, the fat, by virtue of its lighter specific gravity, goes toward the center, while the watery portion, containing the caseine, albuminoids and salts, and whatever dirt or foreign solid matter is present, is found nearest the walls of the cylinder. The dirt, which is sometimes considerable, being next to the periphery. In this way an almost complete separation is effected. The cream is drawn off by a suitable arrangement of tubes leaving the skimmed milk in the cylinder. This is also removed through tubes and the operation is made continuous. A number of applications of this principle for skimming or separating the milk are known, but all are essentially the same. The best known are the following which are named after their inventors: De Laval, Nielsen & Petersen, Lefeldt, Fesca and H. Petersen.

W. Fleischmann* made a series of eight experiments on the influence of temperature on the efficiency of the Nielsen & Petersen separator. He found that on gradually increasing the temperature from five to forty per cent, the fat remaining in the milk decreased from eighty-five per cent in the first case to twenty-two per cent in the last. This demonstrates the beneficial effects of heat on the operation. Fjord† compared three forms of separators with a view of testing their power of separating the cream. He found that the De Laval form always left the most cream in the milk; that is, performed the operation in the least perfect satisfactory manner. The cream yielded by the Nielsen & Petersen separator was found, on comparing with other systems by the same observer, to be the most economical and advantageous for the purpose of making butter.

A comparison‡ of the value of the several styles of separators was made with the following result:

---

* Biedermann's Centralbl. 1883, vol. 411, p. 415.
† Biedermann's Centralbl., 1884, vol. 339, p. 341.
‡ Wagner's Jahresbericht, 1882, p. 920.

Designation.	Mark or number.	Capacity per hour in Litres.	Price in Marks.	Cost for each 100 Litres, hourly capacity Marks.
De Laval................	I	250	7 00	2 80
De Laval................	II	500	16 50	3 30
Neilsen & Petersen.......	I	250	10 00	4 00
Neilsen & Petersen.......	II	500	15 50	3 10
Lefeldt	0	225	6 50	2 89
Lefeldt	II	625	16 50	2 64
Lefeldt	IV	1,250	26 50	2 12
Fesca..................	I	300	11 20	4 57
Fesca..................	II	150	13 70	7 47
Fesca..................	III	75	8 70	11 60
H. Petersen	M. 100	450	21 00	4 67
H. Petersen	100	900	35 00	3 89

From this it appears, as will be readily seen, that the Lefeldt machine is the most economical.

These machines have not come into very extended use in this country, but in Europe they are used with satisfactory results.

The average amount of fat in skimmed milk by the setting systems is given by Vieth as from eight-tenths to one per cent, and the amount by the separator five-tenths per cent.

The following table of analyses from Vieth shows the amount of fat in skimmed milk by the different systems. Also their specific gravity:

Analysis of skimmed milk.

Number.	Specific gravity.	Total solids.	Fat.	Solids not fat.	Remarks.
1	1·0350	9·75	0·55	9·20	Shallow setting system.
2	1·0355	9·90	0·54	9·36	
3	1·0340	10·10	1·00	9·10	
4	1·0355	10·43	0·98	9·45	
5	1·0355	9·68	1·05	8·63	Deep setting system.
6	1·0345	9·70	0·60	9·10	
7	1·0355	9·81	0·43	9·38	
8	1·0350	10·26	0·88	9·38	
9	1·0365	9·96	0·46	9·50	Centrifugal system.
10	1·0350	9·28	0·34	8·94	
11	1·0370	9·94	0·34	9·60	
12	1·0370	9·80	0·35	9·45	

The following is the result of an analysis of the cream separated by centrifugal force, and also of the skimmed milk, by E. W. Martin:

	Cream.	Sk'med milk.
Water	52·21	90·34
Fat	41·16	0·15
Sugar	3·11	3·98
Caseine	3·40	4·80
Salts	0·12	0·78

The following table of analyses shows the difference in the cream and skim milk obtained by the ordinary setting system and Laval's separator, by Vœlcker:

	Ordinary cream.	Cream by separator.	Skimmed milk by Laval separator.	Ordinary skimmed milk.
Water	77·30	66·12	90·82	87·25
Butter fat	15·45	27·69	0·31	1·12
Casein	3·40	2·69	3·31	3·69
Milk sugar	3·15	3·03	4·77	5·16
Mineral matter	0·70	0·47	0·79	0·78

An artificial cream is described by Blyth (p. 277, 1882). It was made of albumen and ordinary cream, and was colored with what was supposed to be the coloring matter of the carrot.

BUTTER.

In milk and cream the fat exists in the form of distinct g u es, which vary in size from 1-2700,* to 1-300 † of an inch in diameter. Much discussion has been going on of late years as to whether or not the milk globules contained an albuminous capsule or envelope. The circumstances affirming the existence of a covering are as follows: Henle ‡ found that the shape of the globules were disturbed by adding acetic acid to the milk. Mitscherlich ascertained that they were not dissolved by shaking with ether, as they should be if they were free globules of fat, unless the milk was first treated with acetic acid, caustic potash, or carbonate of potash, which were supposed to destroy the covering and allow the ether to come in direct contact with the fat and dissolve it. Lehmann observed that merely treating with ether would cause the surface of the globules to become wrinkled, turbid and less transparent, and that the ether dissolved the fat after adding phosphate of sodium. Moleschott extracted the fat with ether after treating the milk with alcohol and acetic acid. He was able to distinguish many empty vesicles which he succeeded in filling with an etherial solution of chlorophyl.

* See Ap Ency. on Milk, body of work.
† See Johnson's New Universal Ency., do.
‡ Vide, Hassall, p. 408 *et sequent*, 1876.

Hoppe-Seyler* based his belief that a caseous membrane existed upon the fact that cream contained a larger proportion of caseine to its water than milk.

Letheby † says "Acherson showed as far back as 1840, that albumen always coagulates from its solution around a fat globule, and this is seen in the little fatty particles of milk which have a covering like a cell-wall of consolidated caseine."

In opposition to this:

Soxhlet‡ denies that the globules possess any covering and accounts for the solidification and agglomeration when agitated by churning by the phenomenon in physics of substances remaining liquid when at rest at a temperature far below their point of solidification.

He believes the globules of fat to be in a liquid state at a temperature below their point of solidification and that the churning causes them to solidify and come together.

Stork § advances the fanciful theory that churning causes the formation of a glue-like substance from the caseine which cements the fat globules together.

James Bell ‖ thinks there is not proof of the existence of a covering except what might be caused by a slight adhesion to the globules of the non-fatty elements of the milk.

He states that as the fat globules are in the same solid condition as butter would be at the same temperature, it is easy to understand how a portion of the milk fluid would adhere to the exterior surface of the fat particles. On this view he attributes the production of butter not to the liberation of the oil globules from an envelope, but to the cohesion of solid particles of fat brought together by the agitation of churning, and adds as proof that the distinctly globular condition of the fat is maintained after being massed into butter.

Blyth** asserts that the fat can be extracted from milk by shaking with ether, if the volume of ether exceeds that of the milk; and assigns as the cause of small quantities of ether dissolving the globules after the addition of acetic acid, to the fact that the acid has the effect of destroying the emulsifying property of the milk.

He also states that the globules are colored by analine red; and that it is difficult to understand on the "membrane" theory how, when milk in thin layers is dried at the ordinary temperature of the air, and under conditions which involve no destruction of the membrane, should this exist, to account for the fact that ether so readily dissolves the butter-fat. In reference to Hoppe-Seyler's belief, he remarks that if the caseous layer does exist it must be so thin as not to be capable of estimation by weight.

S. D. Babcock, of the New York Experimental Station, estimated the number and size of fat globules of a specimen of milk in a certain space under the microscope. He then warmed the milk up to the melting point of butter-fat, shook it and after allowing it to cool examined again with a microscope, and found that the number of globules

---

* Archiv. für path. Anatom., bd. XVII, s. 417, 1859.
† On Foods, page 75. 1872.
‡ Landwirth, Versuchsstat vol. XIX, p. 1.
§ Milchzeitung, 1881, p. 589.
‖ Analysis and Adulteration of Foods, part II, page 4, 1883.
** Foods, Composit. and Anal., 1882, p. 202 *et seqnent.*

had increased and their size diminished. By repeating this he found he could diminish the size of the globules with a proportionate increase of the number.

Edward W. Martin treated cotton-seed oil in the same manner and obtained the same result.

However, modern physiologists teach that fat drops take to themselves a covering when suspended in an albuminous liquid as is observed in the minute fat globules after entering the lacteals.

Foster* says "Each oil globule is invested with an albuminous envelope; this may be dissolved by the aid of alkalies, whereupon the globules run together."

Kirke† asserts that oil always becomes covered when set free in minute drops in all albuminous fluids. This is proved when water or diluted acetic acid is added to chyle which causes many of the molecules, 1-30000 of an inch in diameter, to disappear, and drops of oil are seen in their place, which is due to the albuminous layer around the molecule being dissolved, allowing them to run together.

The best flavored butter is made from "sweet cream," although cream is often allowed to become slightly sour before churning. If the acid fermentation is allowed to go too far it will deleteriously affect the butter. The length of time cream can be kept sweet depends upon the temperature to which it has been exposed. Three days is said to be the longest that cream should be kept for good butter. However, all these conditions will vary somewhat under different circumstances and consequently the individual experience of the dairy farmer will furnish him the best guide.

Great care should be taken to keep cream away from all decomposing matter, or substances giving out disagreeable odors, as it is like milk, a powerful absorbent of volatile substances and readily affected by adventituous influences. In illustration of this may be cited the custom in some parts of Europe‡ of hanging certain herbs in the churn to impart an agreeable odor to the butter.

A writer in the Chemical News, vol. XXXVII, page 50, speaks of a practice of straining milk through a sprig of fir for the purpose of imparting to it an agreeable aroma. Also a practice among the Bedouins in the Hedjaz, mentioned by Smith,§ of boiling herbs with butter for the purpose of changing its flavor.

The effect of churning is generally explained upon the supposition that the fat globules contain an envelope which is broken by the agitation caused by the dasher, which allows the fat drops to come together.

The "grain" of butter is supposed to be due to the globules maintaining their identity when brought together, and it is a desired object to preserve this.

Sometimes the whole milk is churned, but as considerable power is required and the product of inferior quality, the process of churning only cream prevails.

James Bell states that on small farms in England the cream is simply stirred in a shallow tub with the hand, and that a cold hand is considered an indispensable qualification in a Devonshire dairymaid.

---

* Foster's Physiology, Reichert, 1885, p. 363.
† Kirke's hand-book of Physiology, 1885.
‡ Les Mondes, Dec. 6, 1877.
§ Smith's Foods, 3d ed., page 136.

There are a great variety of churns. In looking over the list of patent churns, in the report of the United States patent office, we found that though they number thousands they are nearly limited to two varieties, the type of which is the old upright dasher churn, and the paddle-wheel rotary. The excellency of a churn rests upon its ability to uniformly agitate the cream so that the butter will "come," throughout the whole mass as nearly as possible at the same time; for, if otherwise, the grain of the butter first formed will be injured by being beaten till the last is formed.

Experience has decided that the best temperature, both for the cream and atmosphere, to begin churning is from 54 to 59 degees Farhenheit.

The friction produced by churning, together with some chemical action raises the temperature eight or ten degrees Fahrenheit during the process. The time required for churning the whole milk is from two and one-half to three and one-half hours. That for cream alone from one-half to one hour.

If the churning is done too quickly the butter is apt to be soft and fatty; and prolonged churning injures the flavor.

Many influences affect the time required for churning; as the health of the cow and her food, and many that are unknown.

Mr. John Higgins, of Speedsville, N. Y., a few years ago recommended that a little cold water be poured into the churn at short intervals two or three times in order to lower the temperature, and that as the butter formed to reduce the speed, not allowing the dasher to come to the top of the cream. In this way he claimed to get a purer and harder quality of butter.

In our American factories the dasher in barrel churns is made to make fifty strokes per minute, as that is found to be the best rate.

The speed is kept slow at first until the cream is thoroughly mixed when it is brought up to the desired rate; it is again reduced when the butter begins "to come."

The great object desired is to preserve the grain of the butter, and it is claimed that this is best done with a cylinder churn and dasher and rather low temperature. A temperature at or above the melting point (35·8 degs. C. = 96·44 Fahr. average, Blyth) will cause the globules to lose their identity. Too much force in working butter will also injure the grain.

After churning, butter is washed with pure cold water and worked with a wooden butter ladle in a wooden bowl, or, as in large factories, on an incline slab, to remove the buttermilk. In many places in Europe butter is eaten without any salt, but in this country a little salt, about one-quarter ounce to the pound, is added for flavor, even when intended for immediate use; and, when intended to be kept, from one to two ounces to the pound is added. Much more is used at times which will be referred to under the head of adulterations.

Butter, like cream, is very susceptible to bad odors. We have detected the peculiar flavor of cod-fish in butter kept in a refrigerator with the fish.

Butter contains all the ingredients of milk, with the addition of salt, with the fat largely preponderating, and when it is allowed to stand exposed to the air especially in a warm temperature it undergoes decomposition and is said to be rancid. This is brought about by the caseine acting upon both the sugar and fat. The fat of butter is composed of a mixture of nine different fats, viz.: stearine, palmitine, oleine, myristine, arachidine, butyrine, caproine, capryline and caprine or rutine. These in turn are a chemical combination of glycerl, which is commonly known in its hydrated form as glycerine, and a corresponding fatty acid as stearic acid, palmitic acid, butyric acid, etc.

Now caseine acts as a ferment, under proper conditions, upon the fats, especially butyrine; and breaks it up into butyric acid and glycerine. It also acts as a ferment upon sugar and converts it into lactic acid; rancidity is the result of this decomposition.

According to Lang, the first step in the decomposition of genuine butter is a conversion of lactic into butyric acid, then the butyrine is broken up into butyric acid and glycerl.

Pure butter fat will remain free from rancidity a long time. We have in our laboratory a pure dry butter-fat corked up in a flask, which is over a year old, and perfectly free from rancidity, nowithstanding it has been exposed to a temperature of over 100 degrees Fahrenheit for several months. This teaches that a butter to be preserved must be freed as much as possible of the butter-milk, though butter owes some of its delicate flavor to the presence of butter-milk.

It is to obviate the effects of caseine that salt is used, but it will not, in quantities not to spoil the taste, do what proper working will.

Great care should be taken in the selection of the salt, for its impurities sometimes cause a bad flavor in the butter.

The chlorides of calcium and magnesium are common adulterations of salt. The more evenly the salt is worked through the butter the more protection it affords. For this purpose salt should be finely pulverized.

The Irish method of curing butter as given by Dr. Ure, is one part of sugar, one part of nitre and two parts of Spanish great salt, mixed and finely powdered and thoroughly incorporated with newly made butter in proportion of one ounce of the mixture to one pound of butter.

Melting and washing old butter will remove its rancidity, but it also destroys its flavor.

There is a comparatively new traffic, and fraud as well, being carried on quite extensively in this country, viz.: The old rancid butter is bought up at a low price, five cents a pound, heated and washed and probably deodorized, etc., until its unpleasant taste and smell are removed, and is then churned with new milk, salted and sold as fresh butter.

The secret in packing butter lies in keeping it from the air. For this purpose oak tubs have largely taken the place of glazed earthen jars, probably on account of their being less easily broken when shipped. It is also shipped in sealed tins. Various materials are placed around the butter, as syrup, salt, brine and water. Probably the best is brine.

Brean* recommended a weak solution of tartaric acid. Payen tested this and found that butter had retained its freshness at the end of two months which had been kept at a temperature of from 60 degrees to 68 degrees Fahrenheit.

* Pavy's Food and Dietetics, 2d ed., p. 128.

When a cloth is laid between butter and salt care should be exercised in washing the starch all out, as starch is said to take the color out of butter.

It is generally conceded that the best butter (with exceptions of course) is made by large manufacturers, and the reasons are that all the various steps in the process of manufacture are carried out with more care and precision—the temperature is better regulated, the cream is kept in better condition and, like the churning, is more uniform, the advantage derived by cleaning the utensils with steam, the surroundings are more closely looked after, the cost is less and the evil effects of milk of a cow out of condition would be lessened by an admixture with such large quantities of healthy milk.

The most of these reasons however could be overcome by the farmer's wife.

It has been said that "a man weighs and a woman guesses," and in this lies the secret of man's excelling in performing domestic work.

The composition of good dairy butter according to numerous authorities both in this country and abroad is:

Butter fat	80 to 90	per cent.
Curd	0·2 to 2	"
Water	10 to 12	"
Salt	2 to 6	"

A cow whose milk contains 3·5 per cent of fat will yield about one pound of butter to twelve quarts of milk, allowing 80 per cent of fat for pure butter. Then if she averaged 12 quarts per day for nine months a year she would produce 270 pounds per annum. Of course the milk of many choice cows will contain a very much larger per cent of fat, and also many cows will average a larger yield.

Mr. N. M. Blish, a farmer of Delaware county, N. Y., in his testimony before the Committee on Public Health,* said that the average production of butter from a cow for a year was about 200 pounds, and that the actual cost per pound to the farmer was from twenty to twenty-five cents, nearer latter than former.

The adulterations of butter are increased amount of its natural ingredients—water, salt and curd—at the expense of the fat; and the admixture of foreign substances, as yolks of eggs, animal and vegetable oils, etc., as will be seen by the abstracts of patents. Water is sometimes worked into butter to a great extent, even 50 per cent, which constitutes what is known as "bosh."

Salt and curd are found in butter in varying proportions and constitute adulteration only when in large excess of the average, which is given on page 28 of this report.

Oleomargarine or Margarine Mouries.

This article was originally invented by the French chemist, M. Mége Mouries. In order to make what follows more intelligible to those not familiar with the subject, we will explain the significance of the term oleomargarine. At the time of Mége's experiments it was understood that most animal fats were composed of a mixture of three

* See Report of Committee Public Health, 1884, p. 76.

fats, viz.: stearine, margarine and oleine. The stearine was the hardest of them all not melting below 143 degrees Fahr.; margarine next at 118 degrees Fahr.; and oleine remaining liquid to nearly the freezing point of water. The process of making this article as will be seen from the description, involves the separation of most of the stearine from the oleine and margarine, hence the name oleomargarine. It has since been learned, however, that what was called margarine was a mixture of stearine and palmatine. Mége * was instigated to make his researches by a request of the government for the purpose of obtaining a substitute for butter for the use of the navy and poorer classes, that would cost less and keep from rancidity longer than butter. With this object in view he commenced the following experiments at the farm at Vincennes: He placed several milch cows on a diet which caused them to lose weight and also caused a corresponding decrease in the yield of milk, but as the milk always contained butter he believed it was produced from the fat of the animal in this wise: As the animal lost weight the blood took up the fat from the tissues, and carrying it to the lungs its stearine was used in supplying that organ with material for combustion, the oleomargarine going to the udder in which it came in contact with an animal ferment which converted it into butyric oleomargarine or butter. Under this belief he experimented on fats for the purpose of imitating the natural process through artificial means. The subjoined copy of the patent obtained from the United States government December 30, 1873, will show the methods finally adopted:

" *To all whom it may concern:*

Be it known that I, Hippolyte Mége, of Paris, France, chemist manufacturer, have invented improved means for transforming animal fats into butter, of which the following is a specification.

My invention, which is the result of physiological investigations, consists of artificially producing the natural work which is performed by the cow when it reabsorbs its fat in order to transform the same into butter. The improved means I employ for this purpose are as follows:

I. *Neutralization of the ferments.*— In order to prevent the greasy substance which is settled in the tissues of the animals from taking the disagreeable taste of the fat, it is necessary that the ferments which produce this taste shall be completely neutralized for this effect as soon as possible after the death of the animal. I plunge the raw fats called *graisses en branches* into water containing fifteen per cent of sea salt and one per cent of sulphite of soda. I begin thus the transformation an hour at least after the immersion and twelve hours at most afterward.

II. *Crushing.*— A complete crushing is necessary in order to obtain rapid work without alteration. For this purpose, when the substance is coarsely crushed, I let it fall from the cylinders under millstones, which completely bruise all the cells.

*Vide translation from the Moniteur Scientifique, by Fred H. Hoadley, B. A., of an extract from a report made to the Board of Health of the Department of the Seine on the product presented under the name of artificial butter, by M. Mége Mouries, by M. Felix Baudet, in Amer. Chemist, vol. IV, p. 370.

"III. *Concentrated digestion.*— The crushed fat falls into a vessel which is made of well-tinned iron, or enamelled iron or baked clay. This vessel must be plunged in a water bath of which the temperature is raised at will. When the fat has descended in this vessel, I melt it by means of an artificial digestion so that the heat does not exceed 103 degrees Fahrenheit, and thus no taste of fat is produced. For this purpose I throw into the wash-tub containing the artificial gastric juice about two litres per hundred kilograms of greasy substance. (This gastric juice is made with the half of a stomach of a pig or sheep, well washed, and three litres of water containing thirty grams of biphosphate of lime. After a maceration during three hours, I pass the substance through a fine sieve, and I obtain the two litres which are necessary for a hundred kilograms.) I slowly raise the temperature to about 103 degrees Fahrenheit, so that the matter shall completely separate. This greasy matter must not have any taste of fat. It must, on the contrary, have the taste of molten butter. When the liquid does not present any more lumps I throw into the said liquid one kilogram of sea salt, reduced to powder, per hundred kilograms of greasy matter. I stir during a quarter of an hour and let it sit until obtaining perfect limpidness. This method of extraction has a considerable advantage over that which has been previously essayed. The separation is well made and the organized tissues which do deposit are not altered.

IV. *Crystallization in a mass.*— In order to separate the oleomargarine of the stearine, separated crystallizers or crystallizations at unequal temperatures have been already employed. I have contrived for this purpose the following method, which produces a very perfect separation, and it is as follows: I send the molten fat in a vessel which must be sufficient for containing it. This vessel is placed in a wash-tub of strong wood which serves as a water bath. In this wash-tub I put water at the fixed temperature of eighty-six degrees Fahrenheit for the soft fats proceeding from the slaughter-houses, and ninety-eight degrees for the harder fats, such as mutton fat. Afterward the wash-tubs are covered, and after a certain time, more or less long, according to the fats, the stearine is deposed in the form of teats at the middle of the oleomargarine liquid.

V. *Separation by centrifugal force.*— In order to avoid the numerous inconveniences of the employment of the presses which have been hitherto used, I cause the mixture of stearine and oleomargarine to flow into a centrifugal machine called "hydro-extractor." The greasy liquid passes through the cloth and the stearine is collected. When all the liquid is passed I put the machine in motion, and the crystals of stearine are entirely exhausted without the auxiliary of the presses. However, during certain seasons there are animals which produce crystals of stearine soft enough for rendering necessary the stroke of a press as a last operation; but in this case this operation has little importance, because it is applied only to a fraction of the product. In all cases the oleomargarine is separated from the stearine when it is cold and passed to the cylinder, constitutes (especially if its yellow color has been raised) a greasy matter of very good taste, and which may replace the butter in the kitchen, where it is employed under the "name of "margarine"; but, if it is desired to transform it into more perfect butter, I employ the following means:

VI. In the methods hitherto employed the margarine is transformed into cream, and this latter into butter. This complicated operation has many inconveniences. I obtain the same result by the following manner: I take ten litres of natural and fresh cream of milk; I add ten grammes of bicarbonate of soda and two hundred grammes of the udders of a cow, which must be fresh and well-hacked in order to give all the mammary pepsine. The fresh udder may be replaced by udder collected in slices in sea-salt. After a maceration of an hour, I pass the whole through a very fine sieve. I add the necessary quantity of yellow color which is employed for the ordinary butter, and I put these ten litres into a hundred kilograms of liquid margarine at seventy degrees. I stir or mix until the combination is complete — that is to say, until the pepsine has effectuated its action. At this moment the liquid becomes thick, it takes the taste of cream, and after it has been more thoroughly agitated I let the same become completely cold. When the butter is cold and solid, coarsely scrape it in order to pass it between two large cylinders which give it the homogeneousness and the consistence which are the qualities of the natural butter.

VII. When it is desired to produce butter intended to be preserved, which must contain no animal matter, I plunge the udders into pure water instead of cream, in order to macerate the same. Afterward the water which proceeds from this operation is mixed with the margarine at about eighty-six degrees of temperature — that is to say, to a degree which permits the pepsine to effectuate its action without production of cream. After an hour I let the liquid set, and the margarine, which is decanted, is mixed after it has been reduced in temperature to about seventy-one degrees with an emulsion of butter made with five litres of water, five kilograms of butter, one kilogram of sea-salt, or more, according to the uses, and ten kilograms of bicarbonate of soda. When I add this emulsion with the margarine, which has already been submitted to the pepsine action, I obtain a rapid combination, and all the molecules take the qualities of the ordinary butter. It is a delicate operation, which must be exactly made.

VIII. For the long conservations I only treat the margarine by the mammary pepsine, as before described. I decant it in order to avoid any trace of water or animal matter. If it is desired to add ordinary butter, I do that at the temperature of seventy-one degrees by well diluting it.

IX. The stearine which has been separated from the margarine forms a hard fat, which can be bleached by the known processes in order to produce wax candles of lower quality; but it is preferable to saponify it by any convenient process and crystallize the greasy acids (which are charged with from seventy to eighty per cent of stearic acid, instead of fifty) in a chamber heated to ninety-six degress, so as not to let them become hard by cooling. In this state they can be pressed under heat in order to produce stearic acid much superior to that of the trade, both by its beauty and by its point of melting.

I claim:

1. The improved material herein described, produced by treating animal fats so as to remove the tissues and other portions named, with

" or without the addition of substances to change the flavor, consistency or color as set forth.

2. The process herein described of treating animal fats in the production of oleomargarine.

In testimony whereof I have signed my name to this specification before two subscribing witnesses.

H. MÉGE."

Witnesses:

This process was patented some time before in France, and was the original successful method of making artificial butter. From this beginning has grown one of the most stupendous traffics and colossal frauds of the nineteenth century. The enormous sale of these products is evinced by the fact that in 1883, 10,000,000 pounds were sold in New York city alone and 45,000,000 pounds in the United States.

The rapid development of the traffic in this article is marked by this characteristic: It was not in response to a public demand on account of its merits, for the public has always held it under suspicion, but it was fraudulently foisted upon the people under the guise of genuine butter, and recent events have shown that even now the retail dealers would rather risk the penalty of a misdemeanor than sell it on its merits for what it is.

That Mége's product was at once accepted, on chemical grounds only, as a wholesome article of food is evinced by the fact that the Parisian government imposed upon it the samé tax as upon genuine butter, and also by the following comments made by M. Baudet. After minutely describing the process of manufacture he says:

"This artificial butter presents then this advantage, that it contains much less water and animal substance which makes the ordinary commercial butter rancid; moreover, for the same weight it furnishes more genuine butter. These two circumstances assist without doubt in its preservation, which is much more perfect than that of common butter. They also prevent it from acquiring the odor and the acridity which are soon developed in the latter."

"During warm weather, when ordinary butter can with difficulty be preserved from melting, it is easy to give to the artificial butter a more or less solid consistence by preparing an oleomargarine more or less free from stearin."

"On the other hand, M. Mége has observed that by washing his butter with water at a temperature of only five or six degrees, he is able to leave in it less water and thus to obtain a product capable of being kept a very long time. A specimen of butter thus prepared, and which M. Mége called "butter without water," carried from Paris to Vienna, in Australia, the 29th of October, 1871, has just been brought back on the 5th of April, and is found still, after five months, in a good state of preservation."

"In order to fully appreciate the value of the product of M. Mége, as regards domestic economy and hygiene, I have requested several of my colleagues to try the oleomargarine and the artificial butter; I have submitted this product to the judgment of several breeders and butter merchants of the Auge valley; I have used it myself also in my household, and we all have been of the opinion that the oleomargarine constitutes an excellent *grease* for the kitchen, and that if the artificial butter has not the fine and aromatic taste of the Normandy butter for eating with bread, or use in culinary preparations, it does afford in many other respects the qualities of ordinary butter perfectly."

"The experiments which I have witnessed in the works of M. Mége, those which I have myself made or which have been made at my instance on the new products which he has brought forward, authorize me to believe that he has realized a happy application of his knowledge and his inventive genius in this employment of beef fat, and that he has furnished for consumption two new and important products."

"The first, called cooking grease or oleomargarine, offers a valuable material for cooking purposes, especially for naval vessels during long voyages, by reason of its good quality and of its capability of long and excellent preservation."

"The second, possessed of properties which allow of its close comparison with butter, in a chemical point of view, as well as regards its uses, may take the place of the latter in many instances, and in consequence of the small expense at which it can be made, it has been put in competition with milk butter which will lower necessarily the price of the latter to the benefit of the consumer, which will render the consumption of it less considerable, and will allow the breeders to devote a much greater quantity of milk to the raising of calves, a great advantage to their industry."

"As regards healthfulness, it is evident that the origin and preparation of these two products presented by M. Mége, do not afford any circumstance which can render its employment a matter of suspicion."

"There is then no reason for opposing the sale of these products if we include the proviso that that which M. Mége Mouries compares to butter is not really butter in the usual and true acceptation of the word. It should not be sold under the name of butter, but under a particular designation which will permit it to be distinguished from butter properly so called or true milk butter."

Perhaps it is unnecessary to say that Mége's premises were altogether wrong; the fat of milk is not obtained in any such manner, neither did his product contain the elements peculiar to butter.

The following abstracts of patents issued by the United States government will show the modifications of Mége's process, and also the methods and materials used in making what is claimed to be *perfectly wholesome substitutes* for milk butter. I am informed that there are no less than 180 of these patents on file in the patent office at Washington:

Improvement in Processes of Bleaching and Clarifying Fats.

[Specification forming part of Letters-Patent No. 156,404, dated November 3, 1874.]

"These oils are often dark-colored and offensive in consequence of the presence of decomposing organic matter. My improved process is intended to remove these impurities and render the oil sweet and light-colored, and my process may be also applied to lard, grease and other solid fats.

"To enable others skilled in the art to understand and use my invention, I will proceed to describe the manner in which I have carried it out.

I first add to every hundred (100) gallons of oil three or four gallons of good yeast; or, instead of yeast, about half a pound of putrid cheese, mixed with water to reduce it to a thin paste, may be added to the oil, and the whole well agitated. The oil is kept warm, but at a temperature not exceeding 100 degrees Fahrenheit. Fermentation soon sets in, and the albuminous and mucilaginous matters in the oil are dêcomposed and rendered susceptible to chemical agents, to be used as hereinafter described. The yeast should remain in contact with the oil for about ten days, after which it is allowed to settle, and the oil is then drawn off into another vessel. This treatment with the yeast or other fe ment decomposes all albuminous and mucilaginous matter in the oil. r

I then submit the oil to chemical treatment for removing the residuary products of the decomposition which has been produced by the yeast. For this purpose I employ the following mixture: Permanganate of soda, one ounce; sulphuric acid, two ounces; water, one gallon. The permanganate of soda is dissolved in the water, and the acid added, and one gallon of this solution is agitated with five gallons of oil, or in that proportion for any quantity, namely, twenty per cent of the above solution. After agitation, the oil separates and rises, when it is drawn off, and, after being well washed by agitation with water, is ready for use."

United States Patent Office—Improvement in Processes for Purifying and Preserving Animal Fats.

[Specification forming part of Letters-Patent No. 169,008, dated October 19, 1875.]

"The nature of my discovery or invention consists in purifying and preserving animal fats entirely sweet and odorless, and utilizing and manufacturing the same into butter.

To enable others skilled in the art to make and use my discovery or invention, I will now proceed to describe the same.

First, in ten (10) gallons of water dissolve three (3) pounds of common salt and four (4) ounces of soda-ash. (The proportions of my ingredients may be increased or diminished according to the quantity of fat I desire to treat.) Then boil the admixture in a suitable vessel by hot-air, or steam pipes, or any other suitable means. When the ingredients are thoroughly dissolved by this process, a scum arises at the top, which I then skim off carefully, and add one hundred (100) pounds of animal fat cut in small pieces, keeping the mass well agitated until the whole is thoroughly melted, when I again skim the mass carefully. Then draw the oil off through a filter into cold water, which must be well agitated, until the oil is cool enough to be removed. Second, the fat, as purified by the first step of my process, is then put into a second solution, consisting of about four or five (4 or 5) gallons of water, about two (2) pounds of bicarbonate of potassa, and about two (2) pounds of salt. It is then heated and kept agitated until it becomes thoroughly melted. Then draw the oil off again through a filter into cold water, keeping it well agitated until the fat becomes "cold; then thoroughly remove from it the water contained therein; the fat will then assume an entirely pure, sweet and odorless condition.

It will be observed that my process consists in two separate steps — the first being to purify or remove all impurities from the fatty matter and the second to make it perfectly neutral and sweet.

To utilize and manufacture the fat thus prepared into butter, I place the fat in any suitable churn, with rotary beaters revolving not less than ninety (90) to one hundred (100) times per minute, and agitate the same till it becomes an entire foamy mass, to which add sweet cream in the proportion of one-third, and continue to agitate the same till the whole becomes a foamy mass, when it will have all the general characteristics of natural butter. Then take the the butter from the churn, salt to taste, and pack it in the ordinary manner for market."

Refining oils, etc.

[Specification forming part of Letters-Patent No. 316,668, dated April 28, 1885.]

"The following is a description of my process, and to make it more comprehensible I will take for illustration cotton-seed oil after it is partially refined from crude or "red" oil into "yellow" oil by means of alkalies, as is generally practiced now, and proceed with it in a manner to make of it a white oil, perfectly adaptable to any and all purposes to which best colorless, odorless and tasteless pure oil can be adapted, and which for food is equal to the best Provence olive oil. As a first step, I put the oil to be operated upon into a wooden tank to which a mechanical stirrer is adapted. Any kind of a contrivance will answer. I generally use an upright wooden shaft revolving in the center of the tank, and furnished with horizontal paddles or blades fixed to it at an angle to resemble a propelling screw and revolved in such direction that the contents of the tank are forced by it downward at the center and raised at the periphery. Having started the stirrer and the oil — say four hundred gallons in quantity — being well set in motion, I now let into it a stream about one-half inch in diameter of well-settled and thoroughly clear lime-water until one hundred gallons of it are admitted, the whole being kept in motion continually, and at the same time I let run into this tank another half-inch stream of clear water until a like quantity of this also is admitted, and keep stirring the whole until a perfectly uniform emulsion is made of it, which usually takes place in about two hours. Of course the above proportions of clear and lime water may previously be united together and then let into the oil. The effect will be the same. The lime-water for this purpose should be made fresh every time it is needed, slaking and dissolving the lime a few hours before use by stirring thoroughly, and then letting it settle until the water is perfectly clear. About two pounds of good unslaked lime to one hundred gallons of water will make a thoroughly saturated solution and leave but little waste.

While making the above emulsion, should the contents of the tank be very thick and sluggish, as often happens when an inferior vegetable oil is operated upon, I warm it gently with a steam coil with which the operating tank should be provided, taking good care that the temperature in no case exceeds one hundred and ten degrees Fahrenheit. Well-made emulsion has the appearance of a homogeneous white mass, "and when this point is attained I then add to it (keeping the stirrer still running and shutting off the steam from the steam-coil if this is used) a solution of permanganate of potash in a stream not over one-fourth of an inch in diameter, and falling at or near the center of the tank, whereby it is immediately forced downward by the stirring-screw, and thus brought at once into a more intimate contact with the oil than it would otherwise be. This solution I prepare in a tub placed above the operating-tank by putting into it a quantity of crystallized permanganate of potash in proportion of two and a half pounds of permanganate to each one hundred gallons of oil, filling the tub with water, letting it run into the tank by means of a faucet, and replenishing this tub with water every time it empties until the whole of the permanganate is dissolved with very little, if any, stirring, in order that this solution should always be somewhat below the point of saturation. The emulsion will now look speckled, with minute bright red dots which gradually, as the permanganate deoxidizes, will disappear. The whole of this solution of permanganate being added, I let the stirring continue for from four to six hours longer, and then add to the contents of the tank, also very gradually — that is, in a stream about one-fourth inch in diameter — a solution of sulphuric acid, which solution is in proportion of one of acid to twenty of water, three pounds of acid to every one hundred gallons of oil being necessary in this case. This solution I prepare in a separate vessel lined with lead, and always taking care while making it to pour the acid into water and not the water into the acid. Hydrochloric acid instead of sulphuric acid can be used. It will now be observed that as soon as acid solution is added the mass operated upon becomes more liquid. The stirring should be continued for one or two hours longer, then stopped, and the mass left to separate. In a few hours the oil will be found completely separated and floating on the water, which water, in this case, if tested with litmus paper, should always give a slightly acid reaction to prove a complete neutralization of the lime. The water is drawn out now through a faucet at the bottom of the operating tank, and the oil transferred to a settling tank, which in winter should be kept warm by means of a steam-coil, but not to exceed ninety degrees Fahrenheit. In from three to four days the oil will be found to be perfectly white, transparent, odorless, not liable to become speedily rancid, and for food equal in every way to the best imported olive oil.

Yellow oil equal in every respect, color excepting, to the above white oil is made by diminishing the quantity of permanganate used to about one-half of the proportion given above.

In general the proportion as given already and will be given hereinafter depends on the amount of impurities contained in the material to be operated upon, and hence is variable; but only a short experience with any given hydrocarbon will at once indicate the proportion needed, and neither the excess nor the insufficiency of the agents indicated can result in any great loss, because, in case of in sufficiency, it will only cause the trouble of repeating the operation, and in case of excess the loss will be the quantity in excess of the permanganate, the lime and the acid, which in all cases is used only in proportion but slightly in excess of the quantity necessary to thoroughly neutralize "the lime and no more, while the loss of the material operated upon in either case would be hardly perceptible. Castor oil can also be refined in the above manner, and rendered perfectly white, tasteless and odorless without losing any of its medicinal properties; but it will require a longer setting and at a higher temperature, which, however, must not exceed one hundred and forty degrees Fahrenheit.

Manganate of potash instead of permanganate may also be used, since the former is converted into the latter when dissolved in water; but its use is objectionable because of the uncertainty as to the effect it will produce, and the reaction on the oil of the residuum of the former which may take place, also the difficulty of thoroughly clarifying the oil by settling, because of the finely subdivided residual peroxide of manganese. Other compounds rich in oxygen and parting with it in presence of alkaline earths — such as chromic acids, bichromate of potash, etc.— can be used instead of the manganic combinations with potash. These last, however, I prefer because of their innocuousness, while the former leave in the material operated upon substances deleterious to health, such as green oxide of chromium, etc. The use and proportion of the lime water must of course be varied according to the kind of oil used and the kind to be produced, the main object of using it being that it should in the first place combine with and free the oil of any albuminous and other nitrogenous substances it may hold in suspension; and, secondly, that it should combine with a part or whole, as the operator may desire, of stearine, margarine, etc., contained in the substances operated upon, as it will do this in preference of combining with oleine by reason of the stronger affinity to it of these solid hydrocarbons than of the more liquid ones; hence it is self-evident that if the quantity of lime added to an oil is not large enough to make a combination with the whole mass of the latter, the above-mentioned solid hydrocarbons, by reason of their stronger affinity, will be the first to enter into this combination, leaving oleine free and pure to the same extent, more or less, as the quantity of lime was proportioned to the quantity of those solid hydrocarbons; but even in the preparation of the oil first herein specified, and in which all of the stearine was purposely left, the use of lime-water cannot wholly be dispensed with, because, aside from its reaction specified already, it also assists very materially the deoxidation of the permanganate, on the thoroughness of which deoxidation, the success of the operation and the saving of time and material depend. Besides this, not using lime-water would leave in the material to be refined all of the nitrogenous matter it contained, and hence the defecation of it by the permanganate above-mentioned would last only a very short while, the whole becoming rancid even before it was well settled.

Other alkaline earths — such as baryta, lithia, etc.— can be used instead of lime, and with the same effect. I prefer lime, however, because of its cheapness and the facility of obtaining it.

To make a non-freezing oil (known in trade as "winter oil," made by artificially congealing the oil and pressing the oleine out of it while in that state), a larger quantity of lime-water should be used, and at the end of the operation its combination with oil not neutralized with acid, as specified above, but the whole mass left to settle at a tempera-

"ture of about one hundred and twenty degrees Fahrenheit. In two or three days nearly pure oleine — *i. e.,* winter oil — can be siphoned off, leaving at the bottom of the tank a compact white mass, which, when heated and the lime now neutralized with an acid (hydrochloric acid is preferable in this case) and settled while being kept warm, will free the whole of the hydrocarbons which it held in combination.

Solid hydrocarbons — such as tallow, lard, butter, etc., especially the latter two when rancid — can be thoroughly defecated by the above process, care being taken that the whole operation is carried on at a temperature at which the fats operated upon are in a state of fluidity needed absolutely to produce the desired effect speedily and cheaply. In these last cases the lime-water should also be prepared at the same temperature to insure that there should be no precipitation of lime in the operating tank by the rise of temperature, which precipitation would invariably take place if the lime-water, while being made, was at a lower temperature, and also that its neutralization at the end of the operation should be very complete, as otherwise the loss of margarine in butter would result in converting it into oil.

In case the material operated upon is extremely rancid — that is, in or almost in a state of putrid decomposition — hydrochloric acid for neutralizing the lime is preferable to sulphuric acid, the material, after settling, being thoroughly washed (by means of the same stirring and agitating in the operating-tank) with a light solution in water of crystallized pure hyposulphite of soda.

In purif ing butter by the above means the objectionable feature is that it willy be completely bleached at the same time. This defect, however, can easily be remedied by artificially coloring it again with carrotine or other innocuous substance."

Improvement in Processes for Deodorizing Oils and Fats.

[Specification forming part of Letters-Patent No. 133,322, dated November 26, 1872.]

"The nature of our invention consists in the following treatment: Cocoa-nut oil consists principally of laurine and myristine, and of a small percentage of butyrate, capronate, and caprylate of glycerine. The latter three constituents are the principal cause of the disagreeable smell of the oil and of the soap made from it.

In order to separate these latter three substances from the oil, and to obtain the pure laurine and myristine, we proceed as follows: The cocoa-nut oil is melted at a temperature of 95 to 100 degrees, and an equal volume of alcohol added, when the whole is continually and thoroughly agitated at the above temperature. The alcohol is thrown off carefully from the oil, which settles soon at the bottom. The same process is repeated twice over with fresh alcohol. The alcohol thus dissolves the offensive principle, consisting of the above-named substances, while pure laurine and myristine remain behind, which represent the purified and deodorized cocoa-nut oil.

There will be no loss in alcohol whatsoever if the alcohol is well separated from the oil and redistilled in a suitable apparatus, whereby the alcohol is obtained of the same strength and purity as when applied before. Thus the same alcohol may be reapplied over and over again, so that the cost of the process is a very trifling one.

"The alcohol, after distillation, leaves behind in the still the butyrate, capronate and caprylate of glycerine, forming about three to four per cent of the original oil. These substances may be utilized with much profit by converting them into ethers and so-called flavoring essences, which is done by saponifying the oily residuum, separating the fatty acids by sulphuric or hydrochloric acid, and treating them with alcohol saturated with hydrochloric acid gas, and subjecting it afterward to distillation. Essences of the finest flavor, such as pine-apple, bananas, œnanthic ether, etc., are obtained. These oils have great commercial value for liquors, candies, ice creams, etc. Rancid tallow, when treated in the same manner, becomes white and sweet. Spoiled butter reacquires its original mild taste and flavor.

The same method is applicable to the purification of palm-oil, cod-liver oil, and all fish oils and other fatty substances."

Improvement in Purifying and Separating Fats.

[Specifications forming part of Letters-Patent No. 187,564, dated April 8, 1873.]

"It is physiologically and chemically known that the fresh fat of animals is mainly composed of oleomargarine, stearine and membrane. In the manufacture of tallow the whole mass is heated up at a high temperature, hence its peculiar odor. I found that if the mass is heated at a temperature not surpassing 120 degrees Fahrenheit for about two hours the whole of it will melt the same as at a high temperature and remain perfectly odorless.

The nutritive property of animal fat is most entirely owing to the oleomargarine it contains, the stearine being constantly oxidized by pulmonary respiration. It appeared to me, therefore, of the greatest importance to separate from fresh fat all its nutritive parts, at the same time keeping it odorless, in order to use the same for domestic, cooking, perfumery and medical purposes, as well as the raw material for the manufacture of useful articles from it.

The first part of the operation consists in taking the fresh fat and hashing it as fine as possible in a regular meat-hasher; then to introduce it with its own weight of water in a wooden tank able to be heated by means of a steam-pipe. The whole mass is heated at a temperature varying from 100 to 120 degrees Fahrenheit, and stirring it constantly. After two hours all the stearine and oleomargarine will be separated from the membrane or scraps. The whole mass is allowed to cool. The mixture of congealed oleomargarine, stearine and membrane is separated from the water, which is thrown away, and worked thoroughly, with two per cent of common salt, between two cylinders. This has for its object to extract most of the water from the fatty mass. It is then introduced in cotton or cotton-flannel bags, which are either introduced in a hand-press, cider-press or hydraulic press, or in a hydro-extractor similar to those used in sugar refineries for the purpose of separating the syrup from the crystallized sugar. Either of the above-named operations must functionate in rooms all the time heated at a temperature of 50 to 60 degrees Fahrenheit, which is the melting point of the oleomargarine, but which has no effect on the stearine or membrane. Therefore, at this temperature and by the means of either of the above-named mechanical separations the melted "oleomargarine is separated from the solid stearine and membrane, and, after being congealed, worked again between two cylinders, with two per cent of salt to separate the last trace of water, composes a substance highly suited for domestic and cooking purposes; also, the best base for pomatums and perfumery articles, as well as salves for medicinal purposes."

Improvement in Treating Fatty Matters.

[Specification forming part of Letters-Patent No. 121,162, dated November 21, 1871.]

"The substances which I employ are mixtures of carbonates, of oxide of sodium, or of potassium, or ammonium with some earthy or alkaline chlorides, and as chlorides of sodium, potassium, calcium, magnesium, and aluminium, in solution, and put in contact by ebullition with the fatty matters which it is proposed to melt or treat. Among the various alkaline carbonates (monocarbonate, sesquicarbonate, or bicarbonate) those whose action is the most certain are the sesquicarbonates or bicarbonates of oxide of potassium, used in the proportions of about two pounds and a quarter to a thousand pounds' weight of the fatty matter to be treated, mixed with about five pounds of chlorides of potassium, or sodium, or other alkaline chloride, to a thousand pounds' weight of the fatty matters to be melted. For fatty matters already melted and which it is proposed to restore, the proportion of alkaline carbonates is to be about one pound of the alkaline carbonates to a thousand pounds weight of fatty matters, and when using earthy or alkaline chlorides the proportion must be about two pounds to a thousand pounds' weight of the fatty matters to be treated. The carbonate of oxide of potassium alone, but much better when combined with chlorides of sodium, potassium, or aluminium, are those which are the most suitable for the melting of alimentary fats, such as beef and pork fats, and for the melting of mutton fats to be used for perfumery and pharmaceutical purposes, and for the lubricating of steam and other machinery, and for the manufacture of candles. The carbonates of oxide of sodium alone, but better when combined with chloride of sodium or aluminium and with sugar, produce inoxidizable and hard products, savory, and particularly suitable for summer melting. For the melting of one thousand pounds of raw fat I use a proportion of about two and a half pounds of carbonate of oxide of sodium, about five pounds of chloride of sodium, and about eight to ten pounds of sugar, and all fatty matters thus treated are e fe t sweet and neutral.

I have obtained notable advantages by mixing, in about one-half the proportions above given, the carbonates of oxide of sodium and potassium with chlorides of sodium, potassium, or aluminium.

In conclusion, I will describe the operation of melting and treating one ton of raw fat, either beef or mutton or pork, according to my invention. Being first provided with a vat or vessel containing about one hundred and forty-two gallons of distilled water, I pour in the said water a solution of about four and three-quarters to five pounds of sesquicarbonates or bicarbonates of oxide of potassium or aluminium. The dose is to be about three and three-quarters to four pounds if the monocarbonate of oxide of potassium is used. A solution of about ten pounds of chloride of sodium or potassium, or about four "pounds of chloride of aluminium, is then to be added. When this is done the ebullition is commenced, and when it has attained its full development the ton of fat, after having previously been cut into small pieces, is thrown into the vat gradually, and the operation begins. As the coagulated albumen rises to the surface in scum it is carefully skimmed off. The ebullition must be very active during the first hour; afterward it is to be moderated, and it is left to go on from about five to six hours, when all the fat becomes separated from the membranes and cells which retain it. At this point, and after having added some fresh distilled water two or three times during the operation, the melting is ended, the steam is withdrawn, and the whole is left to rest for two or three hours before passing the melted fat into the refrigerators, from which it is afterward poured into the casks or intended packages.

If it is intended to restore the fats already melted the proportion of the above salts to be used is about half the quantity of what is used for melting raw fats, and the ebullition is to last only from about thirty to sixty minutes."

Purifyiny Fatty Oils and Fats.

[Specification forming part of Letters-Patent No. 278,187, dated May 22, 1883.]

"My invention has for its object the purification of fats and oils which, originally palatable, have acquired an obnoxious taste, odor, or color, either by long-continued storage or by an irrational production.

To this end my invention consists in the following process: I first wash the fats and oils with alcohol of at least 96 degrees Tralles, by preference in iron stirring-vessels. The washing operation may be repeated, and it is conducted most successfully at a temperature of 45 degrees centigrade. The alcohol is separated from the oil and fat by allowing the mass to subside in proper vessels. The alcohol contained in the oil or fat is distilled off, the last traces being driven off by dry steam. The alcohol which contains the oily and fatty acids and such ethereal oils as may be present is separated from these by distillation, the oily and fatty acids forming a valuable by-product. The air which enters the retorts when the same are being emptied must be dry. If, after having been treated with alcohol, as above stated, the oils and fats still show an obnoxious taste or odor, due to the presence of ethereal oils which are not, or not completely, extracted by the alcohol, I then subject the oil to heat, 110 to 175 degrees centigrade, in suitable vessels, wherein it is treated with aqueous vapors or steam heated to the same temperature, the height of temperature depending upon the boiling point of the ethereal oils contained in the fats or oils. This action of heat and steam is continued until the escaping steam no longer gives off any odors of the oils, after which the fats are filtered through bone-black. In most cases this is not required, and the oil or fat, after having been washed with alcohol, can be brought directly upon the bone-black filter, whence it runs off in a palatable and odorless condition."

Purifying and Bleaching Fats and Fatty Oils.

[Specification forming part of Letters-Patent No. 306,324, dated October 7, 1884.]

"I first fuse together in an iron pot or pan the following ingredients, in about the proportions named: Ten (10) pounds of caustic soda or its equivalent in caustic potash, and seven (7) pounds of chlorate of potash. After fusion add in small quantities at various times eight (8) pounds of the fine powder of black oxide of manganese. After the above are thoroughly incorporated I add twenty-five (25) pounds of the bichromate of potash, and while increasing the heat stir well until the fusion and firmness of the mass are complete.

To subject the fats or fatty oils to treatment, I use the following described mixture: Two (2) ounces of the foregoing chemical salts dissolved in one-half (½) of a gallon of hot water, agitate the same with seven and one-half (7½) pounds of melted fat or oil. Then add thereto six (6) fluid ounces of diluted sulphuric acid, and continue the agitation and boil with live steam until the mixture becomes thorough and complete. Then allow the mixture to settle, after which draw off the chemical water. Then wash with water, and finally wash by agitation with a light solution of carbonate of soda. When settled, draw off and wash with water and live steam. The oil or fat is then ready for use.

My product may be distinctly recognized by a chemist who is versed in the examination of oil by its freedom from mucilaginous matter and the products of decomposition of fatty matter, which I have not before succeeded in removing entirely by any other process; and it may also be identified by the absence of the chemicals used to bleach and refine it, which it has hitherto been impossible to eliminate entirely. Such bodies, of which traces have remained behind in my former products, are, first, manganic acid and the binoxide of manganese, which are the products of the reaction of the mucilaginous matter with the permanganate salt used; and, second, the green oxide of chromium, which is the product of the reaction of the mucilaginous matter with the bichromate."

Improvement in Process for Purifying and Bleaching Tallow, Lard, etc.

[Specification forming part of Letter-Patent No. 145,840, dated December 23, 1873.]

"I proceed as follows: Into a tank supplied either with a steam-jacket, or having a coil of steam-pipe placed within the same near the bottom, I place one-fourth as much water as the quantity of tallow or lard to be treated; the water having mixed with it *two to two* and *a half* per centum of strong *sulphuric acid.* I now heat the same to a temperature of about 180 degrees Fahrenheit, and into this I draw the tallow or lard from the rendering-tank, and stir the same, while its temperature is maintained at 200 degrees for half an hour. I now turn off the steam and let the whole settle for one hour or more, but do not allow the temperature to fall below 160 degrees. Near the above tank, and, I prefer, sufficiently lower, so that the contents of the first may be drawn readily into the second, I place a tank having a steam-jacket or a coil of steam-pipe placed within and near the bottom of the same, and also coil of perforated pipe placed near the "bottom and connected *with an air-pump.* Into this second tank I place again from twenty to twenty-five per centum of water having *dissolved in it from one-half to three-fourths of a pound of alum for* each one hundred pounds of tallow or lard to be purified. The tallow or lard is now drawn from the first tank into the second, so as to retain in the first most of the water, and the impurities that have settled in the same. The whole mass is now raised to a temperature of above 200 degrees, but not above 212. The air-pump is started, and air is forced through the whole mass, so as to come in contact with every particle. This is continued from thirty to forty-five minutes, according to the depth of the tallow or lard. The steam is now turned off and the air-pump stopped, and the whole mass allowed to gradually cool for one hour. I now draw off from the bottom of the tank, as near as possible, all the water and settled precipitated impurities, after which I heat up the tallow or lard to 212 degrees, and again start the air-pump, forcing air through the same, and gradually increasing the heat to 230 degrees. When this has continued, according to the quantity, from fifteen to thirty minutes, and when the foam forming on top has a white pearly appearance, the air-blast is stopped and the heat allowed to fall to 200 degrees Fahrenheit, at which temperature it is maintained for from thirty to sixty minutes to allow all impurities to settle, when the same may be drawn off into molds or packages, and the lard or tallow will be found of excellent quality, hard, pure, and of pearly whiteness. It is free from all impurities and well adapted for export, or to keep a long time without injury. It commands a ready sale at the highest prices.

Some tallow-renderers macerate the tallow in water which contains a small percentage of sulphuric acid before rendering the same. Such tallow I place at once into the second tank, and, after heating the same from 220 to 230 degrees, I start the air pump and force air through the same for from thirty to sixty minutes, until the appearance of the foam becomes white and pearly. I then allow it to become quiet by stopping the air-pump and maintaining the temperaure at about 200 degrees for about one hour, allowing all impurities to settle to the bottom, when it is ready for packing, and will be found hard, pure and sweet.

When I wish to impart an extra fine color and taste to the tallow or lard, I draw the air, before it enters the air-pump, through a furnace and over a clear coke fire, in which a small quantity of bay-salt is continually falling, so as to vaporize it; and this chloride of sodium vapor is drawn in with the air and forced through the tallow and lard, while it is maintained at a temperature of from 220 to 230 degrees for half an hour, when the same is allowed to settle as before, the temperature being maintained at about 200 Fahrenheit. A very fine, hard and white tallow or lard of a sweet, pure taste is obtained which will keep for a very long time in perfect condition."

Improvement in Compounds for Culinary Use.

[Specification forming part of Letters-Patent No. 110,626, dated January 3, 1871.]

"This invention relates to a new composition for lard, butter, or shortening, whereby a very cheap, consistent, and coagulate lard or "butter is manufactured, and one superior to ordinary shortening, answering the purpose of lard, butter, or cream for culinary and other uses or purposes.

This invention consists in the application of such ingredients to refined vegetable or fixed oils that the same is changed into a coagulate or consistent state.

My lard or shortening is composed of the following ingredients, in about the following proportions to twelve parts in weight: Beef or mutton suet (tallow), three parts; refined vegetable or fixed oils, seven parts hog's lard (stearine), two parts=twelve parts. In a suitable vessel or tank I place six parts of water, to which I add the above ingredients—suet, oil and lard. The mass is then agitated, washed, and heated for one hour by means of steam injected into the vessel or tank through pipes from an ordinary steam-boiler."

Process of Purifying Fats and Oils.

[Specification forming part of Letters-Patent No. 233,458, dated October 19, 1880.]

"My invention relates to the bleaching, refining, and purifying of tallow, lard, oil, stearic acid, and other dark animal oils and fats, and also of certain vegetable oils, in an economical manner, without the use of acids or any poisonous or deleterious chemicals.

Hitherto it has been customary to treat vegetable and animal oils either by distillation or with chemicals in order to remove the impurities and the coloring matter. These methods are tedious and expensive, and are therefore not economical for the treatment of oils and fats for various purposes and uses.

I have discovered that fuller's earth will remove the impurities from animal oils and fats and from certain vegetable oils, and that it also has an affinity for the coloring matters of these oils and fats, so that by the treatment of them with said fuller's earth they are purified and rendered practically colorless.

My invention, therefore, consists in mixing with such oils and fats reduced to a liquid condition pulverized dry fuller's earth, and then separating the earth from the oil or fat, preferably by allowing it to settle in the same vessel or removing it to another vessel to settle.

In carrying out my invention I take the oil or fat to be purified in any desired quantity, place it in any ordinary vessel suited for the purpose, and warm it until it is in a suitably liquid condition.

The degree of temperature required will, of course, vary with the different kind of oil or fat; and I do not deem it, therefore, necessary to specify any particular degree of heat, it being simply sufficient that the material to be treated should be reduced to a perfectly liquid state. When the fat or oil is in such a proper state of liquefaction I spread over its surface and mix with it a quantity of fuller's earth, or equivalent clay, which should be finely pulverized. After this has been thoroughly agitated and mixed it is allowed to settle. The fuller's earth will then be found at the bottom, and the oil or fat left perfectly free from all impurities and from coloring matter, but in other respects unchanged and ready for use in the making of soap or candles, or for any such purpose where pure, colorless oil or fat is required or desirable. The residuum, consisting of fuller's earth mixed with oil, after "the clear portion has been drawn off, may be put into boiling water, which separates the oil or fat from the earth and permits it to rise to the top, where it can be recovered. The refuse may then be thrown away or utilized in any desired manner.

The amount of fuller's earth which I have found to be necessary varies with the different kinds of fats and oils, but may be stated at from one to fifteen per cent in weight of the fat or oil to be treated thereby.

Obviously no stills or other expensive machinery are needed, the only apparatus required being an ordinary vessel of suitable capacity in which to warm the oil or fat, and, if desired, one or more settling-tanks, separate or connected.

This process is designed mainly for the treatment of animal oils and fats, such as those hereinbefore specified. I have found, however, that the treatment, though not applicable to many of the vegetable oils, may be advantageously applied to the treatment of cotton-seed oil, mustard-seed oil and cocoanut oil, and may also be applied effectively to other oils of like constitution. In no case, however, have I found necessary, nor do I contemplate, the use, with my process, of any acid or other chemical treatment, or the mixture of any other substance.

I am aware that fuller's earth has been heretofore suggested in the English provisional specification No. 3,721 of 1867, in connection with a preliminary sulphuric acid treatment for the purification of paraffine. I do not, therefore, broadly claim the use of fuller's earth in connection with other matters, or as an element in the purification of mineral oils.

I am aware that pumice and other magnesian earths, silica, and silicious earths, in connection with acid treatment or other processes, have been heretofore known in the treatment of oils, and I do not broadly claim such, my process being confined to an argillaceous non-aluminous clay, such as fuller's earth. In the treatment of mineral oils, however, I have found the simple application with fuller's earth as I use it effective for the purpose of removing impurities, and at the same time removing the coloring matter, and I confine my claim, therefore, to the treatment of animal oils and fats, and certain vegetable oils, as specified, by means of fuller's earth taken alone, the fuller's earth having for these specified fats and oils a special fitness by reason of its affinity for the coloring matter contained in them.

By means of this process I am enabled to prepare, in an inexpensive manner, the dark and cheap oils and fats, and to render them practically colorless and fit to be used in the manufacture of soaps, candles, and other articles."

Process of Refining and Hardening Lard.

[Specifications forming part of Letters-Patent No. 251,629, dated December 27, 1881.]

"In carrying out my invention I first take the lard, place it in a proper agitating-vessel, and subject it to just sufficient steam-heat to reduce it to a liquid. I then add the proportion of ten pounds of stearine, in a melted state, to each one hundred pounds of lard, thoroughly mixing the two together until the stearine becomes a component part. I next add to each one hundred pounds of lard the "proportion of two and one-half ounces of saltpetre which has first been dissolved in a small quantity of water. The compound is now agitated again and allowed to cool, when the lard is ready for the market.

The lard prepared in this manner has a firmer texture and will not liquefy from the effect of natural heat, and thereby become rancid and sour from irregular changes of temperature, which is the case when the lard has merely passed through the ordinary rendering process. This process also imparts a much whiter appearance to the lard."

Improvements in Methods of Making a Substitute for Butter from the Oils of Animal Fats.

[Specification forming part of Letters-Patent No. 166,955, dated August 24, 1875.

"Take the pure refined animal oils, such as the oils obtained from beef-suet or caul-fat after they have been prepared for this purpose. I find the oil which is made by the process described in the patent of William E. Andrew, patented August 11, 1874, and numbered 153,999, preferable, as it is pure and sweet and has no objectionable taste or odor, as is the case with kettle or water rendered oils, such as are in general use by most of the oleomargarine or artificial-butter manufacturers. This oil, as there described, is made by taking fresh animal fat, chopping or hashing it fine, placing it in bags, and subjecting it to pressure, in a suitable press, at a *temperature sufficient to start the oil. The product is then allowed to cool, when it is again placed in bags and subjected to a second pressure at a lower temperature, whereby the oil is separated from the stearine and a desirable oil is obtained. I first take the oils, after they have been carefully prepared, and add to the oils a sufficient quantity of annotto to color it to the required shade. To one hundred pounds of this oil I add one pound of Ashton salt. The oil thus prepared is then poured into a churn and the churn set in motion, the stroke of which should be regulated or graduated to from sixty-five to seventy-five strokes per minute, and the churning continued for from twenty to thirty minutes, when the oil-globules will be completely broken up and the mass thoroughly amalgamated. I then stop the churn, and as quickly as possible pour the mass from it on or amid pulverized ice or into very cold water, and stir it briskly until it is hard and firm. I then remove it to an inclined tray for the purpose of draining, and when thoroughly drained, more salt may be added, to flavor. The refrigerating process above described has for its object the rapid changing of the temperature of the oily mass, which, after having been churned as above described, and before it is refrigerated, becomes of a creamy consistency, and by thorough and quick refrigeration it is at once converted into a mass which has the qualities of natural butter, except the flavor. In this condition I find it very useful for culinary purposes, such as cooking, pastry, etc. If desirable to keep it for a length of time, it should be properly worked with a sufficient quantity of salt, as may be desired.

When it is desirable to give it the flavor of natural cream-made butter I take from fifteen to twenty quarts of thick milk or cream, put it in a churn, and agitate it until the butter begins to form. I

* Which is less than 100 degrees Fahrenheit.

"then add about one hundred pounds of the product made from the oils, and formed as before mentioned, to the cream-butter already in the churn. I then agitate the mass until it becomes thoroughly homogeneous; then take the butter from the churn, place it on trays; then add salt to taste, work and pack as practiced in ordinary butter-making. It is then ready for market.

Improvement in treating fats to separate the stearine from the oleine.

[Specification forming part of Letters-Patent No. 158,350, dated July 21, 1874.]

"The parts hereinafter mentioned are by weight.

The fat to be treated should be in the rendered condition and clear.

First step.— One hundred parts of fat are melted and heated to about one hundred and thirty-five degrees Fahrenheit, and turned into a tub (preferably one lined with lead) having the means for thoroughly stirring the contents by hand or machinery.

Second step.— Prepare a solution of two parts of sugar of lead to five of water.

Third step.— Prepare a solution of two parts of alum to five of water.

Fourth step.— Prepare a solution of two parts of bicarbonate of potash to five of water.

Fifth step.— Prepare a solution of two parts of nitrate of soda to five of water.

Sixth step.— Mix together the two solutions first named, to-wit, those of sugar of lead and alum, and allow the mixture to settle; after which draw off the clear liquid, leaving the sediment as waste.

Seventh step.— Mix the last-named liquid with the two others, to-wit, those of bicarbonate of potash and nitrate of soda.

Eighth step.— Turn this mixture of all the solutions into the fat, at the temperature named, and stir for one or two hours until a separation of the stearine and oleine takes place.

Ninth step.— The mass is then put into a tub lined with lead, and having arrangements for boiling with open steam in any of the approved modes. Enough of water is then poured in to cover the steam inlets and form a wash. From three to five parts of commercial sulphuric acid diluted with about ten parts of water are also added. The steam is then admitted until the mass boils, for about half an hour.

Tenth step.— The steam is then shut off and the contents of the tub allowed to settle. The acid-water will be at the bottom. The stearine and oleine are then drawn off and cooled. This may be hastened in any of the usual modes by distributing it into smaller vessels, or putting it into a single shallow one, or by applying ice or cooling mixtures, or otherwise.

Eleventh step.— The stearine and oleine are then pressed in the cold hydraulic press in the usual manner until the oleine is expelled.

Twelfth step.— The stearine is then again boiled with open steam with a little more of the dilute acid, when it is ready to be drawn off into candle-molds or for other use.

The result of this process is a larger percentage of stearine of better "quality and at less cost than by any methoa heretofore known. It resembles wax more nearly, both in color and quality, than any before produced.

'It will be observed that my invention consists mainly in the new materials used in separating the stearine and oleine."

Improvement in processes for purifying and bleaching tallow, lard, etc.

[Specification forming part of Letters-Patent No. 145,840, date December 23, 1873.]

"The object of my invention is to deprive tallow, lard and other fatty matter of all impurities which impair their value and appearance, and particularly of all matter liable to decomposition, and at the same time bleach the same and improve their appearance, as also their value, as an article of commerce, and better adapted for the uses for which they are intended. Tallow and lard, when rendered in the usual manner, contain a large percentage of minute, even microscopic, cellular tissues containing oleine or elaine and albumen which are liable to decomposition, particularly in the tallow and lard rendered during the summer months, imparting offensive smells to the same, while steam-rendered tallow also contains more or less glutinous matter which readily decomposes, producing offensive odors impairing the value of the article. Tallow also contains coloring matter particularly when from lean and unhealthy animals, and when such is mixed with the other tallow, as is always the case when the business is carried on on a large scale, it imparts to it a dingy and a cloudy appearance, and it has to be sold as an inferior article at a less price than prime tallow. The oleine in such tallow easily separates from the stearine in ordinary summer temperature and prevents the tallow from hardening, or if packed in a cooler season it is liable to separate and leak through the packages, causing loss by waste as well as by a reduced value, on account of its unsightly appearance.

In order to remove the said defect and produce tallow or lard free from all impurities, hard and solid at all seasons of the year, sweet in taste, without any offensive smell, and possessing a clear, white color through the whole mass, I proceed as follows: Into a tank supplied either with a steam-jacket or having a coil of steam-pipe placed within the same near the bottom, I place one-fourth as much water as the quantity of tallow or lard to be treated; the water having mixed with it two to two and a half per centum of strong sulphuric acid. I now heat the same to a temperature of about one hundred and eighty degrees Fahrenheit, and into this I draw the tallow or lard from the rendering tank and stir the same while its temperature is maintained at two hundred degrees for half an hour. I now turn off the steam and let the whole settle for an hour or more, but do not allow the temperature to fall below one hundred and sixty degrees. Near the above tank, and, I prefer, sufficiently lower, so that the contents of the first may be drawn readily into the second, I place a tank having a steam-jacket or a coil of steam-pipe p a e within and near the bottom of the same, and also a coil of perforated pipe placed near the bottom and connected with an air-pump. Into this second tank I place again from twenty to twenty-five per centum of water, having dissolved in it from one-half to three-fourths of a pound of alum for "each one hundred pounds of tallow or lard to be purified. The tallow or lard is now drawn from the first tank into the second, so as to retain in the first most of the water and the impurities that have settled in the same. The whole mass is now raised to a temperature of above two hundred degrees but not above two hundred and twelve degrees. The air-pump is started and air is forced through the whole mass so as to come in contact with every particle. This is continued from thirty to forty-five minutes according to the depth of the tallow or lard. The steam is now turned off and the air-pump stopped, and the whole mass allowed to gradually cool for an hour. I now draw off from the bottom of the tank, as near as possible, all the water and settled precipitated impurities, after which I heat up the tallow or lard to two hundred and twelve degrees, and again start the air-pump, forcing air through the same and gradually increasing the heat to two hundred and thirty degrees. When this has continued, according to the quantity, from fifteen to thirty minutes, and when the foam forming on top has a white pearly appearance, the air-blast is stopped and the heat allowed to fall to two hundred degrees Fahrenheit, at which temperature it is maintained for from thirty to sixty minutes to allow all impurities to settle, when the same may be drawn off into molds or packages and the lard or tallow will be found of excellent quality, hard, pure and of pearly whiteness. It is free from all impurities and well adapted for export or to keep a long time without injury. It commands a ready sale at the highest prices.

Some tallow renderers macerate the tallow in water which contains a small percentage of sulphuric acid before rendering the same. Such tallow I place at once into the second tank, and after heating the same from two hundred and twenty degees to two hundred and thirty degrees, I start the air-pump and force air through the same for from thirty to sixty minutes, until the appearance of the foam becomes white and pearly. I then allow it to become quiet by stopping the air-pump and maintaining the temperature at about two hundred degrees for about one hour, allowing all impurities to settle to the bottom, when it is ready for packing and will be found hard, pure and sweet.

When I wish to impart an extra fine color and taste to the tallow or lard, I draw the air, before it enters the air-pump, through a furnace and over a clear coke fire, in which a small quantity of bay-salt is continually falling, so as to vaporize it; and this chloride of sodium vapor is drawn in with the air and forced through the tallow and lard while it is maintained at a temperature of from two hundred and twenty degrees to two hundred and thirty degrees for half an hour, when the same is allowed to settle as before, the temperature being maintained at about two hundred degrees Fahrenheit. A very fine, hard and white tallow or lard of a sweet, pure taste is obtained which will keep for a very long time in perfect condition."

Improvement in Treating Animal Fats.

[Specification forming part of Letters-Patent No. 146,012, dated December 30, 1873.]

"I. *Neutralization of the ferments.*—In order to prevent the greasy substance which is settled in the tissues of the animals from taking the disagreeable taste of the fat, it is necessary that the ferments which "produce this taste shall be completely neutralized for this effect as soon as possible after the death of the animal. I plunge the raw fats called *graisses en branches* into water containing fifteen per cent of sea-salt and one per cent of sulphite of soda. I begin thus the transformation an hour at least after the immersion and twelve hours at most afterward.

II. *Crushing.*— A complete crushing is necessary in order to obtain rapid work without alteration. For this purpose, when the substance is coarsely crushed, I let it fall from the cylinders under millstones, which completely bruise all the cells.

III. *Concentrated digestion.*— The crushed fat falls into a vessel which is made of well-tinned iron or enamelled iron or baked clay. This vessel must be plunged in a water bath, of which the temperature is raised at will. When the fat has descended in this vessel, I melt it by means of an artificial digestion, so that the heat does not exceed 103 degrees Fahrenheit, and thus no taste of fat is produced. For this purpose I throw into the wash-tub containing the artificial gastric juice about two liters per hundred kilograms of greasy substance. (This gastric juice is made with the half of a stomach of a pig or sheep, well-washed, and three litres of water containing thirty grams of biphosphate of lime. After a maceration during three hours I pass the substance through a fine sieve, and I obtain the two litres which are necessary for a hundred kilograms.) I slowly raise the temperature to about 103 degrees Fahrenheit, so that the matter shall completely separate. This greasy matter must not have any taste of fat. It must, on the contrary, have the taste of molten butter. When the liquid does not present any more lumps, I throw into the said liquid one kilogram of sea-salt, reduced to powder, per hundred kilograms of greasy matter. I stir during a quarter of an hour and let it sit until obtaining perfect limpidness. This method of extraction has a considerable advantage over that which has been previously essayed. The separation is well made and the organized tissues which do deposit are not altered.

IV *Crystallization in a mass.*— In order to separate the oleomargarine of the stearine, separated crystallizers or crystallizations at unequal temperatures have been already employed. I have contrived for this purpose the following method, which produces a very perfect separation, and it is as follows: I send the molten fat in a vessel which must be sufficient for containing it. This vessel is placed in a wash-tub of strong wood, which serves as a water-bath. In this wash-tub I put water at the fixed temperature of 86 degrees Fahrenheit for the soft fats proceeding from the slaughter-houses, and 98 degrees for the harder fats, such as mutton fat. Afterward the wash-tubs are covered, and after a certain time, more or less long, according to the fats, the stearine is deposed in the form of teats at the middle of the oleomargarine liquid.

V. *Separation by centrifugal force.*—In order to avoid the numerous inconveniences of the employment of the presses which have been hitherto used I cause the mixture of stearine and oleomargarine to flow into a centrifugal machine called "hydro-extracter." The greasy liquid passes through the cloth and the stearine is collected. When all the liquid is passed I put the machine in motion, and the crystals of stearine are entirely exhausted without the auxiliary of the presses.

"However, during certain seasons there are animals which produce crystals of stearine soft enough for rendering necessary the stroke of a press as a last operation; but in this case this operation has little importance, because it is applied only to a fraction of the product. In all cases the oleomargarine is separated from the stearine when it is cold and passed to the cylinder, constitutes (especially if its yellow color has been raised) a greasy matter of very good taste, and which may replace the butter in the kitchen, where it is employed under the name of "margarine;" but, if it is desired to transform it into more perfect butter, I employ the following means:

VI. In the methods hitherto employed the margarine is transformed into cream, and this latter into butter. This complicated operation has many inconveniences. I obtain the same result by the following manner: I take ten liters of natural and fresh cream of milk. I add ten grams of bicarbonate of soda and two hundred grams of the udders of a cow, which must be fresh and well hacked in order to give all the mammary pepsine. The fresh udder may be replaced by udder collected in slices in sea-salt. After a maceration of an hour, I pass the whole through a very fine sieve. I add the necessary quantity of yellow color which is employed for the ordinary butter, and I put these ten liters into a hundred kilograms of liquid margarine at 70 degrees. I stir or mix until the combination is complete — that is to say, until the pepsine has effectuated its action. At this moment the liquid becomes thick, it takes the taste of cream, and after it has been more thoroughly agitated I let the same become completely cold. When the butter is cold and solid, coarsely scrape it in order to pass it between two large cylinders, which give it the homogeneousness and the consistence which are the qualities of the natural butter.

VII. When it is desired to produce butter intended to be preserved, which must contain no animal matter, I plunge the udders into pure water instead of cream, in order to macerate the same. Afterward the water which proceeds from this operation is mixed with the margarine at about 86 degrees of temperature — that is to say, to a degree which permits the pepsine to effectuate its action without production of cream. After an hour I let the liquid set, and the margarine, which is decanted, is mixed after it has been reduced in temperature to about 71 degrees, with an emulsion of butter made with five liters of water, five kilograms of butter, one kilogram of sea-salt, or more, according to the uses, and ten kilograms of bicarbonate of soda. When I add this emulsion with the margarine, which has already been submitted to the pepsine action, I obtain a rapid combination, and all the molecules take the qualities of the ordinary butter. It is a delicate operation, which must be exactly made.

VIII. For the long conservations I only treat the margarine by the mammary pepsine, as before described. I decant it in order to avoid any trace of water or animal matter. If it is desired to add ordinary butter, I do that at the temperature of 71 degrees by well diluting it.

IX. The stearine which has been separated from the margarine forms a hard fat, which can be bleached by the known processes in order to produce wax-candles of lower quality; but it is preferable to saponify it by any comvenient process and crystallize the greasy acids (which are charged with from seventy to eighty per cent of stearic "acid, instead of fifty) in a chamber heated to 96 degrees, so as not to let them become hard by the cooling. In this state they can be pressed under heat in order to produce stearic acid much superior to that of the trade, both by its beauty and by its point of melting."

Improvement in Rendering and Clarifying Fats.

[Specification forming part of Letters-Patent No. 172,942, dated February 1, 1876.]

"The nature of my invention relates to rendering and clarifying fats for various purposes, among which I especially name the caul or other fat of beeves, rendered and clarified for the purpose of combining the product with other ingredients in the manufacture of artificial butter.

The obiect of my invention, however, is to extract from animal fats an oil free from animal fiber, blood, animal odor or flavor, or any other deteriorating property.

My invention consists, first, in a rendering-vessel provided with a perforated bottom to carry off the liquid fat as fast as rendered, and with two or more induction-apertures in its sides for the induction of dry hot air or steam. Surrounding this rendering-vessel is a cylindrical belt, through which the hot air or steam passes, and it is provided with suitable pipes leading to the induction-apertures, and also with suitable connections to a steam-boiler or hot-air reservoir. Beneath the perforated bottom of the rendering-vessel is a receiving-chamber, made, preferably, in the form of an inverted cone, which receives the rendered fat as fast as it becomes liquid, and passes through the perforated bottom, and conveys it to a conveying-pipe, through which it passes to the clarifying-kettle, which will be hereinafter described.

Second, in a glass outlet-tube situated at the bottom of the clarifying-vessel, by means of which the impurities of the oil may be discerned and the outflow of the same cut off as soon as the clarified oil appears in the glass tube. The clarifying-vessel is a jacket-kettle with a suitable connection, through which steam may be led to the chamber between the kettle and the jacket, and, as usual, it is provided with suitable means for letting off the condensed steam. A portion of this vessel is of the form of an inverted cone, and this portion leads to the glass tube. Beneath the glass tube is a stop cock to allow of removing the impurities which are precipitated by means of my improved process, which will be described. The pure and clear oil may be drawn off through the glass tube, or other means may be employed — for instance, a stop-cock through the body of the jacket-kettle near the bottom thereof. (Not shown.) Third, in a clarifying process which consists in injecting into liquid fat taken direct from the rendering-vessel as fast as rendered, and held at a temperature of about 120 degrees Fahrenheit in the clarifying-vessel, water which is heated to, at or about 140 degrees Fahrenheit, said water having been prepared with chloride of sodium or nitrate of potash, or both, and injected under force in the form of a mist or fine spray.

The approximate infinitesimal particles of the prepared water, heated as described, gravitate through the oil at every point, and take up the portions of blood, tissue, etc., held in the liquid fat, and carry the same to the bottom, from whence they may be removed.

I have discovered, by actual experiment, that in order to prevent the rendering oil from having an animal flavor or taste, it is positively "necessary that it should be removed as fast as rendered. When the oil remains in contact with the tissue in the rendering-vessel, fermentation usually occurs, and decomposition taints the product and depreciates its market value. I have also discovered that in order to procure a pure and clear oil, free from animal odor or taste, it is necessary that the complete operation of rendering and clarifying should be as nearly simultaneous as possible, and, to save time, labor and expense, the processes should follow each other. To this end I have devised means for rendering and conveying the oil as fast as rendered to a clarifying-vessel, and, by a peculiar process separating and removing the impurities, and finally obtaining the pure oil, the whole being one continuous process for this purpose. The elaine or pure oil, if used in the manufacture of artificial butter or cheese, must be perfectly free from animal taste or odor. In such case I select the finest beeves' fat — the caul preferred. It must be used while fresh and dry. It is put in an ordinary meat-hasher, and as fast as it is hashed it is placed in the rendering-vessel and subjected to the action of dry heat or superheated steam, when it will at once begin to melt and pass through the perforated bottom to and through a suitable pipe to a clarifying-vessel, where it is clarified and removed for use.

For the production of tallow or lard, steam may be used in contact with the fat; but for elaine to be used in the manufacture of butter, or as a substitute for olive-oil, no water should be in contact with the fat until it reaches the clarifying-kettle.

It has long been a desideratum to thoroughly clarify and wash rendered fats, and various means have been resorted to for this purpose, such as agitation with water which has been treated with various chemicals; but in all such porcesses the intermingling of the water with the liquid fat has not been sufficiently complete, and a thorough washing consequently not obtained. By my improved process of using water heated to a higher degree of temperature than the oil, treating the same with substances to increase its specific gravity, and then injecting it in the form of a mist into the oil, I obtain very desirable results.

Having thus described my process, I will now explain the mechanical construction of my apparatus.

In the drawings, Figure 1 is an elevation of my invention, and Figure 2 a sectional view.

Referring to the drawings, A represents the rendering-vessel, of any desired dimensions, and it is provided with clusters of small holes or heat-induction apertures, JJ. These apertures may be of any number desired, and located at any place where the most desirable results may be obtained. The bottom of this rendering-vessel A is perforated, so that the fat, as fast as it becomes liquid, may pass through and become separated from the tissue, etc. B is a hollow belt or pipe, surrounding the rendering-vessel A, and it is provided with branch pipes corresponding to the induction-apertures J, and also with connection to the hot-air, superheated-steam or steam reservoir. Through this belt and its branches the heat passes and is introduced directly into the fat in the rendering-vessel. C is the receiver, made preferably in the form of an inverted cone, and this leads to a pipe or carrier, D, which conveys the rendered oil to the clarifier. E is the clarifying-

"vessel, having a chamber surrounding it, into which enters a steam-pipe. This vessel is in the form of a jacket-kettle, and is provided with suitable means for removing condensed steam. T is the rose or other suitable means for forming the mist or fine spray of chemically-prepared water used in my improved process. Any other suitable equivalent device may be used which will atomize the water without departing from the gist of my invention M is the lower portion of the clarifying-vessel, preferably of the form of an inverted cone, and properly attached to its smaller end is the glass tube F, through which are discerned the impurities which have been precipitated with the water and the pure oil when the water has been drawn off in the vessel or tub H through the stop-cock G.

Two o1 more clarifying-vessels may be used, and the carrier be so constructed that it may lead to one while the clarifying process is being carried on in another. The means for injecting the spray may also be arranged so as to be shifted over any one of the series.

The operation is as follows: After hashing a quantity of the caul or other fat of beeves or other animals, say one ton, more or less, is placed in the rendering-vessel A, and heat introduced through the pipes and belt B into the hashed fat. Generally the rendering apparatus is situated in the story above the clarifying apparatus. As fast as the fat is melted it gravitates through the perforated bottom, and by way of the carrier D to the clarifying-vessel E. A stop-cock in the carrier D allows the operator to control the flow of rendered oil. When a sufficient quantity of rendered oil has been conveyed to the clarifying-vessel, a further flow is prevented, and the oil held at about 120 degrees Fahrenheit by means of the steam-jacket. The chemically-prepared water, heated to about 140 degrees Fahrenheit, is then, under force, injected into the mass in a fine spray or mist, which gravitates through the oil, and collects the blood, tissue, etc., and carries the same to the bottom, from whence it is drawn off by means of the cock G, until the pure clear oil is discerned through the glass tube I. The oil is then removed, allowed to cool sufficiently, and placed in bags in a suitable press, and subjected to pressure.

The process of employing this product in the manufacture of artificial butter is no part of this invention, but forms the subject-matter of another application.

The scrap may be removed from the rendering-vessel and subjected to pressure until all the oil is eliminated, and the scrap sold as scrap-cake."

Improvement in processes for separating oleomargarine and stearine from animal fat.

[Specification forming part of Letters-Paten No. 153,999, dated August 11, 1874.)

"This invention relates to certain improvements in processes employed for separating oleomargarine and stearine from animal fat for domestic purposes, whereby a saving of time, labor and expense is effected, a better article is produced and the risk of loss by the non-separation of the oleomargarine, stearine and membrane, as well as other evil effects which result when water is used in the process, are avoided in the accomplishment of the desired object; and the invention consists in a peculiar method of subjecting the fat to heat and "pressure at different temperatures and extracting the desired products without the aid of water or chemicals as will be hereinafter described.

In carrying out my invention I take fresh animal fat and chop or hash it fine in a room at a comfortable temperature which is generally sufficient to start the oil. After being chopped, as before mentioned, the fat is placed in suitable cloths or bags. The bags are placed in series, a little separated from each other, between metal plates in a suitable press and subjected to pressure. This press should be provided with a suitable jacket by means of which the heat used in the process is confined to it and its contents. The temperature within the said jacket should be from one hundred and ten degrees to one hundred and forty degrees Fahrenheit, which is effected by the introduction of dry hot air in any convenient manner, so as to extract the oleomargarine and stearine from the membrane of the animal fat, which membrane remains in the press.

I have also discovered that if the animal fat remains heated any considerable time in a mass or body before the oleomargarine and stearine are separated from the membraneous matter the butter produced from the oleomargarine so extracted will have a disagreeable animal odor or flavor resulting from the cooking or decomposition of said membraneous or fibrous matter. I obviate this by expressing the oleomargarine and stearine from the mass as speedily as possible after it becomes sufficiently heated for that purpose, applying little heat until the mass, in separate parcels as aforesaid, has been placed in the press, and then pressing out the oily matter before the animal fiber has had time to become partially cooked or decomposed.

To facilitate the speedy uniform heating of the mass in the press I so place the separate parcels in bags or cloths, as aforesaid, that the heated air will circulate freely between or among them. The oleomargarine and stearine thus extracted are conveyed away immediately as fast as rendered, through suitable carriers or troughs, to a cool apartment where it becomes a thick mass. Thus it will be observed the oleomargarine and stearine are subjected to heat only sufficiently long to convert the same into a state approximating so nearly to liquid that it will run over an inclined surface in the heated apartment. After cooling the mass is then placed in cloths or bags and again subjected to pressure as before, except that the temperature is reduced to about eighty-five degrees Fahrenheit. This latter pressure at the reduced temperature has for its object the separation of the oleomargarine from the solid stearine which is readily effected at about eighty-five degrees, the membrane having been eliminated by the previous pressure leaving the stearine in the press, the oil passing off and thus being ready for use for any desired purpose. Thus, as is obvious, each element of the animal fat is effectually separated from the others without the aid of water or chemicals of any description, with all deleterious matter eliminated, and the oleomargarine, which is the valuable product sought, may be made into butter by the usual method of intermixing and churning with milk.

After many experiments I have discovered that a free and perfect separation of the oleomargarine and stearine from the membrane, and without animal odor or flavor, can be obtained from the fat taken before it has lost any great degree of animal heat or become tallowed "or set. Experience has demonstrated that the fat is in a better condition for allowing the separation of the oleomargarine and stearine from the membrane while in this condition, and the result is not deleteriously affected in odor or flavor. The heat of the fat should then be increased to, say, from one hundred degrees to one hundred and thirty degrees by the application of dry heat in any suitable manner as before mentioned, and immediately subjected to pressure.

It has been discovered that the application of water or steam or other moisture to the fat while in process of rendition as formerly practiced, is highly injurious to the product of oleomargarine, especially so when the oleomargarine is to be used in the manufacture of butter as it renders the butter liable to granulate and crumble, and in any event when water is used there is much trouble and expense necessary in separating the water from the mass as is obvious. The product of the first pressure, which is the oleomargarine with a large percentage of stearine, is suitable for the manufacture of butter for use in warmer climates as it will be of a harder and firmer nature and not so susceptible to the effects of heat.

Improvement in Treating Animal Fats.

[Specification forming part of Letters-Patent No. 154,251, dated August 18, 1874.]

"The invention relates to compounds used by bakers and cooks for shortening and enriching the flour or meal with which they make bread, cakes, and pastry.

Heretofore cheap butter and stearine have been employed for this purpose, and of late years several other compounds of the latter are the compounds made under patents Nos. 110,626 and 137,564. These have fixed oil as a basis, and use tallow only in small proportions to give stiffness. The resultant in these cases, however, liquifies at the low temperature of 60 to 65 degrees, and when hard is of a crumbling consistency, that does not admit of being cut without fracture. Compounds made under patents Nos. 146,012 and 121,162 have been also somewhat used, but the alkali upon which they depend for effecting their combination, imparts an alkaline taste, to which there is very general objection.

My object in the present invention is to remedy these several objections, and to offer to the public a compound of a waxy consistency, that will always admit of a clean cut there through, will not liquefy under 85 to 95 degrees Fahrenheit, and yet will be entirely without any alkaline ingredient.

In order to carry out my invention I make tallow the basis of my compound, and mix with cream, as follows: I take churned cream (with butter globules appearing) one part; molten tallow (pure) two parts. These are first thoroughly intermixed, the plastic product next suffered to granulate and harden, and the resultant finally subjected to, preferably rotary, beaters, which whip the mass until it assumes a frothy appearance. After hardening, it may be manipulated or molded into any desired form, being of a waxy consistency, and possessing not only the susceptibility of a clean cut, but the property of non-liquefaction in a temperature less than from 85 to 95 degrees.

By my process of incorporating tallow and the butter globules, the "former is enabled quickly to take up the margarine properties of the butter globules as they form, which consist of glycerine, butyric, cupraic, capsylic, and capsic acids. These impart to the tallow a sweet and agreeable buttery smell and taste, and when thoroughly incorporated tend to soften the tallow so that it can be easily cut or manipulated.

In order to produce a thoroughly satisfactory result with tallow compounds, where the main object is to utilize the stearine, the proper preparation of the tallow becomes a very essential condition to success.

I grind the suet with granulated salt, which effectually breaks the animal tissues or cells, and retains the finely-ground suet at a temperature of at least 180 degrees Fahrenheit in a water bath. This separates the animal tissue.

In order to get rid of any impurities that remain, the molten tallow is drawn into a vessel kept at a temperature of about 160 degrees, and subjected to an application of salt and albumen, in about the proportion of three pounds of fine salt and six ounces of dry-blood albumen to every 100 pounds of suet.

After a few minutes' agitation, and a subsequent rest, the undissolved salt will have gathered the more solid impurities and the albumen the lighter ones, while the former is precipitated and the latter caused to rise to the surface as a scum.

The melted tallow being now drawn off into a separate vessel, will be found almost entirely free from organic substances, and to exhibit a remarkable purity.

Improvement in processes of manufacturing products from animal fats.

[Specification forming part of Letters-Patent No. 155,816, dated October 13, 1874.]

"It is a well-known fact that all animal fats contain a valuable nourishing substance together with a valuable article for illuminating purposes.

Many attempts have heretofore been made to produce these articles at a small cost by extracting from the original fatty tissue all the above-named products it may contain.

After repeated experiments I have fully succeeded in solving the problem.

The following is the manner in which I operate: Hashed fat mixed with or without water is heated to a temperature two hundred degrees Fahrenheit, which temperature keeps fatty substances perfectly sweet, providing, however, that a mixture of two-thirds ($\frac{2}{3}$) of caustic potash and one-third ($\frac{1}{3}$) of carbonate of soda are added to said fat in the proportion of from five to ten per cent, which operates as a partial saponifier and separates all the substances which fat contains from the membrane or scrap in the short time of one hour, treating at two hundred degrees Fahrenheit, and obtain these products in a perfectly odorless and sweet state. Further, in order to thoroughly separate from products so obtained in a pure and smooth state, the oily parts from the hard parts, and then entirely prevent the grain or granulation which is so injurious to said oil when not extracted, I add to the hashed fatty substances (three-fourths of an hour after they have been treated with or without water, in addition to the alkaline mixture "above mentioned in proportion of from five to ten per cent) two per cent of extra glycerine, the result being a production of an article which, when subjected to hydraulic pressure, at a temperature of ninety degrees Fahrenheit, gives in liquid from press an entirely new and sweet compound, which never granulates, and which is pure oleopalmitine, and the residuum of the press being pure stearine.

The oleopalmitine can be used for cooking and table use, for lubricating purposes where fine oil is needed, for perfumery and for the manufacture of butter."

Improvement in Methods of Making a Substitute for Butter from the Oils of Animal Fats.

[Specification forming part of Letters-Patent No 166,955, dated August 24, 1875.]

'This invention consists of the following improvements in making butter from the oils of animal fat, whereby a great saving of time, labor, and expense is effected and better results accomplished.

Take the pure refined animal oils, such as the oils obtained from beef suet or caul fat after they have been prepared for this purpose. I find the oil which is made by the process described in the patent of William E. Andrew, patented August 11, 1874, and numbered 153,999, preferable, as it is pure and sweet and has no objectionable taste or odor, as is the case with kettle or water rendered oils, such as are in general use by most of the oleomargarine or artificial butter manufacturers. This oil, as there described, is made by taking fresh animal fat, chopping or hashing it fine, placing it in bags, and subjecting it to pressure in a suitable press at a temperature sufficient to start the oil. The product is then allowed to cool, when it is again placed in bags and subjected to a second pressure at a lower temperature, whereby the oil is separated from the stearine and a desirable oil is obtained. I first take the oils, after they have been carefully prepared, and add to the oils a sufficient quantity of anotto to color it to the required shade. To 100 pounds of this oil I add one pound of Ashton salt. The oil thus prepared is then poured into a churn and the churn set in motion the stroke of which should be regulated or graduated to from sixty-five to seventy-five strokes per minute, and the churning continued for from twenty to thirty minutes, when the oil globules will be completely broken up and the mass thoroughly amalgamated. I then stop the churn, and as quickly as possible pour the mass from it on or amid pulverized ice or into very cold water, and stir it briskly until it is hard and firm. I then remove it to an inclined tray for the purpose of draining, and when thoroughly drained more salt may be added to flavor. The refrigerating process above described has for its object the rapid changing of the temperature of the oily mass, which, after having been churned as above described, and before it is refrigerated, becomes of a creamy consistency, and by thorough and quick refrigeration it is at once converted into a mass which has the qualities of natural butter, except the flavor. In this condition I find it very useful for culinary purposes, such as cooking, pastry, etc. If desirable to keep it for a length of time, it should be properly worked with a sufficient quantity of salt, as may be desired.

When it is desirable to give it the flavor of natural cream-made butter

"I take from fifteen to twenty quarts of thick milk or cream, put it in a churn and agitate it until the butter begins to form. I then add about 100 pounds of the product made from the oils, and formed as before mentioned, to the cream butter already in the churn. I then agitate the mass until it becomes thoroughly homogeneous; then take the butter from the churn, place it on trays, then add salt to taste, work, and pa as practiced in ordinary butter-making. It is then ready for marketck

Improvement in processes for purifying and preserving animal fats.

[Specification forming part of Letters-Patent No. 169,008, dated October 19, 1875.]

"The nature of my discovery or invention consists in purifying and preserving animal fats entirely sweet and odorless and utilizing and manufacturing the same into butter.

To enable others skilled in the art to make and use my discovery or invention, I will now proceed to describe the same.

First, in ten (10) gallons of water dissolve three (3) pounds of common salt and four (4) ounces of soda ash. (The proportions of my ingredients may be increased or diminished according to the quantity of fat I desire to treat.) Then boil the admixture in a suitable vessel by hot air, or steam pipes or any other suitable means. When the ingredients are thoroughly dissolved by this process, a scum arises at the top which I then skim off carefully and add one hundred (100) pounds of animal fat cut in small pieces, keeping the mass well agitated until the whole is thoroughly melted when I again skim the mass carefully. Then draw the oil off through a filter into cold water which must be well agitated until the oil is cool enough to be removed. Second, the fat, as purified by the first step of my process, is then put into a second solution consisting of about four or five (4 or 5) gallons of water, about two (2) pounds of bicarbonate of potassa and about two (2) pounds of salt. It is then heated and kept agitated until it becomes thoroughly melted. Then draw the oil off again through a filter into cold water, keeping it well agitated until the fat becomes cold; then thoroughly remove from it the water contained therein; the fat will then assume an entirely pure, sweet and odorless condition.

It will be observed that my process consists in two separate steps — the first being to purify or remove all impurities from the fatty matter and the second to make it perfectly neutral and sweet

To utilize and manufacture the fat thus prepared into butter, I place the fat in any suitable churn with rotary beaters revolving not less than ninety (90) to one hundred (100) times per minute, and agitate the same till it becomes an entire foamy mass, to which add sweet cream in the proportion of one-third, and continue to agitate the same till the whole becomes a foamy mass, when it will have all the general characteristics of natural butter. Then take the butter from the churn salt to suit taste and pack it in the ordinary manner for market."

Process of refining fats.

[Specification forming part of Letters-Patent No. 280,822, dated July 10, 1883.]

"My invention has reference to the manufacture of a superior lard or fat; and it consists in a process of refining fat from the leaf of the "hog for culinary or other purposes where a fine article of lard is required to be used; further, in deodorizing the peculiar flavor characteristic of the fat of the hog, which peculiar flavor is largely in the tissue of the fat. In the rendering of the fat I find that the fat largely takes up the flavor of the tissue and, therefore, the process of deodorizing should take place, as far as possible, before the fats are rendered.

Hitherto various processes have been used for refining the fat of the hog, but no one has heretofore acted on the fact which I have herein stated and described — a process of deodorizing the fat before it is rendered. For this purpose I have invented an improved process of deodorizing the fat which I will now proceed to describe. I take leaf lard and finely grind it, after which I mix it thoroughly in the proportion of a bag or two bushels of salt to a ton of fat. It is then placed in tanks containing cold water where it is kept very cold for two or three days, when it is worked over with spades or other suitable implements three or four times a day. At the end of this time it is rendered in a water bath at a low temperature and as quickly as possible to remove the fat from the tissue. I am enabled also to further deodorize the pure fat by using a solution made from a half to a pound of nitrate of potassa and the same quantity of bicarbonate of soda dissolved in two gallons of water for about two tons of fat. This solution I sprinkle on the surface of the molten fat which immediately falls leaving the fat entirely odorless. The molten fat is then drawn off from the tissue into settling-basins where the fat is clarified by the use of very finely powdered salt sifted on the top. After resting a sufficient time the pure deodorized and clarified fat is drawn off and cooled as rapidly as possible. By experiment I have found that fat can be deodorized by the process which I have described even if the solution last-named is not used; but I prefer to use the full process as I have described it as I find a better result is obtained."

Improvement in Processes of Treating Fats.

[Specification forming part of Letters-Patent No. 187,327, dated February 13, 1877.]

"My invention relates to a novel process for the conversion of suet and other animal fats into an article suitable for kitchen or table use, and which will retain its agreeable and proper savor and odor in all temperatures.

I have found equal injury to accrue to the fat in the process of rendering by heating, either at too low or too high a temperature. In fact, the only heat that can be used without any risk is from 140 to 145 degrees Fahrenheit. I have also observed that both water and the cellular membranes have a bad effect on the fat while it is being heated.

The fat must be cleansed thoroughly by washing in cold water, but not dried by mechanical pressure or by heat, because necessitating a consumption of time that gives opportunity for putrefaction to set in.

My process is as follows: The crude fat having been subjected to several washings in cold water, and then heated in a jacket kettle to 140 degrees Fahrenheit, my first charge of antiseptics is applied, namely, to each thousand pounds of fat I add thirty pounds of com-

"mon salt, five pounds of saltpetre, two pounds of borax, one-fourth of a pound of boracic acid, and one-fourth of a pound of salicylic acid. These agents becoming dissolved by the watery portions, serve to prevent decomposition during the cooking process, and, combining with the membraneous matters, cause their precipitation; they also, by increasing the specific gravity of the water, enable its complete separation from the melted fat, which is thus enabled to float on top, whence it can be easily drawn off.

The chemicals employed in the above described stage of the process are necessarily sacrificed in performing their duty as precipitants, leaving the fat intact. The fat thus purified is fitted for immediate use, without further treatment of any kind; but for long preservation or shipment to distant places, it is necessary to treat it to a second charge of antiseptics, consisting of common salt, one pound; saltpetre, four ounces; borax, four ounces; boracic acid, one ounce; salicylic acid, one ounce; and benzoic acid, a quarter of an ounce. These ingredients having been well incorporated with the purified fat in a warm kettle, the tallow is ready for package and shipment to any distance or climate.

I am aware that it has been proposed to rectify crude fat by the application thereto of albumen and common salt, in conjunction with heat. In preference to albumen, which is at best a mere precipitant, I prefer to use the agents hereinbefore specified for the following reasons: So far from adding to the decomposable ngredients, they render the ingredients of the fat proper less subject to decomposition, and by superseding the necessity of excessive salting, avoid the deliquescent effects thereof. They in fact preserve intact the natural odor and savor, which the salt alone will not do; besides which, from their powerful antiseptic qualities, a very much smaller relative quantity suffices."

Treating Animal Fats.

[Specification forming part of reissued Letters-Patent No. 10,187, dated June 13, 1882.]

"This invention, which is the result of physiological investigations, consists of artificially producing the natural work which is performed by the cow when it reabsorbs its fat in order to transform the same into butter. The improved means he employed for this purpose are as follows:

I. *Neutralization of the ferments.*— In order to prevent the greasy substance which is settled in the tissue of the animals from taking the disagreeable taste of the fat, it is necessary that the ferments which produce this taste shall be completely neutralized. For this effect, as soon as possible after the death of the animal he plunged the raw fats, called "graisses en branches," into water containing fifteen per cent of sea-salt and one per cent of sulphite of soda. He began thus the transformation an hour at least after the immersion and twelve hours at most afterward.

II. *Crushing.*— A complete crushing is necessary in order to obtain rapid work without alteration. For this purpose when the substance is coarsely crushed he let it fall from the cylinders under millstones which completely bruise all the cells.

"III. *Concentrated digestion.*—The crushed fat falls into a vessel which is made of well-tinned iron, or enameled iron, or baked clay. This vessel must be plunged in a water bath of which the temperature is raised at will. When the fat has descended in the vessel he melted it by means of an artificial digestion, so that the heat does not exceed 103 degrees Fahrenheit, and thus no taste of fat is produced. For this purpose he threw into the wash-tub containing the artificial gastric juice about two litres per hundred kilograms of greasy substance. This gastric juice is made with the half of a stomach of a pig or sheep well washed and three litres of water containing thirty grams of biphosphate of lime. After a maceration during three hours he passed the substance through a fine sieve and obtained the two litres which are necessary for a hundred kilograms. He slowly raised the temperature to about 103 degrees Fahrenheit, so that the matter shall completely separate. This greasy matter must not have any taste of fat. It must, on the contrary, have the taste of molten butter. When the liquid does not present any more lumps he threw into the said liquid one kilogram of sea-salt (reduced to powder) per hundred kilograms of greasy matter. He stirred during a quarter of an hour and let it set until obtaining perfect limpidness. This method of extraction has a considerable advantage over that which has been previously essayed. The separation is well made and the organized tissues which do deposit are not altered.

IV. *Crystallization in a mass.*—In order to separate the oleomargarine of the stearine, separated crystallizers or crystallizations at unequal temperatures have been already employed. He contrived for this purpose the following method, which produces a very perfect separation, and is as follows: He rendered the molten fat in a vessel which must be sufficient for containing it. This vessel is placed in a wash-tub of strong wood, which serves as a water bath. In this wash-tub he put water at the fixed temperature of 86 degrees Fahrenheit for the soft fats proceeding from the slaughter-houses, and 98 degrees for the harder fats, such as mutton fat. Afterward the wash-tubs are covered and after a certain time, more or less long according to the fats, the stearine is deposited in the form of teats at the middle of the oleomargaric liquid.

V. *Separation by centrifugal force.*—In order to avoid the numerous inconveniences of the employment of the presses which have been hitherto used, he caused the mixture of stearine and oleomagarine to flow into a centrifugal machine called "hydro-extractor." The greasy liquid passes through the cloth and the stearine is collected. When all the liquid is passed he puts the machine in motion, and the crystals of stearine are entirely exhausted without the auxiliary of the presses. However, during certain seasons there are animals which produce crystals of stearine soft enough for rendering necessary the stroke of a press as a last operation; but in this case this operation has little importance, because it is applied only to a fraction of the product. In all cases the oleomargarine is separated from the stearine when it is cold and passed to the cylinder, constituting, especially if its yellow color has been raised, a greasy matter of very good taste, and which may replace the butter in the kitchen, where it is employed under the "name of "margarine;" but if it is desired to transform it into more perfect butter, he employed the following means:

VI. In the methods hitherto employed the margarine is transformed into cream, and this latter into butter. This complicated operation has many inconveniences. He obtained the same result by the following manner: He took ten litres of natural and fresh cream of milk. He added ten grams of bicarbonate of soda and two hundred grams of the udders of a cow, which must be fresh and well hacked in order to give all the mammary pepsin. The fresh udder may be replaced by udder collected in slices in sea-salt. After a maceration of an hour he passed the whole through a very fine sieve. He added the necessary quantity of yellow color which is employed for the ordinary butter, and he put these ten litres into a hundred kilograms of liquid margarine at 70 degrees. He stirred or mixed until the combination was complete — that is to say, until the pepsin had effectuated its action. At this moment the liquid becomes thick, it takes the taste of cream, and after it has been more thoroughly agitated he let the same become completely cold. When the butter was cold and solid he coarsely scraped it, in order to pass it between two large cylinders, which gave it the homogeneousness and the consistence which are the qualities of the natural butter.

VII. When it is desired to produce butter intended to be preserved, which must contain no animal matter, he plunged the udders into pure water, instead of cream, in order to macerate the same. Afterward the water which proceeds from this operation is mixed with the margarine at about 86 degrees of temperature — that is to say, to a degree which permits the pepsin to effectuate its action without production of cream. After an hour he let the liquid set, and the margarine which is decanted is mixed, after it has been reduced in temperature to about 71 degrees, with an emulsion of butter made with five litres of water, five kilograms of butter, one kilogram of sea-salt, or more, according to the uses, and ten kilograms of bicarbonate of soda. When he added this emulsion with the margarine, which had already been submitted to the peptic action, he obtained a rapid combination and all the molecules take the qualities of the ordinary butter. It is a delicate operation, which must be very exactly made.

VIII. For the long conservations he only treated the margarine by the mammary pepsin, as before described. He decanted it in order to avoid any trace of water or animal matter. If it is desired to add ordinary butter, he did that at the temperature of 71 degrees by well diluting it.

IX. The stearine which has been separated from the margarine forms a hard fat, which can be bleached by the known process in order to produce wax-candles of lower quality; but it is preferable to saponify it by any convenient process and crystallize the greasy acids (which are charged with from seventy to eighty per cent of stearic acid instead of fifty) in a chamber heated to 90 degrees, so as not to let them become hard by the cooling. In this state they can be pressed under heat, in order to produce stearic acid much superior to that of the trade, both by its beauty and by its point of melting."

Preparing Animal Fats for Culinary Uses.

[Specification forming part of Letters-Patent No. 262,207, dated August 8, 1882.]

"Since the issuing of letters-patent granted to me of date June 6, 1882, and numbered 258,992, for a compound to be used in the place of butter and lard for cooking purposes, I have discovered that the fat obtained from swine can be rendered, purified and deodorized in the same manner as beef-suet by mixing with the fat while being rendered slippery elm bark, as described in said letters-patent, in the manner following, viz.: Take the crude fat, trim and cleanse it in any of the well known ways of easy rendering, place the same in steam-jacket or other vessels for rendering fats, and subject the same to heat at low temperature — say from about 150 degrees to 200 Fahrenheit — and keep the same in constant agitation until the whole is sufficiently rendered. To purify and at the same time take out the offensive odor of this fat while the same is being rendered, and also to hasten the settling of the scraps, sift in about one three-hundredth part of ground or powdered slippery elm bark. The effect of the slippery elm bark thus introduced is to cause the scraps and other impurities to settle almost immediately to the bottom of the kettle and the offensive odor to disappear. The clear oil so obtained may then be drawn off to form a part of the compound. Then take cotton-seed oil or an equivalent vegetable oil purified as described substantially in my said letters-patent, to-wit: Place the oil in vessels to be heated, and while the same is being heated mix in the oil from one-seventieth to one-eightieth part of powdered or ground slippery elm bark. Then subject the same to heat from about 190 to 200 degrees Fahrenheit for about an hour and a half, causing the same to be kept in constant agitation, and let the oil cool for eight or ten hours, and then draw it off. Then take beef-stearine, which is prepared in the following manner: Render the beef-suet in suitable kettles at a low temperature until the suet becomes liquefied, and draw off the clear oil into a cooling-tank, and allow the oil to cool until it becomes granulated or thick. Take folding-bags made of cloth suitable for the purpose, and place one or two pounds of granulated oil upon the cloth unfolded, and then fold the ends of the cloth over the granulated oil so the folded bag shall be about five inches by six inches in size, and lay the bags on the bottom of the press, and thus lay the bags in series one above the other until the press is full. Then, gently at first, bring pressure to bear upon the bags, and increase the pressure until the oleine and margarine constituents of the oil are pressed out through the bags. The remaining constituent of the oil is left in a white, solid cake, and this is stearine.

To form the compound, take about sixty-eight parts of cotton-seed oil or other vegetable oils, about twenty-eight parts of the prepared swine-fat, and about five parts of beef-stearine, and place the same in vessels to be heated, and heat the mass to about 160 degrees Fahrenheit, keeping the mass in constant agitation for about a half hour. Then draw off the compound into a cooling-tank, stirring it all the while until it is cooled to about the consistency of cream. It is then ready for market packages."

Improvement in Manufacture of Butter from Whey.

[Letters-Patent No. 60,656, dated December 18, 1866.]

"After describing suitable apparatus for cooling, proceeds as follows: By this means the whey is rapidly cooled, which is of the first importance, and thus brought to the right temperature. In the process of making cheese, a brine produced from rennet is used to assist in making the curd. A compound known as annotto is dissolved in water and also put into the milk, for the purpose of coloring the cheese. These ingredients give the whey an unwholesome taste, and if the cream should be taken from it in this state and churned, the butter would be worthless and unfit for use. To destroy the deleterious effects of the rennet and annotto left in the whey, and also to cause the cream to rise rapidly, and restore it to a pure and wholesome condition, I use a solution made of saltpetre, borax and saleratus. I take a pail containing about ten quarts of pure water, in which I dissolve three ounces of borax, three ounces of saltpetre and about one ounce of saleratus. To twenty gallons of whey, or thereabouts, I add one quart of this liquid, but if the weather should be cool I use less, say one pint. This liquid is poured in as soon as the whey is placed in the cooler, and it is then thoroughly stirred. The whey is allowed to stand in the cooler, for a period of fifteen to twenty-two hours, or thereabouts, by which time the cream has been fully extracted from the whey, and is ready for skimming. In skimming the cream I use a tin skimmer, made and shaped like an ordinary tin dust-pan. The cream is churned in the ordinary way, the butter taken out and suffered to stand in cold water for ten or fifteen minutes. The water is then poured off, and the butter worked in the usual way. I used about one pound of salt to six pounds of butter. The butter is then ready for use or for packing and will be found equal to the best dairy butter manufactured directly from milk."

"By the old method the whey was placed in zinc tanks and merely kept in a cool place, and a handful of salt thrown in. This method is open to several objections. It was found that butter made in that way was little else than grease and unfit to eat. The zinc communicated a black substance to the whey, making the butter full of black streaks. In warm weather the whey was not cooled with sufficient rapidity and became sour, thus spoiling the cream. The salt did not operate to destroy the effect of the rennet and annotto, consequently the cream and the butter made from it tasted of those ingredients and was unwholesome. Whereas by my process of using a tin vat or cooler, surrounded by cold water, and applying the ingredients I have described, all these objections are obviated and good butter produced with less labor and expense."

Improvement in the Manufacture of Butter.

[Specification forming part of Letters-Patent No. 70,417, dated November 5, 1867; antedated October 29, 1867.]

"To enable others skilled in the art to make and use said invention, I will proceed to describe it.

"To one gallon of sweet milk is added eight pounds (avoirdupois) of butter, one ounce of loaf-sugar, twenty grains (troy) of nitrate of potash, one fluid ounce of liquid rennet, and ten grains (troy) of annotto. These are mixed and churned together in the same manner as cream in the common process of making butter. After the butter is separated from the milk by the process of churning, it is gathered and worked in the usual manner. The chemicals and butter added to the milk cause a speedy separation of the butter globules from it, causing it to yield all or nearly all that it contains, and producing an article of good quality and flavor. The annotto simply gives the butter a yellowish color, and having heretofore been used for a similar purpose, no claim is made to it separately considered."

Improved Process of Treating Milk to obtain Useful Products.

[Letters-Patent No. 78,640, dated June 9, 1868.]

"The object of my invention is to produce marrow from milk, and to utilize the waste in making butter, such as whey, curd, etc., in producing useful substances, viz.: vinegar, cordial, soap, etc.

I place milk in pans, as is usually done, to raise cream. This cream is churned until a substance is produced, which, when properly washed and salted, becomes butter. Instead of churning the cream, I can extract the substance from the same by a process not necessary to describe in this specification, but for which I am about to apply for letters-patent. I have discovered that this substance consists, in greater part, of marrow, such as is found in the bones of animals, and which bone-marrow is used for making pomatum, ointments, and in a great many different ways, but which is comparatively very costly, but a small quantity being obtained from any animal.

I take this substance, and, instead of washing and salting it, place it in a vessel, with a sufficient quantity of water, heat the same to such a degree that the water shall thoroughly permeate all the fatty substance, and extract from the same all milky and other impure articles which remain in the same after churning. I then allow the mass to cool, when the pure marrow will be found on top, while the water and all impure particles remain at the bottom. This process of purification ought to be gone through with at least twice. The marrow is then again subjected to heat without adding any water, so that any water still remaining in the marrow will be evaporated. After cooling the marrow, it is well to beat or work it for a short time, as it thereby becomes finer in grain.

In taking the pure marrow from the vessel, it is unavoidable to leave some on the surface of the water, which is skimmed off the same, but, not being pure enough to be used for ointment, etc., makes an excellent grease for soap; and this is a very important part of my invention. The soap may be made by any of the well-known methods.

I will now describe another part of my invention.

I allow the milk, from which the cream is taken for the above process, to sour. The curd is then separated from the whey. The curd may be used for making cheese, for feeding, or any other purposes to which it can be applied.

By mixing the whey with a proper amount of brown sugar, and allowing it to ferment, an excellent cordial, or, at the pleasure of the "operator, a good strong vinegar, is produced. To make the cordial, a greater quantity of sugar is required than for making vinegar. In the first case, I usually take one pound of sugar to six quarts of the whey, and for the vinegar, one pound of sugar to sixteen quarts of whey; but I desire it to be understood distinctly that I do not confine myself to any given quantities, as climate, temperature, etc., may necessitate different proportions.

The whey obtained in cheese-factories from the sweet milk, and which now is thrown away, can be made use of, in the above-described manner, to make vinegar, cordial, etc., thus changing the waste into a source of great profit."

Improvement in Manufacturing Butter from Whey.

[Specification forming part of Letters-Patent No. 79,078, dated June 23, 1868.]

"This invention relates to a new and improved mode of manufacturing butter from whey; and the invention consists in, first, settling the whey in clear and clean vessels or vats, where, having allowed it to remain for twenty-four (24) hours, skim or take off the cream; second, then add to the cream, for ten (10) pounds of butter, one (1) ounce of saltpetre, first dissolved and filtered, and, having mixed it thoroughly with the cream, churn immediately.

The saltpetre dissolves the membranous coverings of the atoms of cream, and also acts as a preservative.

Butter having thus been produced, wash it in cold water, aud then work it thoroughly; after which add one (1) pound of salt and one-quarter (¼) pound of white sugar, ground fine, and again work it throughly, when it is fit for use or for being packed.

The solution of nitre is mixed with the cream before churning, in order to purify the cream from the cheesy taste and quality derived from the use of rennet in the milk."

Improvement in Compounds for Culinary use.

[Specification forming part of Letters-Patent No. 110,626, dated January 3, 1871.]

"This invention relates to a new composition for lard, butter or shortening, whereby a very cheap, consistent and coagulate lard or butter is manufactured, and one superior to ordinary shortening, answering the purpose of lard, butter or cream for culinary and other uses or purposes.

This invention consists in the application of such ingredients to refined vegetable or fixed oils that the same is changed into a coagulate or consistent state.

My lard or shortening is composed of the following ingedients, in about the following proportions to twelve parts in weight: Beef or mutton suet (tallow), three parts; refined vegetable or fixed oils, seven parts; hog's lard (stearine), two parts = twelve parts. In a suitaole vessel or tank I place six parts of water, to which I add the above ingredients — suet, oil and lard. The mass is then agitated, washed and heated for one hour by means of steam injected into the vessel or tank through pipes from an ordinary steam-boiler."

Improvement in Shortening for Culinary uses.

[Specification forming part of Letters-Patent No. 120,026, dated October 17, 1871; antedated October 3, 1871.]

"Heretofore shortening in culinary use has been confined to hog's lard and butter, while cotton-seed oil was not used for such purpose by reason of its offensive taste and smell, thus rejecting as useless this abundant and rich oil for shortening.

The object of my invention is to deodorize and render palatable cotton-seed oil for culinary use; and I do hereby declare that the following is a full and exact description thereof.

Into any suitable vessel for heating I place cotton-seed oil and water for the purpose of washing, purifying, or deodorizing the oil. For each gallon of oil I use one ounce chlorate of potash and nitre. I then introduce a jet of steam by means of a pipe leading from an ordinary steam boiler and heat and agitate for three hours. The steam is then removed and the oil and water allowed to separate. The oil is then drawn into another vessel and heated to a proper degree, or to 400 degrees Fahrenheit. While thus heated I infuse the same with oxygen. An economical plan is to heat the peroxide of manganese strongly in an iron retort in a reverberatory furnace. One pound of good oxide of manganese will yield seven gallons of oxygen with some carbonic acid. This last is removed by passing the gas through a wash-bottle containg a solution of potash, which absorbs carbonic acid. The oxygen is then passed into the oil through a suitable pipe. The heat and oxygen deodorize and oxygenate the oil, making it sweet and palatable for culinary use.

I do not intend to confine myself to this process alone, as other methods of producing oxygen may be used.

My improved shortening is put up in cans or other suitable packages in an unctious or oleaginous state. I also prepare it to the consistency of common lard by adding to the oil equal parts, weight, suet (tallow). My shortening is also found to be a pepastic and alterative when used in diet."

Improvement in Butter-coloring Compounds.

[Specification forming part of Letters-Patent No. 142,891, dated September 16, 1873.]

"The compound is composed of the following ingredients, in about the following proportions, viz.: Annottoine, five ounces; curcuma, pulverized, six ounces; saffron, one ounce; lard oil, one pint; butter, five pounds.

The butter is first melted and strained through a cloth, and the saffron is made into half a pint tincture of saffron. The tincture of saffron, butter and curcuma are then placed in a boiler and allowed to boil for about fifteen minutes, after which it is all strained through a cloth. This strained compound is then put back into the boiler and the annottoine and lard oil added, when the whole is allowed to boil for about fifteen minutes. Afterward it is strained again through a cloth and stirred until cool.

The amount of coloring thus obtained will be sufficient to color about six thousand pounds of butter.

Improvement in making Butterine.

[Specification forming part of Letters-Patent No. 148,767, dated March 17, 1874.]

"This invention is a process for producing from milk a new article of food, the same being a compound formed of curd or cheese and butter, and which may be used as a substitute for either; and it consists in subjecting the milk, in connection with a base of prepared butter, to certain chemical action and mechanical agitation, by means of which the curd is precipitated, and the oily globules burst and gathered to form butter.

In connection with the process, certain essential apparatus is employed, which, in connection with the foregoing, will be fully described hereinafter. A general statement of my invention may be made as follows:

A suitable quantity of a base of butter, prepared as will be hereinafter described in detail, is properly placed within a churn of peculiar construction with the yelks of eggs. To this a suitable quantity of fresh milk is added, and, by proper instrumentalities, the mass of milk and butter is subjected to galvanic action, and also thoroughly agitated. By this means a compound is formed resembling fresh butter in appearance and taste, which is composed of the curd or cheese formed from the caseine of the milk and butter gathered from its oily globules.

The details of the invention will now be described, with the best method known to me of practicing the same.

The butter employed as the base is prepared in the following manner: Take, first, an earthenware cylinder or bottomless dish, and provide it with a zinc bottom, and also with a similar cover, which latter should be perforated to allow the escape of steam or gas. The zinc plates forming the bottom and cover are first prepared by coating them with saltpetre, which operation may be performed by placing a proper quantity of the substance in a tea-kettle, and directing upon them a steam jet from the spout. When thus prepared, place in the dish a proper quantity of good sweet butter, packing the same loosely, or with a space at the bottom. Then place the dish in a flat bottomed kettle or pan containing a solution of salt in water, blood warm in temperature, in sufficient quantity to make the depth about a quarter of an inch when the kettle is in place. A steam jet from a kettle spout, or some suitable boiler, should then be applied to the outside of the dish at one point of its circumference, and an air-blast from a bellows, or other air-blowing mechanism, at the same time at an opposite point, the dish being at the same time revolved; this operation being continued until, by means of the application of heat and air and the galvanic action arising from the zinc and saltpetre, a certain oil is extracted, which would otherwise interfere with the successful practice of my invention. In treating a pound of butter, about one ounce is extracted, and about five minutes are required to perform the operation. The residuum of this process constitutes the base, which I employ in "connection with the milk. In connection with this base of prepared butter, I employ the yelks of eggs, for the purpose of obtaining a settling action in the milk, and a consequent accumulation of the butter in a mass, when the churning process is being performed. When one pound of butter is used, two eggs will be the proper proportion. The yelks of these should be carefully removed from the white portions, and be well beaten in a half teaspoonful of salt, which is employed for the purpose of cutting them thoroughly. Then strain through a linen cloth with luke-warm water, in equal parts by measure, or in sufficient quantity to cause the yelks to strain readily, so that all stringy substances may be removed. Then mix the egg-yelk with the base of prepared butter; or first apply the butter to the interior of the churn, and then afterward apply the eggs. This base is then placed in the churn, which will now be described.

Its main body portion is made of ordinary tin, but its bottom is composed of a concave zinc plate, as shown in Fig. 1. It is made tapering in form, inclining inward from the bottom upward, and its lower edge forms a perforated flange, as shown in Fig. 1. It is provided near its neck with a flange or rim, having a lip extending downward, so that a chamber is formed around the churn to retain the steam and heat from the water in the heater; and it has, also, suitable cover and a proper convex dasher, having spiral or inclined openings, as shown. This base of prepared butter is placed in this churn, mainly at the bottom, but the side also may be coated by rubbing with the hand. The churn, when thus prepared, is allowed to stand thirty minutes to permit the butter and egg to become thoroughly incorporated.

In the meantime, or previous to coating the churn, the heater is prepared as follows: This heater consists of a cylindrical vessel made entirely of zinc. Previous to using it it is coated with saltpetre, in a manner similar to that before described; then take a little more than a fluid gallon of water, in which has been placed a tablespoonful of saltpetre, and reduce the same by boiling to a gallon; then pour the same into the heater add a teaspoonful of salt, and stir the whole well. When the temperature of this water is about 110 degrees Fahrenheit, the milk (one gallon of which should be used with one pound of butter) may be poured into the churn, which already contains the prepared butter and egg-yelk, and the churn then be placed in the heater containing the hot solution of saltpetre. The dasher of the churn should then be immediately operated, in the usual well-known manner, and continued in operation for about the space of sixty seconds, more or less, according to the indications, after which the churn should be removed and set into the cooler which consists simply of a cylindrical vessel made of zinc and coated, as before described, and contains cold water having a spoonful of salt to the gallon. The dasher being still operated for about the space of thirty seconds, the churn may be removed, when it will be found that the milk has entirely disappeared, and a compound has been produced, resembling, in appearance and taste, fresh butter, which may be formed and molded in glass molds.

That portion of the compound which is formed from the mik con-

"sists mainly of curd or cheese and butter. After the compound has once been made, a portion of it may be employed, if desired, to form the base for a new batch.

If the operation is properly performed, as before described, the results will be as set forth, the theory of the operation being, it is believed, as follows: The galvanic action arising from the saltpetre battery decomposes a portion of the caseine, and consequently, lactic acid is formed from the sugar of milk, by means of which the curd is precipitated, the mechanical agitation at the same time bursting the oily globules, and uniting the compound in one homogeneous mass.

It will be understood that, while this is a compound composed of several elements — butter, cheese, and egg-yelk — it is in no sense an adulteration, all the elements being pure and healthful. It is evident that it can be produced at a small cost, and that it can be used as a substitute for butter.

It should be understood that in producing this invention much depends upon immediate action when the right temperature of the liquid is obtained in the heater.

It will be understood that this substance does not resemble fresh butter so closely as to deceive persons accustomed to examine the same — a marked difference, in fact, upon inspection, distinguishing the one from the other."

Improvement in Butter-Coloring Compounds.

[Specification forming part of Letters-Patent No. 163,610, dated May 25, 1875.]

"I put into a suitable can, which is suspended in a caldron of water, fifty pounds of lard, fresh butter or olive-oil, to serve as a body for the coloring matters and the antiseptic which I employ. I then take three pounds of annotto, and add water to it, and stir until it is about the consistency of thin paste, when I add one-half of a pound of *curcuma* (turmeric), and stir it well with the annotto. When thoroughly mixed, I heat the lard, butter, or oil, whichever one of these substances it is desired to use, to about 110 degrees Fahrenheit, and take about five pounds of it and stir thoroughly with the annotto and *curcuma*, prepared as stated, and put the whole mass into the can above referred to, and stir the substances well together, adding at the same time five pounds of common salt and three ounces of saltpetre. The substances are then brought to a boiling heat, and stirred from time to time until the coloring matter is dark enough for use. It takes ordinarily from twelve to twenty-four hours to properly cook the coloring matter, after which it is poured into cans which can be well stopped, and to every ten pounds five fluid ounces of bromochloralum are added, when it is shaken until cold."

Improvement in Processes for Making Artificial Butter.

[Specification forming part of Letters-Patent No. 173,591, dated February 15, 1876.]

"To obtain the oleine and margarine from animal fats, I take any given quantity of animal fat, and cause the same to be cut or minced by any suitable machine for the purpose, after which I place the fat "in a vessel, and subject it to a heat that will cause the whole mass to become fused. I then place the liquid fat in other vessels, and allow it to remain and cool until the degree of Fahrenheit is from 80 to 90 degrees, and when in this state I place a portion in bags of cloth. These bags are then placed in a press and subjected to a high pressure, which causes the oleine and margarine to free itself from the stearine and fibrous tissue.

The above shall be known and designated as oleine and margarine No. 1.

The process above described will give the same results as described by Chevreaul in Brande's Work of Chemistry, page 482, and published in the year 1829.

To obtain the vegetable oleine and margarine, I use any one of the following articles of commerce, viz.: oil peanut, oil sweet almonds and oil olives.

To produce the lactic acid, I take fourteen parts cane-sugar, sixty parts water, four parts caseine, and five parts chalk. This mixture is kept at a temperature of 80 to 95 degrees Fahrenheit for eight or ten days, or until it becomes a crystalline paste of lactate lime. This is pressed in a cloth, dissolved in hot water and filtered. This solution is then concentrated by evaporation. The acid is obtained from this lactate by treating it with the equivalent quantity of sulphuric acid and filtering from the insoluble gypsum. The solution of lactic acid I make as follows: one dram of lactic acid and sixteen ounces water.

The solution of lactic acid assists digestion, it prevents the product from becoming deteriorated before use, and it assists also in giving the product a butyraceous consistency. By the use of lactic acid all putrefactive and catalytic action is arrested, which action would take place if such acid were not added, and by this means there is prepared an article which is fit for use at any time, and which will preserve its original state and flavor.

To obtain the loppered cream or milk, I take the cream as obtained from the surface of milk, or milk as obtained from the cow, and place it in open vessels, and allow it to remain until the putrefactive and catalytic action has taken place. When in this state it will be ready for use.

To obtain the margarine from the oleine and margarine No. 1, I take a given quantity of oleine and margarine designated as No. 1, and place it in a water bath, and subject it to heat until the same assumes a semi-liquid state; then I place it in bags of cloth. These bags are then placed in a press and subjected to a high pressure, when the oleine will free itself from the margarine.

The above shall be known and designated as oleine No. 2.

I find that the composition of butter made from cows' milk, according to Robin, Verdeil, and described by Dalton in his work of Human Physiology, page 320, and published in the year 1867, is as follows: In one hundred parts — margarine, sixty-eight parts; oleine, thirty parts; butyrine, two parts; and it being a fact that butter made from cows' milk does not contain the same proportion of oleine and margarine in summer as in winter, it having a larger proportion of oleine in winter, and having described the several articles that I use, and to enable others skilled in the art to make and use my discovery or inven-

"tion, I will now give my improved process for making butter for winter and summer use.

To manufacture butter for table use in the winter by my improved process, I take and place in a suitable vessel nine parts of oleine No. 2, one part of fruit or nut oil, one part of solution lactic acid, one part of loppered cream or milk, as hereinbefore described. I then cause the same to be rapidly agitated with a revolving skeleton beater until the whole assumes the consistence of butter made of cream taken from cows' milk, after which I add coloring matter, and then remove the mass upon a table or other suitable receptacle, and then work it until all the fluid portions are expressed. I then add salt to suit the taste, when it will be ready for use.

To manufacture the above for summer use, I take and place in a suitable vessel nine parts of oleine and margarine No. 1, one part of margarine as obtained from No. 1, one part of solution lactic acid, one part of loppered cream or milk, as hereinbefore described, and proceed as I do in making the same for winter use."

Preservation of butter.

[Specification forming part of Letters-Patent No. 226,467, dated April 13, 1880.]

"In carrying out my invention I take of glacial metaphosphoric acid in solution, say, twenty-four grains of acid for every pound of butter. This solution I thoroughly mix, blend and incorporate with the butter by any suitable means. The preservative process is then complete and the butter thus prepared may be placed in any suitable vessel for domestic or commercial purposes.

Instead of applying the glacial metaphosphoric acid in solution, I may apply it in a solid crushed state.

The strength of the solution will vary according to circumstances and requirement — say from three drams to one ounce to the ounce of water.

I would observe that it is preferable that the butter should be treated with the preservative agent as soon after it is taken from the churn as possible, and that the butter be thoroughly freed from buttermilk; also that the flavor will be improved by the addition of a small quantity of salt.

Having now described the nature of my invention and in what manner the same is or may be performed, I would have it distinctly understood that I do not confine myself to the relative proportions hereinbefore given, as such may be varied to suit various kinds of butter, the length of time for which it is desired to preserve the butter, the atmospheric conditions under which it is prepared and other causes. The proportion of the preservative agent will not, however, in any case exceed one dram to one pound of butter.

I am aware that phosphates and phosphites have been long known as antiseptics, and also that metaphosphoric acid in solution has been proposed as a preservative for fish, vegetables and fruits, and for hardening fats by being melted therewith. Such I do not claim as my invention, nor would such means or methods be applicable to the preservation of butter.

The mechanical admixture of the metaphosphoric acid with the butter as carried out in practicing my invention effects the preserva-

"tion of the butter by bringing the reagent in contact with the caseine and other substances which would otherwise putrefy."

Substitute for butter.

[Specification forming part of Letters-Patent No. 236,483, dated January 11, 1881.]

"I first separate the oleine and margarine from the stearine by any known method—for example, by mincing and melting the fat, and then pressing it in bags of open texture. I next place the oleomargarine thus obtained with an alkaline solution, preferably in the following proportions: To eighty pounds of oleomargarine twenty pounds of water and eight ounces of bicarbonate of soda. I next agitate the oleomargarine and the alkaline solution together until the oil globules of the former are thoroughly mixed with the alkaline solution and partly saponified by the action of said alkali. I then add to the oleomargarine thus partly saponified a small quantity of butyric acid, preferably in the proportion of one dram to every hundred pounds. This gives to the article such a fine flavor that even an expert can scarcely distinguish it from excellent dairy butter. Of course the proportion of butyric acid thus added may be varied to suit the requirements of each particular article or the tastes of certain classes of purchasers.

This process, as above described, avoids the use of milk and consequently the presence of caseine. When this latter element is present in any considerable quantities a lactic fermentation often occurs, continuing until the caseine is all converted into butyric acid. Now, a large amount of said acid will make the compound rancid, although a small amount will flavor without injuring it; hence, by dispensing with caseine and substituting therefor a slight flavor of butyric acid, I have greatly improved the article produced."

Process of making artificial butter.

[Specification forming part of Letters-Patent No. 266,568, dated October 24, 1882.]

"I take pure leaf lard that has parted with all its animal heat and divide it minutely in a grinding or hashing apparatus. It is next heated sufficiently to liquefy the lard proper, and is then passed through a series of strainers thereby separating the lard from the tissues intermingled with it in the leaf. It is then treated with coloring-matter such as is commonly used in the making of dairy butter. When it has received a satisfactory color it is poured into tanks containing a strong cold brine, sufficient in quantity to immerse the lard, which is solidified almost immediately by the cold water. In this brine the lard remains for thirty-six hours, and under the operation of the brine parts with a large proportion of those elements that impart to it the characteristic flavor which makes it unacceptable to sensitive palates. After remaining in the brine thirty-six hours it is taken out and placed on tables or shelves of ash which are kept scrupulously clean. Here it is covered with salt and remains in this condition another thirty-six hours, the salt absorbing all that remainder of odorizing matter which was not separated by the brine, and leaving the lard free from peculiar taste or smell. The clarified lard is then heated again in suitable vessels sufficiently to liquefy it, its temperature "being raised to about one hundred and thirty degrees Fahrenheit, and is then thoroughly mixed by stirring with about an equal quantity of lukewarm pure buttermilk and about one-sixth of its weight of clarified tallow, a minute quantity of pepsin (about one ounce to a hundred pounds of the lard) being also added, and having the effect to separate any remainder of nitrogenous matter and leave the fats pure. The mixture having been completely effected, the liquid is left to stand long enough for the buttermilk and foreign substances to precipitate, the temperature being meanwhile maintained at a height sufficient to keep the lard liquid. The buttermilk having completely settled, the mixture of lard and tallow is dipped off and poured into a vessel containing a quantity of pure dairy butter about half as much, by weight, as the lard. This butter has been previously reduced by a chopping apparatus to comparatively fine particles, being warmed just enough to make this reduction practicable. With this comminuted butter the liquid compound of lard and tallow is thoroughly stirred, so as to make of the whole a semi-fluid mass of about the consistency of gruel. When the mixture is complete it is poured into a vessel containing a sufficient quantity of cold water to immerse it. Here it is thoroughly worked with the hands till it becomes of a uniform consistency like that of soft butter, this result being promoted by the already-mentioned admixture of tallow with the lard.

The mass is then put into a butter working apparatus, where it is salted and the water and any remains of buttermilk worked out of it in the usual way. The result is a comparatively inexpensive substance having all the nutritive properties of butter, and being at the same time perfectly sightly and palatable, and necessarily free from any admixture of deleterious matter, nothing having been put in at any point of the process but familiar food substances, except a perfectly neutral coloring, such as is an ingredient of nearly all dairy butter, and pepsin, which is a substance native to the stomach."

Compound to be Used in the Place of Butter for Cooking Purposes.

[Specification forming part of Letters-Patent No. 258,992, dated June 6, 1882.]

"I take what is known as 'caul' and 'kidney' beef suets and trim the same carefully, and then pass the same through a cutter or such process as will soften and prepare the suets for easy rendering. I then place the same in steam-jacket kettles or other vessels to be heated and subject the mass to heat of about 150 degrees Fahrenheit, and keep the same in constant agitation until the whole is sufficiently rendered. To purify and give a flavor to the oil, and also to hasten the settling of the scraps, I introduce during the process of rendering about one three-hundredth part of ground or powdered slippery-elm bark. After the scraps are settled I then draw off the clear oil thus prepared for use in the compound. I also take refined cotton-seed oil or any equivalent vegetable oil and place the same in steam-jacket kettles or other vessels to be heated, and while the same is being heated I mix into the oil about one-seventieth or one-eightieth part of slippery-elm bark powdered or ground. This serves both to purify the oil and to improve its flavor. I then subject the same to heat of about 190 or 200 degrees Fahrenheit for about one and a half hour, and during the pro-

"cess I cause the mass to be constantly agitated. I then remove the steam from the kettle and leave the oil to cool for about eight or ten hours, and then draw off the oil ready for use in the compound. I then take about sixty-eight parts of the prepared cotton-seed oil and twenty-eight parts of the prepared beef suet oil and add thereto about five parts of beef-stearine, and place the mass in steam-jacket or other vessels and heat the same to about 160 degrees Fahrenheit, constantly agitating or stirring the same for about half an hour. It is then prepared to be drawn off into a cooling-tank, and should be kept in constant agitation all the while until it is cooled to about the consistency of cream. The compound may then be drawn off into packages suitable for the market. The compound thus treated will harden to about the consistency of lard.

The compound thus formed is found to be superior to and more economical than butter or lard for all cooking purposes, and this method of purifying and settling the scraps of suets and purifying the oils by means of slippery-elm is new and useful. Slippery-elm as applied also at the same time corrects the disagreeable odors arising from the oils and imparts an agreeable and pleasant odor when the compound is used."

Process in Making Artificial Butter.

[Specification forming part of Letters-Patent No. 263,199, dated August 22, 1882.]

"My invention relates to the manufacture of artificial or oleomargarine butter; and it consists in first purifying what is commercially known as 'leaf-lard,' usually put up in kegs, firkins and casks, which purification is accomplished by first washing the leaf-lard, then cutting and mincing the same in a suitable machine, and then placing the cut and minced lard in a vessel capable of being heated, so as to melt or fuse the lard, which is then strained by being forced through a fine sieve, whereby it is substantially freed of all fiber. The lard which has passed through the sieve is then subjected to the action of cold water to which has been previously added and thoroughly stirred a quantity of borax and nitric acid, about in the proportions hereinafter specified. By treating the lard in this solution, composed of water, borax and nitric acid, the effect is to further cleanse the lard and make it partake of or assume a clear white color free of all odor and almost perfectly tasteless. After being subjected to this treatment the mass is removed and thoroughly rewashed in cold water, preferably in a separate and distinct vessel from that previously employed, whereby the product becomes a purified or deodorized leaf-lard, its characteristics being that it is of a beautiful color — a clear white — perfectly odorless, remarkably solid, and free from the disagreeable taste usually present with lard. Arriving at this stage of the process, a certain minute quantity of nitric acid is added to the water and incorporated with a certain quantity of the purified or deodorized lard to further strengthen the solution, and this mode of treatment and addition of nitric acid are continued as mass after mass of the purified or deodorized lard is prepared, the operation being continued until the product assumes a clear white color, void of odor and taste. The product thus obtained is mixed with oleomargarine, which is now a "commercial article and readily obtained in the market, and when all is thoroughly mixed the mass is subjected to heat—say to about 95 degrees Fahrenheit. After having been so mixed and heated it is ready for the churn, where milk and sugar are added, and after the churning operation it is conveyed to a refrigerated receptacle, where it is instantly solidified, after which it is salted and subjected to a rolling or mixing action for the purpose of impregnating it with a desired quantity of salt, which renders it ready for the market, capable of being made into rolls, blocks, or any desired shape.

In practicing my invention I purchase in open market fresh leaf-lard, and after having thoroughly washed it cause it to be cut up and minced in a suitable machine. The washed and minced lard is then placed into a vessel and subjected to a heat that will cause the mass to become thoroughly fused. The vessels are then prepared, which may be designated as 'Nos. 1, 2 and 3.' In Nos. 1 and 2 there is placed about sixty gallons of ice-water, and in the first named (No. 1) there should be mixed about three ounces pulverized borax and three ounces nitric acid. The washed and melted lard is then run through a fine sieve into the vessel designated 'No. 3,' for the purpose of separating the fiber. Of the lard thus washed there is then taken, say, about one hundred and fifty pounds at a time, which is thoroughly washed in the vessel designated 'No. 1,' which causes it to solidify, from whence it is taken and placed in vessel designated 'No. 2,' where it is rewashed and thoroughly cleansed. When each one hundred and fifty pounds has passed through this process there is added, in addition to the cold water, one ounce nitric acid to maintain the strength of the solution, as the first-treated quantity deprives the solution of a portion of its strength, and this addition of nitric acid is continued until all the lard prepared for the purpose has passed through the process. Should the water become milky-white, the ice-water vessels should be emptied and the solution made over again. The result obtained is a beautiful clear, white, odorless and tasteless product, which I call 'deodorized lard.'

My process of manufacturing oleomargarine butter and its various treatments in various stages are about as follows: Oleomargarine oil—a commercial article—is procured, and about one thousand pounds placed in a tank, to which is added from five to fifty per cent of the purified or deordorized lard, according to the state of the climate, and all is then subjected to a heat of about 95 degrees Fahrenheit. When sufficiently heated it is run into or conveyed to a suitable churn, where milk or cream, or both, are added, together with dissolved sugar and sufficient coloring-matter familiar to dairymen or those skilled in such arts. After the churning operation, which should last about thirty minutes, it is run off into a box containing ice or ice-water, or other suitable refrigerant, which causes the entire mass to solidify almost instantly, and when reduced to this solid condition it is conveyed to tables prepared for the purpose, whereon it is salted to suit the taste, where it is allowed to remain from twelve to twenty-four hours. It is then passed through rollers for the purpose of uniformly mixing the salt. The article is then ready to be packed in any desirable shape.

The result of the above process is the production of a uniform article in every respect resembling the finest quality of butter, which "will not become rancid or crumble or break down in warm or cold weather.

I am aware that crude fat has been treated at a temperature of 140 to 145 degrees Fahrenheit in contact with common salt, saltpetre, borax, and boracic and salicylic acids, withdrawing the separated fat and incorporating therewith a second and smaller charge of the above chemicals, with the addition of boracic acid; and I am also aware that fat has been separated from the stearine and from the oleine by subjecting it successively to the action of solutions of sugar of lead, alum, bicarbonate of potash, and nitrate of soda, and to mechanical pressure; and I am also aware that it is not new to purify and bleach tallow, lard and other fatty matter by subjecting the same in succession to the action of sulphuric acid, alum, and atmospheric air, whether the air be or be not charged with vapor of salt."

Manufacture of Artificial Butter.

[Specification forming part of Letters-Patent No. 263,042, dated August 22, 1882.]

"I have discovered that by the use of the more infusible part of vegetable oils extracted from said vegetable oils by subjecting them to pressure at a temperature sufficient to allow the more fusible portions to run off, leaving the more infusible portions of the vegetable oils, composed mainly of vegetable stearine with some margarine, in the bags. It is this more infusible substance that I mix with the animal oleomargarine obtained by what is called the 'Mége process,' or analogous processes.

I mix my materials above described with animal oleomargarine before the emulsionizing process with the milk, etc., takes place — in the Mége process — by melting it and mixing it into the animal oleomargarine, also in the melted condition. I add from about ten to forty per cent of this material. A larger proportion may be used in winter than in summer. The advantages of this material are that it unites thoroughly with the animal oleomargarine and imparts to the whole compound a more butter-like texture than can be otherwise obtained.

The vegetable stearine to be used can be obtained from any pure vegetable, seed, or nut oils by pressing them at a temperature as above set forth, or it may be obtained in the market at times as vegetable stearine. I prefer that obtained from pressing cotton-seed oil, benne-oil, or mustard-seed oil. The vegetable oil may be purified or refined before pressure, or what I call the 'vegetable stearine,' may be purified after being extracted from the vegetable oil."

Artificial Cream.

[Specification forming part of Letters-Patent No. 264,516, dated September 19, 1882.]

"In carrying out our invention we take olive, lard, oleomargarine, or butter oils, or other animal or vegetable oils, of suitable quality, and with each pound of oil incorporate from two to three pounds of milk. In daily practice we prefer to take one part of oil and two parts of sweet skim-milk. The oil and milk are preferably heated in separate "vessels to from 112 degrees to 150 degrees Fahrenheit. When the proper temperature has been attained a stream is drawn from each vessel to a suitable emulsion or disintegrating-machine in which the mechanical incorporation or admixture of the oil and milk is effected. The machine which we have successfully employed for this purpose is described in Letters-Patent No. 238,091, granted to Wm. Cooley, February 22, 1881; but any other machine adapted to effect a thorough disintegration and admixture of the milk and oil, so as to cause them to blend with each other in the formation of an artificial cream may be used, if desired. The heated milk and oil should be introduced to the machine simultaneously in suitable proportions, and for this purpose the stream of milk drawn from its separate vessel is preferably of about twice the volume of oil drawn from its vessel.

By the mechanical admixture or mingling of the oil and milk, every globule of oil is coated with the caseine contained in the milk, and as caseine is the heaviest constituent of the milk, the specific gravity of the oil thus inclosed therein is practically increased, so that the emulsion or artificial cream thus produced may be put into a large mass of milk without liability of the oil separating and rising to the surface, as would ordinarily be the result from a simple mixture of oil and milk. The oil globules, each separately inclosed in its sack of caseine, thus remain thoroughly incorporated in the artificial cream, of which they form a part, so that the entire mass of milk to which such artificial cream is added may be treated with rennet, coagulated in the usual time without haste, and cheese made in the same manner as when made from milk by ordinary methods.

This artificial cream differs from all other emulsions on account of its contained oil being so finely and evenly divided and so uniformly distributed throughout the milk in such fine particles or infinitestimal atoms that when the compound or 'cream' is put into a thin fluid even it will not separate or rise to the surface quickly, as would be the case with ordinary emulsions. It is obvious that if the atoms or particles of oleaginous material were larger, and not so evenly distributed, their buoyancy would be sufficient to cause them to separate and rise to the surface quickly. By reason of this minute subdivision of an oleaginous material, and its thorough blending with the caseine of the milk, which coats each globule of oil, we are therefore enabled to produce an artificial cream that can be treated and used in every respect the same as natural cream.

It will therefore be seen that the artificial cream can be readily manufactured at one factory and then taken to cheese-factories in other places, not too remote, where it may be added to milk and made into cheese without any separation of the oil.

We have also made butter from this artificial cream by making the cream one day, then setting it aside for twenty-four hours, or until it has become acid, the same as cream is ordinarily treated and then churning the artificial cream for butter.

We have also taken the artificial cream and put it into a quantity of new milk, and then let the mixture 'set' twenty-four to thirty-six hours, the real cream and the artificial cream coming to the surface of the milk together. We have then taken off all the cream thus pro-

"duced and treated it exactly as if it had all been real or natural cream —that is, we have kept it twenty-four hours, until it became acid, and then churned it, obtaining real fine butter that was very difficult to detect from the finest creamery products.

We are aware that cheese has been made from skim-milk and oil mixed together; but not by first producing an artificial cream. Practical success has not attended the putting of oil directly into the entire mass of milk and then trying to confine the oil by the action of something to coagulate the milk quickly, thereby attempting to catch the oil and hold it, so to speak, in the mass of milk. The results of such attempts show that much of the oil is released and cannot be worked into the cheese.

We do not claim the manufacture of cheese from an admixture of milk and fat with rennet added, as that is covered by the patent to H. O. Freeman, No. 136,051, dated February 18, 1873. Neither do we claim butter made instantly from an admixture of milk and oil, as described in the patents of H. Mége, No. 146,012, reissued September 24, 1878, No. 8,424, and G. Cosine, No. 173,591, dated February 15, 1876. Nor do we claim the addition of an oleaginous material — such as cream, melted butter or lard — to a curd prepared from buttermilk, as described in the patent to Wm. Cooley, No. 241,788, dated May 24, 1881. We do believe, however, that we have invented an important and valuable substitute for cream as an article that can be used and treated as real cream; and, further, that by the use of such artificial cream a great deal finer quality of cheese can be made with skim-milk than has heretofore been obtained; and we also claim that by putting this artificial cream into new milk and allowing both the real cream and the artificial to rise to the surface together, setting from twelve to thirty-six hours, and then treating the entire amount of cream so obtained as is usual in making butter, we get a product made up from one-half to three-fourths oil and the balance real butter, which product it is difficult to distinguish from the finest butter made entirely from pure cream; and we can also produce a good, fair article of artificial butter much better than real butter that is so often strong and frowy by making an artificial cream of one part oil and two or three parts milk, treating this cream the same as real cream — that is, keeping it until it is slightly acid and then churning it in any of the ordinary churns employed to make real butter; or the artificial cream may be churned immediately, if desired.

It will be understood that we do not limit ourselves to making an artificial cream from fat or oil and skim-milk, as an excellent article may be made from new milk and animal or vegetable oil; and we have also made the artificial cream from buttermilk and oil; but in this case the product is apt to have the flavor of the buttermilk."

Manufacture of Artificial Butter or Oleomargarine.

[Specification forming part of Letters-Patent No. 265,833, dated October 10, 1882.]

"It is my object to avoid the use of vegetable oils and the like commonly used for the purpose of preventing oleomargarine butter from becoming crumbly at low temperatures, and also to obtain a larger "yield from milk or cream of the creamy substance which should be mixed with oleomargarine oil to produce a good quality of artificial butter. The agent with which I treat both the milk and the oil is sal-soda deprived of its water, and the manner in which I proceed is as follows: I take ordinary sal-soda which I grind or pulverize into a fine powder, then spread out in a room, the temperature of which should be kept at about one hundred degrees Fahrenheit. At the end of about twenty-four hours the sal-soda will become white and the water contained in it will have evaporated.

Preparation of the milk.— Of this prepared sal-soda I take about six ounces to a can of milk containing about ten gallons. The milk containing this proportion of the sal-soda should be kept for about twelve hours in a room of a temperature of sixty-two degrees Fahrenheit. The action of the prepared sal-soda will run the milk to about seventy degrees Fahrenheit. At the end of the twelve hours the milk will appear slightly thickened and will have a slightly salty taste. I then churn this product about five minutes, until it becomes foamy or cream-like, taking care not to carry the operation so far as to allow any butter globules to appear. By this process of treating the milk I obtain a larger percentage out of the milk, and the creamy substance is perfectly sweet. The residue whey or buttermilk is drawn off slowly, leaving the creamy substance entirely free therefrom.

Treatment of the oil.— I take of oleomargarine oil about two hundred pounds in a cold, liquid state. To this I add eight ounces of the prepared sal-soda, and agitate them thoroughly until the mass becomes mushy and of a whitish cream color. The soda is by this time thoroughly intermingled with the oil. I then melt the oil so prepared until it becomes a liquid. It will be perfectly clear and no sediment will be found remaining. I then run the oil thus prepared into the churn containing the prepared creamy substance above described and churn for about thirty minutes. I then add coloring, churn five minutes longer, then remove the mass from the churn, place it on tables and salt. The product will be perfectly sweet and of the consistency of natural butter, and also having the same grain without that greasy and heavy appearance usually to be seen in oleomargarine.

My product will be light in weight and will yield a larger volume in bulk, owing to the action of the sal-soda, and also to the addition of the milk as prepared which would otherwise be lost in whey or buttermilk.

Oleomargarine so prepared will keep sweet and retain the salt longer than ordinary artificial butter. Besides this the granular formation of the oil is broken, and hence it avoids the crumbling which is a great detriment to oleomargarine in winter, and enables me to dispense with use of any substances — such as vegetable oils, etc. — commonly used to prevent its crumbling."

Compound for Culinary Use.

[Specification forming part of Letters-Patent No. 264,545, dated September 19, 1882.]

"In a suitable tank or vessel, preferably a steam-jacket kettle, I place eight parts of water and twelve parts of stearine obtained from animal fats or vegetable oils, and subject the same to a boiling heat "for about one hour. While the mass is boiling I place therein, to each one hundred pounds of stearine, five pounds of fine salt and one-half a pound of powdered orris root. This boiling of the stearine in water serves to cleanse it of all impurities, while the orris root deodorizes it and imparts an agreeable flavor thereto. After boiling the mass the specified time the heat is withdrawn, and it is then allowed to stand for about half an hour, when the stearine is drawn off into a suitable tank or vessel, to be afterward used in the compound. I next place in a suitable tank or vessel, preferably a steam-jacket kettle, refined cotton-seed oil or an equivalent vegetable oil, and subject the same to a heat of 150 degrees to 190 degrees Fahrenheit, keeping it constantly agitated, and during this heating process I stir into the mass and thoroughly incorporate therewith powdered orris-root, in the proportion of one pound of orris-root to each one hundred pounds of oil. This heating process, which serves to deodorize and flavor the oil, is continued for about one hour, after which the heat is withdrawn and the oil allowed to stand until it is thoroughly settled and cooled. The oil thus treated is then drawn off into a suitable tank or vessel, preferably a steam-jacket kettle, and united or mixed with the aforesaid purified and deodorized stearine, in the proportion of seventy-five parts of oil to twenty-five parts of stearine, and the mass is then subjected to a heat of about 150 degrees Fahrenheit and kept constantly agitated for thirty to sixty minutes until the ingredients are thoroughly united or commingled. The compound is then drawn off into smaller tanks or churns, and churned for about half an hour at a temperature of 90 degrees to 120 degrees Fahrenheit, and it is then drawn off into cans, pails, tubs, tierces, or other packages, and is then ready for the market.

I have found that orris-root has a peculiar fitness, and is especially adapted for the purpose of deodorizing and flavoring the compound, as it possesses a much more delicate and agreeable flavor and odor and is more pleasant to the taste than any other substance heretofore used for flavoring compounds of this description, and consequently my improved compound is thereby rendered more appetizing and its market value proportionately enhanced."

Artificial Butter and Method of Making the Same.

[Specification forming part of Letters-Patent No. 266,580, dated October 24, 1882.]

"The ingredients of my improved compound consist of hog's lard, fresh beef-suet, cream butter and glycerine, together with a due proportion of salt and water, and suitable coloring matter. These ingredients enter into the compound in substantially the following proportions, viz. : Lard, fifty to sixty per cent ; butter, thirty to thirty-five per cent ; beef suet, five to ten per cent ; glycerine, one to two per cent ; salt and water together, five to ten per cent, and annotto or other coloring matter in due quantity to give the desired complexion to the product.

Coloring Compound for Artificial Butter.

[Specification forming part of Letters-Patent No. 266,417, dated October 24, 1882.]

"To about four hundred pounds of oleomargarine oil, when mixed with genuine butter and sour milk, or cream, sweet or sour, in the usual proportions, I add, after the usual churning, four pounds of the following mixture, after the same has been thoroughly mixed — that is to say: Fifty parts of sugar, finely powdered; thirty parts of glycerine, the purer the better. If the glycerine be not pure, then must take proportionately more. After the powdered sugar and the glycerine have been thoroughly mixed, add thereto twenty parts of annotto color, and thoroughly mix it with the previous mixture of sugar and glycerine until the mass forms a homogeneous mass or paste-like body. The above proportions are proportions by weight. This mixture is put into the churn containing the previously churned oil and milk or cream, and also at the same time add about three pounds of oil of ben, and the churning is then continued about ten or fifteen minutes longer. The product is then treated as usual.

The results of using my coloring compound are that, owing to the annotto having been thoroughly mixed with the dissolved sugar and glycerine, it has become completely mingled with them, coloring their entire mass. Therefore, when put into the churn, there being such a large body of coloring matter, every portion of the contents of the churn is more completely brought into contact with the coloring matter; also, the sugar adds a more perfect butter-like taste to the product, and the glycerine and oil of ben remove the tendency which these artificial butter-like bodies have to become flaky, brittle or granular, so that when spread with the knife it is hard and of a tallowy texture, instead of smooth like genuine butter."

Substitute for Lard.

[Specification forming part of Letters-Patent No. 266,778, dated October 31, 1882.]

"A new and useful article of manufacture, which we have named 'Oleard,' of which the following is a clear, full and exact description.

We take, say, about seventy parts of cotton-seed, or any suitable vegetable oil, prepared by emulsifying with about ten parts of wheaten or other wholesome farinaceous flour, the latter being first thoroughly cooked with about sixteen parts of water and four parts of salt. If any watery substance should remain after emulsion, it must be separated by freezing, as in an ordinary ice-cooler or in any convenient manner, and the oleard is then canned and ready for the market, and is a cheap and delicious substitute for lard."

Substitute for Butter.

[Specification forming part of Letters-Patent No. 266,777, dated October 31, 1882.]

"Take of crude or refined cotton-seed oil, or any other suitable vegetable oil (preferably the former), say, about twenty gallons, and heat it to 90 degrees Fahrenheit (more or less) by means of steam coils, and then add slowly about one gallon of caustic soda at about 40 "degrees barometer, while the whole is being violently agitated. With specially heavy oils it is not necessary sometimes to add more than half a gallon of the caustic soda to bring the oil to a light straw color. The mixture is then allowed to settle and the supernatant oil is separated by decantation. About seventy-five parts of the oil thus prepared is then emulsified with about twenty parts of corn starch, or any other wholesome farinaceous flour; but we prefer the corn starch, the same having first been thoroughly cooked in salt and water of the strength of five parts of salt to twenty parts of water. Thorough incorporation can be accomplished in a large mortar with heavy pestle, worked by any well known power. The coloring material should be added before working in the mortar. After having been thus prepared, the whole mass should be transferred to a suitable churn and churned in the usual manner. Then place it in a suitable vessel and work the "oil milk" thoroughly out of the mass and add the flavoring, preferably using six drops of butyric ether to the gallon, and the mass becomes what we term "butteroid" — a most excellent and wholesome vegetable substitute for butter."

Manufacture of Artificial Butter.

[Specification forming part of Letters-Patent No. 267,687, dated November 14, 1882.]

"My invention consists in the improved process of combining, under peculiar conditions hereinafter described, certain old ingredients — to-wit, sweet cream, oleomargarine or oil derived from tallow, an oil derived from lard or hog fat, and an oil derived from butter, all of which oils are rendered without cooking from the stock used at the low temperature of about ninety degrees, with the aid of pressure as hereinafter described, and which are mixed with one or more vegetable oils — such as the oil of sesame or benne, or oil of sunflower seed, or cotton seed, together with salt, ice and coloring matter — as, for instance, annotto or annottoine, as used in coloring butter. The proportions of these ingredients will be determined by the temperature of the season in which the article is made as hereinafter set forth.

Before describing the mixing of the ingredients, however, and their treatment after mixing, I will proceed to specify the manner of obtaining the ingredients or certain of them.

To obtain the sweet cream I take fresh milk and allow it to stand until the animal heat is extracted, and then place it in a deep narrow vessel surrounded by water kept at a temperature of about forty degrees or forty-five degrees Fahrenheit, and after leaving it in this vessel for a period of from six to ten hours, more or less, I draw away the milk while still sweet and leave the cream also sweet.

To obtain oil of tallow I take the soft part — or what is known as the "caul" — from rough tallow or intestinal fat of the cow or beef and hash it very fine in a sausage-hasher or other suitable machine, and placing it in small quantities — of about five pounds — in cloths, subject it to heavy pressure in a room heated to about ninety degrees Fahrenheit, when the oil will flow without having resort to cooking.

I obtain the oil of lard by taking leaf lard or the intestinal fat of hogs and treat it in the same way as I do the tallow.

My object in rendering these oils by hashing and pressure is to avoid "cooking and the subjecting of them to any greater heat than milk is subjected to in its elaboration within the cow.

The oil of butter I obtain by subjecting butter to pressure in the same way that I do the oil of tallow, with the exception that I do not hash the material.

The oil of lard is used to soften the lard of tallow while the oil of butter, together with the vegetable oils mentioned, is used to give texture and a butter like appearance to the compound. I use annotto or annottoine to color the product and salt to give it flavor."

Process of Treating Butter.

[Specification forming part of Letters-Patent No. 327,636, dated October 6, 1885.]

"To each gallon of milk used I add certain ingredients in about the proportions named, as follows: One gallon of milk, one teaspoonful of white wine rennet, one teaspoonful of sugar, one teaspoonful of salt, one-fourth teaspoonful bicarbonate of soda, five grains of bicarbonate potassium, ten grains of alum, four pounds of good butter. These ingredients, in about the proportions herein stated, are placed in a churn of any usual or desired construction and agitated in the usual manner, and the butter will be produced in much less time than usual, and all the solid matter withdrawn from the fluid leaving only a thin water as a residue."

Improvement in Making Butter.

[Specification forming part of Letters-Patent No. 68,639, dated September 10, 1867.]

"I take one pint of warm new milk. If not warm from the cow warm it to about that degree. This measure of milk will weigh about one pound. To this are added, while warm, the yolks of two eggs which is then thoroughly beaten together until fully and evenly mixed. To this is now added one pound of good common butter. If the butter is hard soften it to the consistency of that just churned; put the whole into a churn, if the quantity being made is large, or if a small mass, as now under consideration, it may be beaten up in a bowl until the materials have become combined and solid which will take from ten to fifteen minutes. At this time all the milk will have wholly disappeared and become incorporated with the butter, and which will now, in appearance, be like new made butter, possessing the same unadhesive character, so that it will come from the churn freely, leaving nothing behind as a residual product. The butter can now be salted and worked up into rolls or pats in the ordinary way."

Process of Manufacturing Oleomargarine Butter.

[Specification forming part of Letters-Patent No. 264,714, dated September 19, 1881.]

"The feature that characterizes my invention, so far as concerns the preparation of the compound of milk or cream and oleomargarine, is that I preliminarily prepare the soured milk or cream by only half churning it, by which I mean churning until the liquid thickens, but stopping the churning operation before butter comes, then separating the whey or thin liquid from the thicker part, and then mixing the latter substance only with the oil. This substance, of course, is not "curds, nor is it either sour milk or butter. It is a creamy substance resembling the beaten white of an egg, but of somewhat thicker consistency, and is so light that it floats upon the oil. I find that this substance amalgamates with the oil during the churning operation and seemingly becomes an integral part of it, for when the resulting product is heated the two do not separate, nor does the product have a spotted appearance. It also imparts a flavor of natural butter to the product, and does not become rancid.

That my invention may be better understood, I will proceed to describe more in detail the manner in which I now prefer to practice it:

I take the milk or cream, sour it, and after souring it cool it with ice. This lowers its temperature and causes a distinct separation of the fatty portion from the residue or whey. I then place the milk or cream in a churn and churn for about five minutes, until it gets thick, but is nevertheless free from butter particles or globules. It is essential that the churning operation should cease before reaching the latter point. The whey or thin liquid is then drawn off, leaving a thick product — neither sour milk nor butter — having the characteristics above set forth. The oleomargarine, suitably prepared and heated — usually to about 95 degrees Fahrenheit — is then run into the churn in proper quantity, the churn remaining at rest until the oil is all in. The milk or cream product rises and floats on the surface of the oil. I then churn for about forty minutes, and then, after adding flavoring and coloring matter, if desired and working it into the product, remove the latter from the churn, after which I salt, work and pack it in the usual way.

I have found that the quality of the butter-like product can be still further improved by preliminarily treating the oleomargarine in the following way: I place the oil in a suitable tank or vessel, and there, by means of suitable stirring or agitating mechanism, stir and work it until it appears of a fine consistency, like butter, smooth and uniform. I prefer to conduct this operation in the jacketed tank in which the oil is subsequently heated; but during the operation of agitating the oil the tank should be without heat and the oleomargarine should be cold, and if taken into the tank as it comes from the press the stirring operation should continue until it cools and obtains the consistence above referred to.

The agitator in the tank may be of any suitable construction, and should run at a comparatively slow rate of speed — say from twenty to twenty-five revolutions per minute. The agitator which I in practice employ consists of a rotating wooden cage or cylinder surrounding a shaft from which project paddles, the cylinder and shaft revolving in opposite directions. I obtain in this way a more perfect admixture and amalgamation of the component parts of the oil and break up the globular formations which are always observable when ordinary oleomargarine becomes mushy or hard. Oil after being thus treated will not when subjected to cold become as hard as the same oil previous to agitation, and when taken in the mouth it, like butter, dissolves slowly and leaves no globules. The untreated oil, on the contrary, under like conditions, breaks up in the mouth into globules of fat. The oil, when thus treated, becomes opaque — usually of a light yellow color — and perceptibly swells or increases in bulk and resembles butter of the same consistency in smoothness and uniformity."

We now come to the all important aspect of the subject — is artificial butter a wholesome article of food? We answer it in the negative on the following grounds:

First. On account of its indigestibility.

Second. On account of its insolubility when made from animal fats.

Third. On account of its liability to carry germs of disease into the human system.

Fourth. On account of the probability of its containing, when made under certain patents, unhealthy ingredients.

Before entering upon the argument, we wish to state that we have investigated the claim made by the "oleo" makers that the "weight of the testimony of the medical profession was in favor of its being healthy." This, no doubt, was true a few years ago, but we have made it a point of inquiry for nearly two years past, and find that this opinion of the physicians was based, not as a general thing upon investigation, but upon the sanction given to the stuff by such eminent chemists as Profs. C. F. Chandler, R. Ogden Doremus, etc. The opinion was also based upon Mège's product, which must be admitted to be less deleterious to health than most if not all the others. Then, too, these spurious articles were sold so surreptitiously, until those whose personal interests were incidentally affected stirred up the Legislature to investigate, that but little or no attention was given to the subject, and consequently but little known about it. But now, since attention has been so forcibly called to it by the agitation of the Dairy Commissioner in his endeavor to execute the laws prohibiting its manufacture and sale, no difficulty will attend the finding of plenty of eminent physicians who will declare that it may be a very unhealthy article of food. We wish also to state here that the physiology, like the chemistry, of fats, until recently has been studied as a whole, and consequently but little was known of their individual properties.

We read in "Wanklyn's Milk Analysis," published in 1874, that "with regard to the question of admixture of foreign fats with milk-fat we are unable, in the present condition of our knowledge to deal with that part of the problem." We now have no less than four reliable chemical methods for distinguishing butter-fat from other fats. Experimental physiologists are now entering this unexplored field and important discoveries may be confidently expected.

In order to give an appreciable understanding of the indigestibility of artificial butter we must briefly describe the digestive processes. The great variety of foods taken by man is derived from the mineral and organic kingdoms. From the mineral comes water, salts, etc., with which we have no concern at present. The organic foods are the products of living organized bodies and divided into two great classes, viz.: First, Protein principles, also called albuminoid and nitrogenized principles. These are chemically composed of oxygen, hydrogen, carbon and nitrogen. The latter element chemically distinguishes them from the second class, the hydro-carbons, which are composed of oxygen, hydrogen and carbon. In sugar, starch and some other substances belonging to this class the oxygen and hydrogen exist in proper proportions to form water which has given rise to a subdivision of the hydro-carbons into hydro-carbons and carbo-hydrates. Fats and oils belong to the hydro-carbons.

When food is taken into the mouth, its presence stimulates through the nervous arrangement, the salivary glands to produce a copious flow of saliva, which during mastication is (or ought to be) thoroughly mixed with the food. Aside from a slight conversion of starch into sugar, the act of mastication is purely mechanical — the food is broken up, lubricated and gathered into proper form to be swallowed. The temperature in the mouth is 100 degrees Fahrenheit, and of course, any free fat whose melting point is at or below this temperature, will liquefy. The chemical reaction of saliva is alkaline.

When the food reaches the stomach, its presence, as in the mouth, acts as a stimulus and causes an increased secretion, which had already begun when the food was taken into the mouth, of the acid fluid called gastric juice.

The muscular construction of the stomach keeps the food in constant motion so that it is thoroughly mixed with the gastric juice. When the bolus of food mixed with saliva comes in contact with the acid gastric juice the conversion of starch into sugar ceases; the proteids are broken up and dissolved; the proteid cell-walls of the adipose tissue are dissolved, which sets the fat-drops free; and *the free fats which liquefy at or below* 100 *degrees, or perhaps* 101 *degrees Fahrenheit, which is the highest temperature in the stomach, are melted and to some extent, emulsified and split up into fatty-acid and glyceryl.* The acidity of the gastric juice is essential to its activity.

As the food is dissolved or digested (it is now called chyme), it is mostly carried into the intestines by the muscular action of the stomach where it is met by three other digestive fluids; the bile, pancreatic juice and intestinal juice, which are all alkaline in reaction. When the chyme leaves the stomach it is, under normal conditions, acid; but as it is mixed with these alkaline fluids its acidity is neutralized and its reaction becomes alkaline.

In the intestine the conversion of starch into sugar takes place with great rapidity and the proteids or peptones, as they are called after being acted upon by the pepsine of the gastric juice, are still further broken up. The pancreatic juice, so far as is known, is the chief agent in bringing about these changes. The bile does nothing more than to aid in neutralizing the acidity and thus prepare them for the action of the pancreatic juice. But with fat it becomes an important factor. Its salts unite with any free fatty acid and form soaps. It also dissolves soaps which, as we shall see hereafter, materially aid the pancreatic fluid in its action upon fats. Bile also has some emulsifying power on fats. A soap is a fat acid united with a base, as soda, potash, etc.

The pancreatic juice has a powerful emulsifying effect upon fats; that is, divides them into very minute particles. It also has the power, to some extent, of breaking them up into their fatty acid and glycerine; and if an alkali is present the fatty acid unites with it to form soap.

As we have already stated, bile has a slight emulsionizing and solvent effect upon fat, but the fact which is known to be the most important in its relation to the digestion of fat is, that it unites with the free fatty acids which are present in the chyme and forms soaps. It also dissolves soaps that may have been formed before reaching it;

and the presence of soluble soaps are known to aid the emulsion of fats.

Foster* says in reference to this: "Thus a rancid fat, *i. e.*, a fat containing a certain amout of free fatty acid, forms an emulsion with an alkaline fluid more readily than does a neutral fat. A drop of rancid oil let fall on the surface of an alkaline fluid, such as a solution of sodium carbonate of suitable strength, rapidly forms a broad ring of emulsion and that even without the least agitation. As saponification takes place at the junction of the oil and alkaline fluid, currents are set up, by which globules of oil are detached from the main drop and driven out in a centrifugal direction. The intensity of the currents and the consequent amount of emulsion depend on the concentration of the alkaline medium and on the solubility of the soaps which are formed; hence some fats, such as cod-liver oil, are much more easily emulsionized in this way than others. Now, the bile and pancreatic juice supply just such conditions as the above for emulsionizing fats; they both together afford an alkaline medium. The pancreatic juice gives rise to an adequate amount of free fatty acid, and the bile in addition brings into solution the soaps as they are formed. So that we may speak of the emulsion of fats in the small intestine as being carried on by the bile and pancreatic juice acting in conjunction, and as a matter of fact the bile and pancreatic juice do largely emulsify the contents of the small intestine, so that the grayish turbid chyme is changed into a creamy looking fluid, which has been sometimes called chyle."

Now, we believe that butter fat is especially fitted to supply these conditions. Butter, as is well known, readily becomes rancid and no doubt butter contains some free acid very shortly after being made, but we will consider a perfectly fresh specimen. According to Lang, the first step in the decomposition of butter is a conversion of lactic acid into butyric. The second is the breaking up of butyrine into butyric acid and glycerine, the butyrine furnishing by far the most free acid, about seven per cent.† Thus we see that the first fat in the mixture of butter to break up outside of the body is butyrine, and doubtless this is the case inside.

J. Bell,‡ asserts that when a solution of alcohol and an alkali is used in insufficient quantity to saponify all the butter fat treated, the alkaline base unites with the soluble fatty acids and what is left undecomposed are the fats containing the insoluble fatty acids. He also illustrates this by relating an actual experiment. This strongly corroborates the supposition that it is the butyrine that is first broken up in the stomach and intestines.

We have seen in the process of stomach digestion that some fat was emulsionized and broken up into its acid and glycerine constituents. So we have butyric acid set free in the stomach to unite with a base from some of the weaker salts, as the carbonates, for instance, to form a very soluble soap which is dissolved by the bile as soon as it comes in contact with it, and thus furnishing, even a fresh butter, the most favorable conditions for starting the action of the pancreatic juice

---

*Foster's Physiology, by Reichert, 1885, p. 357.
†Blyth, p. 205.
‡Analy. and Adult. of Foods, 1888, p. 45.

upon fats. Indeed Roberts* claims that a small admixture of a free fatty acid in the chyme together with the agitation produced by the movements of the intestines is sufficient to emulsify fats without the aid of pancreatic juice.

Routh† also declares the same. None of the other animal fats contain butyrine.

The large proportion of butyrine in butter and its non-occurrence in any of the other animal fats together with the volatility of its acid, has long impressed us with the belief that it had some important office to perform in the digestive process. Under this belief we began a series of experiments upon the artificial digestion of different fats. Our digestive fluid was composed of five grains of Fairchild Bros. and Foster's "Extractum pancreatis," five grains of bicarbonate of soda dissolved in ten c. c. of distilled water. After the solution was complete we added half a dram of melted fat.

The whole was well agitated in a test tube and placed in an oven at a temperature of from 100 to 101 degrees Fahrenheit. The fats experimented on were cod-liver oil, butter, oleomargarine butter, the commercial oleomargarine oil, lard oil, benne oil, cotton-seed oil, lard, and mutton and beef suet. The cod-liver oil was bought from a reliable drug store.

Both fresh and stale butter was used, and was such as we had made ourselves seeing the milk from which it was made drawn from the cow, or such as we had analyzed ourselves and found to be pure, cow's butter. Fresh and stale "oleo" was used and was also either made by ourselves under the "Nathan" patent, which "oleo" contained some free acid, or was that which we had analyzed. The oils were all obtained from "oleo" makers or dealers in New York city. Both the pure washed dry fats of the butters and "oleos" and the natural products were compared as will be described directly. The contents of the test tubes were examined under a microscope at intervals of one, two, three, four, six, twelve, sixteen and twenty hours.

The cod-liver oil nearly always showed the finest emulsion.

Next, and the difference was often just perceptible came genuine butter. See Plate I. "Oleo" and lard oil came next, there being frequently no appreciable difference between them, but between the butter and the "oleo" there was a marked difference at the end of each period.

Fig. 4, Plate I, and Fig. 1, Plate II, shows the difference between "oleo" and genuine butter after being acted upon by the digestive fluid for one hour. It will be noticed that there is no emulsion at all of the "oleo" while the butter is well advanced.

Fig. 5, Plate I, and Fig. 2, Plate II, shows the same at the end of four hours. It is seen that the "oleo" is not nearly as much emulsified as the butter was at the end of one hour.

Fig. 6, Plate I, and Fig. 3, Plate II, presents the same at the end of twelve hours, which shows that the "oleo" is but a trifle, if at all, further emulsionized than the butter was at the end of the four hours.

---

* Indigest. and Biliousness, Fothergill, page 58.
† Routh, On Infant Feeding, page 131, 3d ed.

ARTIFICIAL DIGESTION.

Plate I.

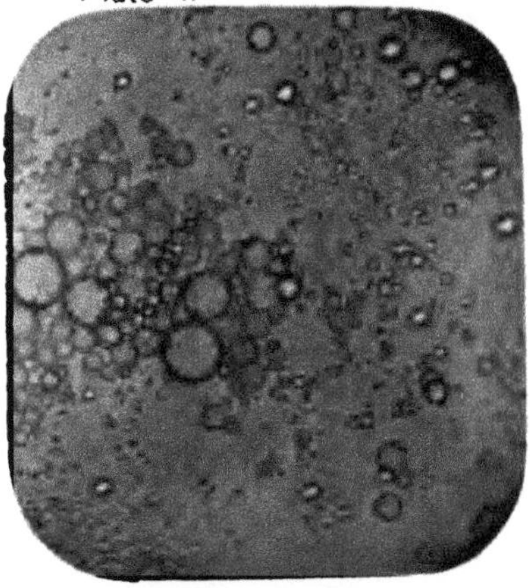

FIG. 1 COD LIVER OIL

At the end of 1 Hour X 250.

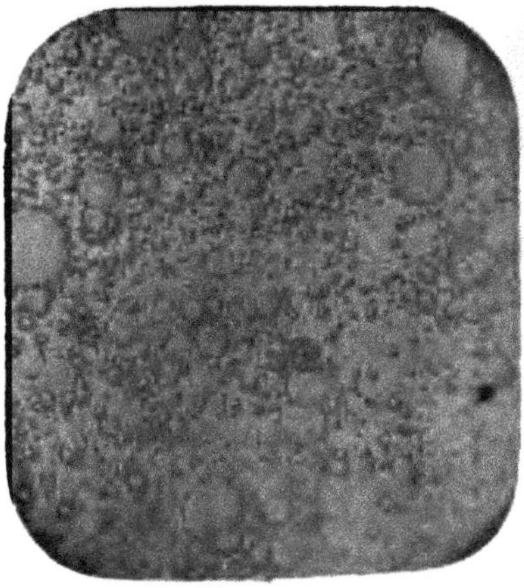

FIG. 4 BUTTER

At the end of 1 Hour. X 250.

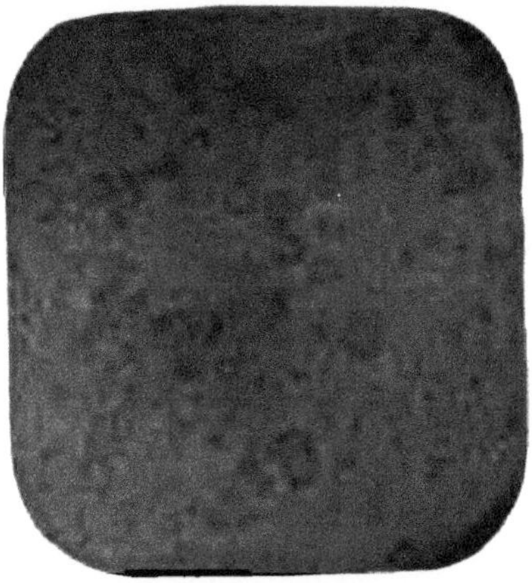

FIG. 2 COD LIVER OIL.

At the end of 4 Hours X 250.

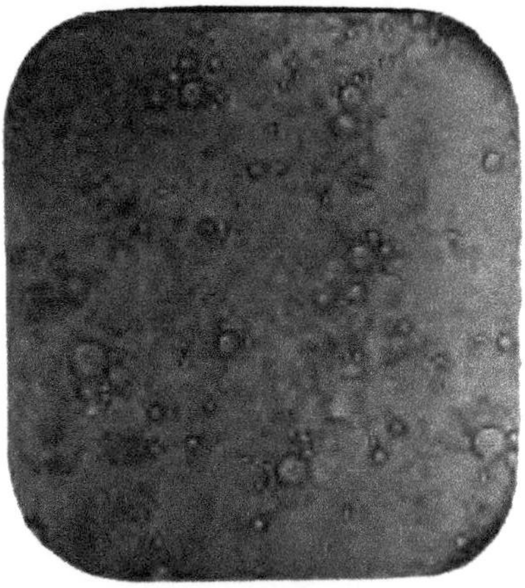

FIG. 5 BUTTER.

At the end of 4 Hours, X 250.

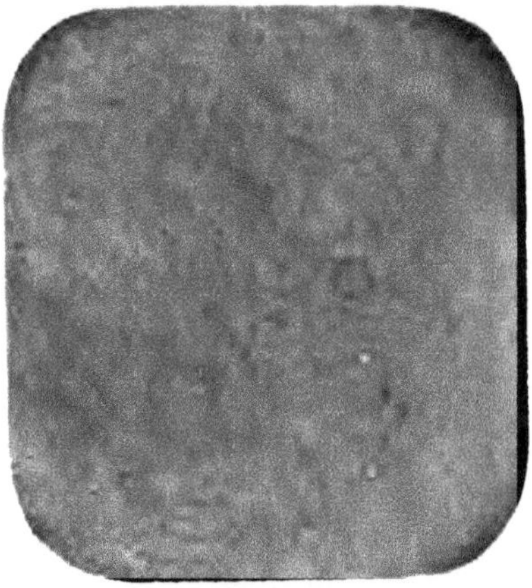

FIG. 3 COD LIVER OIL.

At the end of 12 Hours x 250

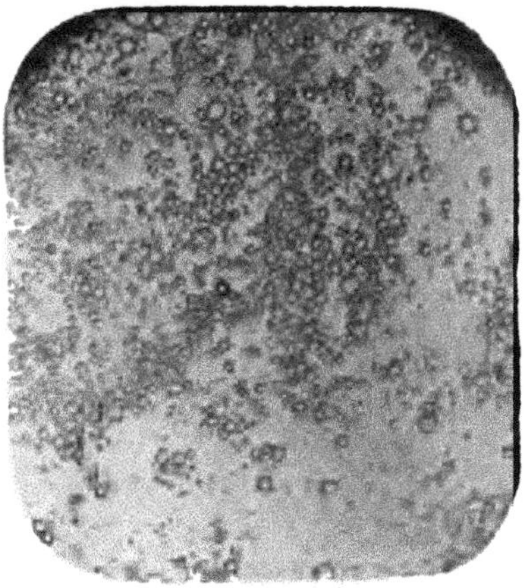

FIG 6 BUTTER.

At the end of 12 Hours x 250

ARTIFICIAL DIGESTION.

Plate II.

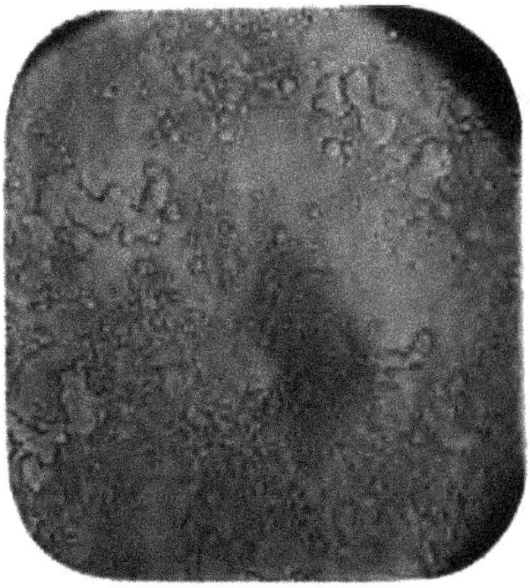

FIG 1 OLEOMARGARINE
At the end of 1 Hour X 250.

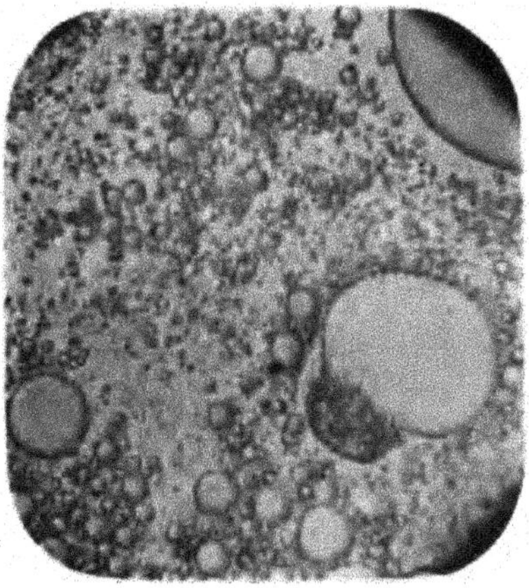

FIG, 2 OLEOMARGARINE
At the end of 4 Hours. X 250

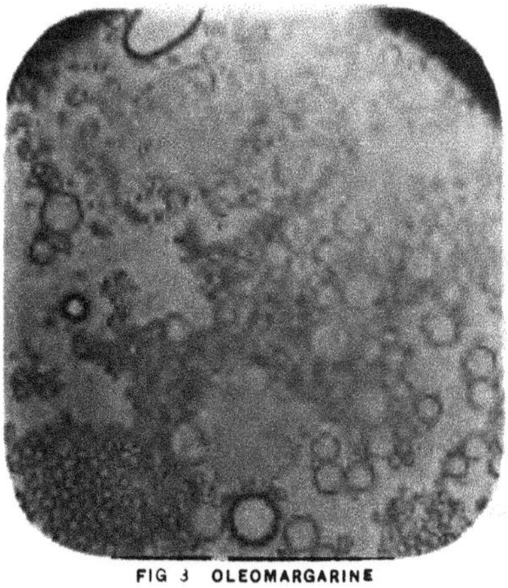

FIG 3 OLEOMARGARINE
At the end of 12 Hours X 250

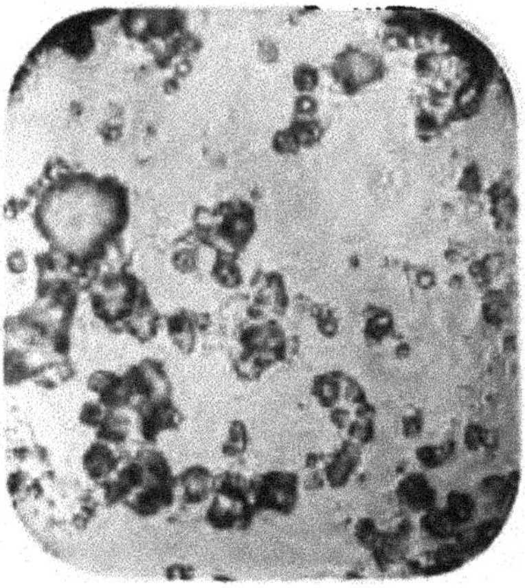

FIG. 4 BEEF SUET.
At the end of 1 Hour. X 250.

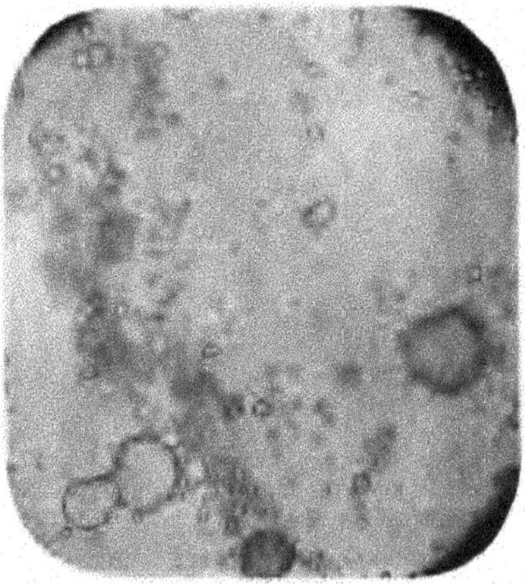

FIG 5 BEEF SUET
At the end of 4 Hours. X 250

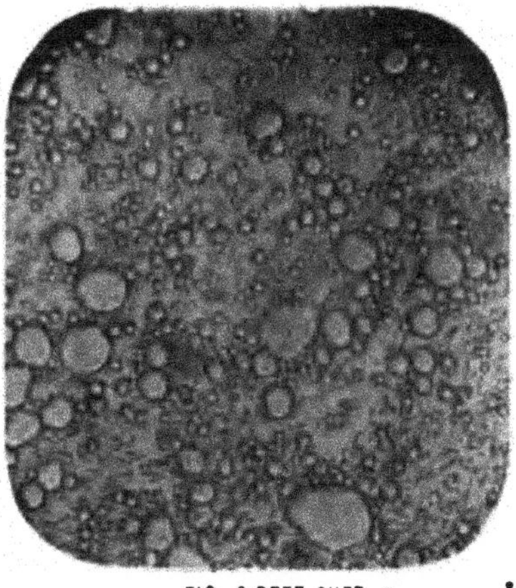

FIG. 6 BEEF SUET.
At the end of 12 Hours X 250

FIG. 1 LARD OIL.

At the end of 1 Hour. X 250

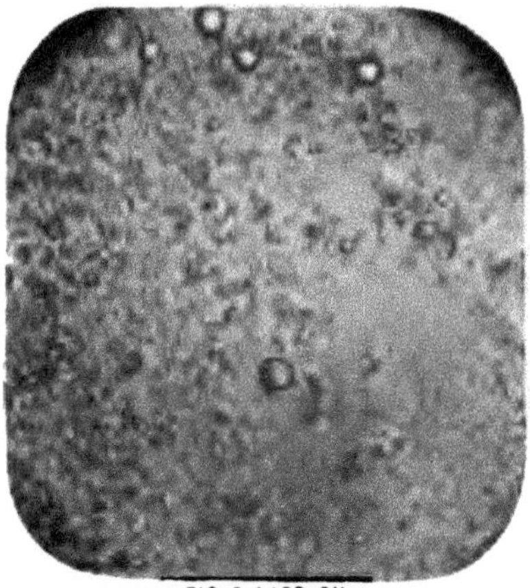

FIG 2 LARD OIL.

At the end of 4 Hours. X 250.

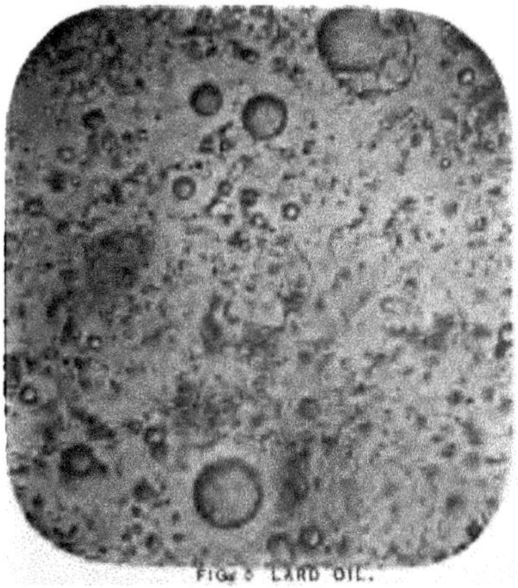

FIG. 3 LARD OIL.

At the end of 12 Hours. X 250.

FIG. 4 MUTTON SU

At the end of 1 Hour. 0.

At the end of 4 Hours X 2

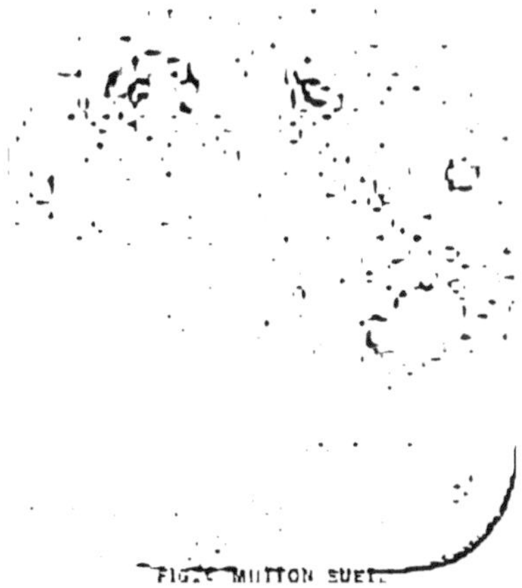

FIG. 6 MUTTON SUET.

At the end of 12 Hours. X 250

ARTIFICIAL DIGESTION.

Plate IV

FIG. 2 COTTON SEED OIL
At the end of 4 Hours X 250

FIG. 3 COTTON SEED OIL.

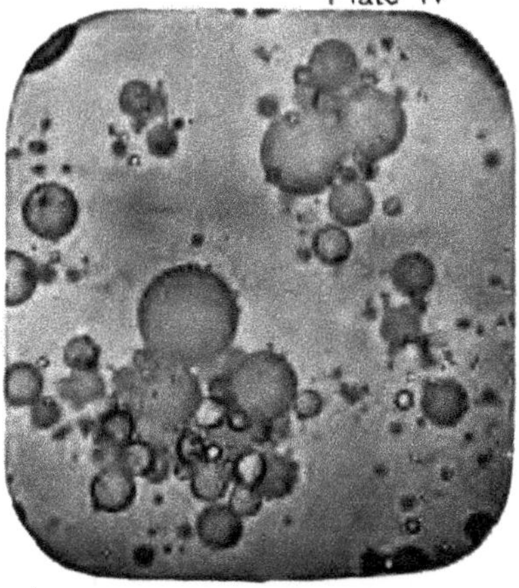

FIG 4 SESAME OIL
At the end of 1 Hour X 250.

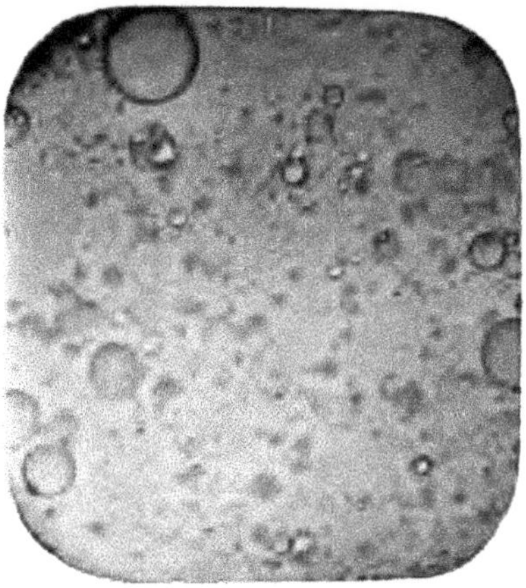

FIG 5 SESAME OIL.
At the end of 4 Hours. X 250

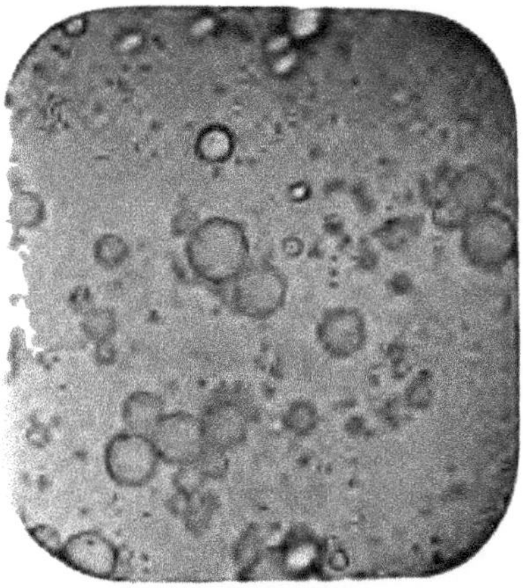

FIG 6 SESAME OIL

It will be further noticed that the globules of butter are finer, more uniform, containing very few large globules, and what is particularly conspicuous is the clearness and distinctness of the butter globules. They are well defined, sleek-looking and have a clean-cut outline which strongly intimates that they would go through an animal membrane — which they are required to do as will be seen later on — than oleo, which has a rough, course, ill-defined appearance. This holds true until the whole is saponified. The best results were obtained after exposing the fats to the digestive fluids for five or six hours at a temperature of one hundred degrees Fahrenheit, then allowing the whole to stand over night at a temperature of about sixty degrees Fahrenheit, and in the morning adding an equal bulk of warm water. The butter then presents under the microscope a most perfect emulsion. The globules are all very minute, grading off into almost imperceptible granules.

By examining the corresponding figures on the different plates, the comparative digestibility of the various fats and oils used in making artificial butter may be seen.

That butyric acid does have some important role to play in the alimentary canal is evident from the fact that sugar undergoes butyric fermentation in the small intestines. Yeo* says in reference to this: "Some of the sugar in the intestines, moreover, undergoes fermentation by which it is converted into lactic and butyric acids. How much of the sugar is absorbed as lactic and butyric acid has not been determined, but the amount of sugar found in the portal vessels or lacteals does not at all correspond with the amount that disappears from the cavity of the intestines."

Foster† says: "This suggests the possibilit of the sugar of the intestinal contents undergoing the butyric acid fermentation (during which, as is well known, carbonic anhydride and hydrogen are evolved), and thus, so to speak, put on its way to become fat; * * * moreover it is probable that by other fermentative changes a considerable quantity of sugar is converted into lactic acid, since this acid is found in increasing quantities as the food descends the intestines."

No doubt the lactic acid is converted into butyric acid which in turn is converted into soluble soaps and which may perform, and we believe do perform, important offices. As will be seen further on, fat is often covered with soap when absorbed and soaps are found in the chyle as well as some fatty acids. Furthermore it is shown that fats undergo still further emulsion after being absorbed while passing through the lacteals to enter the general circulation. Now, these soaps may be and very likely are the chief agents in accomplishing this. One of the arguments always advanced by the advocates of artificial butter is that it possesses better keeping qualities and does not become rancid and is therefore more wholesome than rancid butter. Now, it is true that it does not set free butyric acid (as it contains no butyrine), which gives the rancidity to butter, but as it contains some cellular tissue (in our specimen considerable), it undergoes a different decomposition which is liable to develop the septic material peculiar to dead animal matter and which is often very poisonons to

* Yeo's Physiology, 1884, p. 201.
† Foster's Physiology, 1885, p. 360.

human beings. On the other hand rancid butter is probably more readily digested than fresh and is not poisonous, the repugnance to it being simply one of taste as will be seen from the following taken from Roberts by Fothergill:* "The different behavior of two specimens of the same oil, one perfectly neutral and the other containing a little free fatty acid, is exceedingly striking. I have here before me two specimens of cod-liver oil, one of them is a fine and pure pale oil, such as is usually dispensed by the better class of chemists; the other is the brown oil sent out under the name of DeJongh. I put a few drops of each of those into two beakers and pour on them some of this solution, which contains two per cent of bi-carbonate of soda. The pale oil, you see, is not in the least emulsified; it rises to the top of the water in large clear globules; the brown oil, on the contrary, yields at once a milky emulsion. The pale oil is a neutral oil, and yields no acid to water when agitated with it — in other words it is quite free from rancidity; but the brown oil when treated in the same way causes the water with which it is shaken, to redden litmus paper. '(When the inhabitant of Arctic regions prefers his fat rancid, probably he is only following out what experience has taught him is good in his liberal consumption of fat)'. The bearing of these observations on the digestion of fat is plain. When the contents of the stomach pass the pylorus they encounter the bile and pancreatic juice, which are alkaline, from the presence in them of carbonate of soda so that the fatty ingredients of the chyme, if they only contain a small admixture of free fatty acids, are at once placed under favorable circumstances for the production of an emulsion without the help of any soluble ferment, the mere agitation of the contents of the bowels by the peristaltic action being sufficient for the purpose." (Roberts.)

"Possibly some fats containing a large proportion of oleine emulsionize more readily than others. But the whole subject is in its infancy so far as our acquaintance with it is concerned."

Cod-liver oil contains about one and four-tenths per cent of volatile fatty acid, some of which is butyric acid. This, together with its fluidity, accounts for its easy digestion and absorption.

The following is what some of the standard authors say about the digestibility of butter and other fats:

"Like† other fats and oils it (lard) is difficult of digestion, and therefore is sometimes used as a laxative for children and for its protective power in diarrhœa, dysentery, etc. * * * It has been proposed as a substitute for cod-liver oil in the treatment of phthisis (consumption), but its indigestible nature unfits it for this purpose."

"Apart,‡ however, from the deficiency in flavor it is doubtful whether 'butterine' (artificial butter) can be said to fully supply the place of butter as an article of diet. When the highly complex and peculiar character of the constitution of butter is considered, and that it is the fat derived from or natural to milk which for a time at least is the principle food of the young, it is probable that butter performs some more specific office in the system than ordinary fats."

---

*Indigestion and Biliousness, 1881, p. 58.
† National Dispensatory, 1874, page 102.
‡ Bell's Analy. and Adults. of Foods, 1883, page 62.

"As* before stated, fats consist of a fatty acid and oxide of lipyl. In the adult it is the pancreas which effects this separation into these approximate constituents. We all know that if this change does not occur the fat passes off unchanged by the bowels; and, as Bernard has shown, the expulsion of fat is one of the surest indications of diseased pancreas. In the infant, judging from the want of development of the salivary glands, the pancreas probably does not suffice to the complete performance of this function.

"*It is here that we remark one of those wonderful adaptations of nature. First, in butter we have excess of a free fatty acid; therefore rendering the assimilation of it possible without the assistance of the pancreas.*" * * *

"Another way in which this emulsion of fat can be accomplished is by giving the patient, not fats, properly so called, but the fatty acids of which they are composed, and which are very readily absorbed into the system. The good effects of cod-liver oil are probably in some measure due to the excess of fatty acids present. *So also, those of butter, it is indeed a matter of popular observation, that many children grow fat upon bread and butter. They appear to thrive on it when other means fail. This good effect cannot be due simply to the bread, for reasons before stated* (see page 176 same book) *but to the free acid which is also in excess in butter.*"

"It† (butter) is the best known of all this class of substances (fats) but it is eaten in very different quantities; from the large cup full before breakfast, as drank by the Bedouins near the Red Sea and the Persian Gulf, to the scarcely perceptible layer on the bread eaten by the needle women of London, and the supply is limited by pecuniary means rather than desire. It is also the form of separated fat which is less frequently disliked by consumptive people and invalids generally, as was shown by me in an inquiry into the state of 1,000 patients at the Hospital for Consumption, Brompton."

In answer to a letter of ours, Prof. Stelle, of Philadelphia, says: "If you care for my personal opinion it is that fresh butter and fresh olive oil are the most digestible of fatty bodies; next to them comes lard and finally tallow."

Finally, it is a matter of common observation among physicians that natural butter is taken by invalids, especially consumptives, when other fats, even cod liver oil, cannot be tolerated.

It is important to know that the approval‡ given to Mége's oleomargarine as an article of food by the council of health of Paris, in 1872, on the strength of the favorable report made by M. Felix Baudet (an abstract of which is given on page 30 of this report) was morally, at least, withdrawn in consequence of a report of an investigation made by a commission of the Academy of Medicine for the Prefect of the Seine, disapproving of the article for use except to a limited extent in cooking, on the ground of its comparative indigestibility. It was never allowed to be sold in the public markets of Paris except *under its own name.* Its sale is now prohibited in the public markets.

---

* Routh on Infant Feeding, 3d ed., 1879, page 130.
† Smith's Foods, 3d ed., page 128.
‡ Wagner Jahr. 1880, p. 711; Dingl. Polyt. Jour., vol. 237, p. 478; Revue de Medicine, 1880.

The insolubility of those artificial butters made from animal fats is another potent quality for rendering them indigestible. In man the digestive process is carried on with greater rapidity than in any of the lower animals; and the gastric juice acts upon food from the outside toward the centre; that is, it does not soak the material and exert its solvent action upon the whole of it at the same time; consequently the greater amount of surface of food directly exposed, the more rapid its digestion. It is for this reason that it is so necessary for man to carry out the process of mastication thoroughly. It is for this reason also that some people experience distress after eating eggs boiled just hard, but none after eating them soft-boiled or after being boiled for some time when they become "mealy." The difference in the digestion of an egg is again felt when eaten raw without beating and when it is beaten. The beating mixes the albumen with the air rendering it porous.

The artificial butters made from animal fats, although the oleine and palmitine are separated as much as possible by pressure, will not liquefy at the stomach temperature as is demonstrated by the following experiments: We placed in an oven kept at a temperature of from one hundred to one hundred and four degrees Fahrenheit, four beakers containing respectively pure butter, oleomargarine butter, oleomargarine oil (commercial) and lard oil, about twenty drams of each and which were all of the temperature of about sixty degrees Fahrenheit when taken. At the expiration of thirty-five minutes and the temperature at one hundred degrees Fahrenheit, the butter presented a clear, limpid appearance, but the others remained solid being but very little affected; and at the end of five hours, the temperature being from one hundred and one to one hundred and four degrees Fahrenheit, they were in a semi-solid condition. The oleomargarine oil being most softened, the oleo butter next and the lard the least softened.

These insoluble fats then must interfere with digestion in two ways; first, by not being acted upon themselves by the gastric juice; and second, by being thoroughly mixed with the other foods in the mouth they form an impervious covering to them, thereby preventing the gastric juice from coming in direct contact with them.

Randolph* says that "a further reason that the fats, especially when cooked with other foods, are frequently found to be unwholesome, is that in the process of cooking, they so surround and saturate the tissues of the substance with which they are combined, that it is rendered nearly inaccesible to the action of the saliva and gastric juice and at times digestion is in so far delayed that the fried substance does not become entirely freed from this more or less impervious coating of fat until subjected to the action of the pancreatic juice."

This retards digestion and prevents that increased flow of gastric juice which follows the absorption in the stomach of the first portion of food digested as is shown to be the case by Heidenhains† experiment, and also deprives the proteids of that aid in their digestion which fats are declared to render.

---

*Carbohydrates and Fatty Foods, by N. A. Randolph, M. D., Therapeutic Gazette, vol. IX, p. 732.

†Foster's Physiology, p. 283.

"In* experimenting with gastric festulæ on different dogs, for example, we have found in one instance, like Dr. Beaumont, that the gastric juice was always entirely absent in the intervals of digestion; the mucous membrane then presenting invariably either a neutral or slightly alkaline reaction. In this animal, which was a perfectly healthy one, the secretion could not be excited by any artificial means, such as glass rods, metallic catheters, and the like; but only by the natural stimulus of ingested food. Tough and indigestible pieces of tendon introduced through the fistula, were expelled again in a few minutes, one after the other, without exciting the flow of a single drop of acid fluid; while pieces of fresh meat introduced in the same way produced at once an abundant supply."

After food has been changed by the act of digestion it is required to enter the current of blood before it can fulfill its office of nourishing the body. In order to do this it must pass through the walls of the alimentary canal, which passage constitutes the process of "absorption."

While absorption may take place through any part of the body containing blood and lymph vessels, and not covered with a hard thickened cuticle like the palms of the hand and soles of the feet, yet the locality especially adapted to it is the upper part of the small intestine. Here the lining membrane is thrown into numerous folds in order to increase the amount of surface and covered with myriads of minute projections resembling the pile of velvet which are technically called villi. Each little villus constitutes an absorbent gland. Its surface is covered with columnar epithelial cells containing protoplasm and also little rod-like projections extending from their free extremities.

These cells rest upon a basement membrane which contains muscular tissue so arranged as to aid in carrying along the solid particles of food on their passage to the lacteals and blood-vessels.

This membrane encloses a framework of connective tissue in which are contained the blood-vessels and lacteals. The blood-vessels are arranged in the form of lattice-work around the lacteals which latter contain no perceptible openings. Now, fat is the only element of food that is absorbed in the form of solid particles, at least to any extent, and therefore, would seem to be the most difficult of absorption. This absorption of solid particles of fat has, indeed, always been a puzzle to physiologists. The peptones and sugar are almost wholly liquefied and cannot be recognized by the microscope after entering the lacteals, but fat is seen after reaching the lacteals in a very minute state of division. On the principle of osmosis, it is easy to understand how liquid foods are absorbed. Some physiologists believe that the epithelium covering the villus is prolongated, so to speak, into the central lacteal vesicle and that the fat granules pass not through, but between the epithelial cells along this prolongation of protoplasm and so reach the lacteal. Others believe that they pass through the cell by being taken up by the protoplasm in the manner in which an amoeba takes its food, and passed on to the lacteals by this protoplasmic agent, being aided by contraction of the muscular element in the villus. The latter theory is the most satisfactory, and probably the most modern. It is also believed that the

*Dalton's Human Physiology, 1875, p. 162.

layer of rods or pores projecting from the free surface of the epithelium has to do with the absorption of fats. Whichever theory is correct it seems plain to us that the finer the particles of fat the more readily will they be absorbed. Moreover it is well known that an animal membrane moistened with water will not allow the passage of emulsionized fat but when moistened with bile fat passes through it. From this fact it is quite probable that the soaps formed as previously described perform important work in connection with the absorption of fat.

Yeo* says in reference to this: "It has therefore been suggested that the epithelial cells of the mucous membrane are more or less moistened with bile, and the particles of fat in the emulsion are also coated with a film of bile or soap. Thus they are enabled to pass into the epithelial cells, in which they can be detected during digestion. The bile or soapy coating of the fat particles may no doubt aid in their transit through the various obstacles on their way to the lacteal radicles."

I know of but few actual experiments upon human beings as to the comparative absorptivity of butter and other fats, but it is fair to assume from the foregoing circumstances that butter is much more readily absorbed than its sham congeners. Rubner† ascertained that butter was much more readily absorbed than ham fat. Randolph says that cod-liver oil is absorbed with the greatest ease and to a greater degree than any of the other fats, and that on the other hand, the vegetable oils are the least readily absorbed.

A. Mayer‡, experimented to determine whether natural or artificial butter was the easiest absorbed by the system. He took a man and a boy and fed them for three days on various mixtures of bread, milk, eggs and vegetables together with natural butter. Then followed two days rest, they being fed on ordinary diet; after which for three days they were given precisely the same food as on the first three days, except artificial was substituted for natural butter. Each successive day of the experiment the solid evacuations were collected and analyzed; commencing twenty-four hours after the beginning of the experiment. The amount of fat in the excrements was estimated which determined the amount of fat that had been absorbed. The following is the percentage of the amount absorbed:

Man.	1st day.	2d day.	3d day.
Natural butter	97·0	99·4	98·7
Artificial butter	94·6	97·9	96·7
Boy.			
Natural butter	97·8	94·8	98·7
Artifical butter	93 3	94·6	97·6

It will be seen, therefore, that the average was about one and six-tenths per cent less of the artificial absorbed than of the natural. The greatest difference was two and five-tenths per cent less of the artificial. The experimenter concludes that except in sickness this trifling difference may be overlooked with safety.

---

* Yeo's Manual of Physiol., page 203.
† An article by N. A. Randolph, M. D., Therapeutic Gazette, November 16, 1885, p. 735.
‡ Landwirthschaftliche Versuchsstation, vol. 29, p. 215.

Of course, these experiments were not carried on long enough to be of much value, but as far as they go, they harmonize exactly with our idea of the difference in the absorption of these two articles. If this difference was manifest in three days we would expect a very much greater difference in three months.

Magendie's experiments on dogs for the purpose of testing the effect of feeding nothing but fat, incidentally shows a striking difference in the life-sustaining power between butter and lard. He used two dogs for the experiment. One he fed butter and the other lard. The first lived sixty-eight, the second fifty-six days; that is, the dog fed on butter lived twelve days longer than the other, or one-fourth of the whole time which the other dog lived.

The liability of conveying disease germs into the human system through artificial butter is, in our opinion, greater than is supposed by those not familiar with the subject. In the first place investigations are showing that many more diseases than was formerly supposed are communicable from animal to man. The following are some of those known to be such: Consumption, anthrax, trichinosis, tape-worm, glanders, foot and mouth disease, cow-pox, hydrophobia, etc. Many more as epidemic pleuro-pneumonia, small-pox of sheep, splenic apoplexy, braxy of sheep, typhus, etc., have, when the flesh of animals suffering from them was eaten, produced serious sickness in human beings.

We would like to give the history of these diseases and also of the cases of the sickness resulting from consumption of the flesh of these diseased animals; for we think the effect would be to startle the populace and to induce it to lend a heartier support to those public officers to whom has been assigned the duty of preventing unwholesome food being sold to it; but the want of time prevents. We must content ourselves with a brief reference to some points bearing directly upon the subject in hand. The manner in which trichinæ can get into artificial butter can easily be seen from the following: When the animal takes a cyst containing a trichina into its stomach the cyst is dissolved by the gastric juice which sets the trichina free when it passes out of the stomach into the intestine where it developes in from a week to ten days, and the female deposits her embryos — from 60* to 2,000 for each female trichina. The young trichinæ then make their way through the connective tissue† to the muscles. Trichenæ are found in hogs, cattle and sheep. Now, if those animals are killed during the migratory stage the caul fat would doubtless contain the parasite. Dr. Billings‡ says he has frequently found encysted trichinæ in the adipose tissue between muscular tissue of very fat hogs, but not in the fat lying upon the muscles. He states, however, that Prof. Taylor, of the Department of Agriculture at Washington, has seen in the journal of the Microscopical Association that they have been found in fat. Everyone is aware of the dangerous character of this disease.

A tape-worm is developed from a kind of germ called a cysticercus.

---

*W. C. W. Glazier, M. D. A pamphlet published by the Illustrated Med. Jour. Co., Detroit, Mich.

†Ibid.

‡Relation of Animal Diseases to the Public Health, by F. S. Billings, D. V. S., 1884, p. 7.

These are of different varieties and are found in the solid parts of hogs, cattle and sheep. Animals infested with these germs are said to have the measles. A cysticercus is developed from the egg of a tape-worm. The fully-matured tape-worm is developed in two separate stages as follows: The eggs of the worm pass out of the body and are eaten by a man, or another animal. They then find their way into the solid tissues of this animal when they develope into cysticerci and so remain until the cysticerci are again taken into the intestines of another animal or man, where they reach their full development as a tape-worm. Now, the heat applied to the fats employed in making artificial butter is not sufficient to destroy these germs as most of them are treated at a temperature below 140 degrees Fahrenheit, as is seen from the abstracts. One patent for making a compound to substitute butter for cooking purposes requires a temperature of 190 degrees to 200 degrees Fahrenheit. One other for "improvement in shortening for culinary uses" uses a heat of 400 degrees Fahrenheit. Six for purifying and bleaching tallow, lard, etc., heats to 140, 150, 200, 200, 200 and 347 degrees Fahrenheit, respectively.

Much interest is manifested at the present time in regard to germs and their destruction, and as is always the case with new subjects, there is some difference of opinion in regard to the efficacy of different disinfecting agents. The following will give some idea of the amount of heat required to kill disease germs:

Toussaint,* showed by experiment that the tuberculous element was not confined to the diseased localities, but were diffused through all the tissues, and that the juice of the flesh of a consumptive animal had produced a disease in others after having been heated to fifty degrees or sixty degrees Cent., (122 degrees to 140 degrees Fahrenheit) the temperature of roasting beef, and that when given in very small doses.

Referring to these experiments, Bartley says: "Considering the facts in this light we ought to establish no degrees in tuberculosis; when it exists it renders the consumption of the flesh dangerous."

In reference to trichinæ, some observers as Vallin† state that a temperature of 129 degrees to 133 degrees Fahrenheit, kills most of them, and that 140 degrees Fahrenheit is safe; but Collin‡ found living trichinæ in half a pound of steak that had been boiled for ten minutes, presenting a white appearance when cut, having no red points—and discovered trichinæ in the intestines of a bird after having been fed upon it.

Pasteur § asserts that an exposure for ten minutes to a temperature of 129·2 degrees Fahrenheit, will kill anthrax rods, but spores resist prolonged boiling. The spores develop in the rods rapidly after the death of the animal, under proper conditions, and will remain active for years. They are not destroyed by drying or putrefaction when exposed to oxygen (Maguire).

* Report of the Bureau of Animal Industry of the Department of Agriculture, 1884, p. 863.

† Ibid, 848.

‡ Ibid, 848.

§ Bacteria by Dr. A. Magnin, translated by S. M. Sternberg, M. D., F. R. M. S., 1884, p. 270.

Klein* also affirms that the anthrax spores will resist prolonged boiling.

Vantieghem is quoted by Magnin as saying that a temperature of 121 degrees Fahrenheit is fatal to most bacteria; but he has studied the bacillus that is able to multiply and form spores in a culture fluid at 165.2 degrees Fahrenheit; but which cease to multiply at 171.5 degrees Fahrenheit. Magnin also states as coming from Lebedeff that septic blood does not lose its virulence at the end of forty days; or by being heated to the boiling point (212 degrees Fahrenheit), for from three to twenty-four hours, and that the bacteria in it are capable of multiplying after such exposure.

Arloing and Chauveau† have found what they consider to be the bacillus causing gangrenous septicæmia. When fresh it is destroyed by a temperature of from 194 to 212 degrees Fahrenheit, but when dried it required 248 degrees Fahrenheit.

The heat to be trusted for destroying pathogenic germs in practice will be seen from the following:

Dr. Van Bush, of Berlin, used a temperature of 149 degrees to 167 degrees Fahrenheit, for the destruction of puerperal fever contagion. The late Dr. Elisha Harris,‡ in 1859, employed a temperature at and above 212 degrees Fahrenheit to disinfect clothes of yellow fever subjects. He quotes Dr. William Henry as saying: "That the infectious matter of cow-pox is rendered inert by a temperature not below 140 degrees Fahrenheit, from whence it is inferred that more active contagion is probably destructible at temperatures not exceeding 212 degrees Fahrenheit."

Dr. Henry could not communicate typhus after exposing flannel shirts to 204 degrees Fahrenheit; same with scarlet fever. He says: "The experiments which we have related appear to be sufficiently numerous to prove that *by exposure to a temperature not below* 200 *degrees Fahrenheit during at least one hour, the contagious matter of scarlatina is rather dissipated or destroyed.*"

The following circular,§ issued to the customs officers December 22, 1884, shows what temperature is considered safe by the government: "All circulars of the department concerning the importation of old rags are modified as follows: No old rags except afloat on or before January 1, 1885, on vessels bound directly to the United States, shall be landed in the United States from any vessel, nor come into the United States by land, from any foreign country, except upon disinfection at the expense of the importers, as provided in this circular or may hereafter be provided.

Either of the following processes will be considered a satisfactory method of disinfection of old rags, and will entitle them to entry and to be landed in the United States upon the usual permit of the local health officer, viz.:

1. "Boiling in water for two hours under a pressure of fifty pounds per square inch."

2. "Boiling in water for four hours with pressure."

---

* Micro-organisms and Disease by E. Klein, 1885, p. 187.
† British Medical Jour., January, 1884.
‡ Utility and application of heat as a disinfectant, by Elisha Harris, M. D., 1860.
§ Sanatarian, January, 1885, p. 69.

3. "Subjection to the action of confined sulphurous acid gas for six hours, burning one and a-half or two pounds roll brimstone in each one thousand cubic feet of space, with the rags well scattered upon racks."

4. "Disinfection in the bale by means of perforated screws or tubes through which sulphur dioxide or superheated steam at a temperature of not less than 330 degrees shall be forced under a pressure of four atmospheres for a period sufficient to insure thorough disinfection," etc., etc.

James A. Russell, in Quain's Dictionary of Medicine, says: "It is extremely improbable that any contagium can withstand a temperature of 220 degrees Fahrenheit (104·5 C.), maintained during two hours. When contagium is shielded by thick material into which heat penetrates slowly, the time necessary to reach the disinfection temperature may be long and hence the necessity for spreading clothing and opening out bedding in special hot air-chambers where the heat ought not to be less than 220 degrees Fahrenheit (104·5 C.), nor more than 250 degrees Fahrenheit (112·1 C.)"

The following is an abstract from the report* of the committee of disinfectants of the American Public Health Association: "The experimental evidence recorded in these reports seems to justify the following conclusions: The most useful agents for destruction of spore-containing infectious materials are:

1. Fire, complete destruction by burning.
2. Steam under pressure, 230 degrees Fahrenheit, for ten minutes.
3. Boiling in water for one hour. For the destruction of infectious material which owes infecting power of micro-organisms not containing spores, the committee recommended:
1. Fire, complete destruction by burning.
2. Boiling in water half an hour.
3. Dry heat, 230 degrees Fahrenheit, for two hours," etc.

It is alleged by the makers of artificial butter that the fats from animals† dying from disease could not be used in making these articles as they would "stink" and taint the product, and the deoderization would not remove said stink, etc. This is false, for we have tasted and smelled of oil made from horses and dogs picked up in the streets of New York and Brooklyn, dead of disease, and it had no unpleasant taste or appearance, in fact, tastes as sweet as pure dried butter fat. And, too, the suspicion is growing stronger and stronger among those who are cognizant of the facts, that those oils go into the artificial butters. Why should so much pains be taken to render a sweet clear oil from dead horses and dogs? This would be adding unnecessary expense if it was intended for lubricating purposes, and we do not hear of its being commonly used in soap-making.

The following letter in answer to one from us will tell its own story:

"BROOKLYN, N. Y., *January* 18, 1886.

"DEAR DOCTOR — In reply to yours of the 12th inst., I would say that all I can say of the oil I showed in New York was that it was

* Medical News, December 12, 1885, p. 654.

† *Vide* Report of Committee on Public Health, transmitted to Legislature March 21, 1884, p. 88.

manufactured on Newtown creek, by Mr. Henry Beran. Mr. Beran has the contract for the dead animals and offal of the city of Brooklyn. The oil in question was made from the *comb-fat* (so called) of horses. That is, from the top part of the neck of horses, which were obtained from this city and tried out by the contractor. The horses were such as die in every city from both accident and disease. There were a large number of horses killed in Brooklyn last year that were suffering with glanders. Whether any of these horses helped to make up this oil I do not know; nor does Mr. Beran. The specimen I had in New York was a very fine oil, and it shows that an oil can be made from dead horses which in taste and naked eye appearances is as palatable as the best 'oleo' oil.

"Mr. Beran has told me that he is satisfied that some of his oil has been used for the manufacture of 'oleo' butter. He has always been very careful about telling me to whom he sells it and he evidently *thinks* it is used for that purpose; in fact, he says he *knows* it has. I give this as his own statement, and for what it is worth. I could not prove it. From the odor, taste, etc., of this oil I am of the opinion that it can be used to make 'oleomargarine,' and that its use for that purpose ought to be strongly condemned. I also hold that the use of lard, tried out at a temperature below 130 degrees Fahrenheit, should be prohibited. Hoping this will answer your questions I am

"Very sincerely yours,

"E. H. BARTLEY, M. D."

It might be asked if natural butter was not exposed to the same contamination? We answer that it is not; for in the first place, the fat of milk is doubtless manufactured in the gland by the metabolic action of the protoplasmic cells; and consequently would not be apt to contain disease germs even if they were in the cow's system unless the udder itself was diseased. Then, too, it is difficult to make good butter from a diseased cow; and but few farmers would risk their reputation by selling butter made from sick cattle. Furthermore, I am unable to find a single authentic instance where milk-butter has produced any serious sickness, which, in consideration of the length of time it has been known, is significant.

Dr. Alfred Hill,* on account of assertions being made that the milk quickly became rancid, and produced typhoid fever, and that the butter was very offensive which came from cows that had been partly fed on sewage grass, made a thorough examination of the milk and its butter which came from the Birmingham Sewage-Farm† and found that the keeping and other qualities of the milk were not in the least inferior to ordinary milk. In regard to the butter he says: "In order to test the quality of the butter made from it, I requested the wife of the farm manager, who thoroughly understands butter-making (although no butter is ordinarily made on the sewage-farm) to make a churning for me, which she was kind enough to do.

The resulting butter was excellent in quality, and retained its sweetness and other properties as well as other fresh butter, although the

* The Analyst, August, 1885, p. 186.

† A farm manured with sewerage.

weather at the time was excessively hot; so that the conditions of the experiment were as unfavorable as possible."

When we look over the ingredients used in making artificial butter or preparing the fats and oils for the same and find such powerful acids as sulphuric, nitric, benzoic, salicylic, etc., and such alkalies as caustic soda, bicarbonate of soda, carbonate of ammonia, saleratus, sal-soda, etc.; and such drugs as sugar of lead, alum, carbonate of potash, nitrate of soda, sulphate of soda, borax, nitre, etc.; and such easily decomposed material as slippery elm bark, rennet, yolk of eggs, cow's udder, fresh vegetable pulps, etc., mixed with it; and after having prepared this stuff according to the specifications of certain patents, we cannot repel the conviction that the greatest care must be exercised or they will contaminate the product. By referring to patent No. 263,199, it will be seen that about 150 pounds of melted lard is thoroughly "washed"—that is, mixed—with sixty gallons of ice-water holding in solution three ounces of nitric acid (strong) and borax. The lard solidifies in this solution and while solid is washed in sixty gallons of ice-water. Every time this quantity of fat is washed in the acid water one ounce more of nitric acid is added, which shows that this amount of nitric acid is considered to be taken up by the lard. In the manufacture of "oleo" under this patent from five to fifty per cent of this deodorized lard is added to commercial oleomargarine oil.

The whole is then subjected to a heat of ninety-five degrees Fahrenheit (which is not sufficient to melt it) and churned with milk or cream, sugar and coloring matter. It is then treated with ice-water which causes it to rapidly and completely solidify. After mixing thoroughly and salting, it is ready for market.

It will be seen by this process that the fat, after being treated with nitric acid, is never again subjected to a thorough washing, and in view of the fact that fats possess the property of retaining free acids with remarkable tenacity, it is difficult to believe that the marketed product does not contain nitric acid.

The following is the conclusion of Nothnagel and Rossbach* concerning the effect of small, greatly diluted doses of acids: "When acids are used for too long a time the appetite and digestion are finally injured and a series of pathological conditions result."

"It is readily supposable that the long-continued administration of diluted mineral acids to the living organism leads to the decomposition of the alkaline combinations with the weaker acids, *e.g.*, carbonic acid, or with the albuminoids; the stronger acids uniting with these alkalies and being excreted with the urine, as mineral salts; so that not only the blood, but the whole body, would become poorer in alkalies and salts."

"Salkowski and Lasar proved directly that the alkalesence of the blood is diminished by the internal administration of dilute mineral acids."

We now return to the question is artificial butter a wholesome article of food? It seems to us from the facts set forth in the foregoing pages that there can be but one answer to this question.

We do not mean to say that every individual who eats artificial but-

* Materia Medica by Drs. Nothnagel and Rossbach, New York, 1884, p. 822 and 823, from 4th German ed.

ter will sicken and die any more than every man who uses ardent spirits, tobacco or narcotics to excess would do so, but what we do mean to say is that it, like them, possesses physiological properties "unfavorable to health" and are very liable to possess ingredients very dangerous to health. Dyspepsia is a prevalent disease in this country and is not acquired in a day; for a strong stomach will stand much abuse before it will permanently rebel.

Several instances are on record where pennies and other metallic substances have been swallowed and digested — even jack-knives have been swallowed and their bone handles completely digested, but no person would consider these healthy articles of diet.

Strong, vigorous men and those whose habits are invigorating to the digestive powers might substitute a food hard of digestion for an easy one, for a long time, with apparent impunity, but weaker men and those whose habits are sedentary and whose labors are mental, which tend to debilitate digestion, would soon be injured.

Fats as a whole are considered by medical men to be difficult of digestion; and to substitute those hard of digestion for one that is easy, and too, for one which we believe is endowed by nature with properties that not only render it, *per se*, easily digested and assimilated but which also render important aid in these processes to other fats, must eventually produce sickness. The little genuine butter added to these spurious articles helps as far as it goes, but the amount in most of them is very small indeed.

It is true we eat fats which when raw are more difficult of digestion than some of the artificial butters, but it must be borne in mind that they are eaten in conjunction with natural butter, and the cooking process to which they are subjected no doubt renders them much more easily digested. As is well known, "drippings" are much easier digested than the fats from which they come.

That cooking renders fats much more easily emulsionized by artificial means is demonstrated by the following experiments:

We subjected a portion of oleomargarine butter placed in a frying-pan to the heat of a cook-stove, the same as would be employed to fry a piece of meat, for about five minutes. (Our thermometer registered 200 degrees Cent., and the heat went above this somewhat.)

The fat was then poured off and equal quantities of it and the same specimen of "oleo" uncooked were exposed to the action of artificial digestive fluid, the two specimens being placed under exactly the same conditions.

At the end of four hours the microscope showed that the cooked oleo was decidedly the best emulsion — approaching in appearance natural butter uncooked under the same circumstances. It was intended to have artotypes to show this, but the experiments were not completed in time, and we would add here that we are carrying on various experiments with a view to demonstrating the differences between natural and artificial butters, which we hope to publish in our next annual report.

As the fusing points of the cooked and uncooked "oleos" remained identical, the difference in the emulsions must have been due to chemical changes produced by the heat, as the separation of the fatty acids and glycerine which again gives us a free fatty acid.

After pouring off the cooked fat, there remained in the frying-pan a considerable quantity of scrap.

Fothergill* says: "But heat does liquefy fat and separates (we believe) olein, from stearin and margarin. The liquid portions of fried bacon is digested by many who cannot digest the solid portion of bacon fat. This is a well known fact."

Furthermore, the great heat to which fats are subjected in frying is probably sufficient to set free considerable quantities of fatty acids, and also to cause partial breaking up of the whole fat.

The friends of the bogus butter ask us, in a spirit of defiance, to show any cases of sickness produced by it. This is in fact a demand for a complete demonstration, and may be answered by stating that we have seen a great many cases of sickness, and much of it dyspepsia, during the period in which the bogus butter has been sold for which we have been unable to assign a cause. This may have been artificial butter, but the deceptive manner in which it has been handled has prevented physicians from ascertaining its effects; consequently we must judge it by its qualities.

No person would gainsay that these articles if they contained germs of disease, or such materials as enumerated above, were unwholesome. We have pointed out the liability and great probability of their containing them; and many things have been publicly condemned on less liability to produce sickness; for instance, the water of Albany has been used by nearly 100,000 people for several years, and no serious results can be shown; yet the conditions are present which render it *liable* to produce disease, and this circumstance has agitated the public mind to such an extent that some of the best medical and other men of the city have devoted themseves to finding a better supply; and they have finally decided that it is expedient to obtain it from another source than the present which will necessitate the expenditure of $450,000.

"Bob veal" produces sickness in comparatively few cases, yet on account of its *liability* to produce disease its sale is prohibited.

Dr. Fox† says, in connection with anthracic diseases, "that large quantities of this meat have been eaten with apparently no injurious effects, but so many disastrous occurrences have followed its employment as to warrant the medical officer of health in condemning such meat."

MILK.

The following table shows the result of analyses of samples of milk as brought in by the inspector and was sold as pure, whole milk:

---

* Indigestion and Biliousness, 1881, p. 51.

†Fox's Sanitary Examinations of Water, Air and Food, London, 1878, p. 412

Analyses of Milk — Below Standard.

Sample number.	TEST.		ANALYSIS.					Remarks.
	Lactometer.	Thermometer.	Per cent of cream by volume.	Per cent of water.	Per cent of fat.	Per cent of caseine and sugar.	Per cent of ash.	
1..........	120	60	4	88.237	2.259	....	.683	
2..........	....	...	....	88.38	2.281	....	....	Sample No. 1, after having been frozen.
3..........	103	65	4½	89.176	2.047	....	.598	
4..........	.. .	....	....	89.249	2.571	....	....	Sample No. 3, after having frozen.
5..........	102	58	5	89.310	2.645	....	.542	
6..........	118	57½	6½	88.096	2.656	....	.678	
7..........	114	58	4½	88.566	2.355	....	...	
8..........	97	60	7	89.567	2.921	....	....	
9	100	58	7	89.106	3.08	....	.. .	
10..........	117	59	7	88.287	3.199	....	....	
11...	101	60	7	89.156	2.798	....	....	
12..........	105	60	6	89.006	2.500	....	....	
13..........	122	58	7	88.15	2.591	....	....	
14..........	123	60	5	87.408	2.013	....	....	
15..........	107	60	6	88.784	2.381	....	....	
16...... ...	121	58	6	88.402	2.170	....	....	
17..........	110	62	6	88.764	2.694	....	....	
18..........	107	60	6	89.000	2.869	....	....	
19..........	140	60	7	84.840	1.529	....	. ..	
20...... ...	....	...	....	88.698	2.648	....	....	Duplicate of No. 17.
21..........	107	60	6	89.068	2.802	...	....	
22..........	116	60	6	88.518	2.588	....	....	
23.	109	60	6½	88.743	2.806	....	....	
24..........	116	60	6½	88.296	2.763	....	....	
25..........	116	60	7	88.754	1.851	....	....	
26..........	106	62	7	88.781	2.415	....	.. .	
27..........	99	62	7	88.995	3.074	....	....	
28..........	124	60	4½	87.799	2.267	....	....	
29..........	115	62	6	88.38	1.88	....	....	
30.	121	60	6	88.33	2.127	....	....	
31..........	124	60	7	88.01	2.471	....	....	
32	124	60	6	87.718	1.859	....	....	
33..........	114	60	6½	88.15	3.081	....	....	
34...... ...	....	....	8	88.173	2.390	....	.688	From a creamery in Schenectady county.
35..........	....	....	6	89.154	1.951	....	.696	
36...... ...	....	....	6	89.620	1.475	....	.689	
37..........	....	....	5	88.461	2.709	....	.620	
38...... ...	101	60	6	89.721	2.820	....	.557	
39..........	121	62	7	88.155	1.640	....	.716	
40..........	110	58	7	89.267	2.875	....	.606	

Analyses of Milk — Below standard — (Continued).

Sample number.	Test.		Analysis.					Remarks.
	Lactometer.	Thermometer.	Per cent of cream by volume.	Per cent of water.	Per cent of fat.	Per cent of caseine and sugar.	Per cent of ash.	
41	120	62	6	88.874	1.579	....	.675	
42	102	58	4	90.229	1.824	....	.888	
43	....	....	....	88.465	2.551	....	.724	This is a duplicate of number 39. It was slightly sour when analyzed.
44	124	58	5	89.249	.767	....	.491	
45	102	58	6	89.041	2.659	....	.597	
46	117	58	6	88.138	2.844	...	.685	
47	108	58	7	89.494	1.686	....	.608	
48	120	62	6	88.412	2.458	....	.724	
49	118	60	6	89.085	1.987	....	.610	
50	....	....	....	89.268	2.080	....	.561	
51	112	60	4	90.222	1.818	....	.674	
52	120	62	6	87.778	2.810	....	.887	
53	105	62	5½	90.160	1.638	....	.649	
54	119	68	6½	88.455	1.400	....	.725	
55	98	66	6	90.166	1.944	....	.575	
56	110	68	7	89.170	2.421	....	.597	
57	122	66	6	87.661	1.641	....	.682	
48	114	65	7	88.217	2.699	....	.627	
59	114	62	6	88.566	2.899	....	.658	
60	118	58	6	89.171	2.286	....	.688	
61	104	60	5	90.414	1.687	....	.539	
62	99	65	7½	89.818	3.076	....	.555	
63	97	66	8½	89.076	2.850	....	.587	
64	108	65	6	89.456	2.357	....	.596	
65	96	66	9½	88.691	3.216	....	.539	
66	108	56	6½	89.894	1.844	....	.559	
67	101	58	8	90.017	1.740	....	.547	
68	94	65	4	91.291	1.807	....	.580	
69	102	64	7	89.318	2.478	....	.520	
70	117	62	7½	89.552	1.784	...	.594	
71	....	....	....	90.169	2.086	....	.558	
72	....	....	....	89.508	1.474	....	.678	
				Above Standard.				
73	116	58½	5½	87.470	3.300	....	.620	
74	124	55	7	86.910	3.066	....	.685	
75	115	60	7	87.580	3.148	....	....	
76	122	60	7½	84.970	3.119	....	....	
77	121	60	12	86.620	3.500	....	....	
78	119	62	8	87.291	3.371	....	....	
79	120	60	7	86.550	3.759	....	....	
80	120	60	7½	87.366	3.001	....	....	
81	111	60	6	87.269	3.694	....	....	
82	117	60	6	87.160	3.536	....	....	
83	118	60	....	86.966	3.755	....	....	
84	118	60	12	86.480	3.966	....	....	
85	118	60	12	86.620	3.985	....	....	
86	108	58	....	83.680	7.870	....	....	
87	116	60	....	86.890	3.640	....	....	
88	112	62	14	86.490	3.368	....	....	
89	115	60	10	86.780	3.430	....	....	
90	121	62	11	86.650	3.419	....	....	
91	112	62	9	86.605	3.155	....	....	
92	115	60	7	87.702	3.172	....	....	
93	121	55	7	87.618	3.098	....	.672	

The milkmen, from whom Nos. 6, 23, 24, 25 and 28 in the table were taken, declared that the milk was given to the inspector just as it came from the cows; and that if it was below the standard it was the fault of the cows and not their's. In order to ascertain the truth of the matter, the Dairy Commissioner directed us to go to the dairies of these gentlemen and procure samples which *we knew* to be "just as they came from the cows." This we did and the result is shown in the table by Nos. 83, 84, 87, 88, 89 and 90.

Nos. 83 and 84 were taken at different times from the same dairy from which came No. 6. Nos. 89, 90, 88 and 87, came from the same dairies as Nos. 23, 24, 25 and 28 respectively.

The following table gives the results of analyses of cheese.

Number.	Per cent of water.	Per cent of fat.	Per cent of curd.	Per cent of ash.	State brand No.	Remarks.
1..........	23.78	29.18	42.17	4.87	188	
2..........	28.96	27.95	38.84	4.25	162	
3..........	26.86	29.98	38.69	4.47	364	
4..........	27.77	23.84	43.94	4.45	250	
5..........	27.33	27.34	41.53	3.80	410	
6..........	23.32	32.89	39.28	4.51	218	
7..........	23.50	35.62	36.77	4.11	127	
8..........	32.33	28.78	35.60	3.29	90	
9..........	29.70	26.49	39.83	3.98	105	
10..........	31.27	27.00	37.44	4.29	10	
11..........	28.37	31.28	36.52	3.83	......	Cheese exhibited at the State Fair, 1885, belonging to S. G. Bartlett, full cream.
12..........	30.59	27.05	38.27	4.09	118	
13..........	27.11	32.16	37.12	3.61	122	
14..........	28.62	28.58	39.21	3.59	187	
15..........	28.62	29.90	37.66	3.82	......	Took half of the spec'l prize at State Fair, 1885, made by Seth Bonfoy.
16	15.67	40.07	39.79	4.47	189	
17..........	18.36	38.62	37.87	5.15	188	Duplicate of No. 1, made 3 weeks after.
18..........	33.75	28.95	33.70	3.60	......	Full do r e a m cheese ma e by G. Merry, Verona, N. Y.
19..........	29.80	27.28	38.60	4.16	228	
20..........	38.35	19.93	38.48	3.24	......	Made by George M. Crill, Holland Pat'nt, "night-skim."
21..........	30.17	27.85	38.73	3.25	120	
Average..	27.82	28.61	38.10	4.39	......	

The analysis No. 17 was of the same specimen of cheese as that of No. 1, but was made about three weeks later, during which time the cheese dried out so that it lost 5·42 per cent of its water. This increases the relative proportion of the solids and demonstrates that the fat of cheese should be calculated from the solids only.

The following table shows the results of the analyses of the fat of the cheese above tabulated.

Number.	Insoluble fatty acid.	Soluble fatty acid.	State brand No.	Remarks.
1	87.14	6.80	188	
2	87.78	6.38	162	
3	87.31	6.29	364	
4	85.60	6.44	250	
5	87.81	5.66	410	
6	87.98	6.42	218	
7	88.31	5.70	127	
8	85.64	5.84	90	
9	88.71	5.72	105	
10	87.30	6.82	228	
11	88.87	6.67	10	Cheese on exhibition at the State Fair, 1885; belonging to S. G. Bartlett.
12	88.40	5.90	....	
13	83.53	9.34	113	
14	86.34	6.39	122	
15	84.16	7.93	187	Took half of the special prize at State Fair, 1885; made by Seth Bonfoy.
16	86.26	6.71	....	This analysis was made with K. H. O., purified by barium.
17	88.95	6.13	189	
18	86.68	8.57	189	This analysis was made with K. H. O., purified in the usual way with alcohol.
19	86.46	8.63	....	Full cream cheese made by G. Merry, Verona, N. Y.

The high percentage of soluble fatty acids in some of the analyses is probably due to the free acid in the cheese. The lower percentage in some is due to the fact that the samples were washed before being analyzed.

Respectfully submitted,
R. D. CLARK, M. D.,
Albany, N. Y.

THE STATE CHEESE BRAND.

UTICA, N. Y., *January* 2, 1886.

Hon. JOSIAH K. BROWN, *State Dairy Commissioner:*

DEAR SIR — In answer to your inquiry as to the effect of the statute passed last spring, creating a State brand for "full-cream cheese," it will be necessary to refer briefly to the position of New York State cheese in 1884. The practice of skimming the milk in spring and fall, among the so-called full-milk factories, had been carried to such excess that June and July were practically the only months during which full-cream cheese was made in the majority of our factories. The evil was insidious for this reason, that when only the night's milk was skimmed and mixed with that of the morning, it took an expert judge to detect it in the cheese. The argument among advocates of the practice was, that it was so difficult to tell cheese that was thus made, no great harm could be done by it. For a certain length of time, perhaps four or five months, the cheese would keep and be apparently fine. Then suddenly it would get off-flavored, and the buyer who held it at this period was the loser. English buyers who had previously been caught in this way, grew shy of our fall cheese in 1884, and Canadian cheese of the same dates sold at one-half to one cent a pound higher than ours. In trying to overreach others, our makers had overreached themselves, and when this became evident, the more prudent ones among them began to devise some means for regaining their lost reputation.

Robert McAdam of the Utica Board of Trade, was the first to suggest the creation of a State brand, which there should be a penalty for using on any but full-cream goods. At a meeting of the Utica Board and afterward at a meeting of the Rome Board, his resolutions were adopted. Soon after, the State Dairymen's Association approved and indorsed the plan. Lists of signatures were sent to Albany and the statute as it now stands was framed and enacted, very properly placing the issue of the brands in the hands of the State Dairy Commissioner. During the season of 1885, I understand that 452 numbers were issued, which means that 452 factories adopted the brand in full or in part. To my own knowledge several creameries also applied for and obtained the brand to use during a part of the season when the milk was not skimmed. It is evident that the brand would be even more valuable to them than to full-milk concerns, for it would remove the suspicion that might naturally taint all their product, and enable them to sell their genuine full cream goods for what they really were.

The establishment of this brand has worked marvels in the course of a single season. It has freed our cheese from all suspicion of being skimmed whenever the brand is on it. So far as I am aware, no case has occurred in which it was believed that the brand had been illegally used. Instead of selling below Canadian cheese, our make has sold one-half to three-quarter cent above it for the same dates. And strange to say, the brand has largely increased our home trade in the best quality of our make. So popular has the brand become, and so thoroughly has its value been recognized, that home dealers all over the country, both in this and other States, will now purchase no New York cheese that is not branded with the State mark. Instances have occurred in which cheese were carelessly forwarded to New York without being branded. The consignee has written back to learn if the goods were not full-cream; and upon receiving the brand and marking it on them, the cheese at once sold for one-half cent a pound more than any offer he could get for them before.

It may be objected that cheese was never so low and depressed as it was in the past season. That is true, but it is not the whole truth. The proper way to put it is this: if the State brand had not been adopted, cheese would have sold one-half to one cent lower than it did. It has created confidence in the goods of New York State, and has made them in demand all over the Union. Its effect has been such that no factory making full-cream goods can hereafter expect to command top prices unless its cheese bears this brand. It is doubtful if any other plain and simple law on our statute books has ever brought such immediate and valuable assistance to one of our chief industries, as this little restrictive paragraph has brought to the cheese makers of this State.

The salesmen of the Utica Board are entirely satisfied with the result thus far; and another season the brand will be used by every factory in this region, without exception, in which full-cream cheese is made.

Respectfully yours,

BENJAMIN D. GILBERT,

Secretary of Utica Board of Trade.

LAWS.

Chapter 215, Laws of 1882.

AN ACT to regulate the manufacture and sale of oleomargarine or any form of imitation butter and lard or any form of imitation cheese, for the prevention of fraud, and the better protection of the public health.

Section 1. No person, persons, firm or corporation manufacturing with intent to sell any article or substance in semblance of natural butter or natural cheese not the legitimate product of the dairy, and not made exclusively from milk or cream, or both, with salt or rennet, or both, and with or without coloring matter or sage, but into which any animal intestinal or offal fats, or any oils or fats of any kind whatosever not produced from milk or cream, or into which melted butter, lard or tallow shall be introduced, shall add thereto or combine therewith any annatto or compounds of the same, or any other substance or substances whatsoever, for the purpose or with the effect of imparting thereto a color resembling that of yellow (or any shade of the same) butter or cheese; nor shall they introduce said coloring matter into any of the articles of which the same is composed.

§ 2. No person, persons, firm or corporation shall deal in, sell, expose for sale or give away any article or substance in semblance of natural butter or natural cheese described in the first section of this act, and known as oleomargarine or imitation butter and lard or imitation cheese; and no keeper of any hotel, restaurant, boarding-house or other place of public entertainment shall keep, use or serve either as food for their guests or for cooking purposes, any such imitation butter or cheese which shall contain any of the coloring matter therein prohibited, or be colored contrary to the provisions of this act.

§ 3. No person, persons, firm or corporation shall manufacture, with intent to sell, deal in, sell or expose for sale any article or substance in semblance of natural cheese not the legitimate product of the dairy, and not made exclusively of milk or cream, or both, but into which any animal, intestinal or offal fats or oils of any kind whatsoever, not produced from milk or cream, shall be introduced, unless the words "imitation cheese," shall be plainly stenciled, in plain Roman letters at least one-half inch in length with durable paint, upon the sides of each and every cheese and also upon the outside of top, and opposite sides of each and every box containing the same, in letters and with paint as before mentioned and described.

§ 4. For the purposes of this act the terms "natural butter" and "natural cheese" shall be understood to mean the products usually known by these names, and which are manufactured exclusively from milk or cream, or both, with salt and rennet, and with or without coloring matter or sage.

§ 5. Every person, firm or corporation violating the provisions of this act shall be deemed guilty of a misdemeanor, and upon conviction thereof shall be punished by a fine of not less than $50 nor more than $200, or by imprisonment in the county jail for not less then ten nor more than thirty days, or by both such fine and imprisonment for each and every offense, in the discretion of the court, one-half of such fine to be paid to the complainant, the other half to be paid to the officer or officers having charge of the poor fund of the town where such prosecution occurs, for the support of the poor, or if the poor of such town are supported by the county, then said money shall be paid to the officer or officers having charge of the poor fund of the county in which said town is located, to be used for the support of the poor of such county. But nothing in this act shall be so construed as to interfere with or to abridge any right obtained, secured or guaranteed by any law of congress, or by any patent duly granted by the United States government.

§ 6. This act shall take effect sixty days after it becomes a law.

CHAPTER 238, LAWS OF 1882.

AN ACT for the protection of dairymen, and to prevent deception in the sales of butter and cheese.

SECTION 1. Every person who shall manufacture for sale, or who shall offer or expose for sale, or who shall export to a foreign country, by the tub, firkin, box or package, or any greater quantity, any article or substance in semblance of butter or cheese not the legitimate product of the dairy, and not made exclusively of milk or cream, but into which any oil, lard or fat not produced from milk or cream enters as component part, or into which melted butter, or any oil thereof, has been introduced to take the place of cream, shall distinctly and durably stamp, brand or mark upon the side of every cheese and also upon the top, and side of every such tub, firkin, box or package of such article or substance the words "oleomargarine butter," or if containing the cheese, the words "imitation cheese," only where it can be plainly seen, in Roman letters, which shall be burned in or painted thereon with permanent black paint, in a straight line, and shall not be less than one-half inch in length, and if for export, shall also invoice the same, and clear the same, through the custom-house as "oleomargarine butter," or if cheese, as "imitation cheese;" and in case of retail sales of such articles or substances in parcels, the seller shall in all cases set, sell, or offer or expose the same for sale from a tub, firkin, box or package stamped, branded or marked as herein stated, and shall also deliver therewith to the purchaser, printed label bearing the plainly printed words, "oleomargarine butter," or if cheese, the words "imitation cheese," only, in Roman letters not less than one-half inch in length, which shall be printed in a straight line; and every sale of such article or substance, or export of the same, by the tub, firkin, box or package, or in any greater quantity, not so stamped, branded or marked, and if exported, not invoiced and cleared through the custom-house as "oleomargarine butter," or if cheese, as "imitation cheese," and every sale of such article or substance at retail, in parcels, that shall not be sold from a tub, firkin, box or package so stamped, branded or marked, or without delivery of a label therewith, as above stated, is declared to be unlawful and void, and no action upon any contract shall be maintained in any of the courts of this State to recover upon any contract for the sale of any such article or semblance not so stamped, branded, marked, labeled or sold.

§ 2. Every person who shall sell, or offer or expose for sale, or export to a foreign country, or have in his or her possession, with intent to sell by the tub, firkin, box or package, or in any greater quantity, any of the said article or substance required by the first section of this act to be stamped, branded, marked, and if exported, invoiced and cleared through the custom-house as "oleomargarine butter," or "imitation cheese," as therein stated; that shall not be so stamped, branded, marked, and if exported, invoiced according to the provisions of this act,or in case of retail sales in parcels every person who shall sell, or offer or expose for sale, any of said article or substance, without selling, offering or exposing for sale, the same from a tub, firkin, box or package stamped, branded or marked, as in said first section stated, or without delivery of a label, as required by section one of this act, shall for every such offense forfeit and pay a fine of $100, to be recovered with costs, in any of the courts of this State having cognizance thereof, in an action to be prosecuted by any district attorney, in the name of the people, and the one-half of such recovery shall be paid to the informer, and the residue shall be applied to the support of the poor in the county where such recovery is had.

§ 3. Every person who shall sell or offer or expose for sale, or export to a foreign country, or who shall cause or procure to be sold, offered or exposed for sale by the tub, firkin, box or package, or in any greater quantity, any article or substance required by the first section of this act to be stamped, branded, marked, and if exported, invoiced and cleared as therein stated, not so stamped, branded, marked, and if exported, invoiced and cleared; or in case of retail sales in parcels, every person who shall sell, or offer or expose for sale, or who shall cause or procure to be sold, offered or exposed for sale, any article or substance required by the first section of this act to be sold, offered or exposed for such sale, from a tub, firkin, box or package, stamped, branded or marked, and labeled as therein stated, contrary to the provisions of said section, shall be guilty of a misdemeanor, and upon conviction shall be punished by a fine of not less than $50, nor more than $200, or by imprisonment in the county jail for not less than ten nor more than thirty days, or by both such fine and imprisonment, for each and every offense.

§ 4. All acts or parts of acts inconsistent with the provisions of this act are hereby repealed.

§ 5. This act shall take effect immediately.

Chapter 246, Laws of 1882.

AN ACT to prevent fraud in the sale of oleomargarine, butterine, suine or other substances not butter.

Section 1. Any person who shall hereafter sell, either at wholesale or retail, any oleomargarine, butterine, suine or other substance not butter, and represent the same to be butter, shall be deemed guilty of a misdemeanor, and upon conviction shall be liable to a fine of not less than $25, nor more than $100, for each and every offense, or by imprisonment in the county jail not to exceed thirty days, or by both such fine and imprisonment.

§ 2. The sale by any person of such oleomargarine, butterine, suine or other substance not butter, representing the same to be butter, shall be deemed presumptive evidence of the guilt of such person.

§ 3. This act shall take effect immediately.

CHAPTER 202, LAWS OF 1884.

AN ACT to prevent deception in sales of dairy products.

PASSED April 24, 1884; three-fifths being present.

The People of the State of New York, represented in Senate and Assembly, do enact as follows:

SECTION 1. No person or persons shall sell or exchange, or expose for sale or exchange, any unclean, impure, unhealthy, adulterated or unwholesome milk, or shall offer for sale any article of food made from the same or of cream from the same. This provision shall not apply to pure skim cheese made from milk which is clean, pure, healthy, wholesome and unadulterated, except by skimming. Whoever violates the provisions of this section is guilty of a misdemeanor and shall be punished by a fine of not less than twenty-five nor more than two hundred dollars, or by imprisonment of not less than one or more than six months, or both such fine and imprisonment for the first offense, and by six months' imprisonment for each subsequent offense.

§ 2. No person shall keep cows for the production of milk for market, or for sale or exchange, or for manufacturing the same, or cream from the same, into articles of food, in a crowded or unhealthy condition, or feed the cows on food that is unhealthy, or that produces impure, unhealthy, diseased or unwholesome milk. No person shall manufacture from impure, unhealthy, diseased or unwholesome milk, or of cream from the same, any article of food. Whoever violates the provisions of this section is guilty of a misdemeanor and shall be punished by a fine of not less than twenty-five nor more than two hundred dollars, or by imprisonment of not less than one or more than four months, or by both such fine and imprisonment for the first offense and by four months' imprisonment for each subsequent offense.

§ 3. No person or persons shall sell, supply or bring to be manufactured to any butter or cheese manufactory, any milk diluted with water, or any unclean, impure, unhealthy, adulterated or unwholesome milk, or milk from which any cream has been taken (except pure skim milk to skim cheese factories), or shall keep back any part of the milk commonly known as "strippings," or shall bring or supply milk to any butter or cheese manufactory that is sour (except pure skim milk to skim cheese factories). No butter or cheese manufactories, except those who buy all the milk they use, shall use for their own benefit, or allow any of their employees or any other person to use, for their own benefit, any milk, or cream from the milk, or the product thereof brought to said manufactories, without the consent of the owners thereof. Every butter or cheese manufacturer, except those who buy all the milk they use, shall keep a correct account of all the milk daily received, and of the number of pounds and packages of butter, the number and aggregate weight of cheese made each day, the number of packages of cheese and butter disposed of, which shall be open to inspection to any person who delivers milk to such manufacturer. Whoever violates the provisions of this section shall be guilty of a misdemeanor and shall be punished for each offense by a fine of not less than twenty-five or more than two hundred dollars, or not less than one or more than six months' imprisonment, or both such fine and imprisonment.

§ 4. No manufacturer of vessels for the package of butter shall sell or dispose of any such vessels without branding his name and true weight of the vessel or vessels on the same with legible letters or figures not less than one-fourth of an inch in length. Whoever violates the provisions of this section is guilty of a misdemeanor and shall be punished for each offense by a fine of not less than fifty nor more than one hundred dollars, or by imprisonment of not less than thirty or more than sixty days, or by both such fine and imprisonment.

§ 5. No person shall sell or offer or expose for sale any milk except in the county from which the same is produced, unless each can, vessel or package containing such milk shall be distinctly and durably branded with letters not less than one inch in length, on the outside, above the center, on every can, vessel or package containing such milk, the name of the county from which the same is produced, and the same mark shall be branded or painted in a conspicuous place on the carriage or vehicle in which the milk is drawn to be sold, and such milk can only be sold in or retailed out of a can, vessel, package or carriage so marked. Whoever violates the provisions of this section shall be guilty of a misdemeanor, and shall be punished by a fine of not less than twenty-five nor more than two hundred dollars, or not less than two months' or more than four months' imprisonment, or both such fine and imprisonment for the first offense, and by four months' imprisonment for each subsequent offense.

§ 6. No person shall manufacture out of any oleaginous substance or substances, or any compound of the same, other than that produced from unadulterated milk, or of cream from the same, any article designed to take the place of butter or cheese produced from pure, unadulterated milk or cream of the same, or shall sell, or offer for sale, the same as an article of food. This provision shall not apply to pure skim-milk cheese made from pure skim milk. Whoever violates the provisions of this section shall be guilty of a misdemeanor, and be punished by a fine of not less than one hundred nor more than five hundred dollars, or not less than six months' or more than one year's imprisonment, or by both such fine and imprisonment, for the first offense, and by imprisonment for one year for each subsequent offense.

§ 7. No person shall offer, sell or expose for sale in full packages, butter or cheese branded or labeled with a false brand or label as to county or State in which the article is made. Whoever violates the provisions of this section is guilty of a misdemeanor, and shall be punished by a fine of not less than twenty-five or more than fifty dollars, or imprisonment of not less than fifteen or more than thirty days, for the first offense, and fifty dollars or thirty days' imprisonment for each subsequent offense.

§ 8. No person shall manufacture, sell, or offer for sale, any condensed milk unless the same shall be put up in packages, upon which shall be distinctly labeled or stamped the name or brand by whom or under which the same is made. No condensed milk shall be made or offered for sale unless the same is manufactured from pure, clean, healthy, fresh, unadulterated and wholesome milk, from which the cream has not been removed; or unless the proportion of milk solids contained in the condensed milk shall be in amount the equivalent of twelve per centum of milk solids in crude milk, and of such solids twenty-five per centum shall be fat. When condensed milk shall be sold from cans or packages not hermetically sealed, the vendor shall brand or label such cans or packages with the name of the county or counties from which the same was produced, and the name of the vendor. Whoever violates the provisions of this section shall be guilty of a misdemeanor, and be punished by a fine of not less than fifty or more than five hundred dollars, or by imprisonment of not more than six months, or both such fine and imprisonment for the first offense, and by six months' imprisonment for each subsequent offense.

§ 9. The Governor, by and with the advice and consent of the Senate, shall appoint a commissioner, who shall be known as the New York State Dairy Commissioner, who shall be a citizen of this State, and who shall hold his office for the term of two years, or until his successor is appointed, and shall receive a salary of three thousand dollars per annum and his necessary expenses incurred in the discharge of his official duties under this act; said commissioner shall be appointed within ten days after the passage of this act, and shall be charged, under the direction of the Governor, with the enforcement of the various provisions thereof. Said commissioner may be removed from office at the pleasure of the Governor, and his successor appointed as above provided for.

The said commissioner is hereby authorized and empowered to appoint such assistant commissioners and to employ such experts, chemists, agents and such counsel as may be deemed by him necessary for the proper enforcement of this law. Their compensation to be fixed by the commissioner.

The said commissioner is also authorized to employ a clerk at an annual salary of not to exceed twelve hundred dollars.

The sum of thirty thousand dollars is hereby appropriated to be paid for such purpose out of any moneys in the treasury not otherwise appropriated. All charges, accounts and expenses authorized by this act shall be paid by the Treasurer of the State, upon the warrant of the Comptroller. The entire expenses of said commissioner shall not exceed the sum appropriated for the purposes of this act.

The said commissioner shall make annual reports to the Legislature, not later then the fifteenth day of January of each year, of his work and proceedings, and shall report in detail the number of assistant commissioners, experts, chemists, agents and counsel he has employed, with their expenses and disbursements. The said commissioner shall have a room in the New Capitol, to be set apart for his use by the Capitol Commissioner.

§ 10. The said commissioner and assistant commissioners, and such experts, chemists, agents and counsel as they shall duly authorize for the purpose, shall have full access, egress and ingress to all places of business, factories, farms, buildings, carriages, cars, vessels and cans used in the manufacture and sale of any dairy products or any imitation thereof. They shall also have power and authority to opon any package, can or vessel containing such articles which may be manufactured, sold or exposed for sale in violation of the provisions of this act, and may inspect the contents therein and may take therefrom samples for analysis.

§ 11. Courts of special sessions shall have jurisdiction of all cases arising under this act, and their jurisdiction is hereby extended so as to enable them to enforce the penalties imposed by any or all of the sections hereof.

§ 12. In all prosecutions under this act the costs thereof shall be paid out of the fine, if one is collected; if not, the same shall be paid in the manner now provided for by law, and the rest of the fine shall be paid to the State Treasurer.

§ 13. In all prosecutions under this act, relating to the sale and manufacture of unclean, impure, unhealthy, adulterated or unwholesome milk, if the milk be shown to contain more than eighty-eight per centum of water or fluids or less than twelve per centum of milk solids which shall contain not less than three per centum of fat, it shall be declared adulterated, and milk drawn from cows within fifteen days before and five days after parturition, or from animals fed on distillery waste, or any substance in the state of putrefaction, or fermentation, or upon any unhealthy food whatever, shall be declared unclean, impure, unhealthy and unwholesome milk. This section shall not prevent the feeding of ensilage from silos.

§ 14. The doing of any thing prohibited being done, and the not doing of any thing directed to be done in this act shall be presumptive evidence of a willful intent to violate the different sections and provisions hereof.

§ 15. Chapters four hundred and sixty-seven of the laws of eighteen hundred and sixty-two, five hundred and forty-four and five hundred and eighteen of the laws of eighteen hundred and sixty-four, five hundred and fifty-nine of the laws of eighteen hundred and sixty-five, four hundred and fifteen of the laws of eighteen hundred and seventy-seven, two hundred and twenty and two hundred and thirty-seven of the laws of eighteen hundred and seventy-eight, four hundred and thirty-nine of the laws of eighteen hundred and eighty, and two hundred and fourteen of the laws of eighteen hundred and eighty-two, are hereby repealed.

§ 16. This act shall take effect on the first day of June, eighteen hundred and eighty-four, except as otherwise provided therein.

CHAPTER 183, LAWS OF 1885.

AN ACT to prevent deception in the sale of dairy products, and to preserve the public health, being supplementary to and in aid of chapter two hundred and two of the laws of eighteen hundred and eighty-four, entitled "An act to prevent deception in sales of dairy products."

PASSED April 30, 1885; three-fifths being present.

The People of the State of New York, represented in Senate and Assembly, do enact as follows:

SECTION 1. No person or persons shall sell or exchange, or expose for sale or exchange, any unclean, impure, unhealthy, adulterated or unwholesome milk, or shall offer for sale any article of food made from the same, or of cream from the same. The provisions of this section shall not apply to skim milk sold to bakers or to housewives for their own use or manufacture, upon written orders for the same, nor to skim milk sold for use in the county in which it is produced. This provision shall not apply to pure skim cheese made from milk which is clean, pure, healthy, wholesome and unadulterated, except by skimming. Whoever violates the provisions of this section is guilty of a misdemeanor, and shall be punished by a fine of not less than $25, nor more than $200, or by imprisonment of not less than one month or more than six months, or both such fine and imprisonment for the first offense, and by six months' imprisonment for each subsequent offense.

§ 2. No person shall keep cows for the production of milk for market, or for sale or exchange, or for manufacturing the same, or cream from the same, into articles of food, in a crowded or unhealthy condition, or feed the cows on food that is unhealthy, or that produces impure, unhealthy, diseased or unwholesome milk. No person shall manufacture from impure, unhealthy, diseased or unwholesome milk, or of cream from the same, any article of food. Whoever violates the provisions of this section is guilty of a misdemeanor and shall be punished by a fine of not less than $25, nor more than $200, or by imprisonment of not less than one month or more than four months, or by both such fine and imprisonment for the first offense, and by four months' imprisonment for each subsequent offense.

§ 3. No person or persons shall sell, supply or bring to be manufactured to any butter or cheese manufactory, any milk diluted with water or any unclean, impure, unhealthy, adulterated or unwholesome milk, or milk from which any cream has been taken (except pure skim milk to skim cheese factories), or shall keep back any part of the milk commonly known as "strippings," or shall bring or supply milk to any butter or cheese manufactory that is sour (except pure skim milk to skim cheese factories). No butter or cheese manufactories, except those who buy all the milk they use, shall use for their own benefit, or allow any of their employees or any other person to use for their own benefit, any milk, or cream from the milk, or the product thereof, brought to said manufactories without the consent of the owners thereof. Every butter or cheese manufacturer, except those who buy all the milk they use, shall keep a correct account of all the milk daily received, and of the number of packages of butter and cheese made each day, and the number of packages and aggregate weight of cheese and butter disposed of each day, which account shall be open to inspection to any person who delivers milk to such manufacturer. Whoever violates the provisions of this section shall be guilty of a misdemeanor, and shall be punished for each offense by a fine of not less than $25 or more than $200, or not less than one month or more than six months' imprisonment, or both such fine and imprisonment.

§ 4. No manufacturer of vessels for the package of butter shall sell or dispose of any such vessels without branding his name and the true weight of the vessel or vessels on the same, with legible letters or figures not less than one-fourth of an inch in length. Whoever violates the provisions of this section is guilty of a misdemeanor, and shall be punished for each offense by a fine of not less than $50 nor more than $100, or by imprisonment of not less than thirty days or more than sixty days, or by both such fine and imprisonment.

§ 5. No person shall sell, or offer or expose for sale, any milk except in the county from which the same is produced, unless each can, vessel or package containing such milk shall be distinctly and durably branded with letters not less than one inch in length, on the outside, above the center, on every can, vessel or package containing such milk, the name of the county from which the same is produced; and the same marks shall be branded or painted in a conspicuous place on the carriage or vehicle in which the milk is drawn to be sold; and such milk can only be sold in, or retailed out of a can, vessel, package or carriage so marked. Whoever violates the provisions of this section shall be guilty of a misdemeanor, and shall be punished by a fine of not less than $25 nor more than $200, or not less than two months or more than four months' imprisonment, or both such fine and imprisonment, for the first offense, and by four months' imprisonment for each subsequent offense.

§ 6. No person shall manufacture out of any oleaginous substance or substances, or any compound of the same, other than that produced from unadulterated milk, or of cream from the same, any article designed to take the place of butter or cheese produced from pure unadulterated milk or cream of the same, or shall sell, or offer for sale, the same as an article of food. This provision shall not apply to pure skim-milk cheese made from pure skim milk. Whoever violates the provisions of this section shall be guilty of a misdemeanor, and be punished by a fine of not less than $200 nor more than $500, or not less than six months or more than one year's imprisonment, or both such fine and imprisonment for the first offense, and by imprisonment for one year for each subsequent offense.

§ 7. No person by himself or his agents or servants shall render or manufacture out of any animal fat or animal or vegetable oils not produced from unadulterated milk or cream from the same, any article or product in imitation or semblance of or designed to take the place of natural butter or cheese produced from pure unadulterated milk or cream of the same, nor shall he or they mix, compound with, or add to milk, cream or butter any acids or other deleterious substance or any animal fats or animal or vegetable oils not produced from milk or cream, with design or intent to render, make or produce any article or substance or any human food in imitation or semblance of natural butter or cheese, nor shall he sell, keep for sale, or offer for sale any article, substance or compound made, manufactured or produced in violation of the provisions of this section, whether such article, substance or compound shall be made or produced in this state or in any other state or country. Whoever violates the provisions of this section shall be guilty of a misdemeanor and be punished by a fine of not less than $200 nor more than $500 or not less than six months' or more than one years' imprisonment for the first offense, and by imprisonment for one year for each subsequent offense. Nothing in this section shall impair the provisions of section six of this act.

§ 8. No person shall manufacture, mix or compound with or add to natural milk, cream or butter any animal fats or animal or vegetable oils, nor shall he make or manufacture any oleaginous substance not produced from milk or cream, with intent to sell the same for butter or cheese made from unadulterated milk or cream, or have the same in his possession, or offer the same for sale with such intent, nor shall any article or substance or compound so made or produced, be sold for butter or cheese, the product of the dairy. If any person shall coat, powder or color with annatto or any coloring matter whatever butterine or oleomargarine, or any compounds of the same or any product or manufacture made in whole or in part from animal fats or animal or vegetable oils not produced from unadulterated milk or cream whereby the said product, manufacture or compound shall be made to resemble butter or cheese, the product of the dairy, or shall have the same in his possession, or shall sell or offer for sale or have in his possession any of the said products which shall be colored or coated in semblance of or to resemble butter or cheese, it shall be conclusive evidence of an intent to sell the same for butter or cheese, the product of the dairy. Whoever violates any of the provisions of this section shall be guilty of a misdemeanor, and be punished by a fine of not less than two hundred dollars nor more than one thousand dollars. This section shall not be construed to impair or affect the prohibitions of sections six and seven of this act.

§ 9. Every manufacturer of full-milk cheese may put a brand upon each cheese indicating "full-milk cheese," and the date of the month and year when made; and any person using this brand upon any cheese made from which any cream whatever has been taken shall be guilty of a misdemeanor, and shall be punished for each offense by a fine of not less than one hundred dollars nor more than five hundred dollars.

§ 10. No person shall offer, sell or expose for sale in full packages, butter or cheese branded or labeled with a false brand or label as to county or state in which the article is made. Whoever violates the provisions of this section is guilty of a misdemeanor, and shall be punished by a fine of not less than twenty-five dollars or more than fifty dollars, or imprisonment of not less than fifteen days or more than thirty days for the first offense, and fifty dollars or thirty days' imprisonment for each subsequent offense.

§ 11. No person shall manufacture, sell or offer for sale any condensed milk, unless the same shall be put up in packages upon which shall be distinctly labeled or stamped the name, or brand, by whom or under which the same is made. No condensed milk shall be made or offered for sale unless the same is manufactured from pure, clean, healthy, fresh, unadulterated and wholesome milk, from which the cream has not been removed, or unless the proportion of milk solids contained in the condensed milk shall be in amount the equivalent of twelve per centum of milk solids in crude milk, and of such solids twenty-five per centum shall be fat. When condensed milk shall be sold from cans, or packages not hermetically sealed, the vendor shall brand or label such cans or packages with the name of the county or counties from which the same was produced, and the name of the vendor. Whoever violates the provisions of this section shall be guilty of a misdemeanor, and be punished by a fine of not less than fifty dollars or more than five hundred dollars, or by imprisonment of not more than six months, or by both such fine and imprisonment for the first offense, and by six months' imprisonment for each subsequent offense.

§ 12. Upon the expiration of the term of office of the present commissioner, the governor, by and with the advice and consent of the senate, shall appoint a commissioner, who shall be known as the New York state dairy commissioner, who shall be a citizen of this state, and who shall hold his office for the term of two years, or until his successor is appointed, and shall receive a salary of three thousand dollars per annum, and his necessary expenses incurred in the discharge of his official duties under this act. Said commissioner shall be charged, under the direction of the governor, with the enforcement of the various provisions thereof, and with all laws prohibiting or regulating the adulteration of butter, cheese or milk. The said commissioner is hereby authorized and empowered to appoint such assistant commissioners and to employ such experts, chemists, agents and such counsel as may be deemed by him necessary for the proper enforcement of this law, their compensation to be fixed by the commissioner. The said commissioner is also authorized to employ a clerk at an annual salary not to exceed twelve hundred dollars. The sum of fifty thousand dollars is hereby appropriated, to be paid for such purpose out of any moneys in the treasury not otherwise appropriated. All charges, accounts and expenses authorized by this act shall be paid by the treasurer of the state upon the warrant of the comptroller, after such expenses have been audited and allowed by the comptroller. The entire expenses of said commissioner shall not exceed the sum appropriated for the purposes of this act. The said commissioner shall make annual reports to the legislature, on or before the fifteenth day of January of each year, of his work and proceedings, and shall report in detail the number of assistant commissioners, experts, chemists, agents and counsel he has employed, with their expenses and disbursements. The said commissioner shall have a room in the new capitol, to be set apart for his use by the capitol commissioner. The said commissioner and assistant commissioners and such experts, chemists, agents and counsel as they shall duly authorize for the purpose, shall have full access, egress and ingress to all places of business, factories, farms, buildings, carriages, vessels and cans used in the manufacture and sale of any dairy products or any imitation thereof. They shall also have power and authority to open any package, can or vessel containing such articles which may be manufactured, sold or exposed for sale, in violation of the provisions of this act, and may inspect the contents therein and may take therefrom samples for analysis. This section shall not affect the tenure of the office of the present commissioner.

§ 13. Upon the application for a warrant under this act, the certificate of the analyst or chemist of any analysis made by him shall be sufficient evidence of the facts therein stated. Every such certificate shall be duly signed and acknowledged by such analyst or chemist before an officer authorized to take acknowledgments of conveyances of real estate.

§ 14. Courts of special sessions shall have jurisdiction of all cases arising under this act, and their jurisdiction is hereby extended so as to enable them to enforce the penalties imposed by any or all sections thereof.

§ 15. In all prosecutions under this act, one-half of the money shall be paid by the court or clerk thereof to the city or county where the recovery shall be had, for the support of the poor, except in the city and county of New York shall be equally divided between the pension funds of the police and fire departments, and the residue shall be paid to the dairy commissioner, who shall account therefor to the treasury of the State, and be added to any appropriation made to carry out the provisions of this act. All sums of money expended by the dairy commissioner under the provisions of this act, shall be audited and allowed by the comptroller of the State. Any bond given by any officer shall be subject to the provisions of this section.

§ 16. In all prosecutions under this act relating to the sale and manufacture of unclean, impure, unhealthy, adulterated or unwholesome milk, if the milk be shown to contain more than eighty-eight per centum of water or fluids, or less than twelve per centum of milk solids, which shall contain not less than three per centum of fat, it shall be declared adulterated, and milk drawn from cows within fifteen days before, and five days after, parturition, or from animals fed on distillery waste, or any substance in the state of putrefaction or fermentation, or upon any unhealthy food whatever, shall be declared unclean, unhealthy, impure and unwholesome milk. This section shall not prevent the feeding of ensilage from silos.

§ 17. The doing of any thing prohibited being done, and the not doing of any thing directed to be done in this act, shall be presumptive evidence of a willful intent to violate the different sections and provisions thereof. If any person shall suffer any violation of the provisions of this act by his agent, servant, or in any room or building occupied or controlled by him, he shall be deemed a principal in such violation and punished accordingly.

§ 18. Chapters four hundred and sixty-seven of the laws of eighteen hundred and sixty-two, five hundred and forty-four and five hundred and eighteen of the laws of eighteen hundred and sixty-four, five hundred and fifty-nine of the laws of eighteen hundred and sixty-five, four hundred and fifteen of the laws of eighteen hundred and seventy-seven, two hundred and twenty and two hundred and thirty-seven of the laws of eighteen hundred and seventy-eight, four hundred and thirty-nine of the laws of eighteen hundred and eighty, and two hundred and fourteen of the laws of eighteen hundred and eighty-two, are hereby repealed.

§ 19. If any person shall, by himself or other, violate any of the provisions of sections one, two, three, four or five of this act, or knowingly suffer a violation thereof by his agent, or in any building or room occupied by him, he shall in addition to the fines and punishments therein described for each offense, forfeit and pay a fixed penalty of $100. If any person, by himself or another, shall violate any of the provisions of section six, seven or eight of this act, he shall, in addition to the fines and penalties herein prescribed for each offense, forfeit and pay a fixed penalty of $500. Such penalties shall be recovered with costs in any court of this State having jurisdiction thereof, in an action to be prosecuted by the Dairy Commissioner or any of his assistants in the name of the people of the State of New York.

§ 20. This act and each section thereof is declared to be enacted to prevent deception in the sale of dairy products, and to preserve the public health which is endangered by the manufacture, sale or use of the articles or substances herein regulated or prohibited.

§ 21. This act shall take effect immediately. Sections six and seven shall not apply to any product manufactured, or in process of manufacture at the time of the passage of this act; but neither this exemption nor this act shall impair the power to prosecute any violations heretofore committed of section six of the act of which this act is supplemental.

Chapter 193, Laws of 1885.

AN ACT to amend chapter two hundred and two of the laws of eighteen hundred and eighty-four, entitled "An act to prevent deception in sales of dairy products."

Passed April 30, 1885; three-fiths being present.

The People of the State of New York, represented in Senate and Assembly, do enact as follows:

Section 1. Section seven of chapter two hundred and two of the laws of eighteen hundred and eighty-four, entitled "An act to prevent deception in sales of dairy products," is hereby amended to read as follows:

§ 7. No person shall offer, sell or expose for sale butter or cheese branded or labeled with a false brand or label as to the quality of the article or the county or State in which the article is made. The New York State Dairy Commissioner is hereby authorized and directed to procure and issue to the cheese manufactories of the State, upon proper application therefor and under such regulations as to the custody and use thereof as he may prescribe, a uniform stencil brand bearing a suitable device or motto, and the words "New York State Full Cream Cheese." Every brand issued shall be used upon the outside of the cheese and also upon the package containing the same, and shall bear a different number for each separate manufactory, and the commissioner shall keep a book in which shall be registered the name, location and number of each manufactory using the said brand, and the name or names of the persons at each manufactory authorized to use the same. It shall be unlawful to use or permit such stencil brand to be used upon any other than full cream cheese or packages containing the same. Whoever violates the provisions of this section is guilty of a misdemeanor, and for each and every cheese or package so falsely branded shall be punished by a fine of not less than $25 or more than $50, or imprisonment of not less than fifteen or more than thirty days.

§ 2. This act shall take effect immediately.

CHAPTER 427, LAWS OF 1885.

AN ACT to protect butter and cheese manufacturers.

PASSED June 8, 1885; three-fifths being present.

The People of the State of New York, represented in Senate and Assembly, do enact as follows:

SECTION 1. Whoever shall with intent to defraud, sell, supply or bring to be manufactured to any butter or cheese manufactory in this State, any milk diluted with water, or in any way adulterated, unclean or impure, or milk from which any cream has been taken, or milk commonly known as skimmed milk, or whoever shall keep back any part of the milk as strippings, or whoever shall knowingly bring or supply milk to any butter or cheese manufactory, that is tainted or sour, or whoever shall knowingly bring or supply to any butter or cheese manufactory, milk drawn from cows within fifteen days before parturition, or within three days after parturition, or any butter or cheese manufacturers who shall knowingly use or allow any of his or her employees or any other person to use for his or her benefit, or for their own individual benefit, any milk or cream from the milk brought to said butter or cheese manufacturer, without the consent of all the owners thereof, or any butter or cheese manufacturer who shall refuse or neglect to keep or cause to be kept a correct account, open to the inspection of any one furnishing milk to such manufacturer, of the amount of milk daily received, or of the number of pounds of butter and the number of cheese made each day, or of the number cut or otherwise disposed of, and the weight of each, shall for each and every offense forfeit and pay a sum not less than $25 nor more than $100, with costs of suit to be sued for in any court of competent jurisdiction for the benefit of the person or persons, firm or association, or corporation or their assigns upon whom such fraud or neglect shall be committed. But nothing in this act shall affect, impair or repeal any of the provisions of chapter two hundred and two of the laws of eighteen hundred and eighty-four, or of the acts amendatory thereof or supplementary thereto.

§ 2. This act shall take effect immediately.

CHAPTER 458, LAWS OF 1885.

AN ACT to amend chapter one hundred and eighty-three of the laws of eighteen hundred and eighty-five, entitled "An act to prevent deception in the sale of dairy products, and to preserve the public health, being supplementary to and in aid of chapter two hundred and two of the laws of eighteen hundred and eighty-four, entitled 'An act to prevent deception in sales of dairy products.'"

PASSED June 9, 1885; three-fifths being present.

The People of the State of New York, represented in Senate and Assembly, do enact as follows:

SECTION 1. Section six of chapter one hundred and eighty-three of the laws of eighteen hundred and eighty-five, entitled "An act to prevent deception in the sale of dairy products, and to preserve the public health, being supplementary to and in aid of chapter two hundred and two of the laws of eighteen hundred and eighty-four, entitled 'An act to prevent deception in sales of dairy products,'" is amended so as to read as follows:

§ 6. No person shall manufacture out of any oleaginous substance or substances, or any compound of the same, other than that produced from unadulterated milk, or of cream from the same, any article designed to take the place of butter or cheese produced from pure unadulterated milk or cream of the same, or shall sell, or offer for sale, the same as an article of food. This provision shall not apply to pure skim-milk cheese, made from pure skim milk. Whoever violates the provisions of this section shall be guilty of a misdemeanor, and be punished by a fine of not less than $100 nor more than $500, or not less than six months or more than one year's imprisonment, or both such fine and imprisonment for the first offense, and by imprisonment for one year for each subsequent offense.

§ 2. Section seven of said act is amended so as to read as follows:

§ 7. No person by himself or his agents or servants shall render or manufacture out of any animal fat or animal or vegetable oils not produced from unadulterated milk or cream from the same, any article or product in imitation or semblance of or designed to take the place of natural butter or cheese produced from pure unadulterated milk or cream of the same, nor shall he or they mix, compound with, or add to milk, cream or butter any acids or other deleterious substance or any animal fats or animal or vegetable oils not produced from milk or cream, with design or intent to render, make or produce any article or substance or any human food in imitation or semblance of natural butter or cheese, nor shall he sell, keep for sale, or offer for sale any article, substance or compound made, manufactured or produced in violation of the provisions of this section, whether such article, substance or compound shall be made or produced in this State or in any other State or country. Whoever violates the provisions of this section shall be guilty of a misdemeanor and be punished by a fine of not less than $100 nor more than $500, or not less than six months or more than one year's imprisonment for the first offense, and by imprisonment for one year for each subsequent offense. Nothing in this section shall impair the provisions of section six of this act.

§ 3. Section eight of said act is amended so as to read as follows:

§ 8. No person shall manufacture, mix or compound with or add to natural milk, cream or butter any animal fats or animal or vegetable oils, nor shall he make or manufacture any oleaginous substance not produced from milk or cream, with intent to sell the same for butter or cheese made from unadulterated milk or cream, or have the same in his possession, or offer the same for sale with such intent, nor shall any article or substance or compound so made or produced, be sold for butter or cheese, the product of the dairy. If any person shall coat, powder or color with annatto or any coloring-matter whatever butterine or oleomargarine, or any compounds of the same or any product or manufacture made in whole or in part from animal fats or animal or vegetable oils not produced from unadulterated milk or cream whereby the said product, manufacture or compound shall be made to resemble butter or cheese, the product of the dairy, or shall have the same in his possession, or shall sell or offer for sale or have in his possession any of the said products which shall be colored or coated in semblance of or to resemble butter or cheese, it shall be conclusive evidence of an intent to sell the same for butter or cheese, the product of the dairy. Whoever violates any of the provisions of this section shall be guilty of a misdemeanor, and be punished by a fine of not less than $100 nor more than $1,000. This section shall not be construed to impair or affect the prohibitions of sections six and seven of this act.

§ 4. Section seventeen of said act is amended so as to read as follows:

§ 17. The doing of any thing herein prohibited being done, shall be evidence of a violation of the provisions of this act relative to the thing so prohibited; and the not doing of any thing herein directed to be done, shall be evidence of a violation of the provisions of this act relative to the thing so directed to be done. If any person shall suffer any violation of the provisions of this act by his agent, servant, or in any room or building occupied or controlled by him, he shall be deemed a principal in such violation and punished accordingly.

§ 5. This act shall take effect immediately.

A BRIEF SUMMARY OF THE LAWS OF THE DIFFERENT STATES AND TERRITORIES OF THE UNITED STATES RELATIVE TO OLEOMARGARINE, BUTTERINE, SUINE AND KINDRED PRODUCTS.

STATES.

Alabama.

This State has no law upon this subject.

Arkansas.

This State has no law upon this subject.

California.

"An act to prevent the sale of oleomargarine, under the name and pretense that the said commodity is butter."

This law is restrictive, requires the word "oleomargarine" to be branded on the package. The penalty is from $50 to $200, or imprisonment from 50 to 200 days, or both.

"An act to prevent fraud and deception in the sale of butter and cheese."

This law is restrictive, requiring the article to be manufactured and sold under its appropriate name. Penalty is from $10 to $500 or imprisonment from ten to ninety days, or both. Approved, March 2, 1881.

"An act to prevent the sale or disposition as butter of the substance known as 'oleomargarine,' or 'oleomargarine butter,' and when 'oleomargarine' or 'oleomargarine butter' is sold or disposed of requiring notice thereof to be given."

This law is restrictive, requiring branding, also requiring hotel-keepers, etc., to keep posted up in their places of business in three places, the words "oleomargarine sold here." Penalty from $5 to $500, or imprisonment for not more than three months, or both such fine and imprisonment, approved March 1, 1883.

"An act to protect and encourage the production and sale of pure and wholesome milk, and to prohibit and punish the production and sale of unwholesome or adulterated milk."

This law makes it a misdemeanor to sell or expose for sale adulterated or unwholesome milk, or to keep cows for producing the same in an unhealthy condition, or feeding them on feed that wiil produce impure milk, etc. Penalty is $100 for the first offense, and double that amount for each subsequent offense. Approved March 12, 1870.

Colorado.

"An act to encourage the sale of milk, and to provide penalties for the adulteration thereof."

This law makes it a misdemeanor to sell adulterated milk or milk from which the cream has been taken, or for withholding the strippings without the purchasers being aware of the fact. Penalty is from $25 to $100, or imprisonment for six months, or by both such fine and imprisonment. In force May 20, 1881.

"An act to regulate the manufacture and sale of oleomargarine, butterine, suine or other substances made in imitation of, or having the semblance of butter, and to provide penalties for the violation of the provisions hereof."

This law requires that a license shall be necessary to manufacture, import, or sell oleomargarine or kindred products within the State. License to manufacture or import not less than 1,000; license to sell not less than 500; penalty from $50 to $500, or imprisonment not to exceed one year or both. Approved April 6, 1885.

Connecticut.

"An act concerning the sale of oleomargarine and other articles."

This law requires that the article shall be properly branded, and that the seller shall keep a sign posted up in his place of business that such commodity is sold there. Penalty $7, or imprisonment from ten to thirty days or both. Approved April 4, 1883.

Delaware.

"An act to regulate the manufacture and sale of oleomargarine."

This law is restrictive in its nature, penalty $50, commitment until the fine is paid. Approved February 10, 1879.

"An act to amend chapter 154, volume 16, Laws of Delaware."
This amendment has reference to the fact that the substance manufactured is "artificial butter." Passed March 21, 1883.

Florida.

Chapter 80, sections 34–35, McClellans' Digest, 1881.
Section 34 makes it a misdemeanor to sell spurious preparations as butter; section 35 has reference to hotels and boarding-houses. Penalty, not to exceed $100, or imprisonment not to exceed thirty days, or both.

Georgia.

This State has no law upon this subject.

Illinois.

"An act to prevent and punish the adulteration of articles of food, drink and medicine, and the sale thereof when adulterated."
Section 3 of this law has reference to coloring matter in food, drink or medicine. Section 4 of this law has reference to mixing oleomargarine with butter, cheese, etc., requiring the seller to inform the buyer of the fact and the proportion of the mixture. Penalty, first offense, $25 to $200; second offense, $100 to $200, or imprisonment from one to six months or both; third offense, from $500 to $2,000 and imprisonment not less than one year nor more than five years. Approved June 1, 1881.
"An act to require operators of butter and cheese factories on the co-operative plan to give bonds, and to prescribe penalties for the violation thereof."
This law requires the filing of a bond in the penal sum of $6,000 that certain reports will be made on the first of each month and a copy filed with the town clerk, etc. Penalty, from $200 to $500, or imprisonment from thirty days to six months, or both. Approved June 18, 1883.

Indiana.

Section 2071, Revised Statutes, "Selling unwholesome milk."
This section provides against the sale of unwholesome milk, whether from adulteration or from the feed given the cows; also against the use of poisonous or deleterious material in the manufacture of butter and cheese. Penalty, from $50 to $500.
"An act to prevent the sale of impure butter, and the keeping on any table at any hotel or boarding-house of impure butter, providing penalties declaring an emergency."
This law requires the branding with the word "oleomargarine." Penalty, from $10 to $50. Approved March 3, 1883.

Iowa.

Section 4042, Code.
This section provides against the adulteration of milk in any way. Penalty, $25 to $100, and makes the offender liable in double that amount to the party injured.

"An act to protect the dairy interests and for the punishment of fraud connected therewith.

This law requires that "oleo" and kindred products shall be branded with the word "oleomargarine."

Penalty from $20 to $100 or imprisonment from ten to ninety days.

"An act to prevent and punish the adulteration of articles of food, drink and medicine, and the sale thereof when adulterated."

' This law provides that skimmed milk cheese shall be so branded, and when oleomargarine is mixed with any other substance for sale it shall be distinctly branded with the true and appropriate name. Penalty, first offense from $10 to $50; second, from $25 to $100 or confined in the county jail not more than thirty days; third, from $500 to $1,000 and imprisonment not less than one year nor more than five years.

Kansas.

Has no law upon this subject.

Kentucky.

Has no law upon this subject.

Louisiana.

Has no law upon this subject.

Maryland.

"An act to repeal the act of 1883, chapter 493, entitled 'An act for the protection of dairymen, and to prevent deception in the sale of butter and cheese, and to re-enact new sections in lieu thereof.'"

This law requires that substances made in semblance of butter and cheese not the true product of the dairy shall be branded with the word "Oleomargarine" so as to be conspicuous, and that the buyer shall be apprised of the nature of the article that he has bought. Penalty, $100 or imprisonment not less than thirty or more than ninety days for the second offense and not less than three months nor more than one year for the third offense. Approved April 8, 1884.

Mississippi.

This State has no law upon this subject.

Maine.

"An act to amend chapter 128 of the Revised Statutes, relating to the sale of unwholesome food."

This law is prohibitive as to oleomargarine and kindred products. Penalty, for the first offense $100, and for each subsequent offense $200, to be recovered with costs.

Massachusetts.

This State has a law against selling adulturated milk.

Penalty for first offense $50 to $100; for the second offense $100 to

$300, or by imprisonment from thirty to sixty days, and for subsequent offenses $50 and imprisonment from sixty to ninety days.

Michigan.

"An act to prevent deception in the manufacture and sale of dairy products and to preserve the public health."

This law prohibits the manufacture and sale of oleomargarine and kindred products.

Penalty, $200 to $500, or not less than six months nor more than one year's imprisonment or both for the first offense, and by imprisonment for one year for each subsequent offense.

Approved June 12, 1885.

Minnesota.

"An act to prohibit and prevent the sale or manufacture of unhealthy or adulterated dairy products."

This law prohibits the sale of impure or adulterated milk.

Penalty, $25 to $200 or imprisonment from one to six months or both for the first offense and six months' imprisonment for each subsequent offense. This law also prohibits the manufacture and sale of oleaginous substances or compounds of the same.

Penalty from $100 to $500, and from six months to one year's imprisonment or both such fine and imprisonment for the first offense and by imprisonment one year for each subsequent offense.

Approved March 5, 1885.

Missouri.

This State passed the first prohibitory law.

Penalty, confinement in the county jail not to exceed one year or fine not to exceed $1,000 or both.

Nebraska.

Section 2345, "Skimmed milk or adulterated milk." This section provides against the sale of adulterated milk and makes a penalty of from $25 to $100, and be liable to double the amount to the person or persons upon whom the fraud is perpetrated.

Nevada.

This State has no law upon this subject.

New Hampshire.

"An act relating to the sale of imitation butter."

This law provides that no artificial butter shall be sold unless it is colored PINK.

Penalty for the first offense $50, and for a second offense a fine of $100. "A certificate of the analysis sworn to by the analyzer shall be admitted in evidence in all prosecutions."

"The expense of the analysis not exceeding $20, included in the costs."

New Jersey.

Law similar to the New York law.

North Carolina.

This State has no law upon the subject.

Ohio.

This State has a law that is prohibitory except as to oleomargarine made of beef suet and milk.

Penalty, $100 to $500, or from three to six months' imprisonment or both for the first offense, and by such fine and imprisonment for one year for each subsequent offense.

Passed April 27, 1885.

Oregon.

The law in this State provides against adulterated and unwholesome milk, against keeping cows in an unhealthy condition, and against feeding them upon unhealthful food.

It also provides that oleaginous substances sold upon the market shall be so branded as to distinguish them from the true dairy product; and that in hotels, boarding houses, restaurants, etc., where such substances are used as an article of food the bill of fare shall state the fact and that the name of the said substance shall be posted up in the dining-room in a conspicuous place. Passed February 20, 1885.

Pennsylvania.

"An act to protect dairymen, and to prevent deception in sales of butter and cheese.

This act requires the branding of imitation butter and cheese. Penalty, $100. Violations of this act by exportation to a foreign country are punished by a fine of from $5, to $200, or by imprisonment from ten to thirty days, or by both such fine and imprisonment. Approved, May 24, 1883.

"An act for the protection of the public health and to prevent adulteration of dairy products and fraud in the sale thereof."

This law prohibits the sale of oleomargarine and kindred products. Penalty, $100 to $300, or by imprisonment from ten to thirty days for the first offense and by imprisonment for one year for each subsequent offense. Approved, May 21, 1885.

Rhode Island.

"Of the sale of butter, potatoes, onions, berries, nuts and shelled beans."

This law provides that artificial butter shall be stamped "Oleomargarine," and that the retailer shall deliver to the purchaser a label upon which shall be the word "Oleomargarine." Penalty, $100.

South Carolina.

There is no law in this State relative to this subject.

Tennessee.

Code of 1884, chapter 14, sections 2682, 2683, 2684.

This law requires that the substance shall be manufactured under its true and appropriate name and that it shall be distinctly branded with the true and appropriate name. Penalty, from $10 to $300, or imprisonment from ten to ninety days.

Texas.

This State has no law upon this subject.

Vermont.

"An act to prevent fraud in the sale of oleomargarine and other substances as butter."

This law provides that oleomargarine and kindred products shall not be sold as butter. Penalty, $500. Approved, November, 1884.

Chapters 192, Laws of 1874, 76 of 1870, 51 of 1855, provide against the adulteration of milk.

Virginia.

Code of Virginia, 1873, chapter 865, title 26, section 56.

"Provision against adulterating milk intended for the manufacture of cheese."

This law provides against the adulteration of milk carried to cheese manufactories, etc. Penalty, from $25 to $100, with costs of suit.

West Virginia.

Chapter 41, Acts of West Virginia, 1885.

"An act to prevent the manufacture and sale of mixed and impure butter and cheese and imitations thereof."

This law requires that the true and appropriate name of the substance shall be printed thereon, etc. Penalty, from $10 to $100, or imprisonment.

Wisconsin.

Section 1494, chapter 61, Revised Statutes.

This act provides that no cream shall be taken from the manufactory where it is being worked up, also that the persons manufacturing cheese at factories shall keep certain records.

Chapter 361, R. S.

"An act to prevent the manufacture and sale of oleaginous substances or compounds of the same in imitation of the pure dairy products, and to repeal sections 1 and 3 of chapter 49 of the laws of 1881."

This law prohibits the manufacture and sale of oleomargarine and kindred products. Penalty, not to exceed $1,000, or imprisonment not to exceed one year, or by both such fine and imprisonment. Published April 13, 1885.

Territories.

Alaska.

Have not heard from this Territory as yet.

Arizona.

"An act to regulate the sale and manufacture of oleomargarine or other substitutes for butter in the Territory of Arizona."

This law requires that oleomargarine and kindred substances sold in the Territory shall be appropriately branded with the word "oleomargarine." And that the seller shall deliver to the purchaser a printed label on which is the word "oleomargarine." Also that dealers shall keep posted up in their places of business this sign, "Oleomargarine sold here." Penalty for the first offense not less than $5, for the second offense not less than $100 or imprisonment for sixty days, and for each succeeding offense $500 and imprisonment for ninety days. Approved, March 8, 1883.

Dakota.

"An act to secure the public health and safety against unwholesome provisions."

This law requires that all oleaginous substances shall be branded with their true and proper names. Costs of analyses, not exceeding $20, shall or may be included in the costs of prosecutions. Penalty, first offense, $100, and every subsequent offense $200. Passed at the session of 1883.

Idaho.

This Territory has no law upon this subject.

Montana.

This Territory has no law upon this subject.

New Mexico.

This Territory has no law upon this subject.

Utah.

This Territory has no law upon this subject.

Washington.

This Territory has no law upon this subject.

Wyoming.

This Territory has no law upon this subject.

INDEX.

Lightning Source UK Ltd.
Milton Keynes UK
UKHW052300080219
336748UK00008BA/879/P

9 780365 313588